Handbook of Himalayan Ecosystems and Sustainability, Volume 2

Volume 2: Handbook of Spatio-Temporal Monitoring of Water Resources and Climate is aimed to describe the current state of knowledge and developments of geospatial technologies (remote sensing and Geographic Information Systems) for assessing and managing water resources under climate change. It is a collective achievement of renowned researchers and academicians working in the Hindu Kush Himalayan (HKH) mountain range. The HKH region is a part of the Third Pole due to its largest permanent snow cover outside of polar regions. Importantly, the Himalayan belt is geologically fragile and vulnerable to geohazards (e.g., landslides, land subsidence, rockfalls, debris flow, avalanches, and earthquakes). Therefore, critical assessment and geospatial solutions are indispensable to safeguard the natural resources and human beings in the Himalayas using space-borne satellite data sets. This book also showcases various remote sensing techniques and algorithms in the field of urban sprawling, urban microclimate, and air pollution. The potential impacts of climate change on the cryosphere and water resources are also highlighted. This comprehensive Handbook is highly interdisciplinary and explains the role of geospatial technologies in studying the water resources of the Himalayas considering climate change.

KEY FEATURES

- This book is unique as it focuses on the utility of satellite data for monitoring snow cover variability, snowmelt runoff, glacier lakes, avalanche susceptibility, and flood modeling.
- It also explains how remote-sensing techniques are useful for mapping and managing the morphology and ecology of the Himalayan River.
- Addresses how geospatial technologies are valuable for understanding climate change impact on hydrological extremes, the potential impact of land use/land cover change (LULC) on hydrology and water resources management.
- It highlights the impact of LULC changes on land surface temperature, groundwater, and air pollution in urban areas.
- Includes contributions from global professionals working in the HKH region.

READERSHIP

The Handbook serves as a valuable reference for students, researchers, scientists, hydrologists, hydro-ecologists, meteorologists, geologists, decision makers, and all others who wish to advance their knowledge on monitoring and managing water resources and urban ecosystem using remote sensing in the HKH region considering climate change.

Handbook of Himalayan Ecosystems and Sustainability, Volume 2

Spatio-Temporal Monitoring of Water Resources and Climate

Edited by
Bikash Ranjan Parida,
Arvind Chandra Pandey,
Mukunda Dev Behera,
and Navneet Kumar

CRC Press
Taylor & Francis Group
Boca Raton London New York

CRC Press is an imprint of the
Taylor & Francis Group, an **informa** business

First edition published 2023
by CRC Press
6000 Broken Sound Parkway NW, Suite 300, Boca Raton, FL 33487-2742

and by CRC Press
4 Park Square, Milton Park, Abingdon, Oxon, OX14 4RN

CRC Press is an imprint of Taylor & Francis Group, LLC

Library of Congress Cataloging-in-Publication Data
Names: Parida, Bikash Ranjan, editor. | Pandey, A. C. (Arvind Chandra), editor. |
    Behera, Mukunda Dev, editor. | Kumar, Navneet (Senior researcher), editor.
Title: Handbook of Himalayan ecosystems and sustainability / edited by Bikash Ranjan Parida,
    Arvind Chandra Pandey, Mukunda Dev Behera, and Navneet Kumar.
Description: First edition. | Boca Raton, FL : CRC Press, 2023. | Includes bibliographical
    references and index. | Contents: Volume 1. Spatio-temporal monitoring of forests and
    climate—Volume 2. Spatio-temporal monitoring of water resources and climate.
Identifiers: LCCN 2022024150 (print) | LCCN 2022024151 (ebook) | ISBN 9781032203140
    (volume 1 ; hbk) | ISBN 9781032214306 (volume 1 ; pbk) | ISBN 9781032203157
    (volume 2 ; hbk) | ISBN 9781032207735 (volume 2 ; pbk) | ISBN 9781003268383
    (volume 1 ; ebk) | ISBN 9781003265160 (volume 2 ; ebk)
Subjects: LCSH: Environmental monitoring—Himalaya Mountains Region—Handbooks,
    manuals, etc. | Ecosystem management—Himalaya Mountains Region. | Remote sensing. |
    Geographic information systems.
Classification: LCC QH541.15.M64 H364 2023 (print) | LCC QH541.15.M64 (ebook) |
    DDC 363.7/063095496—dc23/eng/20220622
LC record available at https://lccn.loc.gov/2022024150
LC ebook record available at https://lccn.loc.gov/2022024151

ISBN: 978-1-032-20315-7 (hbk)
ISBN: 978-1-032-20773-5 (pbk)
ISBN: 978-1-003-26516-0 (ebk)

DOI: 10.1201/9781003265160

Typeset in Times
by Apex CoVantage, LLC

Contents

Foreword ix
Preface.. xi
Editors.. xv
Contributors xix

SECTION I   *Geospatial Approaches for Monitoring Cryosphere and Hydro-climatic Disasters*

Chapter 1   Monitoring Spatiotemporal Snow-Cover Dynamics with a
Focus on Runoff in the Zanskar Valley, Ladakh, India 3

Munizzah Salim and Arvind Chandra Pandey

Chapter 2   Comparative Assessment on Glacier Velocity Estimation
Using Optical and Microwave Satellite Data over Select
Large Glaciers across the Himalaya...19

*Nishant Tiwari, Arvind Chandra Pandey, and Chandra
Shekhar Dwivedi*

Chapter 3   Monitoring Spatiotemporal Patterns of Glacial Lakes in the
Eastern Himalayas Using Satellite Data and Nonparametric
Statistical Testing Techniques 35

Deepali Gaikwad, Supratim Guha, and Reet Kamal Tiwari

Chapter 4   Modeling Avalanche Susceptible Zones across the Indo-China
Border around the Galwan Valley, Ladakh (India):
Geoinformatics-based Perspective in High-Altitude
Battlefield 53

*Arvind Chandra Pandey, Shubham Bhattacharjee, Munizzah
Salim, and Chandra Shekhar Dwivedi*

Chapter 5   Flood Monitoring and Assessment over the Himalayan
River Catchment.. 69

*Kavita Kaushik, Arvind Chandra Pandey, Bikash Ranjan
Parida, and Navneet Kumar*

Chapter 6   Flood Inundation and Floodwater Depth Mapping Using
Synthetic Aperture Radar Data in the Gandak River Basin 89

*Gaurav Tripathi, Bhawani Shankar Phulwari, Bikash Ranjan
Parida, Arvind Chandra Pandey, and Mukunda Dev Behera*

SECTION II   Geospatial Approaches for Monitoring Water Resources and Climate Change

Chapter 7   Climate Change Impact on the Hydrological Extremes of a
River Basin in the Hindu Kush Himalayan Region: A Case
Study of the Marsyangdi River Basin, Nepal 111

Prajwal Neupane, Sangam Shrestha, and Suwas Ghimire

Chapter 8   Climate-Change Projections in the Himalayan River Basin
Using CMIP6 GCMs: A Case Study in the Koshi River
Basin, Nepal 137

Pragya Pradhan and Sangam Shrestha

Chapter 9   Impact of Structural Barriers on the Morphology and
Ecology of the Himalayan Rivers 165

Gaurav Kailash Sonkar and Kumar Gaurav

Chapter 10 Modeling the Potential Impact of Land Use/Land Cover
Change on the Hydrology of Himalayan River Basin:
A Case Study of Manipur River, India 189

Vicky Anand and Bakimchandra Oinam

Chapter 11 The Application of Remote Sensing for Water Resources
Management in Data-Scarce Watersheds in the Hindu Kush
Himalaya Region: A Case of Kabul River Basin 205

*Fazlullah Akhtar, Abdul Haseeb Azizi, Usman Shah,
Christian Borgemeister, Bernhard Tischbein, and
Usman Khalid Awan*

SECTION III   Geospatial Approaches for Monitoring Urban Ecosystem

Chapter 12 Urban Sprawl and Future Growth Projection vis-à-vis Groundwater Resource Availability in Hill Township in Shillong (Meghalaya), India 225

Arvind Chandra Pandey, Sudipta Hansda, and Navneet Kumar

Chapter 13 Land Surface Temperature Responses to Urban Landscape Dynamics 249

Nimish Gupta and Bharath Haridas Aithal

Chapter 14 Effects of Land Use/Land Cover Changes on Surface Temperature and Urban Heat Island over Kathmandu District in Nepal 275

Sourav Kumar, Aniket Prakash, Sandeep Kumar, and Bikash Ranjan Parida

Chapter 15 Assessing Aerosol and Nitrogen Dioxide Concentration in Major Urban Cities over the Himalayan Region during the COVID-19 Lockdown Phases 293

Shyama Prasad Mandal, Avinash Kumar Ranjan, Bikash Ranjan Parida, and Sailesh Narayan Behera

Chapter 16 Development of Spatially Distributed GIS-based Emission Inventory of Particulate Matter from Anthropogenic Sources over India and Assessment of Trends of Pollution 317

Sailesh N. Behera, Bikash R. Parida, Jitendra K. Tripathi, and Mukesh Sharma

Chapter 17 Spatial Distribution of Particulate Organic Carbon over India and the Prediction of Its Deposition in the Himalayas through the GIS-WRF-CAMx Modeling System 343

Ajay K. Singh, Sailesh N. Behera, Mukesh Sharma, and Bikash Ranjan Parida

Index 377

Foreword

It is my pleasure to write a Foreword to the two-volume Handbook on the topic of Himalayan Ecosystems and Sustainability. It comprises Volume 1: Spatio-Temporal Monitoring of Forests and Climate and Volume 2: Spatio-Temporal Monitoring of Water Resources and Climate, edited by Dr.rer. nat. Bikash Ranjan Parida, Prof. Arvind Chandra Pandey, Prof. M.D. Behera and Dr.-Ing. Navneet Kumar. I believe that the topics covered will attract broad attention to many sectors of society as the Hindu Kush Himalaya (HKH) mountain ecosystem is fragile and vulnerable to climate change. It needs critical assessment using Remote Sensing and Geographic Information Systems (GIS) which offers unique opportunities in mapping and monitoring the Himalayan mountain ecosystem. The Remote Sensing based measurement with the integration of Geospatial pathways bridges the earth's surface data and science to unfold the complex geo-environmental processes and their interconnections. The repository of geospatial data cube has been rapidly expanding, from an array of sources, namely satellite sensors, unmanned aerial vehicles (UAVs) and LiDAR to monitor the earth's surface, subsurface and atmosphere.

The HKH (known for Third Pole) is the source of the ten largest Asian river systems that provide water, ecosystem services (e.g., water, food, energy), and also the source for livelihoods to a population of 250 million people and nearly 1.65 billion people rely on its rivers. A significant volume of freshwater resources is deposited as snow and glacier ice. Over the past several decades, temperature and rainfall patterns are changing considerably. The volume of snow-cover was projected to be reduced due to climate variability that will cause adverse impacts on forest and agro-ecosystems, geosciences, hydrology, water resources, cryosphere, and the socio-economic systems of HKH. Global warming can further increase rapid glacial melting, formation of glacial lakes, and scarcity and uneven distribution of water resources owing to higher runoff in major river basins. Extreme intensity rainfall events are also posing challenges in urban areas of the Himalayas as it abruptly causes flash floods. Hence, the remote sensing-based assessment using geospatial pathways plays a vital role to assess the vulnerability of Himalayan communities to climate change and ways to improve crop diversity, food security, water resources, mountain ecology, climate resilience, incomes and associated risks.

Despite HKH including the Indian Himalayan Region (IHR) known for the fragile mountain ecosystem, relatively less effort has been made to understand its ecosystems and impacts of climate change on the livelihood of people of the Himalayas. The scientific outcomes over these regions are limited and most importantly, the knowledge gathered from science does not reach decision makers and other practicing professionals. This handbook contains a collection of chapters covering the utility of geospatial technologies for monitoring forests, biodiversity and agricultural ecosystems as well as water resources considering impacts of climate change and

land use/land cover. Thus each chapter of this book provides information from different perspectives about the monitoring and management of natural resources in mountainous regions. The publication of this book is an important milestone that aims to convey an insight into advanced space-based technologies for studying and offering progress and developments of geospatial technology in natural resources management. Therefore, a critical assessment of this mountain environment using geospatial technologies would be useful to protect and manage biodiversity, ecosystem services, livelihoods, and water resources of the Himalayas which is consistent with the Sustainable Development Goals (SDGs). With contributions from several international scholars and professionals, and edited by four outstanding scientists with the help of reviewers, this comprehensive handbook is a valuable accession to the cutting-edge science and research on natural resources management considering impacts of anthropogenic activities and climate change in the Himalayas.

I wish all the very best!!

Dr. Raj Kumar
Director, National Remote Sensing Centre (NRSC)
Indian Space Research Organisation (ISRO)
Government of India

Preface

The Himalayas are the youngest mountain range in the world with the highest mountain systems and prominent peaks. The Hindu Kush Himalayan (HKH) mountain range passes through eight countries, Afghanistan, Pakistan, India, China, Nepal, Bhutan, Bangladesh, and Myanmar. The HKH region is also part of the third pole due its largest permanent snow cover outside the polar regions. The HKH region comprises the source of ten major river basins with an area of over 4.2 million km². The region provides water to a large part of the Indian subcontinent, including ecosystem services (e.g., water, food, energy) that directly sustain the livelihoods of 250 million people. It has been estimated that more than 3 billion people benefited from the food produced in its river basins. The major portion of the Himalayan range lies in the Indian subcontinent, called the Indian Himalayan Region (IHR). This mountain ecosystem comprises ten hill states in India, namely, Jammu and Kashmir, Himachal Pradesh, Uttaranchal, Sikkim, Arunachal Pradesh, Manipur, Meghalaya, Mizoram, Nagaland, and Tripura, and two partial hill states, Assam and West Bengal. The IHR comprises nearly 17% of the geographical area of India (0.537 million km²), with 52 million people, and extends over 2,500 km between the Indus and the Brahmaputra River systems. Most of the areas are covered by snow-clad peaks and glaciers of the higher Himalayas, whereas dense forests cover in mid-Himalaya. Since the Himalayan Mountain range is geologically fragile and vulnerable to geohazards (e.g., landslides, land subsidence, rockfalls, debris flow, avalanches, and earthquakes), it is necessary to perform critical assessments and geospatial solutions to safeguard the natural resources and population in the Himalayas. Glacial lakes and snowmelt runoff are the main water sources of rivers for the people living in the Himalaya mountain belt. This Handbook aims to compile the research outcomes on the use of advanced space-based technologies to monitor and manage water resources and offer new insights into the development of this region. Moreover, the region is undergoing rapid change over the past few decades due to climate change and human intervention, such as urbanization, infrastructure development, migration, and tourism, among others. Therefore, the outcome of the interplay of these complex factors of change is challenging to predict but a critical assessment of this mountain environment would be useful to protect biodiversity, ecosystem services, livelihoods, and water resources of the Himalayas.

Variability in temperature and rainfall patterns has been observed over the past six decades in the Himalayas. Temperature is continuously increasing at the rate of 0.1–0.2°C per decade, whilst the projected changes in the temperature over the HKH region are larger (4.2–6.5°C for RCP8.5) than the global mean temperature by the end of the 21st century. The precipitation pattern was quite irregular in the Himalayas region with an increasing trend in the intensity of the precipitation. A reduction in snow-cover area and snow volume has been projected due to a rise in temperature that will cause adverse impacts on the cryosphere including the socioeconomic condition of people living in the region. Furthermore, global

warming could trigger an increase in glacial melting, the formation of glacial lakes, and uneven distribution and scarcity in water availability. Glacial lake outburst floods (GLOFs) in the Himalayas are becoming frequent and erratic, which is causing fatalities, damage to infrastructures, and risks to fragile landscapes. Moreover, climate change helps to form supra-glacial lakes and small water bodies on top of glaciers that tend to exacerbate the instability of the ice. With the increase of glaciers melting, the water from supra-glacial lakes can pass through the glacier ice/glacier bed and could potentially result in glacier surges (i.e., abrupt advances of glacier ice). Glaciers in the Himalayas have been retreating significantly and causing higher stream flows and runoff in major river basins. Additionally, extreme rainfall events (i.e., cloudbursts) are posing challenges as they abruptly cause flash floods in urban areas of the Himalayas. Heavy rainfall over the complex topography also poses a greater risk in the downstream areas because of hydrometeorological hazards (cloudbursts, floods, flood discharge) including rainfall-induced landslides, rockfalls, debris flows, and avalanches in the region. Remote-sensing and Geographic Information System (GIS) techniques offer continuous data freely in higher temporal and spatial resolutions (i.e., optical and synthetic aperture radar [SAR]) and are utilized for various applications. Thus, the remote sensing–based assessment using geospatial pathways plays a vital role in monitoring the vulnerability of the Himalayan cryosphere to climate change and ways to improve water resources, mountain ecology, and associated risks.

This two-volume book (*Handbook of Himalayan Ecosystems and Sustainability, Volume 1: Spatio-Temporal Monitoring of Forests and Climate* and *Volume 2: Spatio-Temporal Monitoring of Water Resources and Climate*) is a collection of chapters from several academicians mostly from South/Southeast Asia working in the domain of monitoring and management of forest, biodiversity and agricultural ecosystems, cryosphere as well as water resources and climate change using geospatial technologies. This two-volume Handbook aims to provide an insight into space-based technologies for studying and offering progress and developments of geospatial technology in the Himalayan Mountain ecosystem.

Volume 2 is divided into three sections: (I) geospatial approaches for monitoring and managing cryosphere and hydro-climatic disasters, (II) geospatial approaches for monitoring and managing water resources and climate change, and (III) geospatial approaches for monitoring and managing the urban ecosystem. This book is a multiauthored book, divided into several broader segments, such as cryosphere studies, geohazards, hydrometeorological hazards, urban sprawling, air pollution, climate change, and its impact on the water resources of the Himalayas. This book has summarized the recent progress and developments of geospatial technology (remote sensing and GIS) for studying the fragile Himalayan ecosystems and their dynamics in the spectrum of climate change.

Section I, Geospatial Approaches for Monitoring Cryosphere and Hydro-climatic Disasters, includes six chapters covering snow cover variability, snowmelt runoff, glacier lakes, avalanche susceptibility, and flood modeling using remote sensing techniques across the IHR. Chapter 1 focuses on the dynamics of snow cover and glacier melt runoff during 2000–2020 using remote-sensing-derived data set in Zanskar Valley, Ladakh, India. The authors highlighted that the discharge due to glacier melt

has increased rapidly from 2000 to 2020 and that precipitation exhibits more control on increasing runoff than temperature. The loss of snow cover is significantly higher in recent years, and such unprecedented changes will have adverse impacts on Himalayan cryospheric ecosystems along with climate change that needs serious attention from glaciologists, environmentalists, and policymakers. In Chapter 2, the authors employed both optical- and SAR-based techniques for deriving glacier velocity in major glaciers in the Himalayas (Siachen, Gangotri, and Bara Sigri, etc.). In Chapter 3, the authors have applied nearly four decades of satellite data to generate spatiotemporal patterns of glacial lakes in the Eastern Himalayas and concluded that the size of glacial lakes has been increasing. In Chapter 4, the authors use satellite data, GIS, and multi-criteria decision-based Analytical Hierarchy Process (AHP) for modeling avalanche susceptible zones across the Indo-China border region in the Himalayas. Their results demonstrated that north- and northeast-facing slopes with the convex surface within the relief of 5000–5500 m are more susceptible to avalanche events. In Chapter 5, the authors highlight the utility of the MODIS near-real-time (NRT) flood product for identifying the flood inundation extent and water stagnancy over the Himalayan River catchment. Their results reveal that a stagnancy period of more than 30 days causes adverse impacts on cropland and buildup in lower catchment areas. Similarly, Chapter 6 discusses the spatio-temporal distribution of inundation along with the floodwater depth. Here the authors suggest that the floodwater characterization information is useful for flood risk planning and management.

Section II, Geospatial Approaches for Monitoring Water Resources and Climate Change, comprises five chapters covering the morphology and ecology of the Himalayan River, the climate-change impact on hydrological extremes, the potential impact of land use/land cover (LULC) change on the hydrology, and water resources management. In Chapters 7 and 8, the authors evaluate the impact of climate change on hydrological extremes of a river basin in the HKH region. The CMIP6-based General Circulation Models (GCMs) predicted an increase of the annual maximum temperature, rainfall, hydrological extremes, and runoff under the SSP5-8.5 scenario by 2100. The increasing rainfall pattern indicates the risk of extreme events including floods, flash floods, landslides, and soil erosion in the future. These findings could assist in providing policy implications and adaptation measures against natural hazards. In Chapter 9, the authors discuss the impact of structural barriers on the morphology and ecology of the Himalayan River. Their results reveal that the structural barriers not only modify the hydraulic habitat of rivers but also induce habitat fragmentation, which, in turn, influences the spatial distribution and the health of aquatic inhabitants. Chapter 10 aims to predict the impact of changes in LULC pattern on the hydrology of the Himalayan River Basin using the Soil and Water Assessment Tool model. Their results indicate an increase in surface runoff and actual evapotranspiration (AET) with a decrease in water yield associated with the expansion of buildup. These findings could help in formulating environmental preservation policies and urban planning for the sustainable use of resources. In Chapter 11, the authors discuss the application of remote sensing for water resources management in data-scarce watersheds in the HKH region of the Kabul River Basin. Their results indicate a decreasing trend in the snow-covered

area with a significant increasing trend of AET that consequently result in a relative shortage of water availability in the basin.

Section III, Geospatial Approaches for Monitoring Urban Ecosystem, includes six chapters, all focusing on the utility of remote sensing and geospatial technologies for understanding LULC changes and their impact on land surface temperature (LST), groundwater, and air pollution in urban areas. In Chapter 12, the authors analyze the changes in land use patterns between 2000 and 2020 using Landsat-7/8 satellite data for assessing groundwater potential. Their results were pivotal for discovering the groundwater availability aspects while planning the town's future. In Chapter 13, the authors discuss changing land use patterns and their impact on urban LST and urban microclimate. Their results support the development of urban policies for the well-being of the city environment and sustainable urban growth. Chapter 14 applies satellite data to discuss the increasing urbanization in the central Himalayas in the last two decades. The study reveals that urban expansion has taken place at the cost of agricultural and barren lands, altering LST and urban heat islands. In Chapter 15, the authors discuss the aerosol and nitrogen dioxide concentrations in major urban cities over the Himalayan region using Sentinel–5P satellite data. The study indicates a significant reduction in air pollution during the lockdown phases. Chapter 16 provides a quantification of particulate matter (PM_{10}) from anthropogenic sources based on a GIS-based emission inventory. The authors demonstrate an increase in PM_{10} concentration significantly over the ecologically sensitive Himalayan region. Finally, in Chapter 17, the authors discuss particulate organic carbon (POC) and the prediction of its deposition in the Himalayas through the GIS-WRF-CAMx modeling system. Their results indicate prominent deposition rates and ambient concentrations of POC at glaciers that are located relatively at lower elevations than surrounding peaks. Therefore, the transported air mass containing pollutant would first settle at the glacier, and the remaining air mass successively would settle at peak in that region.

Editors
Dr.rer.nat. Bikash Ranjan Parida
Central University of Jharkhand, India

Prof. Arvind Chandra Pandey
Central University of Jharkhand, India

Prof. M.D. Behera
Indian Institute of Technology Kharagpur, India

Dr.-Ing. Navneet Kumar
University of Bonn, Germany

How to cite: B.R. Parida, A.C. Pandey, M.D. Behera, N. Kumar (eds) (2023). Handbook of Himalayan Ecosystems and Sustainability, Volume 2: Spatio-Temporal Monitoring of Water Resources and Climate, CRC Press/Taylor & Francis Group.

Editors

Dr.rer.nat. Bikash Ranjan Parida is an assistant professor at the Central University of Jharkhand (CUJ) in India since September 2016. He earned a BSc in Agricultural Sciences (2004) from UAS, Dharwad in Karnataka, and an MSc in Geoinformatics (2006) from the Faculty Geo-Information Science and Earth Observation, University of Twente, the Netherlands. He earned a PhD in Natural Sciences (Specialization: Meteorology & Earth System Modelling) from University of Hamburg/Max Planck Institute for Meteorology in Germany (2011). Prior to joining CUJ, he was employed at various premium institutes such as Max Planck Institute for Meteorology in Hamburg and the University of California, Los Angeles in the US. He has been involved in earth and environmental studies, specifically solving environmental problems using space-based technologies, machine learning techniques, and climate modeling. He is interested in crosscutting activities related to atmosphere-land surface exchange processes (surface water, energy, carbon fluxes, etc.), carbon cycling modeling, application areas in agriculture, forest and natural resources monitoring and assessment, and climate change using remote-sensing and the Geographic Information System. He has contributed to several national-level research projects such as the Science and Engineering Research Board under the Department of Science and Technology, Start-Up Research Grant by University Grants Commission (UGC), NASA-ISRO SAR Mission project by Space Applications Centre (SAC), Ahmedabad, Indian Space Research Organisation (ISRO). He has more than 50 research publications in peer-reviewed journals. Website Link: https://sites.google.com/view/bikash-parida

Prof. Arvind Chandra Pandey is a professor in the Department of Geoinformatics, Central University of Jharkhand (CUJ), Ranchi. He was former Head, Department of Geoinformatics (2013–2020) and Dean, School of Natural Resource Management (2013–2016) in CUJ. He is coordinator (CUJ) for the ISRO, EDUSAT programme since 2013. He previously served as an associate professor (remote sensing) at the Birla Institute of Technology, Mesra, Ranchi, for a decade (2004–2013) and as a research scientist in the Department of Science & Technology (DST), Govermnet of India, Chandigarh, for seven years (1997–2004). He earned an MSc (Applied Geology) and a PhD in Geology from the Department of Geology, University of Delhi in 1993 and 2001, respectively. He has been working in diverse application areas of geoinformatics, namely, water resources, glaciology, natural hazards, urban environment, forestry, and others. Nine PhD and 60 MTech/MSc theses have been completed under his guidance. He has more than 70 publications in refereed international/national journals and two edited books to his credit.

He is recipient of NASA-SERVIR Fellowship in 2013 to work on Himalayan glaciers in Zanskar Valley, Jammu and Kashmir. He also obtained training at Ocean Teacher Global Academy, University Malaysia Terengganu in 2016. He has completed many national projects as principal investigator and co–principal investigator from ISRO, DST, and Ministry of Environment, Forest and Climate on aspects of the Himalayan Glacier Study. Website Link: http://cuj.ac.in/Arvind.php

Prof. Mukunda Dev Behera has made outstanding contributions to the fields of forest remote sensing and ecological climatology through theorizing, modeling, and conducting innovative experiments and field-based measurements. His research innovations have fundamentally transformed the study of phytogeography, with attention to ecological functioning of three ecosystem components, plant diversity, water, and energy. His research highlighted the spatial variations of climate with plant diversity and quantified the contribution of climate drivers in Indian context. He has developed innovative protocols for inclusive biodiversity assessment and new methods for estimating a range of vegetation photosynthetic/structural variables using a suite of satellite data products and modeling protocols.

Dr. Behera has guest edited six special issues in journals and serves on the editorial boards of three Springer journals (*Biodiversity and Conservation*, *Environmental Monitoring and Assessment*, and *Tropical Ecology*). With more than 20 years of research and teaching experience, he has published more than 100 papers in journals and supervised 11 PhD and three postdoc researchers. Website Link: www.iitkgp.ac.in/department/CL/faculty/cl-mdbehera

Dr.-Ing. Navneet Kumar works as a Senior Researcher at Center for Development Research (ZEF), University of Bonn, Germany and currently acts as a disciplinary course coordinator and member ethical committee for the doctoral program at ZEF. He earned his PhD in Engineering (water resource management) from University of Bonn. His main research themes are water and natural resources management and geoinformatics. Particular topics include remote sensing and GIS applications in water management and agriculture, hydrological modeling, climate change, land degradation, risk and impact assessment, irrigation, etc. Dr. Kumar has contributed to several projects in Africa and Asia (Algeria, Ethiopia, India, Mali, Niger, Uzbekistan). He has also been involved in capacity and network building projects with Pan African University – Institute of Water and Energy Sciences (PAUWES) in Algeria and West African Science Service Center on Climate Change and Adapted Land Use (WASCAL). He has contributed to the development of eLearning courses on water management and is currently involved in conducting joint global classroom semester programs on water management with

partner institutions from USA and Brazil. He has conducted several lectures, summer schools and workshops in Africa, India and Germany. Dr. Kumar has presented his work at several international conferences and received several awards and travel grants. He has published in several peer-reviewed journals and is a reviewer for several journals and grant proposals. Website Link: https://www.zef.de/staff/navneet_kumar

Contributors

Aithal, Bharath Haridas
Indian Institute of Technology (IIT)
  Kharagpur
Kharagpur, West Bengal, India

Akhtar, Fazlullah
University of Bonn
Bonn, Germany

Anand, Vicky
National Institute of Technology
  Manipur
Imphal, Manipur, India

Awan, Usman Khalid
International Water Management
  Institute (IWMI)
Lahore, Pakistan

Azizi, Abdul Haseeb
University of Bonn
Bonn, Germany

Behera, Mukunda Dev
Indian Institute of Technology (IIT)
  Kharagpur
Kharagpur, West Bengal, India

Behera, Sailesh Narayan
Shiv Nadar University
Greater Noida, Uttar Pradesh, India

Bhattacharjee, Shubham
Central University of Jharkhand
Ranchi, Jharkhand, India

Borgemeister, Christian
University of Bonn
Bonn, Germany

Dwivedi, Chandra Shekhar
Central University of Jharkhand
Ranchi, Jharkhand, India

Gupta, Nimish
Indian Institute of Technology (IIT)
  Kharagpur
Kharagpur, West Bengal, India

Gaikwad, Deepali
Indian Institute of Technology (IIT)
  Ropar
Rupnagar, Punjab, India

Gaurav, Kumar
Indian Institute of Science Education
  and Research (IISER), Bhopal
Bhopal, Madhya Pradesh, India

Ghimire, Suwas
Asian Institute of Technology
Bangkok, Thailand

Guha, Supratim
Indian Institute of Technology (IIT)
  Ropar
Rupnagar, Punjab, India

Hansda, Sudipta
Central University of Jharkhand
Ranchi, Jharkhand, India

Kaushik, Kavita
Central University of Jharkhand
Ranchi, Jharkhand, India

Kumar, Navneet
University of Bonn
Bonn, Germany

Kumar, Sandeep
Indian Institute of Technology Bombay
Mumbai, Maharashtra, India

Kumar, Sourav
Indian Institute of Technology Delhi
New Delhi, Delhi, India

Mandal, Shyama Prasad
Central University of Jharkhand
Ranchi, Jharkhand, India

Neupane, Prajwal
Asian Institute of Technology
Bangkok, Thailand

Oinam, Bakimchandra
National Institute of Technology
    Manipur
Imphal, Manipur, India

Pandey, Arvind Chandra
Central University of Jharkhand
Ranchi, Jharkhand, India

Parida, Bikash Ranjan
Central University of Jharkhand
Ranchi, Jharkhand, India

Pradhan, Pragya
Asian Institute of Technology
Bangkok, Thailand

Prakash, Aniket
Indian Institute of Technology (ISM)
    Dhanbad
Dhanbad, Jharkhand, India

Ranjan, Avinash Kumar
National Institute of Technology Rourkela
Rourkela, Odisha, India

Salim, Munizzah
Central University of Jharkhand
Ranchi, Jharkhand, India

Shah, Usman
Independent Researcher

Shankar, Bhawani
Central University of Jharkhand
Ranchi, Jharkhand, India

Sharma, Mukesh
Indian Institute of Technology
    Kanpur
Kanpur, Uttar Pradesh, India

Shrestha, Sangam
Asian Institute of Technology
Bangkok, Thailand

Singh, Ajay K.
Shiv Nadar University
Greater Noida, Uttar Pradesh, India

Sonkar, Gaurav Kailash
Indian Institute of Science Education
    and Research (IISER), Bhopal
Bhopal, Madhya Pradesh, India

Tischbein, Bernhard
University of Bonn
Bonn, Germany

Tiwari, Nishant
Central University of Jharkhand
Ranchi, Jharkhand, India

Tiwari, Reet Kamal
Indian Institute of Technology (IIT)
    Ropar
Rupnagar, Punjab, India

Tripathi, Gaurav
Central University of Jharkhand
Ranchi, Jharkhand, India

Tripathi, Jitendra K.
Shiv Nadar University
Greater Noida, Uttar Pradesh, India

Section I

Geospatial Approaches for Monitoring Cryosphere and Hydro-Climatic Disasters

1 Monitoring Spatiotemporal Snow-Cover Dynamics with a Focus on Runoff in the Zanskar Valley, Ladakh, India

Munizzah Salim and Arvind Chandra Pandey*
Department of Geoinformatics, School of Natural
Resource Management, Central University of
Jharkhand, Ranchi-835222, Jharkhand, India

CONTENTS

1.1 Introduction 3
1.2 Study Area 5
1.3 Materials and Methods 5
    1.3.1 Methodology 7
        1.3.1.1 Runoff from the SCA 7
        1.3.1.2 Runoff from SFA 8
        1.3.1.3 Subsurface Runoff 8
        1.3.1.4 Total Runoff 8
1.4 Results and Discussion 9
    1.4.1 Estimation of Glacier Melt Runoff Using SNOWMOD 9
    1.4.2 Relation between Melt Discharge and Climate Forcing 14
1.5 Conclusion 14
References 16

1.1 INTRODUCTION

The livelihoods of millions of people depend on Himalayan glaciers as they provide a major amount of the freshwater supply to many perennial river systems emerging from the Himalayas. The three major river systems, namely, the Ganga, the Brahmaputra, and the Indus contribute 50% of the usable water (Immerzeel

DOI: 10.1201/9781003265160-2

3

et al. 2010; Singh and Jain 2009). Studies related to growth/shrinkage and the mass exchange of glaciers are important (Kaser et al. 2003) as not the only Himalayas but also the Alps, the Andes, and the Rocky Mountains have shown wide-scale glacier retreat (Kulkarni et al. 2002). Global climate change impacted the regional or local hydrological process (Arnell 1999). Thus, the estimation of melt runoff and glacier mass balance is essential to understanding the spatiotemporal changes over glacier area, length, and volume for marking the impact of changing climate.

Climate changes over the glaciological domain made a variation in runoff (Song et al. 1999). All glaciers counter climate perturbations by their mass fluctuations. Climate changes can disturb the atmospheric process, have the potential to increase the precipitation and, by increasing the temperature, also influence evaporation and evapotranspiration. Therefore, the discharge rate of glacier melts and timing also were affected (Foster 2007; IPCC 2007). To advocate the future relationship between climatic parameters and runoff, it is very important to understand the future climatic trend and pattern of glacier changes. As the Earth's climate is changing rapidly, the glaciers become an indicator of climate change, and it provides a glimpse of future issues (Krajik 2002; Hinkle 2003). Therefore, it becomes necessary to understand the relation between climate and Earth (IPCC 2007; Marston 2008). The receding rate of the Himalayan glacier is very high, and about 67% of glaciers are retreating at an alarming rate. Khromova et al. (2003) found that from 1993 to 2001, the glaciers of the Al-Shirak range retreated by 23%. During the 20th century, about 30.35% of glaciers retreated as observed in the Ladakh region (Yablokov 2006). According to Kulkarni (2007), nearly 466 glaciers of the Chenab, Parbati, and Baspa basins faced retreat whereas in Parbati Glacier in Kullu district receded at a rate of 52 m/year from 1990 to 2001 (Kulkarni et al. 2005). According to Dobhal et al. (2004), the Himalayan glacier recession varies from 16 m to 35 m annually. Bradley et al. (2006) opine that with the increase in altitude in the troposphere, the rate of warming also increases, which directly results in glacier retreat. In the recent study of Himalaya–Karakoram glacier changes by Bahuguna et al. (2021), it was estimated that in the Karakoram region about 2143 glaciers witnessed a gain of about 0.026% in contrast to this Himalayan region experienced decrement in glacier area by 1.44% over 17 years. The part of Chenab and Satluj basins experienced the highest amount of loss of about 2.2%, whereas the sub-basin present on the left side of the Indus basin witnessed the lowest loss in glacier area, about 0.76%, and one of its key findings is that the glaciers that are debris-free are more prone to melting due to global warming.

Mountain areas are at high risk due to climate change because there is a large difference in altitude over a small distance (Beniston et al. 1997). There were many instances in which global warming not only is responsible for retreat but also helped in glacier advancement (Winkler et al. 1997; Casassa et al. 2002; Chinn et al. 2005). The changing climate has a direct impact on the snowmelt runoff as well as glacier mass balance (Agarwal 2011). Global climate warming is a serious threat that abates the glaciers continuously (Prasad et al. 2009). Stocker et al. (2013) gave important information on temperature rises of combined land and ocean surface temperature by 85°C from 1820 to 2012. In the context of hydropower generation, snow- and glacier melt runoff have a wide base as in Rathore et al. (2009); a runoff model has been generated over Malana nala in Prabati river

basin, Kullu. In this study, meteorological data sets in conjunction with remote sensing satellite data have been used.

The Moderate Resolution Imaging Spectroradiometer (MODIS) data have been widely used for studying spatiotemporal dynamics of glaciers as it has a high spatial and temporal resolution. But the major constraint in optical data sets is the interference of clouds, and hence, the use of snow-cover products obtained by optical data sets could be erroneous. To overcome this issue, passive microwave products have been used that can easily penetrate the clouds and remain unaffected by the weather, but it is not good for regional snow-cover monitoring as it limits the accuracy. Therefore, cloud removal and downscaling are better approaches for monitoring snow cover using both optical and passive microwave products (Huang et al. 2016).

There are two approaches to compute glacier melt runoff, namely, the energy balance approach and the temperature degree day. The energy balance includes the snowpack surface and constitutes four major components, such as shortwave radiation exchange, longwave radiation exchange, convective heat transfer, and advective heat transfer. The degree-day method is intended to simulate the daily streamflow of the Himalayan basin. Here, air temperature is expressed in degree day and is used in snowmelt computation. It is a unit that expresses the amount of heat in terms of temperature for a period of 24 hr.

Generally, snowmelt runoff is calculated by using two methods, namely, the physical model and the conceptual index model. Physical models include models like VIC (variable infiltration capacity), SHE (Systeme Hydrologique European), the LSM (land surface model), whereas the conceptual index models comprise models like the SRM (Snowmelt Runoff Model), the PRMS (precipitation runoff modeling system), and TOPMODEL (hydrologic simulation model), among others. The present study deals with the estimation of trends in snow cover variability using Landsat satellite data along with glacier melt runoff for Zanskar Valley using the SNOWMOD technique.

1.2 STUDY AREA

The study area is located within the longitudinal range from 77° 10′ 12″ N to 75° 59′ 23″ E and the latitudinal range from 33° 5′ 24″ N to 34° 5′ 24″ N (Figure 1.1). It comprises parts of Zanskar Valley between the Padam and Parkachik regions in Ladakh, India. The valley lies at an elevation of 3500–7135 m above the mean sea level. The region of Ladakh is known as a cold desert with an average precipitation of 80 mm estimated by the weather station at Leh. The majority of this precipitation is in the form of snowfall. The southernmost part of the Ladakh lies in a higher Himalayan range that comprises granitoids and metamorphics. The main river in Ladakh is the Indus river and the other two major rivers are Zanskar and Shyok Rivers, and they serve as sources of water for irrigation and cultivation (Zgorzelski 2006). The region consists of sparse vegetation due to its harsh conditions. Xerophytic shrubs and bushes are common types of plants present in Ladakh.

1.3 MATERIALS AND METHODS

In this study, three types of data sets were utilized, namely optical satellite data sets, digital elevation model (DEM) and meteorological data sets (Table 1.1). Optical data

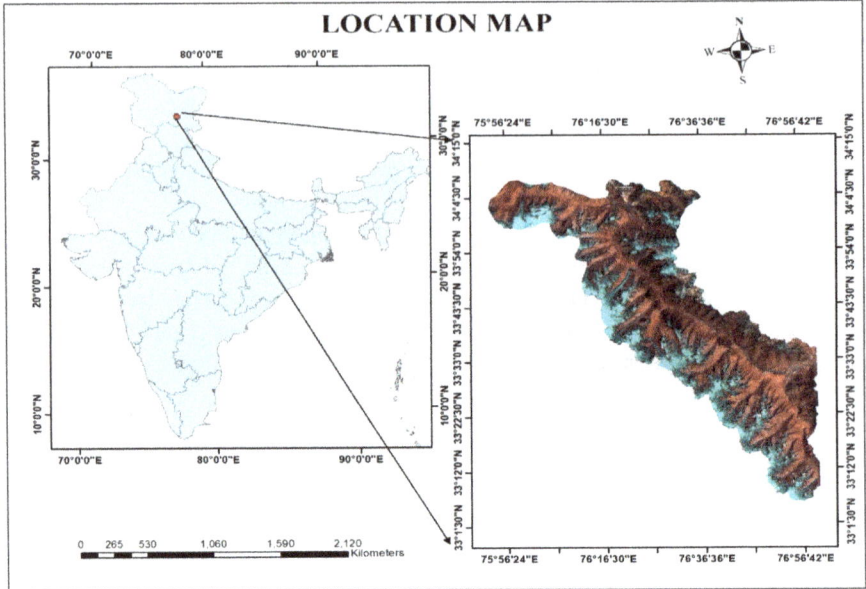

FIGURE 1.1   The study area map of Zanskar Valley.

TABLE 1.1
Data Used in the Present Study

Data Sets	Acquisition Date	Sensor Characteristics	Purpose	Source
Landsat-7	July 2000 July 2005 July 2010	Multispectral Scanner and Thematic mapper	Snow cover monitoring and NDSI estimation	USGS Earth explorer
Landsat-8	July 2015 July 2020	Operational land imager and Thermal infrared sensor	Snow cover monitoring and NDSI estimation	USGS Earth explorer
DEM from ALOS	2012	–	Terrain Analysis	Alaska satellite facility
Meteorological parameters (i.e., precipitation, temperature)	July 2000 – July 2020	–	Parameters used for runoff estimation and correlation with runoff	NASA Power

sets included Landsat series data from 2000 to 2020, whereas DEM data (ALOS PALSAR) were used for the estimation of morphometric parameters and elevation. For meteorological parameters, temperature and precipitation data available with NASA power (https://power.larc.nasa.gov) was acquired.

1.3.1 METHODOLOGY

For this study, the Landsat series data were analyzed from 2000 to 2020 for the peak ablation period (July–August). The Normalized Difference Snow Index (NDSI) has been incorporated to extract snow cover area. NDSI is a measure of the relative difference between visible (green) and short-wave infrared (SWIR) reflectance values, and it ranges from 0 to 1.

Runoff was estimated using the SNOWMOD technique, which incorporates various glacier domain components for runoff calculation (Figure 1.2). The three basic components include runoff from the snow-cover area (SCA), runoff from the snow-free area (SFA), and runoff from the subsurface area. Runoff from SCA comprises three subclasses, such as runoff triggered by an increase in air temperature, melt caused by heat transfer to snow surface due to rain, and runoff from precipitation itself falling over SCA. In contrast, the runoff from SFAs comprises runoff triggered due to ice, debris cover, and hilly terrain. Various equations used in the study for runoff computation follow.

1.3.1.1 Runoff from the SCA

Various factors that triggered the SCA runoff are as follow:

a. Increase in air temperature above melting temperature.

$$M_{s(i,j)} = C_{s(i,j)} D_{i,j} T_{i,j} S_{c(i,j)} \qquad (1.1)$$

Here s represents snow whereas S_c represents snow cover,
where $M_{s(i,j)}$ = snowmelt on ith day for jth band (mm),
   $C_{s(i,j)}$ = the coefficient of runoff for snow on ith day for jth band,
   $D_{i,j}$ = the degree-day factor on ith day for jth band,
   $S_{c(i,j)}$ = the ratio of SCA to total area of jth band for ith day, and
   $T_{i,j}$ = the surface temperature on ith day for jth band.

b. Melt caused by the heat transferred to the snow surface due to rain

$$M_{r(i,j)} = 4.2\, T_{i,j}\, P_{i,j}\, S_{c(i,j)} \div 325 \qquad (1.2)$$

Here r represents rain,
where $M_{r(i,j)}$ = the snowmelt due to rain on snow on ith day for jth band and
   $P_{i,j}$ = the rainfall on snow on ith day for jth band.

c. Runoff due to rain itself falling over the SCA.

$$R_{s,i,j} = C_{s,i,j}\, P_{i,j}\, S_{i,j} \qquad (1.3)$$

$S_{i,j}$ = the snowfall on ith day for jth band.

$$Q_{sca} = \alpha \sum_{j=1}^{n} (M\,s,i,j + M\,r,i,j + R\,s,i,j) A_{sca} \qquad (1.4)$$

where n = the total number in elevation.

A_{sca} = the SCA on ith day for jth band, and

α = factor (0.0116) was used to convert the runoff depth (mm day^{-1}) into discharge (m^3 sec^{-1}).

1.3.1.2   Runoff from SFA

$$\mathbf{R_{f(i,j)} = C_{r(i,j)}\, P_{i,j}\, S_{f(i,j)}} \tag{1.5}$$

where $\mathbf{C_{r(i,j)}}$ = the coefficient of runoff for rain on ith day for jth band,

$\mathbf{P_{i,j}}$ = the rainfall on snow on ith day for jth band, and

$\mathbf{S_{f(i,j)}}$ = the ratio of SFA to the total area of jth band on ith day.

The total runoff from SFA is calculated by the following expression:

$$\mathbf{Q_{sca}} = \alpha \sum_{j=1}^{n} \left(R_{f(i,j)} \mathbf{A}_{SFA\ i,j} \right) \tag{1.6}$$

where $\mathbf{A_{SFA}}$ = the SFA in different zones.

1.3.1.3   Subsurface Runoff

It represents runoff from the unsaturated zone of the catchment to the streamflow. After runoff from the SCA and the SFA, the remaining water infiltrates into the groundwater and appears as a subsurface flow.

$$\mathbf{R_{b(i,j)} = \beta[(1 - C_{r(i,j)})) R_{f(i,j)} + (1 - C_{s(i,j)})\, M_{t(i,j)}} \tag{1.7}$$
$$\text{where } \mathbf{M_{t(i,j)} = M_{s(i,j)} + M_{r(i,j)} + R_{s(i,j)} \text{ and } \beta = 0.50} \tag{1.8}$$

Hence, the base flow is computed by the following formula:

$$\mathbf{Q_b} = \sum_{j=1}^{n} \left(R_{b(i,j)} \mathbf{A}_{i,j} \right) \tag{1.9}$$

where $\mathbf{A}$ is the total area representing the sum of the area of SCA and the area of SFA.

1.3.1.4   Total Runoff

The total runoff was calculated by adding the previously mentioned glacier runoff contributing components:

$$\mathbf{Q = Q_{SCA} + Q_{SFA} + Q_B} \tag{1.10}$$

For the completion of the above study, we have taken into account the following three data sets: satellite image (Landsat series), DEM (ALOS PALSAR), and meteorological data sets (temperature and precipitation). The SCA has been delineated by applying the NDSI over the Landsat image, and with the help of DEM, terrain analysis took place; it includes slope and elevation range analysis through which we

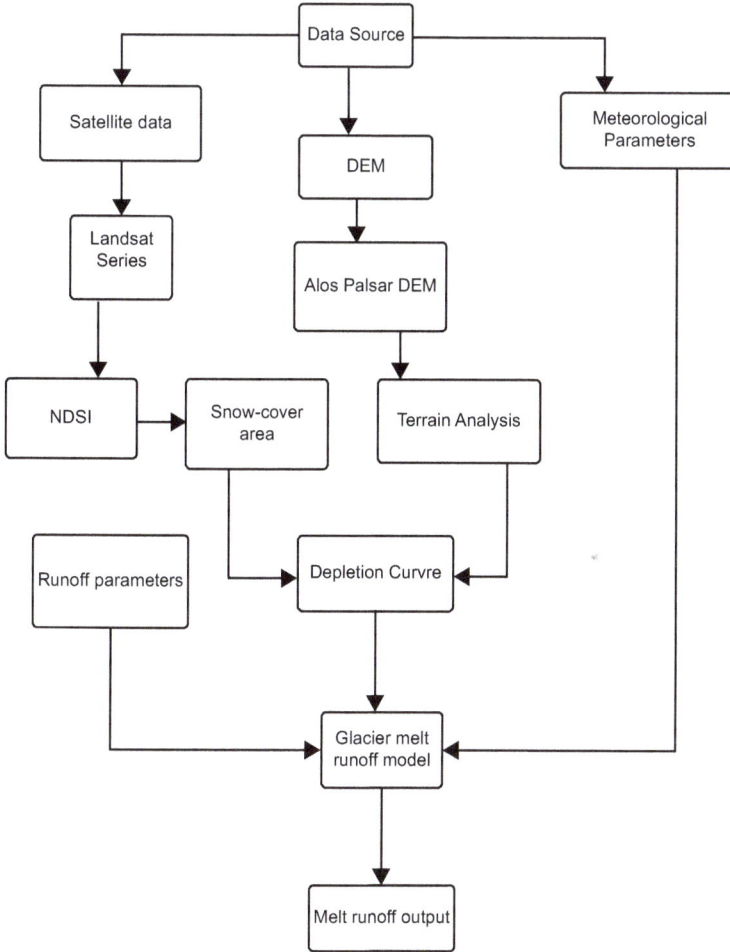

FIGURE 1.2   Methodology flowchart adopted in this study.

can modify the temperature in accordance with the temperature lapse rate with the increasing elevation. Afterward, the snow depletion curve has been estimated, which indicates the change in SCA with time at different elevation levels. And finally with the help of various runoff parameters along with the depletion curve, a glacier melt runoff model was set up, and the melt runoff output was analyzed.

1.4   RESULTS AND DISCUSSION

1.4.1   ESTIMATION OF GLACIER MELT RUNOFF USING SNOWMOD

Terrain mapping of the Zanskar basin was performed along with the generation of its catchment boundaries. The elevation zones (Figure 1.3) are delineated to derive contributing factors of runoff in different elevation zones. The overall area was divided

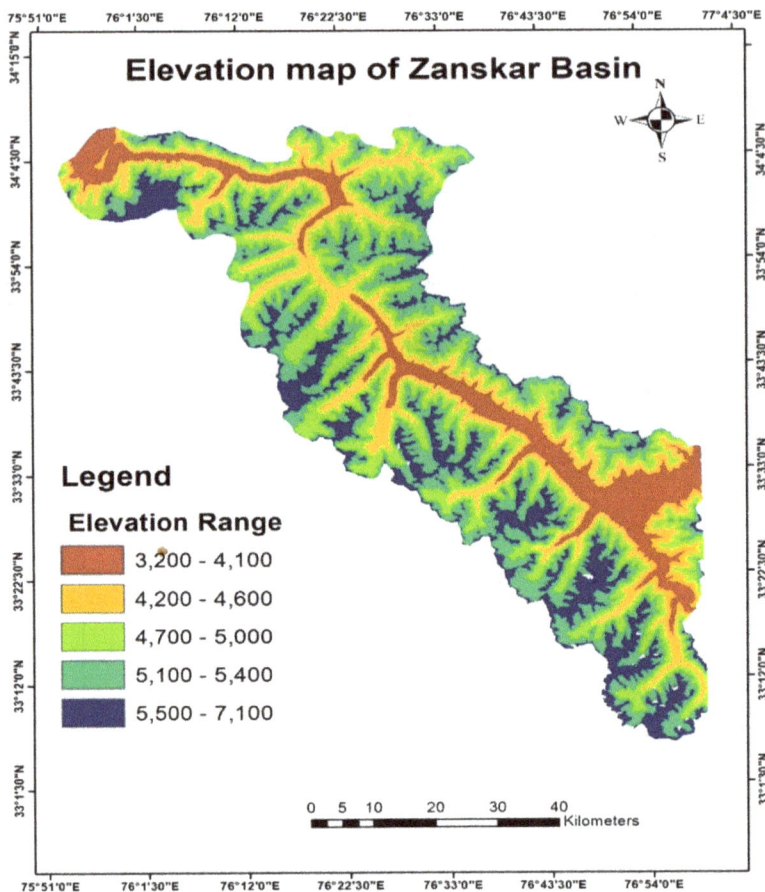

FIGURE 1.3   Elevation map of Zanskar Basin.

into five elevation zones extending from 3200 m to 7100 m. Zone 1 (3200–4100 m) comprises an area of approximately 675 km², zone 2 (4200–4600 m) extends up an area of about 1086 km², zone 3 (4700–5000m) covers an area of approximately 1129 km², zone 4 (5100–5400 m) consists of an area of about 791 km², and zone 5 (5500–7100 m) comprises an area of about 338 km².

The snow-cover map (Figure 1.4) and snow depletion curve (Figure 1.5) showed a continuous increase in snow cover from 2000 (14.41%) to 2005 (73.80%) in the glacier catchment area, while a declining trend in snow cover was observed from 2005 (73.80%) to 2010 (23.09%). An abrupt rise in snow cover was found in 2015 (44.10%) as compared to a major retreat of snow cover in 2020 (37.47%). Various glacier domain components have also been incorporated (Figure 1.6).

The input parameters such as SCA, SFA, subsurface flow, and climatic parameters (i.e., temperature and precipitation) across elevation zones were obtained, and then all the parameters were summed to derive total discharge as glacier melt runoff

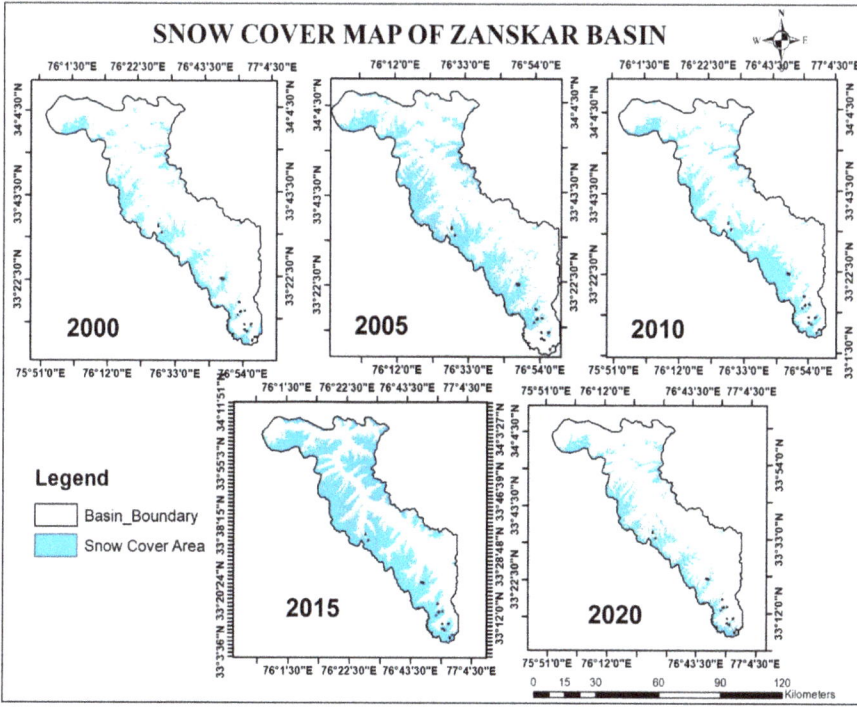

FIGURE 1.4   Snow-cover map.

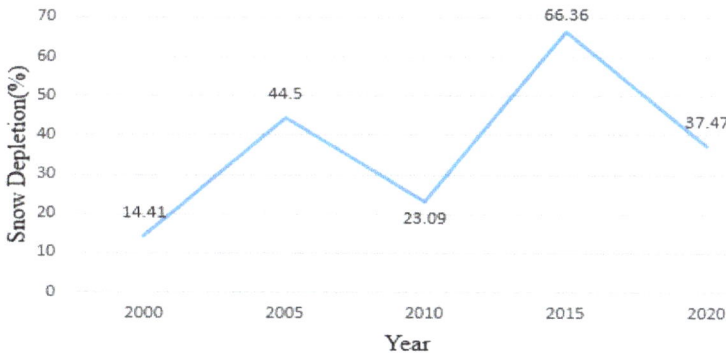

FIGURE 1.5   Snow depletion curve.

(Table 1.2). The calibration and validation of these parameters used in the aforementioned equations have been accomplished by comparison with other studies. The
degree-day factor of 6 (for snow) has been considered in the present study. The SFA
has been further divided into three categories, ice, debris cover, and hilly terrain
(rugged slope), and their runoff is mentioned in Table 1.3.

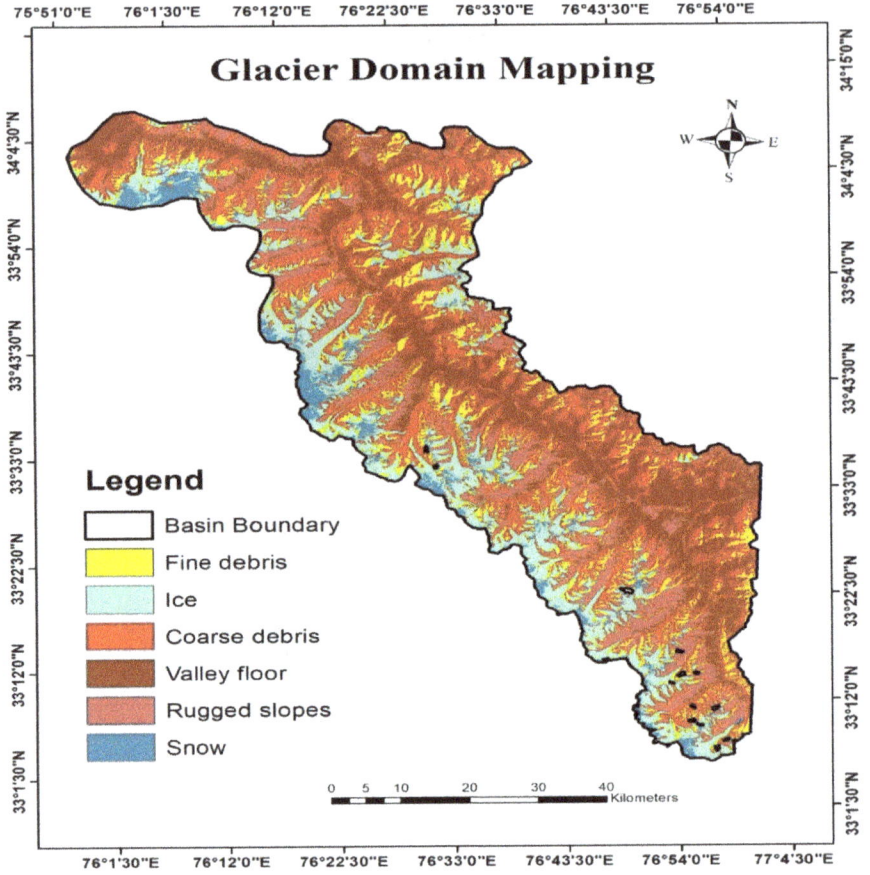

FIGURE 1.6 Glacier domain map of the Zanskar Valley showing the distribution of different types of material across the Zanskar Valley in and around glaciers.

TABLE 1.2

Calculated Parameters for Estimation of Glacier Melt Runoff

Years	SCA Runoff (m³sec⁻¹)	SFA Runoff (m³sec⁻¹)	Subsurface Runoff (m³sec⁻¹)
2000	5.66	92.34	95.63
2005	28.23	64.84	273.42
2010	74.07	115.41	81.75
2015	186.40	176.92	198.36
2020	356.43	270.47	340.75

TABLE 1.3
Categorization of SFA Runoff

Years	Ice Cover Runoff (m³ sec⁻¹)	Debris Cover Runoff (m³ sec⁻¹)	Hilly Terrain Runoff (m³ sec⁻¹)
2000	65.28	16.65	10.41
2005	47.16	11.68	6.1
2010	76.32	26.32	12.77
2015	148.23	22.23	6.46
2020	195.26	43.61	31.6

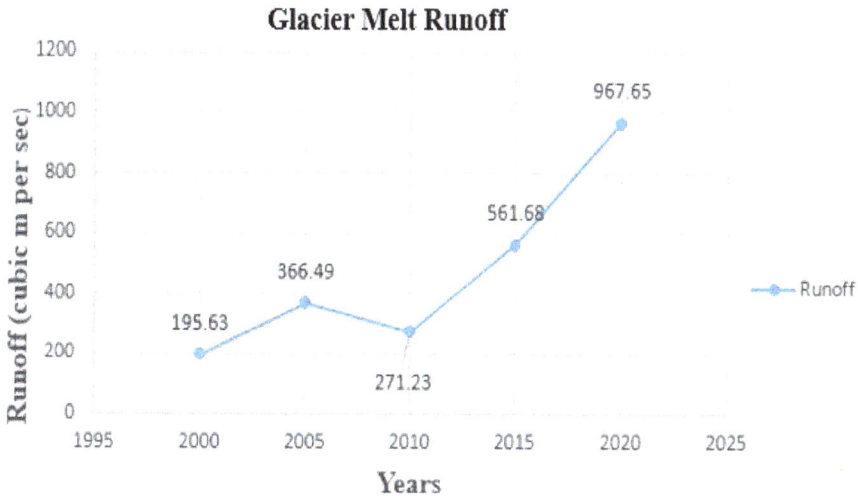

FIGURE 1.7 Glacier melt runoff during the period from 2000 to 2020.

Temporal glacier melt runoff showed that 2020 exhibited the peak melt discharge of 967.65 m³/sec (Figure 1.7), while the lowest melt discharge of 80.65 m³/sec was seen in 2000. Melt discharge increased rapidly from 2000 (193.63 m³/sec) to 2005 (366.49 m³/sec) and exhibited a rapid decrease in 2010 (271.23 m³/sec), followed by an increment in consecutive years from 2015 (561.68 m³/sec) to 2020 (967.65 m³/sec). In the Zanskar Valley, from 1962 to 2001, almost 14 km² of snow area was lost (Schmidt and Nusser 2012). In the Zanskar basin, the glacier cover was reduced to 259.11 km² from 291.8 km² between 1980 and 2000 (Ali et al. 2016). A glacier area loss of about 18.16% was experienced in the Zanskar Valley from 1962 to 2001, with a snout recession of 33 m/year (Nathawat et al. 2008). From 1901 to 2003, the temperature increased, and it possibly caused adverse impacts on snow accumulation and directly affected the snowmelt runoff, which is an essential part of human livelihood either in terms of hydrology or climatology (Beniston 1997). Shakil et al.

(2015) simulated the runoff in the Lidder catchment of the Indus River basin by using the Win SRM model for 2007 (11.94 m³ sec⁻¹) and 2011 (13.51 m³ sec⁻¹). A study carried out in the Gangotri glacier basin between 2010 and 2019 dealing with the snowmelt runoff estimation witnessed that the basin experienced the highest runoff in 2010 (256.79 m³ sec⁻¹), whereas the lowest runoff was experienced in 2019 (80.65 m³ sec⁻¹); the possible reason for this decrement is increased insolation and less precipitation (Salim and Pandey 2021).

In the Zanskar Valley, the highest runoff was experienced in 2020 (967.67m³/sec), whereas the lowest runoff was witnessed in 2000 (195.63 m³/sec). Runoff increased in 2005 (366.45 m³/sec) and again showed a decrement in 2010 (271.23 m³/sec), and afterward, runoff increased till 2020.

1.4.2 RELATION BETWEEN MELT DISCHARGE AND CLIMATE FORCING

In the Zanskar Valley, during the peak ablation period, the highest temperature was found in 2000 (15.83°C), whereas the minimum temperature was witnessed in 2020 (12.88°C). A decreasing temperature trend has been noticed from 2000 to 2020. In the case of precipitation, 2020 (216.54 mm) recorded the highest amount of precipitation in contrast to 2000 (58.67 mm) with minimum precipitation (Figure 1.8). Precipitation showed a sudden increase in 2010 (168.76 mm), and afterward, it turned down in 2015 (125.42 mm) and then again increased in 2020.

From 2000 to 2020, the temperature decreased whereas runoff showed increasing and decreasing behavior. Both runoff and precipitation followed the same trend from 2000 to 2005. The runoff increased with an increase in precipitation but contrasting behavior has been experienced in 2010 in which with an increase in precipitation, the runoff has decreased. From 2015 to 2020, a similar positive trend has been followed like with an increase in precipitation, runoff also increased. In the Zanskar Valley, the coefficient of variation (R^2) between precipitation and runoff exhibited 0.76, which showed a good positive correlation between the two variables (Figure 1.9).

1.5 CONCLUSION

Through this study, the relevance of geospatial techniques for glacier melt runoff estimation has been demonstrated over a part of the Zanskar Valley in India. The total period of study has been divided into four timeframes: 2000–2005, 2005–2010, 2010–2015, and 2015–2020. It was observed that the SCA showed both increasing and decreasing behavior. The utility of SNOWMOD technique accurately demonstrated the trend of glacier melt runoff in the valley, and it exhibited, in general, a continuous increment in runoff from 2000 to 2020. Temperature and precipitation were the prime contributing factors for runoff generation, but precipitation has shown a good correlation with runoff, indicating its major influence on glacier melt runoff. The increasing melt discharge trend possibly reveals the deteriorating health of the glacier system largely in parts of its ablation zones, although increasing snow cover possibly reflects the improving health of accumulation zones in the glaciers of the Zanskar Valley.

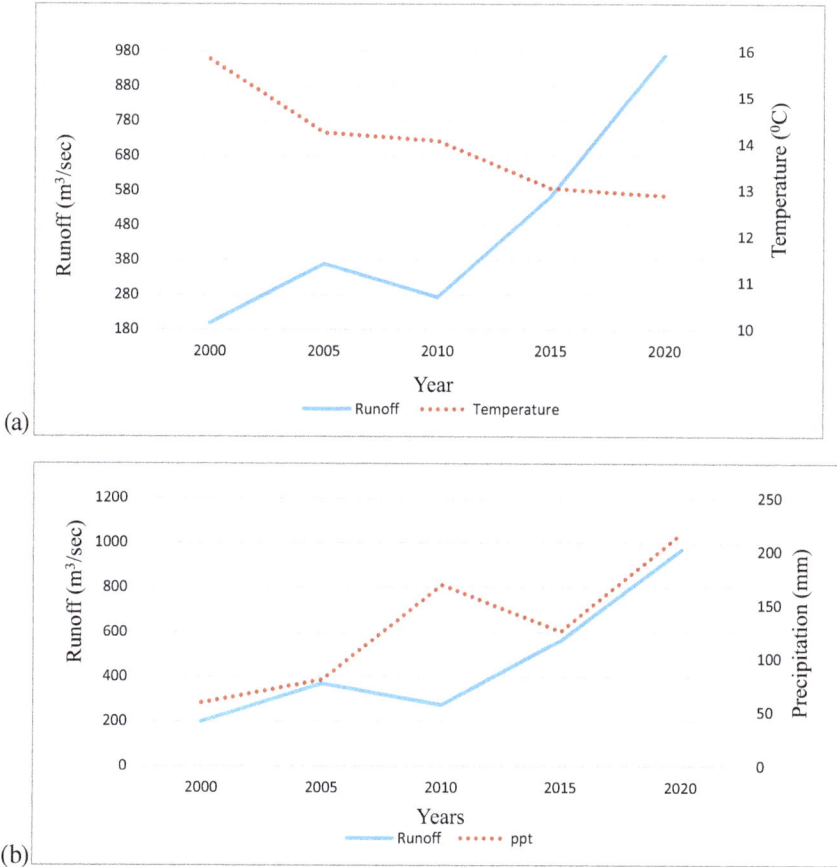

(a)

(b)

FIGURE 1.8 Relation between glacier melt runoff with climatic forcing in the Zanskar Valley where (a) shows with average temperature and (b) with precipitation.

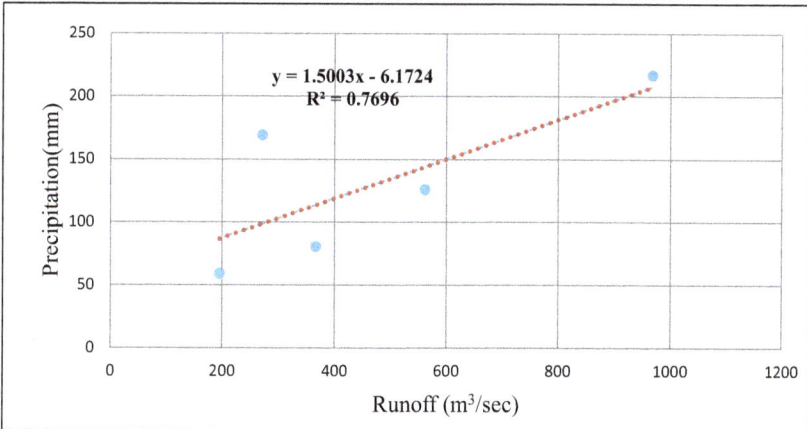

$y = 1.5003x - 6.1724$
$R^2 = 0.7696$

FIGURE 1.9 Correlation between precipitation and glacier melt runoff.

REFERENCES

Agarwal, N. K. 2011. "Remote sensing for glacier mapping and monitoring." *Geological Survey of India Special Publication*, 53, 201–206.

Ali, I., Shukla, A., and Qadir, J. 2016. "Monitoring glacier parameters in parts of Zanskar Basin." In *Geospatial and Geospatial Approach for the Characterization of Natural Resource in the Environment*. Springer International Publishing http://doi.org/10.1007/978-3-319-18663.

Arnell, N. W. 1999. "Climate change and global water resources." *Global Environment Change*, S31–S49.

Bahuguna, I., Rathore, B. P., Jasrotia, A. S., Randhawa, S. S., Yadav, S. K. S., Ali, S., Gautam, N., Poddar, J., Srigyan, M., Dhanade, A., Joshi, P., Singh, S. K., Rajak, D. R., and Sharma, S. 2021. "Recent glacier area changes in Himalaya–Karakoram and the impact of latitudinal variation." *Pagination*, 121(7), 929–940. http://doi.org/10.18520/cs/v121/i7/929-940

Beniston, M., Diaz, H. F., and Bradley, R. S. 1997. "Climatic change at high elevation sites: An overview." *Climatic Change*, 36, 233–251.

Bradley, S. B., Vuille, M., Diaz, H. F., and Vergara, W. 2006. "Threats to water supplies in the tropical Andes." *Science*, 312, 1755–1756. http://doi.org/10.1126/science.1128087

Casassa, G., Smith, K., Rivera, A., Araos, J., Schnirch, M., and Schneider, C. 2002. "Inventory of glaciers in Isla Riesco, Patagonia, Chile, based on aerial photography and satellite imagery." *Annals of Glaciology*, 34, 373–378.

Chinn, T. J., Winkler, S., Salinger, M. J., and Haakensen, N. 2005. "Recent glacier advances in Norway and New Zealand: A comparison of their glaciological and meteorological causes." *Geografiska Annaler*, 87A, 141–157.

Dobhal, D. P., Gergan, J. T., and Thayyen, R. J. 2004. "Recession and morphogeometrical changes of Dokriani glacier (1962–1995) Garhwal Himalaya, India." *Current Science*, 86(5), 692–696.

Foster, J., Hall, D., Eylander, J., Kim, E., Riggs, G., Tedesco, M., Nghiem, S., Kelly, R., Choudhury, B., and Reichle, R. 2007. "Blended visible, passive microwave and scatterometer global snow products." *Proc. 64th Eastern Snow Conf.*, St. Johns, Newfoundland, Canada, 27–36.

Hinkle, K. M., Ellis, A. W., and Mosley, T. 2003. "Cryosphere." In *Geography in America at the Dawn of 21 Century*. Oxford: Oxford University Press, 47–55.

Huang, X., Deng, J., Ma, X., Wang, Y., Feng, O., and Hao, X. 2016. "Spatiotemporal dynamics of snow cover based on multi source remote sensing data in China." *The Cryosphere*, 10(5), 2453–2463.

Immerzeel, W. W., Droogers, P., de Jong, S. M., and Bierkens, M. F. P. 2010. "Satellite derived snow and runoff dynamics in the Upper Indus River Basin." *10th International Symposium on High Mountain Remote Sensing Cartography Band 45/2010*, 303–312.

IPCC. 2007. "Summary for Policy Makers." In M. L. Parry, O. F. Canziani, J. P. Palutikof, P. J. van der Linden, and C. E. Hanson (eds.), *Climate Change 2007: Impacts, Adaptation and Vulnerability. Contribution of Working Group II to the Fourth Assessment Report of the Intergovernmental Panel on Climate Change.* Cambridge: Cambridge University Press, 7–22.

Kaser, G., Fountain, A., and Jansson, P. 2003. "A manual for monitoring the mass balance of mountain glaciers." *Technical Documents in Hydrology*, No. 59. Paris: UNESCO.

Khromova, T. E., Dyurgerov, M. B., and Barry, R. G. 2003. "Late-twentieth century changes in glacier extent in the Ak-shirak Range, Central Asia, determined from historical data and ASTER imagery." *Geophysical Research Letters*, 30, 1863.

Krajik, K. 2002. "Ice man: Lonnie Thompson scales the peaks for science." *Science*, 298, 518–522.

Kulkarni, A. V. 2007. "Effect of global warming on the Himalayan cryosphere." *Jalvigyan Sameeksha*, 22, 93–108.

Kulkarni, A. V., and Bahuguna, I. M. 2002. "Glacial retreat in the Baspa Basin, Himalayas, monitored with satellite stereo data." *Journal of Glaciology*, 48(160), 171–172.

Kulkarni, A. V., Rathore, B. P., Mahajan, S., and Mathur, P. 2005. "Alarming retreat of Parbati Glacier, Beas basin, Himachal Pradesh." *Current Science*, 88, 1844–1850.

Marston, R. A. 2008. "Land, life, and environmental change in mountains." *Annals of the Association of American Geographers*, 98, 507–520.

Nathawat, M. S., Pandey, A. C., Rai, P. K., Ahmad, S., and Bahuguna, I. M. 2008. "Spatiotemporal dynamics of glaciers in Doda valley, Zanskar Range, Jammu & Kashmir, India." In *Proceedings of the International Workshop on Snow, Ice, Glacier and Avalanches*, IIT Bombay, 256–264.

Prasad, A. K., Yand, K. H. S., El-Askary, H. M., and Kafatos, M. 2009. "Melting of major glaciers in the western Himalayas: Evidence of climatic changes from long term MSU derived tropospheric temperature trend (1979–2008)." *Annales Geophysicae*, 27, 4505–4519.

Rathore, B., Kulkarni, A., and Sherasia, N. K. (2009). "Understanding future changes in snow and glacier melt runoff due to global warming in Wangar Gad Basin, India." *Current Science*, 97, 1077–1081.

Salim, M., and Pandey, A. C. 2021. "Estimation of temporal snowmelt runoff using geospatial technique in Gangotri glacier basin, Uttarakhand, India." *Remote Sensing Applications: Society and Environment*, 24, 100660. https://doi.org/10.1016/j.rsase.2021.100660.

Schmidt, S., and Nüsser, M. 2012. "Changes of high altitude glaciers from 1969 to 2010 in the Trans-Himalayan Kang Yatze Massif, Ladakh, Northwest India." *Arctic, Antarctic, and Alpine Research*, 44(1), 107–121.

Shakil, A. R., Dar, A. R., Rashid, I., Marazi, A., Ali, N., and Zaz, S. N. 2015. "Implications of shrinking cryosphere under changing climate on the streamflow in the Lidder catchment in the upper Indus basin, India." *Arctic, Antartic and Alpine Research*, 47, 627–644.

Singh, P., and Jain, S. K. 2009. "Snow and glacier melt in the Satluj River at Bhakra Dam in the western Himalayan region." *Hydrological Sciences*, 47, 93–106.

Song, C., Woodcock, C. E., Seto, K. C., Lenney, M. P., and Macomber, A. S. 1999. "Classification and change detection using Landsat TM data: When and how to correct atmospheric effects." *Remote Sensing of Environment*, 75, 230–244.

Stocker, T. F., Qin, D., Pattner, G. K., Tignor, M., Allen, S. K., Boschung, J., Nauels, A., Xia, Y., Bex, V., and Midhley, P. M. 2013. "The physical science basis." *Contribution of Working Goup 1 to the Fifth Assessment Report of IPCC 1535*. ISBN 978-92-9169-138-8

Winkler, S., Haakensen, N., Nesje, A., and Rye, N. 1997. "Glaziale Dynamik in Westnorwegen: Ablauf und Ursachen des aktuellen Gletschervorstoßes am Jostedalsbreen." *Petermanns Geographische Mitteilungen*, 141, 43–63.

Yablokov, A. 2006. "Climate change impacts on the glaciation in Tajikistan." In *Assessment Report for the Second National Communication of the Republic of Takikistan on Climate Change*. Dushanbe: Tajik Meteorological Service (In Russian).

Zgorzelski, M. 2006. "Ladakh and Zanskar." *Miscellanea Geographia*, 12, 13–24.

2 Comparative Assessment on Glacier Velocity Estimation Using Optical and Microwave Satellite Data over Select Large Glaciers across the Himalayas

Nishant Tiwari, Arvind Chandra Pandey*, and Chandra Shekhar Dwivedi

Department of Geoinformatics, Central University of Jharkhand, Ranchi-835222, India

CONTENTS

2.1 Introduction 19
2.2 Study Area...21
2.3 Data and Methodology 22
2.4 Results... 24
    2.4.1 Morphometric and Terrain Parameters 24
    2.4.2 Glacier Velocity (Optical).. 25
    2.4.3 Glacier Velocity (Microwave)... 26
    2.4.4 Comparison between Optical- and Microwave-Derived Glacial Velocity 27
    2.4.5 Glacier Velocity and Morphometric Parameters 30
2.5 Conclusion 30
Acknowledgment 31
References...31

2.1 INTRODUCTION

Glacier studies, with the help of remotely sensed data, have become widely accepted because of the difficulties in measuring the properties of glaciers from the field itself.

DOI: 10.1201/9781003265160-3

Satellite remote sensing has proved to be a practically useful tool because the area of interest is often inaccessible (König et al., 2001). Remote-sensing techniques have helped in studying supraglacial lake formation, determining the equilibrium line of altitude, monitoring the annual mass balance change, and estimating the various movement rates (Berthier et al., 2007). Demarcating the glacier boundary from remotely sensed data is standard, although debris can cause some serious complications (Bhambri et al., 2011; Shukla et al., 2010; Paul et al., 2007). The Equilibrium Line Altitude (ELA) has been determined with the use of remote sensing, with the use of Landsat-4 (TM), 5 (TM), 7 (ETM+), SPOT 1 to 5, and ASTER (Rabatel et al., 2013). The changes in surface elevation can be accurately done with the help of different periods of the digital elevation model (DEM; Berthier et al., 2007). Mass balance, velocity, and volume assessments have also been done through satellite and airborne data, such as the use of airborne laser altimetry for the estimation of mass balance changes in Alaska (Arendt, 2002). Kumar et al. (2011) utilized a one-day temporal difference ascending and descending pass of the European Remote Sensing satellite (ERS-1/2) Tandem mission satellite data to derive the three-dimensional (3-D) velocity of Siachen Glacier.

Snow cover insulates the ground, and fresh snow reflects up to 81% of the incoming solar energy, which compares drastically with only 20% of bare ground (König et al., 2001; Weller and Holmgren, 1974). The Himalayan glaciers are more sensitive to climatic fluctuations than other glaciers as they are located near to Tropic of Cancer and they receive more heat from solar radiation (Ahmad et al., 2004); this sensitivity to climatic fluctuations has been made worse with the help of climate change, as reflected by the glaciers showing continuous retreat since 1850 (Kumar et al., 2011). Studies have shown an accelerated depletion in glacier areas and an increase in areas of supraglacial ponds and lakes all over India over a few decades (Bhambri et al., 2011; Shukla et al., 2009; Kulkarni et al., 2007). However, recent studies have shown that this is not the case for every region of the Himalayas, as Pandey et al. (2011) reported an advancement and an increase in the area of the glaciers in the Zanskar region of the Great Himalayan range.

Understanding the importance of glaciers as a climate change indicator increases the interest of scientists to monitor glaciers, especially for their surface velocities and retreat rate. Smaller glaciers respond quickly to climatic variations due to lower response times. Methods have been created to estimate the ice thickness of the glacier using surface velocity (McNabb et al., 2012), which helps a lot in understanding the mass balance of the glacier. Multiseasonal data revealed a large difference between summer and winter flow characteristics of the glacier (Quincey et al., 2009). Geomorphological features of the Gangotri glacier had been estimated to access the retreat of the glaciers (Naithani et al., 2001). The availability of satellite data through different sensors renders the selection of the right kind of data sets and methodology for the glacier studies. Studies indicate that different tools led to various kinds of results in terms of magnitude and direction of the flow of glaciers (Jawak et al., 2018). A comparison between the different tools and techniques helps ascertain the best technique for obtaining the glacier surface velocity.

The use of offset tracking and interferometry are widely accepted ways for velocity estimation using Synthetic Aperture Radar (SAR) data sets. Surface motion can

be estimated from InSAR data within the data repeat interval (Goldstein et al., 1993). The remote-sensing technique was used to estimate the velocity of Nabesna Glacier in North America using interferometric techniques using ERS-1/2 SAR images (Li et al., 2008). Zongli et al. (2012) used SAR feature tracking based on ALOS/PALSAR L-band to derive the surface velocity of the Yengisogat Glacier in the Karakoram Range. The use of COSI-Corr (co-registration of optically sensed images and correlation) has been validated in several case examples of horizontal ground displacements (Herman et al., 2011; Ayoub et al., 2009) using the methodology provided by Leprince et al. (2007) and Scherler et al. (2008). Copland et al. (2011) have used the optical remote-sensing satellites, mainly Landsat, to estimate the glacier velocity of the glacier in the Karakoram range; apart from Landsat, JERS-1 and ASTER data sets were also used. Optical data have also been used to estimate the velocity of the debris-covered glaciers using the COSI-Corr technique (Sam et al., 2016).

The estimation of glacier velocities using optical and SAR data sets amount to various errors. While using SAR data sets, the error arises from the SAR image geometry, leading to incomplete spatial coverage due to layover and shadowing (Huang and Li, 2011; Strozzi et al., 2002), the optical data sets rely heavily on the illumination of the sun and are severely limited by the cloud cover (Huang and Li, 2011). Therefore, a proper comparison is needed to understand the utility and accuracy of different types of data sets available for the glacier velocity estimation. In this chapter, the glaciers chosen across the Himalayan belt were studied for the same period (2015–2019) for a comparative assessment of their glacier velocity using the optical and SAR data sets.

2.2   STUDY AREA

The study area encompasses select glaciers from the entire Himalayan range (Figure 2.1). The glaciers lie in Jammu and Kashmir (Siachen Glacier), Himachal Pradesh (Bara Sigri Glacier), Uttarakhand (Gangotri Glacier), Nepal (Langtang Glacier), and Sikkim (Zemu Glacier). The Siachen Glacier is in the Karakoram range; it is the longest glacier in the Himalayan Region. The glacier's climate is heavily influenced by the Indian monsoon and the westerlies, and the microclimatic activities are influenced by the terrain (Sun et al., 2017; Hewitt et al., 1989). The Bara Sigri Glacier is in the Chandra valley of Lahaul–Spiti in the western Himalayan district. The Chandra valley has the monsoon-arid transition zone and is climatically influenced by both the Indian summer monsoon (ISM) and western disturbances (Yellala et al., 2019).

The Gangotri Glacier is one of the longest glaciers in the Himalayan region and is the source for one of the important rivers, the Ganga. The Gangotri Glacier originates from the Chaukhambha group of peaks, which has unique characteristics, namely, an abundance of supraglacial moraines, supraglacial lakes, numerous transverse crevasses, and active and inactive tributary glaciers (Singh et al., 2017). The Langtang Glacier is one of the longest glaciers in Nepal, and it is heavily covered in debris. About 35% of the glacier is covered in debris, and the climate of the glacier is managed by the monsoon period (more than 70% annual precipitation)

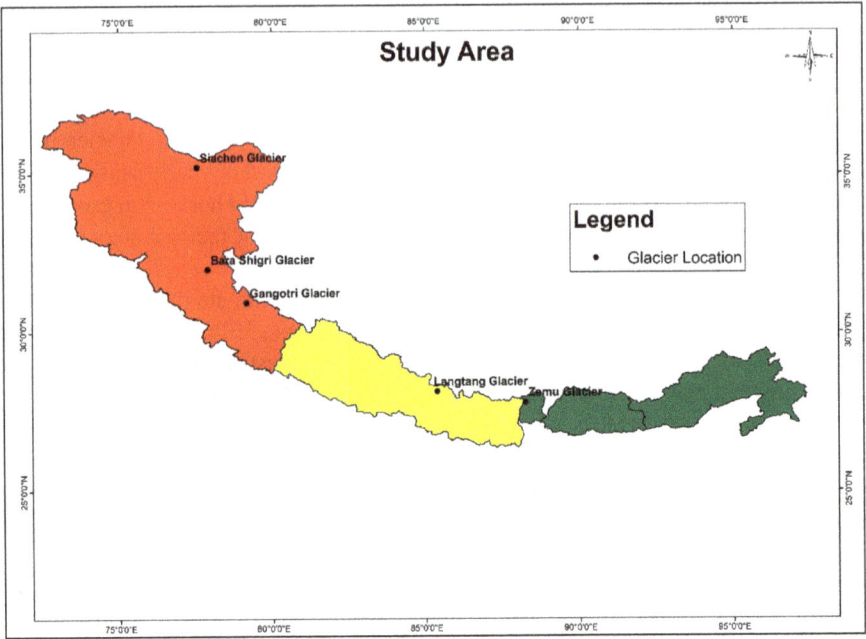

FIGURE 2.1   Study area map displaying glacier location.

and by the westerlies in the winter (Wijngaard et al., 2019; Collier and Immerzeel, 2015; Immerzeel et al., 2012). Finally, the Zemu Glacier lies at the foot of Mount Kangchenjunga; it is fairly covered in debris and receives a large amount of precipitation from the Asian monsoon (Garg et al., 2019; Bookhagen and Burbank, 2010).

2.3   DATA AND METHODOLOGY

Various kinds of data sets are used in this chapter; the data were selected based on the parameter under consideration, that is, morphometric parameters and glacier velocity with optical and microwave data. For physical parameters like area and length, Sentinel-2A data were used, and ALOS W30D data were used for terrain parameters like elevation, slope, and aspect. For glacier velocity estimation through optical data, Landsat-8 data were used, while Sentinel-1A data were used for velocity estimation through SAR data (Table 2.1).

Manual delineation is used for demarcating the glacier boundary in 2019 using ArcMap and Sentinel-2A data. To increase the glacier boundary accuracy, DEM was used. Morphometric parameters like glacier area, length, snout position, and terrain parameters like elevation, aspect, and slope were mapped using ALOS-derived DEM and Sentinel-2A data sets. COSI-Corr software was used for glacier velocity estimation using optical data sets, and an offset-tracking method was used for velocity estimation using microwave data sets. The flowchart of the methodology used in the study is given in Figure 2.2.

TABLE 2.1

Satellite Data Used and Their Specifications

Data Set	Glacier	Acquisition Date (2019)	Acquisition Date (2015)	Purpose	Resolution (m)
Sentinel-2A	Siachen	08-06-2019	–	Glacier Boundary Delineation	10m
	Bara Shigri	28-08-2019	–	Glacier Boundary Delineation	10m
	Gangotri	01-06-2019	–	Glacier Boundary Delineation	10m
	Langtang	28-08-2019	–	Glacier Boundary Delineation	10m
	Zemu	24-06-2019	–	Glacier Boundary Delineation	10m
Landsat-8	Siachen	08-06-2019	07-04-2015	Glacier Surface Velocity Estimation	30m
	Bara Shigri	28-08-2019	15-09-2015	Glacier Surface Velocity Estimation	30m
	Gangotri	01-06-2019	08-09-2015	Glacier Surface Velocity Estimation	30m
	Langtang	28-08-2019	23-10-2015	Glacier Surface Velocity Estimation	30m
	Zemu	24-06-2019	09-10-2015	Glacier Surface Velocity Estimation	30m
Sentinel-1A	Siachen	14-08-2019	11-08-2015	Glacier Surface Velocity Estimation	30m
	Bara Shigri	14-08-2019	23-08-2015	Glacier Surface Velocity Estimation	30m
	Gangotri	27-08-2019	27-08-2015	Glacier Surface Velocity Estimation	30m
	Langtang	20-08-2019	17-08-2015	Glacier Surface Velocity Estimation	30m
	Zemu	02-08-2019	23-08-2015	Glacier Surface Velocity Estimation	30m
ALOS-DEM	All	2016		Terrain Parameters	30m

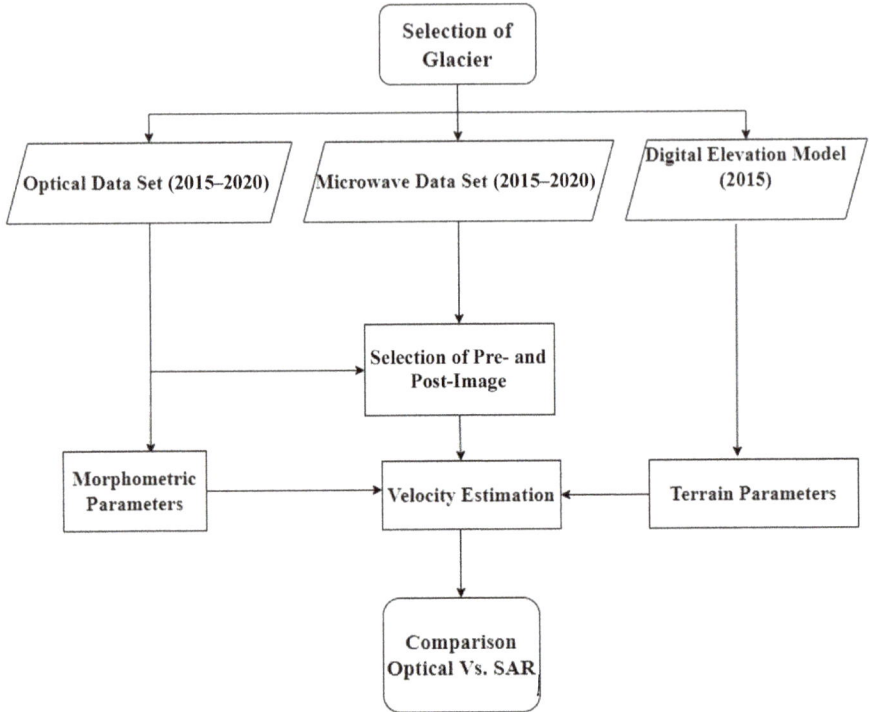

FIGURE 2.2   Methodology flowchart adopted in this study.

2.4   RESULTS

2.4.1   MORPHOMETRIC AND TERRAIN PARAMETERS

Morphometric parameters are needed to study glaciers for the aspects of glacier dynamics. Morphometric parameters taken into consideration in the present
study include glacier length, area, snout altitude, minimum and maximum elevation, aspect, and slope. The highest elevation is represented by the Siachen Glacier,
Western Himalaya, at approximately 7395 masl, and the lowest elevation was also
observed at the snout of the Siachen glacier (3600 masl). Large glaciers under study
exhibit a northwest–southeast orientation as represented by glaciers in the western
and central Himalayas, whereas in the eastern Himalayas, select glaciers show east–
west orientation. The accumulation zone in all the glaciers exhibits a slope of about
7 degrees, whereas in the ablation zone slope ranged from 7–15 degrees. Some parts
of glaciers (near the summit) showed higher degrees of slope; therefore, the slope
was considered along the center line. Table 2.2 shows the data of the morphometric
parameters obtained through various satellite data.

Siachen Glacier is southeast-facing compared to northwest-facing Gangotri and
Bara Sigri Glaciers, whereas Langtang and Zemu are south- and east-facing, respectively. The length of the glacier showed that the western Himalayas had longer glaciers

TABLE 2.2
Morphometric and Terrain Parameters

Glacier Name	Max. Altitude (m)	Min. Altitude (m)	Orientation/ Facing	Dominant Slope Class (degrees)	Area (km²)	Perimeter (km)	Length (km)
Siachen	7395	3723	NW–SE/SE	<7	778	1022	69.71
Gangotri	6974	4038	NW–SE/NW	<7	122	205	31.63
Bara Sigri	6336	3970	NW–SE/NW	<7	121	306	27.68
Langtang	6712	4456	N–S/S	7–15	39	135	18.69
Zemu	6558	4065	E–W/E	7–15	74	184	23.95

than the central Himalayas, and the eastern Himalayas had shorter glaciers. A visual analysis of the satellite images showed that glaciers in the eastern Himalayas had poor health, with the presence of large number of pro-glacial lakes.

2.4.2 GLACIER VELOCITY (OPTICAL)

Landsat-8 images were used for the generation of the surface velocity output using COSI-Corr software, using the SWIR band images. The windows used for the velocity estimation varied from 1024 to 256, depending on the estimated displacement by the glaciers, as they amount to different velocities and sizes. The highest velocity estimated was of Siachen Glacier, which showed an average velocity of about 60 m per year over the period of the study. The maximum velocity exhibited by the Siachen Glacier, which was about 96.64 m per year, and the minimum velocity was 20.53 m per year (Table 2.3). The Bara Sigri Glacier showed the second-highest average velocity at 30 m per year, with the highest velocity of 55 m per year and the lowest velocity of 7.13 m per year. Langtang Glacier has the highest velocity of 20 m per year, with an average velocity over a major part of the glacier of 13 m per year and the lowest velocity being only 0.78 m per year. All these velocities were taken along the center line of the glacier and then were averaged.

The Siachen Glacier had a high-velocity zone (more than 70m/year) in the upper part of the ablation zone, while the accumulation zone had some areas under high velocity. The snout had a lower velocity compared to the upper ablation zone; one of the tributaries (the tributary on the east of the glacier) had a high velocity near the junction. In the Bara Sigri Glacier, the lower accumulation zone and the upper ablation zone were under the high-velocity zone (more than 35m/year), after which the velocity gradually decreased. The Gangotri Glacier had a high-velocity zone (more than 30m/year) in the lower ablation zone and near the snout; meanwhile, its tributaries had lower velocity compared to the trunk. The Langtang Glacier had a high-velocity zone (more than 15m/year) in the mid-trunk at the upper ablation zone and near the snout of the glacier. One of its tributaries (the southernmost) also showed a high velocity in the upper ablation zone. The Zemu Glacier had a high-velocity zone (more than 18m/year) at the lower accumulation zone and the upper ablation zone,

TABLE 2.3

Velocity Using Optical Data Sets

Glacier Name	Maximum Velocity (m/year)	Minimum Velocity (m/year)	Average Velocity (m/year)
Siachen	96.64	20.53	63.12
Bara Shigri	57.19	7.13	34.28
Gangotri	46.12	4.32	30.97
Langtang	21.08	0.78	13.98
Zemu	25.41	2.38	16.82

which was dominant in the southern part of the main trunk; meanwhile, its tributaries showed low velocity.

The velocity maps for Siachen and Gangotri Glaciers showed a lot of noise that was due to the presence of clouds in the image from 2019; therefore, the tributaries in these glaciers showed poor output, although the main trunks of both glaciers were free from clouds/shadows; therefore, the images were taken into consideration.

2.4.3 Glacier Velocity (Microwave)

The velocity map has been produced with high accuracy and then clipped in ArcGIS to match the glacier boundary. The highest velocity was again shown by the Siachen Glacier, where the average velocity along the center line was about 72 m/year. The Bara Sigri Glacier and the Gangotri Glacier had the second- and third-highest velocities at 38 m/year and 35 m/year, respectively, for each glacier. The lowest velocity was recorded over the Langtang Glacier, with a velocity of 13 m/year, whereas the Zemu Glacier showed a velocity of about 18 m/year. All the velocities discussed here are the average velocities of the glaciers and are taken along the center lines of the glaciers. The maximum and minimum velocities are given in Table 2.4.

A much larger high-velocity zone is seen in Siachen Glacier using the SAR data sets; the upper ablation zone showed a much higher area under high velocity, with some patches in the accumulation zone too. The Bara Sigri Glacier had a high-velocity zone toward the lower side of the ablation zone, where the lower accumulation zone showed decent velocity (less than 30m/year), while the upper accumulation zone and the tributaries showed exceptionally low velocity. The Gangotri Glacier showed a large area under the high-velocity zone, near the snout and the lower ablation areas. Some areas in the upper ablation zone also had high velocity, but these regions were small. The Langtang Glacier showed a significant area under the high-velocity zone in the middle part of the trunk and in the upper ablation area, where its tributaries also showed high velocity in the accumulation region. The Zemu Glacier showed high velocities under the upper ablation zone, just after the ELA, gradually decreasing as it reached the lower ablation zone.

TABLE 2.4
Velocity Using SAR Data Sets

Glacier Name	Maximum Velocity (m/year)	Minimum Velocity (m/year)	Average Velocity (m/year)
Siachen Glacier	98.13	7.04	72.05
BaraShigri Glacier	42.97	3.28	38.26
Gangotri Glacier	38.26	2.05	35.49
Langtang Glacier	20.01	1.51	13.06
Zemu Glacier	22.35	1.71	18.22

2.4.4 Comparison between Optical- and Microwave-Derived Glacial Velocity

Comparing the outputs, the images acquired through SAR had more area under a high-velocity zone than the optical counterparts. It is clearly seen that almost all the glaciers had most of their ablation zone in a high-velocity region, whereas in optical-derived maps, only in the Zemu Glacier did more regions lie in the high-velocity zones (Figure 2.3). The Bara Sigri and Gangotri Glaciers mostly showed low velocity compared to the SAR-derived outputs. Most of the glaciers showed a similar pattern of velocity. The regions around where the velocity was high are similar in all the glaciers, and the only difference was the extent of these velocity regions, which was found more in the SAR-derived velocity product.

There are many differences found between the obtained velocity maps owing to different data types and their inherent errors. The optical data sets were not that much clearer and were subjected to various types of errors and defects. Partial cloud cover is present in almost all the optical data sets; therefore, the velocity output generated with these images was not exactly accurate and might show less or more velocity than the actual value. Microwave data, however, is not generally affected by these errors; in fact, the ability to differentiate the wet and dry ice helps it more often. As can be seen in the velocity images that were created by the optical data sets, some places that were covered with cloud gave wrong velocity values. The images also were not very continuous and had many patches with different velocities banded together. The velocities were different for both data sets, and the difference in those values is given in Table 2.5, along with the data from published literature.

Velocity estimates derived from SAR images were close to the data obtained in earlier studies, although the velocity estimates from the optical images were also similar. The difference obtained with reference to published glacier velocity results could be attributed to different periods of observation, the window size used for the estimation of the velocity, the data used, and the inherent errors in the individual satellite data. Since the use of any technique for the velocity estimation depend heavily on the previous knowledge of the glacier's velocity and surface ice movement data, the smaller and less studied glaciers usually give a large difference, as in this case

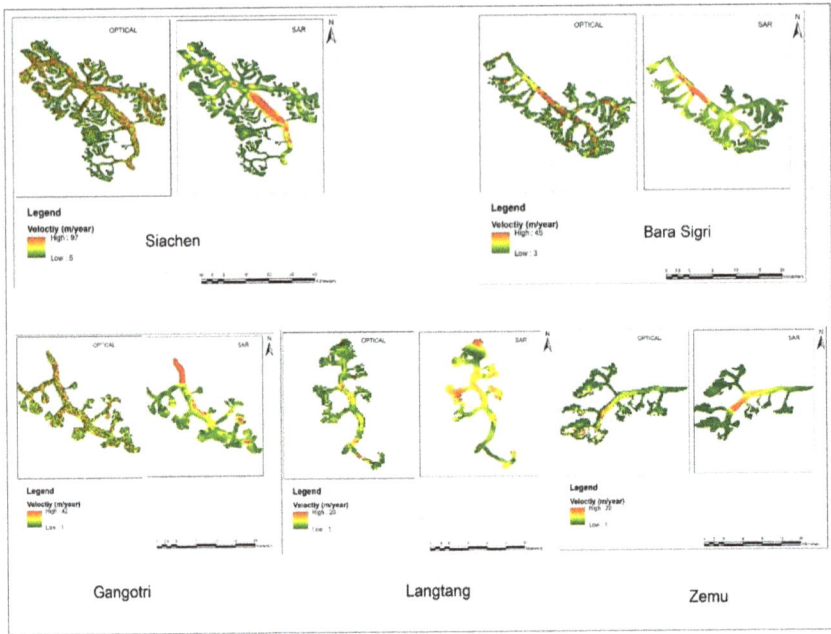

FIGURE 2.3   Optical- and SAR-based velocity maps across different glaciers.

TABLE 2.5

Velocity Comparison between SAR and Optical. The maximum (max) value has been also provided

Glacier Name	Average Velocity Optical (m/year)	Average Velocity SAR (m/year)	Difference in Average Velocity (%)	Previous Study (m/year)
Siachen Glacier	63.12	72.05	12.39%	91.06 (Sun et al., 2017)
Bara Sigri Glacier	34.28	38.26	10.40%	32.5 (2003/04) and 25.3 (2013/14) (Garg et al., 2017)
Gangotri Glacier	30.97	35.49	12.73%	15–20 (cm/year) (max) (Lal et al., 2018)
Langtang Glacier	13.98	13.06	−7.04%	7 (Wijngaard et al., 2019)
Zemu Glacier	16.82	18.22	7.68%	69.71 m/year (max) (Garg et al., 2019)

with the Langtang Glacier. In the case of the Siachen Glacier, Sun et al. (2017) used Landsat-7 images from 1999 to 2003 and used the COSI-Corr for the estimation of the glacier surface velocity. It is important to mention that they used the lower 30% of the glacier length for the center-line average of the velocity. In contrast, in the present study, the images were taken at a 5-year difference, and the whole center

line average (excluding the start and end points of the glacier) was taken. Similarly, in Bara Sigri Glacier, Garg et al. (2017) used TERRA-based ASTER images for the surface velocity estimation using the feature tracking method (Leprince et al., 2007), and they measure the velocity for two periods, 2003–2004 and 2013–2014, and found a significant slowing of the surface velocity. For the Gangotri Glacier, Lal et al. (2018) used short-interval Terra SAR X-band and C-band images to calculate the velocity of the Gangotri and Siachen Glaciers. Satyabala (2016) noted a large difference in the surface velocity of the Gangotri glacier in the summer and winter periods. Therefore, a long-term assessment done in the present study would account for this difference in velocity. Wijngaard et al. (2019), who studied the Langtang Glacier, found the velocity of the glacier about 6 m/year using ASTER VNIR Band 2 and COSI-Corr software for the estimation of surface velocity; however, again, the temporal gap was not as high as attempted in the present study, which accounted for a bigger window size during the velocity estimation. In the case of Zemu Glacier, Garg et al. (2019) used Landsat-5 (TM) and Landsat-7 (ETM+) images with the COSI-Corr technique to calculate the surface glacier velocity. A large temporal difference in data sets would lead to a large difference in velocity (~10 cm/day) as compared to shorter interval data (Lal et al., 2018). Therefore, the observed difference in glacier velocity estimates obtained in the present study in comparison to previous studies could be attributed to the large period of observation taken.

Figure 2.4 shows the difference between the velocity outputs by SAR and optical data sets. The lowest difference in the average velocities was obtained in the Langtang Glacier, where only a 7% difference was seen. Bigger glaciers, like the Siachen, the Bara Sigri, and the Gangotri, did not have much difference (around 10–12%). The Zemu Glacier also showed only a 7% difference in the study. SAR and optical data through this study seem to be complementary, but many other factors heavily affect the selection of these data sets. Since SAR data sets are generally bigger and more complex, they therefore require powerful computational processors. Meanwhile, the optical data sets are lighter in comparison to SAR; therefore, they require less computational power. Again, the atmospheric parameters affect optical data sets heavily but not SAR data sets. Terrain parameters, like relief and aspect, affect the SAR data sets more than they would affect optical data sets.

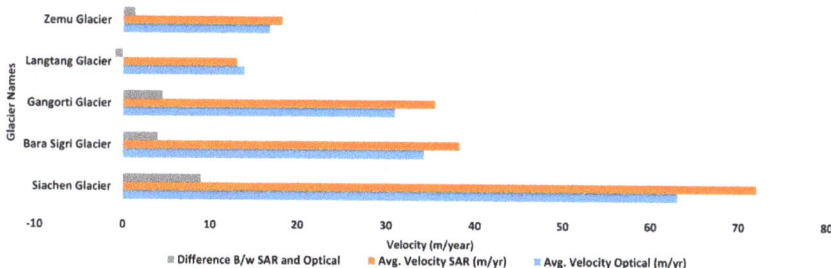

FIGURE 2.4   Difference between velocity across the glaciers between SAR and optical data.

2.4.5   GLACIER VELOCITY AND MORPHOMETRIC PARAMETERS

Glacier area is one of the major factors influencing the glacier velocity (since mass is directly proportional to area and length). A larger size of the glacier means a higher glacier velocity (Shukla et al., 2020) as represented by the Siachen Glacier, with the highest velocity, followed by Bara Shigri and Gangotri, Zemu, and the smallest Langtang Glacier, with the least velocity obtained.

The Langtang Glacier, having a higher elevation than Bara Shigri Glacier, exhibited the lowest velocity largely on account of its small area and mass compared to other glaciers under observation. Both the Langtang and Zemu Glaciers, being located at higher elevations (than Bara Shigri and comparable to Gangotri and Siachen Glaciers), exhibit lower velocities than the other glaciers, possibly reflecting more stability of high-relief small glaciers as compared to lower-relief large glaciers. From the Siachen to the Gangotri Glacier (western to central Himalayas), the average slope of the glacier kept on increasing, but the velocity exhibited a decreasing trend.

Although higher velocities are associated with a steep slope, there is only an insignificant correlation between slope and glaciers' surface velocity (Shukla et al., 2020). Tiwari et al. (2016) also suggested that slope does not affect the glacier velocity significantly as the size does up to a certain area threshold. Terrain factors, such as aspect, also play an important role in the retreat velocity of the glaciers as they control the amount of solar energy the glacier receives. North-facing slopes receive minimum solar radiation while south-facing slopes receive comparatively higher solar radiation, whereas east-facing glaciers receive the highest amount of incoming solar radiation (Shukla et al., 2020). Thus, a glacier facing the north can exhibit lower velocity than east-, south-, and west-facing glaciers. The glaciers in the present study that are southwest-facing (Siachen Glacier) showed the highest velocity, followed by northwest-facing glaciers (Gangotri and Bara Sigri) as compared to the east-facing Zemu Glacier, with the lowest velocity over the south-facing Langtang Glacier. This clearly reflects that among the glacier morphometric parameters, only area has a dominant control on glacier velocity.

2.5   CONCLUSION

Glaciers are very dynamic in nature, the changes in the glacier need to be regularly monitored not only because of climate change but also due to the many hazards associated with the glaciers. From the results, it can be concluded that the SAR images are better for the glacier surface velocity assessment than the optical images, since the microwave images are not subjected to atmospheric error. Also, SAR works on the principle of dielectric constant, which allows the signal to clearly identify the ice from snow and dry ice from wet ice as, in winters, the glacier is generally fully covered with snow.

It was found that the SAR-derived velocity product exhibit similarity to glacier velocity estimation obtained in earlier studies employing high-resolution optical data sets. Additionally, the SAR-based outputs were far more continuous than the optical counterparts where velocity variation among adjacent pixels is more prominent. Although the SAR-derived products showed more area under high velocity, as the

average velocity comes under the accurate reading, it is safe to conclude that the surface velocity image output from SAR images shows accurate regions under the different velocity zones. As glaciers such as Siachen, Gangotri, Bara Sigri, Zemu, and Langtang showed almost similar velocities with only up to a 4-m-per-year difference at the maximum, the optical images can be used as a complementary data set considering shadow effects in a high-relief region, cloud coverage, and the availability of data sets.

ACKNOWLEDGMENT

The authors would like to acknowledge the ESA and the USGS to the access of the Sentinel and Landsat data sets. Furthermore, we would like to thank the California Institute of Technology for the access to the COSI-Corr plugin and the Japan Aerospace Exploration Agency (JAXA) for ALOS-derived DEM.

REFERENCES

Ahmad, S., Hasnain, S., and Selvan, M. T. (2004). Morpho-metric Characteristics of Glaciers in the Indian Himalayas. *Asian Journal of Water Environment and Pollution*, 1, 109–118.

Arendt, A. A. (2002). Rapid Wastage of Alaska Glaciers and Their Contribution to Rising Sea Level. *Science*, 297(5580), 382–386.

Ayoub, F., Leprince, S., and Avouac, J. P. (2009). Co-registration and Correlation of Aerial Photographs for Ground Deformation Measurements. *ISPRS Journal of Photogrammetry and Remote Sensing*, 64(6), 551–560.

Berthier, E., Arnaud, Y., Kumar, R., Ahmad, S., Wagnon, P., and Chevallier, P. (2007). Remote Sensing Estimates of Glacier Mass Balances in the Himachal Pradesh (Western Himalaya, India). *Remote Sensing of Environment*, 108(3), 327–338.

Bhambri, R., Bolch, T., and Chaujar, R. K. (2011). Mapping of Debris-covered Glaciers in the Garhwal Himalayas Using ASTER DEMs and Thermal Data. *International Journal of Remote Sensing*, 32(23), 8095–8119.

Bookhagen, B., and Burbank, D.W. (2010). Toward a Complete Himalayan Hydrological Budget: Spatiotemporal Distribution of Snowmelt and Rainfall and Their Impact on River Discharge. *Journal of Geophysical Research: Earth Surface*, 115(3), 1–25.

Collier, E., and Immerzeel, W. W. (2015). High-resolution Modelling of Atmospheric Dynamics in the Nepalese Himalaya. *Journal of Geophysical Research: Atmosphere*, 120, 9882–9896.

Copland, L., Sylvestre, T., Bishop, M., Shroder, J., Seong, Y., Owen, L., Bush, A., and Kamp, U. (2011). Expanded and Recently Increased Glacier Surging in the Karakoram. *Arctic, Antarctic, and Alpine Research*, 43(4), 503–516.

Garg, P. K., Shukla, A., and Jasrotia, A. S. (2019). On the Strongly Imbalanced State of Glaciers in the Sikkim, Eastern Himalaya, India. *Science of the Total Environment*, 691, 16–35.

Garg, P. K., Shukla, A., Tiwari, R. K., and Jasrotia, A. S. (2017). Assessing the Status of Glaciers in Part of the Chandra Basin, Himachal Himalaya Multi-parametric Approach. *Geomorphology*, 284, 99–114.

Goldstein, R. M., Engelhardt, H., Kamb, B., and Frolich, R. M. (1993). Satellite Radar Interferometry for Monitoring Ice Sheet Motion: Application to an Antarctic Ice Stream. *Science*, 262(5139), 1525–1530.

Herman, F., Anderson, B., and Leprince, S. (2011). Mountain Glacier Velocity Variation During a Retreat/Advance Cycle Quantified Using Sub-pixel Analysis of ASTER Images. *Journal of Glaciology*, 57(202), 197–207.

Hewitt, K., Wake, C. P., Young, G. J., and David, C. (1989). Hydrological Investigations at Biafo Glacier, Karakoram Range, Himalaya; an Important Source of Water for the Indus River. *Annuals of Glaciology*, 13, 103–108.

Huang, L., and Li, Z. (2011). Comparison of SAR and Optical Data in Deriving Glacier Velocity with Feature Tracking. *International Journal of Remote Sensing*, 32(10), 2681–2698.

Immerzeel, W. W., van Beek, L. P. H., and Konz, M. (2012). Hydrological Response to Climate Change in a Glacierized Catchment in the Himalayas. *Climate Change*, 110, 721–736.

Jawak, S. D., Kumar, S., Luis, A. J., Bartanwala, M., Tummala, S., and Pandey, A. C. (2018). Evaluation of Geospatial Tools for Generating Accurate Glacier Velocity Maps from Optical Remote Sensing Data. *Proceedings*, 2(7), 341.

König, M., Winther, J., and Isaksson, E. (2001). Measuring Snow and Glacier Ice Properties from Satellite. *Reviews of Geophysics*, 39(1), 1–27.

Kulkarni, A.V. (2007). Effect of Global Warming on the Himalayan Cryosphere. *Jalvigyan Sameekhsha*, 22, 93–108.

Kumar, V., Venkataramana, G., and Høgda, K. A. (2011). Glacier Surface Velocity Estimation Using SAR Interferometry Technique Applying Ascending and Descending Passes in Himalayas. *International Journal of Applied Earth Observation and Geoinformation*, 13(4), 545–551.

Lal, P., Vaka, D., and Rao, Y. (2018). Mapping Surface Flow Velocities of Siachen and Gangotri Glaciers Using TERRASAR-X and Sentinel-1A Data by Intensity Tracking. *ISPRS Annals of Photogrammetry, Remote Sensing and Spatial Information Sciences*, IV-5, 325–329.

Leprince, S., Barbot, S., Ayoub, F., and Avouac, J.-P. (2007). Automatic and Precise Orthorectification, Coregistration, and Subpixel Correlation of Satellite Images, Application to Ground Deformation Measurements. *IEEE Transactions on Geoscience and Remote Sensing*, 45(6), 1529–1558.

Li, S., Benson, C., Gens, R., and Lingle, C. (2008). Motion Patterns of Nabesna Glacier (Alaska) Revealed by Interferometric SAR Techniques. *Remote Sensing of Environment*, 112(9), 3628–3638.

McNabb, R. W., Hock, R., O'Neel, S., Rasmussen, L. A., Ahn, Y., Braun, M., Conway, H., Herreid, S., Joughin, I., Pfeffer, W. T., Smith, B. E., and Truffer, M. (2012). Using Surface Velocities to Calculate Ice Thickness and Bed Topography: A Case Study at Columbia Glacier, Alaska, USA. *Journal of Glaciology*, 58(212), 1151–1164.

Naithani, A. K., Nainwal, H. C., Sati, K. K., and Prasad, C. (2001). Geomorphological Evidences of Retreat of Gangotri Glacier and Its Characteristics. *Current Science*, 80, 87–94

Pandey, A. C., Ghosh, S., and Nathawat, M. S. (2011). Evaluating Patterns of Temporal Glacier Changes in Greater Himalayan Range, Jammu & Kashmir, India. *Geocarto International*, 26(4), 321–338.

Paul, F., Kääb, A., and Haeberli, W. (2007). Recent Glacier Changes in the Alps Observed by Satellite: Consequences for Future Monitoring Strategies. *Global and Planetary Change*, 56(1–2), 111–122.

Quincey, D. J., Copland, L., Mayer, C., Bishop, M., Luckman, A., and Belò, M. (2009). Ice Velocity and Climate Variations for Baltoro Glacier, Pakistan. *Journal of Glaciology*, 55(194), 1061–1071.

Rabatel, A., Letréguilly, A., Dedieu, J.P., and Eckert, N. (2013). Changes in Glacier Equilibrium-line Altitude in the Western Alps from 1984 to 2010: Evaluation by Remote Sensing and Modelling of the Morpho- Topographic and Climate Controls. *The Cryosphere*, 7, 1455–1471.

Sam, L., Bhardwaj, A., Singh, S., and Kumar, R. (2016). Remote Sensing Flow Velocity of Debris-covered Glaciers Using Landsat 8 Data. *Progress in Physical Geography*, 40(2), 305–321.

Satyabala, S. P. (2016). Spatiotemporal Variations in Surface Velocity of the Gangotri Glacier, Garhwal Himalaya, India: Study Using Synthetic Aperture Radar Data. *Remote Sensing of Environment*, 181, 151–161.

Scherler, D., Leprince, S., and Strecker, M. (2008). Glacier-surface Velocities in Alpine Terrain from Optical Satellite Imagery—Accuracy Improvement and Quality Assessment. *Remote Sensing of Environment*, 112(10), 3806–3819.Shukla, A., Arora, M. K., and Gupta, R. P. (2010). Synergistic Approach for Mapping Debris-covered Glaciers Using Optical–Thermal Remote Sensing Data with Inputs from Geomorphometric Parameters. *Remote Sensing of Environment*, 114(7), 1378–1387.

Shukla, A., and Garg, P. (2020). Spatio-temporal Trends in the Surface Ice Velocities of the Central Himalayan Glaciers, India. *Global and Planetary Change*, 190, 103187.

Shukla, A., Gupta, R., and Arora, M. (2009). Estimation of Debris Cover and Its Temporal Variation Using Optical Satellite Sensor Data: A Case Study in Chenab Basin, Himalaya. *Journal of Glaciology*, 55(191), 444–452.

Singh, D. S., Tangri, A. K., Kumar, D., Dubey, C. A., and Bali, R. (2017). Pattern of Retreat and Related Morphological Zones of Gangotri Glacier, Garhwal Himalaya, India. *Quaternary International*, 444, 172–181.

Strozzi, T., Luckman, A., Murray, T., Wegmuller, U., and Werner, C. L. (2002). Glacier Motion Estimation Using SAR Offset-tracking Procedures. *IEEE Transactions on Geoscience and Remote Sensing*, 40(11), 2384–2391.

Sun, Y., Jiang, L., Liu, L., Sun, Y., and Wang, H. (2017). Spatial-Temporal Characteristics of Glacier Velocity in the Central Karakoram Revealed with 1999–2003 Landsat-7 ETM+ Pan Images. *Remote Sensing*, 9(10), 1064.

Tiwari, R., Garg, P., Shukla, A., Ahluwalia, R., Singh, N., and Chauhan, P. (2016). A Geomorphic and Morphometric Analysis of Surface Ice Velocity Variation of Different Valley Type Glaciers. *SPIE Proceedings*.

Weller, G., and Holmgren, B. (1974). The Microclimates of the Arctic Tundra. *Journal of Applied Meteorology*, 13, 854–862.

Wijngaard, R. R., Steiner, J. F., Kraaijenbrink, P. D. A., Klug, C., Adhikari, S., Banerjee, A., Pelliciotti, F., van Beek, L. P. H., Bierkens, M. F. P., Lutz, A. F., and Immerzeel, W. W. (2019). Modelling the Response of the Langtang Glacier and the Hintereisferner to a Changing Climate Since the Little Ice Age. *Frontiers in Earth Science*, 7, 1–24.

Yellala, A., Kumar, V., and Høgda, K. A. (2019). Bara Shigri and Chhota Shigri Glacier Velocity Estimation in Western Himalaya Using Sentinel-1 SAR Data. *International Journal of Remote Sensing*, 40(15), 5861–5874.

Zongli, J., Shiyin, L., Sichun, L., and Xin, W. (2012). Estimate Yengisogat Glacier Surface Flow Velocities Using ALOS PALSAR Data Feature-tracking, Karakoram, China. *Procedia Environmental Sciences*, 12, 646–652.

3 Monitoring Spatiotemporal Patterns of Glacial Lakes in the Eastern Himalayas Using Satellite Data and Nonparametric Statistical Testing Techniques

Deepali Gaikwad, Supratim Guha,
and Reet Kamal Tiwari*
Geomatics Engineering Laboratory, Department
of Civil Engineering, Indian Institute of
Technology Ropar-140001, India

CONTENTS

3.1   Introduction 35
3.2   Study Area 38
3.3   Data Used and Methods...39
3.4   Results...41
3.5   Discussion and Conclusion 45
Acknowledgment 46
Authors' Contributions 47
References...47

3.1   INTRODUCTION

The Himalayan glaciers have been shrinking and retreating in recent decades owing to global warming (Kulkarni and Karyakarte, 2014; Mitkari et al., 2017). Retreating glaciers promote the formation and enlargement of glacial lakes, depending on the geomorphological condition (Yao et al., 2012; Maanya et al., 2016). These enlarging lakes have the potential for outburst floods, which are an extremely dangerous glacial

DOI: 10.1201/9781003265160-4

hazard due to their size, potential to devastate infrastructure and cropland, and the extent they threaten the downstream communities (Eamer et al., 2007; Wilson et al., 2018; Duan et al., 2020). In 1926, the first lake outburst flood occurred in the Shyok Glacier, in Jammu and Kashmir in the Indian Himalayas, which devastated Abudan village and its surroundings (Mason, 1929). A flood with heavy rainfall in Kedarnath in 2013 caused 6600 human deaths, which occurred probably due to the outburst of the Chorabari Lake (Ahluwalia et al., 2016). Other events in Kinnaur Valley and Himachal Pradesh were also reported between 1981 and 1988 (Philip and Sah, 2004).

In various mountain regions, glacial retreat has been observed by global-level studies across the world, predominantly sensitive to global climate agitations (Garg et al., 2017) during the recent decades, but Karakoram and western Kunlun are exceptional (Shukla et al., 2018). The glacier melting has led to a drastic rise in surface area and in the frequency of lake formation in the glaciated regions across the world, such as in Greenland (Livingstone et al., 2013; How et al., 2021; Tomczyk and Ewertowski, 2020), the Himalaya–Karakoram–Tibet (HKT) region (Ashraf et al., 2015, 2017, 2021; Baig et al., 2020; Chen et al., 2020; Muneeb et al., 2021), High Mountain Asia (Wang et al., 2020; Chen et al., 2020), Central Asia (Janský et al., 2010; Mergili et al., 2013), the Tibetan Plateau (Sun et al., 2018; Liu et al., 2020; Zhang et al., 2021; Cheng et al., 2021; Duan et al., 2020), the Alps (Huggel et al., 2002; Emmer et al., 2015), and the Andes (Cook et al., 2016; Kougkoulos et al., 2018; Vilca et al., 2021).

About 15% of the Himalayas' area is covered by glaciers; therefore, it is known as a "storehouse of world's largest ice mass outside poles" (Brun et al., 2017; Garg et al., 2017). Many studies and observations have concluded that climate-warming impacts are worst in the Himalayas region (Garg et al., 2017; Mitkari et al., 2017), resulting in the formation of high-altitude glacial lakes (Gardelle et al., 2011; Zhang et al., 2015). Glacial lake expansion was recently studied mainly in the Himalayan–Everest region (Salerno et al., 2012; Watson et al., 2016), the central Himalayas (Jha and Khare, 2017; Shrestha et al., 2017; Begam and Sen, 2019; Pandey et al., 2021; Sattar et al., 2020), the eastern Himalayas (Basnett et al., 2013; Govindha et al., 2013; Debnath et al., 2017; Remya et al., 2019; Begam and Sen, 2019), the western Himalayas (Patel et al., 2017; Jain and Mir 2019), and Trans-Himalaya (Kumar et al., 2020; Majeed et al., 2020). Gardelle et al. (2011) mapped glacial lakes in seven sites from Bhutan to the Hindu Kush and the east–west spread of the mountain range from 1990 to 2010. They observed that lakes were more extensive in the eastern part than the western part, whereas lakes have contracted in the Karakoram and the Hindu Kush during the same period.

A recent study in the Sikkim by Shukla et al. (2018) observed 425 high-altitude glacial lakes in 1975, and this number increased to 466 in 2017. An overall incremental increase of 9% and 24% in the number and area of the lakes, respectively. The study estimated that the lake number (6%) and lake area (10%) increased from 1975–1991, recorded as the highest increment.

Another inventory was prepared by Aggarwal et al. (2017) for glacial lakes by employing high-resolution satellite data in Sikkim. The study reported 1104 lakes located above 3500 msl (mean sea level), out of which 472 lakes have an area greater than 0.01 km². Also, glacial lake outburst floods (GLOFs) susceptibility was assessed on these 472 lakes, of which 21 lakes were reported to be GLOF susceptible. Five

of these lakes were low, 14 were medium, and 2 were highly susceptible to GLOFs (Aggarwal et al., 2017). Some more studies on glacial lake dynamics for other regions are presented in Table 3.1.

The glacier wastage in the eastern Himalayas is slightly higher due to global climate change (Brun et al., 2017). Along the Himalayan arc, most of the glaciers are covered with debris and are relatively stagnant. These factors probably influence the enormous growth of lakes in the eastern Himalayas, which can harm the downstream communities (Shukla et al., 2018). However, the magnitude and trend of lake change and its expanding rate in the Sikkim Himalaya are still unknown, which is vital for

TABLE 3.1
Literature Review

Authors	Study Area	Period	Purpose	Findings
Ashraf et al. (2021)	Hindu Kush–Karakoram–Himalaya	2013	Preparation of glacial lake inventory and identification of critical GLOF lakes	The study identified 3044 lakes in HKH ranges. The highest increase in lake area was observed in the lowest range of elevation (2500–3500 m). In addition, they categorized 36 lakes as potentially dangerous glacial lakes to GLOF.
Khadka et al. (2021)	Mahalangur region in Himalaya	1998–2018	Inventory of Glacial Lake and GLOFs susceptibility assessment.	In 2018, they reported 345 lakes (>0.001 km^2) covering a total area of 18.80 ± 1.35 km^2. Out of these, 64 lakes (≥0.045 km^2) were assessed and found seven lakes were highly dangerous to GLOF.
Pandey et al. (2021)	Part of the central Himalayas, Uttarakhand State, India.	1994–2017	To explore the interaction between glacial lakes and glaciers	In 2015, 1353 glacial lakes were observed, covering an area of 7.96 km^2. An overall increase of 57% in the area of glacial lakes lying within 2 km periphery of the mother glacial was observed during 1994–2017. On the other hand, glaciers were reduced by 7 km^2, accounting for an overall loss of 1.92% during the same period.
Wang et al. (2020)	High Mountain Asia	1990–2018	Investigation of glacial lake inventory	They observed that the number of lakes (44.12%) increased more than double the lake area (15.31%).
Kumar et al. (2020)	Nubra and Shyok Basins, Karakoram range	2002–2017	Evaluation and growth assessment of moraine and bedrock dammed glacial lakes.	The number of glacial lakes raised from 215 to 255 during 2002–2017, and the area expanded from ~9.0 km^2 to ~11.27 km^2 from 2003–2017. Noted the highest expansion between 2013 and 2017.

regular monitoring of glacial lake dynamics and further adopting the necessary measures against any GLOF event. Therefore, the present study attempts to bridge this
research gap by estimating the expansion rate of glacial lakes and understanding the
trend in variation using inferential statistics for the last four decades in the Sikkim
Himalaya.

3.2   STUDY AREA

The study was conducted in Sikkim, eastern Himalayas, India, which falls 27°–28°
N latitude and 88–89° E longitude (Figure 3.1). The area of Sikkim is 7093 km^2 and
has a steep slope in more than 43% of the area and escarpments with rugged terrain.
The number and area covered by glaciers were reported to be 449 and 705.54 km^2,
respectively, by the Glacier Atlas of India (Raina and Srivastava, 2008). Sikkim has
more than 1000 lakes that originate from glaciers. The Teesta and Rangit are two
main rivers in Sikkim, which are sources of many lakes and have the potential for
hydropower generation. The elevation range in Sikkim varies from 213 m above msl
at the Teesta River outflow in the south to Mount Kangchendzonga (known as "the
third-highest mountain in the world") at 8586 m msl in the west. The average annual
precipitation received by Sikkim is about 2500 mm, and the mountain region has
heavy rain and snowfall (GSI, 2012). Sikkim comes under seismic zone IV and is
considered high risk during earthquake events (Murty, 2004; Sharma et al., 2012).

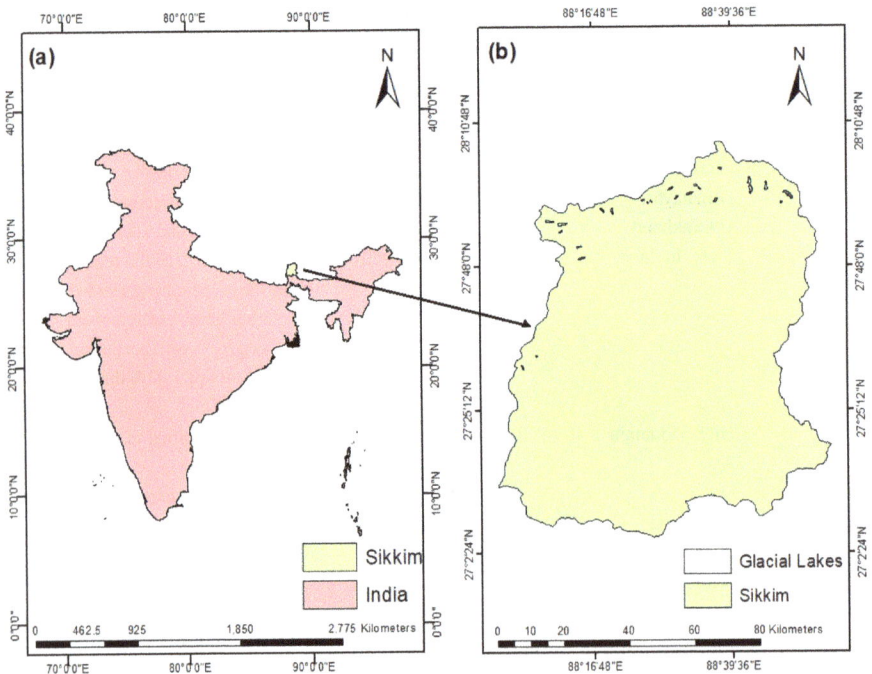

FIGURE 3.1   Sikkim State, Eastern Himalaya, and the study area (green) located in India.

The economy of Sikkim mainly depends on agriculture and tourism, which may be affected by GLOFs.

Figure 3.1 shows the location of Sikkim state (green) in India and samples of the glacial lakes selected in the study area.

3.3 DATA USED AND METHODS

Remote-sensing data are suitable for extracting spatial information because of its synoptic view, repetitive coverage, real-time data acquisition, and temporal coverage (Belal and Moghanm, 2011; Guha et al., 2018; Guha et al., 2019). In this study, Landsat multitemporal images were used to estimate the rate of glacial lake changes. Landsat Thematic Mapper (TM) images were used for 1988 and 2008, Landsat enhanced thematic mapper (ETM+) for 2000, and Landsat operational land imager (OLI) for 2014 and 2020 for glacial lake boundary delineation. All these Landsat data sets were acquired from the USGS platform (https://earthexplorer.usgs.gov/) and projected in the UTM coordinate system (Zone 45N) with WGS 84 datum (Table 3.2). The elevation information of these glacial lakes was acquired from SRTM DEM (Shuttle Radar Topography Mission Digital Elevation Model) which was downloaded from http://srtm.csi.cgiar.org/srtmdata/.

The Landsat data were collected for the ablation period (October–December), with minimal cloud cover in the region. Before mapping, Landsat ETM+ and Landsat OLI data have been converted into 15m resolution by panchromatic (PAN)–sharpening using ArcGIS vs 10.2, for better visualization purposes (Figure 3.2). The PAN sharpening method increases the resolution of the images by combining the low-resolution multispectral bands with the high-resolution panchromatic band. The ArcGIS uses the five image fusion methods that include the Brovey transformation, the intensity–hue–saturation transformation, the Esri pan-sharpening transformation, the simple mean transformation, and the Gram–Schmidt spectral sharpening method. After that, on-screen digitization technique has been used to delineate the lake boundaries.

TABLE 3.2
Details of Acquired Data for the Study

Satellite/ Sensor	Path/ Row	Date	Resolution	Spectral Bands	Land Cloud Cover	Purpose
Landsat-5/TM	139/41	01/12/1988	30m	2, 3, 4	3%	Glacial Lake Mapping
Landsat-7/ETM	139/41	26/12/2000	30/15m (PAN)	2, 3, 4, 8	4%	Glacial Lake Mapping
Landsat-5/TM	139/41	08/12/2008	30 m	2, 3, 4	4%	Glacial Lake Mapping
Landsat-8/OLI	139/41	07/11/2014	30/15m (PAN)	3, 4, 5, 8	4.23%	Glacial Lake Mapping
Landsat-8/OLI	139/41	25/12/2020	30/15m (PAN)	3, 4, 5, 8	2.24%	Glacial Lake Mapping
SRTM DEM	139/41	2000	90 m	NA	NA	Elevation information

Normal Image   Enhanced Image

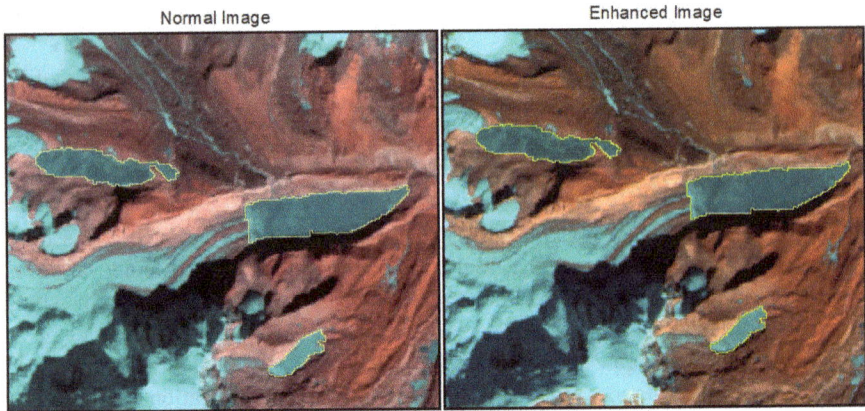

FIGURE 3.2   Pan-sharpened image to advance visualization.

The delineated glacial lake boundaries were validated using the Google Historical Imagery.

The uncertainty associated with glacial lake area calculation is generally due to spatial resolution of satellite imagery, high mountain shadow effect, and cloud cover (Gardelle et al., 2011). The standard method was used to estimate glacial lake mapping error:

$$U = N \times \frac{A}{2}, \tag{3.1}$$

where U is the uncertainty of the lake area, A is the area of a pixel, and N is the number of pixels covered along the lake boundary. The uncertainty error in the total area of glacial lakes has been reported for 2020 was 0.712 km^2, for 2014 was 0.876 km^2, for 2008 was 1.659 km^2, for 2000 was 0.819 km^2, and for 1988 was 1.274 km^2.

Twenty-four glacial lakes of various types, such as proglacial lake (PGL), periglacial lake (PEGL), supraglacial lake (SGL), and cirque lake (CL), have been delineated from different parts of Sikkim. These glacial lakes were different in type, size, shape, and situated at different altitudes. Therefore, these 24 glacial lakes were considered representative data sets for Sikkim. Furthermore, the rate of glacial lake area changes (RGLAC) has been calculated by Equation 3.2

$$Rate\, of\, glacial\, lake\, area\, changes\, (\%) = \left(\frac{A_1 - A_2}{A_1} \right) \times 100 \times \frac{1}{\Delta T}, \tag{3.2}$$

where A_1 is the area of the initial year (km^2), A_2 is the area of the final year (km^2), and ΔT denotes difference in period (in years)

Finally, inferential statistics have been adopted to understand the temporal variation in the RGLAC in the entire Sikkim Himalaya using the 24 representative glacial lakes. A nonparametric Friedman statistical testing method compares two or more

groups of samples selected from the same population. This test is appropriate for nonnormally distributed samples, but it assumes that the sample variance is roughly equal and that all data points are independent (Smalheiser, 2017). The Friedman test statistics were estimated by using Equation 3.3

$$F_r = \frac{12}{nk(k+1)}\left(\sum T^2\right) - 3n(k+1), \tag{3.3}$$

where n is the samples, k is the number of groups (timeframes), and T is the sum of ranks of each group. Here, it was used to identify whether the RGLAC is equal or not in all timeframes. The RGLAC was estimated between 1988–2000 tagged as Timeframe-1 (TF_1), 2000–2008 tagged as TF_2, 2008–2014 tagged as TF_3, and 2014–2020 tagged as TF_4. The null hypothesis (H_0) and the alternative hypothesis (H_a) of the statistical test follow:

H_0: Rate of glacial lake area variations are equal in all four timeframes
H_a: Rate of glacial lake area variations are not equal in all four timeframes

If the null hypothesis is rejected in favor of an alternative hypothesis, that is, if the RGLAC was not equal in all the timeframes, then a post hoc test needs to be done. In our case, the Dunn–Bonferroni post hoc test has been executed to evaluate the significant differences in the RGLAC for all timeframes.

3.4 RESULTS

The surface area fluctuations on a decadal scale of the sample lakes were observed in 1988, 2000, 2008, 2014, and 2020. The location, elevation, surface area changes, and type of all these glacial lakes have been discussed in detail (Table 3.3). We observed the areal expansion in most lakes, but some lakes have been seen to be contracting in these decadal timeframes (Figure 3.3).

Furthermore, we noted that the total area of sample lakes has expanded consistently from 9.617 km² to 10.973 km² (14.10%) between 1988 and 2000, further increased from 10.973 to 12.096 km² (9.55%) between 2000 and 2008, from 12.096 km² to 13.036 km² (8.44%) between 2008 and 2014, and finally increased from 12.590 km² to 13.036 km² (4.25%) between 2014 and 2020 (Figure 3.4).

The individual lake analysis revealed that area changes for each of the 24 sample lakes in different periods were heterogeneous. The maximum area of most glacial lakes was observed in 2014 and 2020, while some lakes were at their largest extent in 2008. However, two lakes showed contraction and had the highest area in 1988 and 2000 (Figure 3.5). Also, the three lakes (GL-05, GL-13, GL-17) appeared as new lakes after 1988 (Figure 3.3 and Table 3.3).

Temporal variations of some glacial lakes were clearly seen in all decades. The expansion of the surface area of few lakes (GL-01, GL-05, GL-07, GL-09, GL-12, GL-17, GL-18, GL- 20) are presented. However, GL-10 represents the contraction from 1988 to 2008 and again showed the expansion from 2014 to 2020 (Figure 3.6).

TABLE 3.3
Lake ID, Location, Area, and Type of the Lake for All Periods

Lake ID	Latitude (N)	Longitude l	Elevation (m)	Area (km²) 1988	2000	2008	2014	2020	Type
GL01	27°59.743'	88°32.742'	5162.83	0.52	0.79	0.88	0.69	0.67	PGL
GL02	27°51.179'	88°14.444'	5219.39	0.09	0.13	0.25	0.32	0.31	SGL
GL03	28°0.491'	88°34.326'	5000.34	0.24	0.26	0.25	0.24	0.23	PGL
GL04	28°0.208'	88°42.795'	5259.10	1.16	1.03	1.20	1.21	1.34	PGL
GL05	27°58.944'	88°30.527'	4935.34	0.00	0.16	0.20	0.28	0.34	PGL
GL06	28°0.781'	88°45.367'	5094.83	1.00	1.00	0.91	1.04	1.06	PGL
GL07	27°59.409'	88°48.906'	5308.91	1.57	1.62	1.61	1.72	1.79	PGL
GL08	27°58.558'	88°36.971'	4961.74	0.54	0.55	0.53	0.58	0.56	PGL
GL09	27°32.019'	88°5.132'	4870.65	0.19	0.30	0.34	0.39	0.31	PGL
GL10	28° 3.535'	88°37.742'	4921.5	0.48	0.28	0.07	0.12	0.35	CL
GL11	27°58.126'	88°47.821'	5358.9	0.15	0.18	0.19	0.20	0.17	PGL
GL12	27°49.280'	88°14.870'	5423.89	0.07	0.30	0.44	0.69	0.78	SGL
GL13	27°58.539'	88°25.050'	5234.96	0.00	0.05	0.15	0.16	0.21	PEGL
GL14	27°58.225'	88°25.787'	4985.22	0.12	0.13	0.15	0.13	0.14	PEGL
GL15	27°58.390'	88°26.311'	5068.45	0.10	0.10	0.11	0.10	0.09	PEGL
GL16	27°33.760'	88°7.375'	4719.79	0.09	0.10	0.12	0.10	0.11	PGL
GL17	27°53.608'	88°11.436'	5524.71	0.00	0.13	0.18	0.22	0.24	PGL
GL18	27°54.853'	88°11.751'	5217.98	0.38	0.72	0.99	1.24	1.43	PGL
GL19	28°1.624'	88°42.615'	5148.46	1.07	1.10	1.05	1.15	1.07	PGL
GL20	27°55.267'	88°9.624'	5462.94	0.44	0.55	0.71	0.83	0.80	PGL
GL21	27°56.910'	88°18.306'	5054.49	0.28	0.33	0.36	0.33	0.36	PGL
GL22	27°56.704'	88°20.022'	5036.39	0.54	0.58	0.63	0.58	0.57	PGL
GL23	28° 0.374'	88°29.540'	5024.3	0.40	0.38	0.44	0.45	0.43	PGL
GL24	28° 0.936'	88°33.722'	5060.11	0.21	0.24	0.26	0.27	0.26	PGL

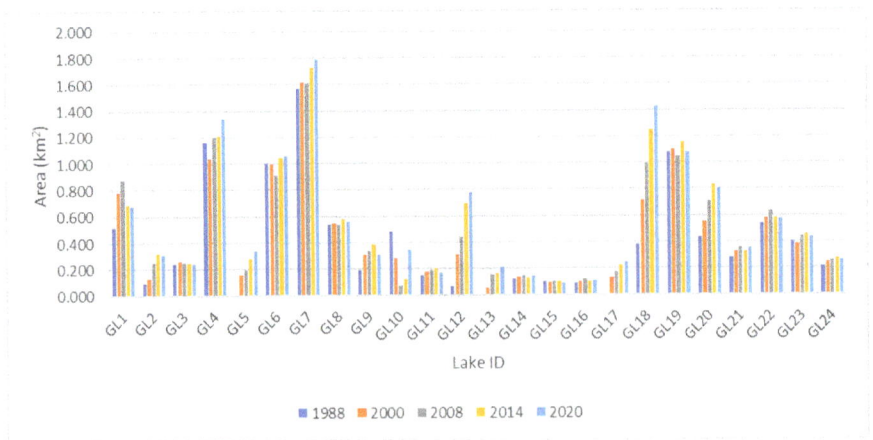

FIGURE 3.3   Areal variation of all representative glacial lakes.

Average area of sample glacial lakes

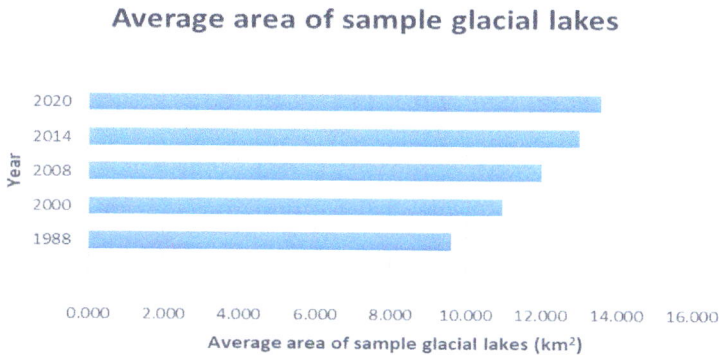

FIGURE 3.4   Average area of all sample glacial lakes from 1988 to 2020.

Number of lakes having the lagerst area

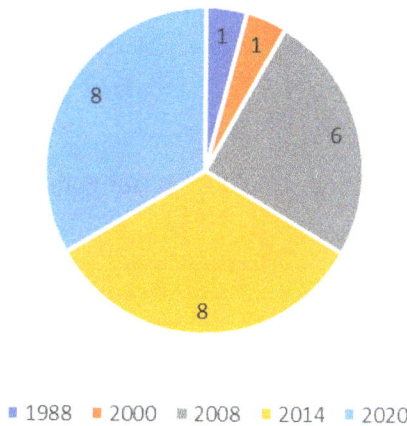

■ 1988   ■ 2000   ▥ 2008   ■ 2014   ■ 2020

FIGURE 3.5   Number of glacial lakes have largest area between 1988 and 2000.

The nonparametric tests have been utilized to understand the temporal variations of the RGLAC for all study periods in the entire region. First, the nonparametric Friedman statistical test has been carried out to compare the RGLAC in different timeframes with a confidence interval of 95%. The test rejected the H_0 as a p-value (0.007) lesser than the significant level (0.05; Table 3.4). This showed that the RGLAC has a heterogeneous trend and was not equal in one timeframe at least. Afterward, a Dann–Bonferroni post hoc test was executed to identify exactly which timeframes have different RGLAC. Finally, we found that TF_1 and TF_4 have significantly differ-ent RGLAC, as the p-value (0.005) is less than the critical level (0.05). These time-frames showed a considerable difference (orange line) in their mean ranks in Dunn's

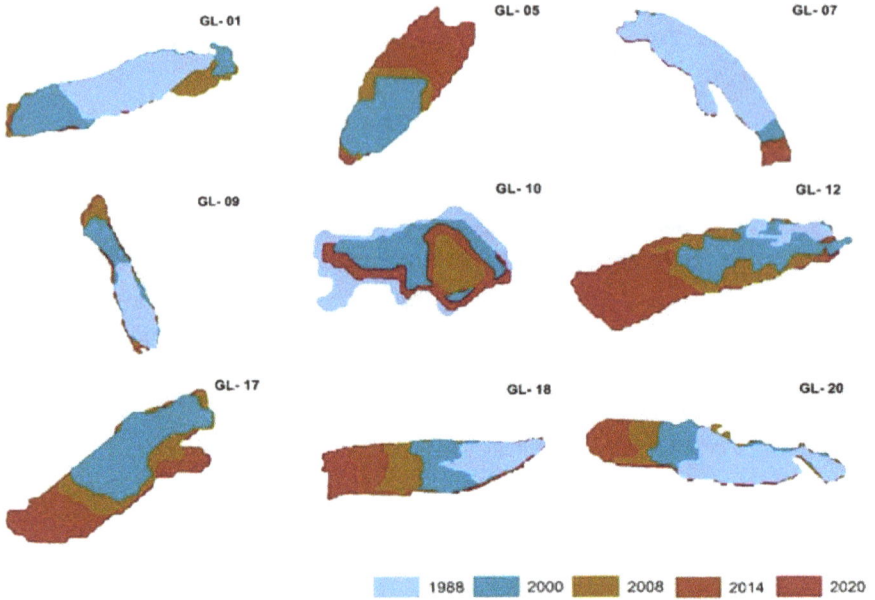

FIGURE 3.6   Decadal variation of some (9 out of 24) representative lakes.

TABLE 3.4

Friedman Statistical Test Analysis of RGLAC for Four Timeframes

Null Hypothesis	Test	Sig.	Decision
The distributions of TF1, TF2, TF3, and TF4 are the same.	Related- samples Friedman's Two-Way Analysis of Variance by Rank	.007	Reject the null hypothesis

---

pairwise post hoc tests (Figure 3.7). Hence, the rate of area changes of all glacial lakes present in Sikkim Himalaya has remarkably changed between 1988 and 2000 and between 2014 and 2020.

The test statistics is the difference between the mean of two timeframes, which is further standardized to compute the p-value. Afterward, Bonferroni correction needs to be done when multiple testing is conducted on the same data sets (Eisinga et al., 2017). It adjusts the p-value, or α-value (significant level), to reduce the probability of Type 1 error, which rejects the null hypothesis when it is true. In the present case, the observed p-value is already adjusted by multiplying the number of tests being carried out on the samples (Figure 3.7).

Dunn's pairwise comparison of the mean rank of RGLAC showed the highest RGLAC in the entire Sikkim Himalaya for TF_1; after that, it slowed down to moderate in TF_2 and TF_3. Finally, the minimum RGLAC has been observed in the last timeframe (TF_4). A decreasing RGLAC has been observed in past decades (Figure 3.7: Pairwise comparison).

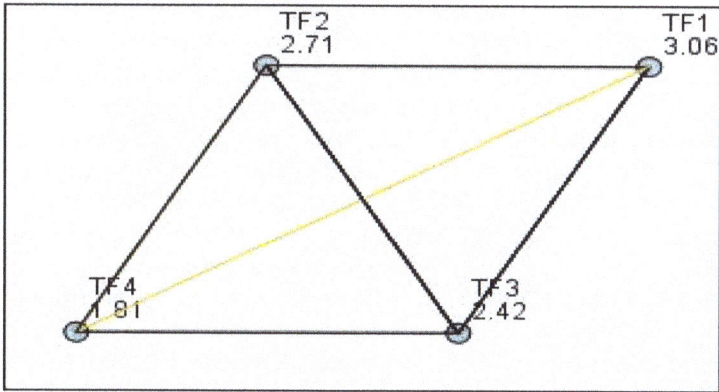

Each node shows the sample average rank.

Sample1-Sample2	Test Statistic	Std. Error	Std. Test Statistic	Sig.	Adj.Sig.
TF4-TF3	.604	.373	1.621	.105	.630
TF4-TF2	.896	.373	2.404	.016	.097
TF4-TF1	1.250	.373	3.354	.001	.005
TF3-TF2	.292	.373	.783	.434	1.000
TF3-TF1	.646	.373	1.733	.083	.499
TF2-TF1	.354	.373	.950	.342	1.000

Each row tests the null hypothesis that the Sample 1 and Sample 2 distributions are the same.
Asymptotic significances (2-sided tests) are displayed. The significance level is .05.

FIGURE 3.7   Pairwise comparison of means of the RGLAC and Dunn–Bonferroni statistical test assessment between 1988 and 2020.

3.5   DISCUSSION AND CONCLUSION

Glacial lakes are mainly fed by meltwater from the nearest glacier; hence, the dynamics of glaciers are responsible for the variation in glacial lakes. The rate of temporal variation of lakes is not constant in any timeframe due to changing climate conditions. Therefore, based on multitemporal remote sensing data, we analyzed the RGLAC for the four timeframes: 1988 to 2000, 2000 to 2008, 2008 to 2014, and 2014 to 2020. Twenty-four glacial lakes were selected from these five timeframes such that they appeared in all periods; however, a few lakes were not present in 1988

(GL-05, GL-13, GL-17). All sample glacial lakes were different in size and shape and were situated at different altitudes. Therefore, using these representative sample lakes, we evaluated the trend of RGLAC by inferential statistics in the entire Sikkim Himalayas. Furthermore, we observed the variation of 24 sample lakes and found that lakes were growing with time. The average area of all sample lakes has changed from 9.617 km^2 to 13.590 km^2 (a 14.31% increase) between 1988 and 2020.

To identify the variation in RGLAC, nonparametric tests (Friedman and Dann-Bonferroni test) have been used. These tests clearly showed that RGLAC for the 1988–2000 (TF$_1$) and 2014–20 (TF$_4$) timeframes are statistically unequal; the mean rank of the RGLAC in TF$_1$ is much higher than TF$_4$. That means that the rate of expansion of the lake has been remarkably reduced between the 1988–2000 period and the 2014–2020 period. The decreasing rate of expansion is probably due to a declination in precipitation, an increase in temperature, a disconnection of the lake from its host glaciers, and the draining of large glacial lakes, as suggested by the studies (Shukla et al., 2018; Garg et al., 2019).

The impacts of climate change are the main reason behind the dynamics of the glacier and glacial lake changes. The climatic parameters observed between 1987 and 2011 by Basnett et al. (2013); noted the rise in mean annual temperature by 1.03 °C in Sikkim Himalayas. However, the precipitation varied highly but did not show any clear trend. Although, on average, a slight declination in winter precipitation had been observed. The study by Garg et al. (2019) also reported a significant increase in summer temperatures than winter temperatures in the study area between 1990 and 2016. Moreover, glacier recession in the Sikkim Himalaya has been consistently investigated, and a reduction in the total area of glaciers by 11.2 ± 2.4% between 1975 and 2014 was reported (Bhattacharya et al., 2018).

This study employed the simple method of finding the status of area change of glacial lakes for the entire region by extracting some samples of glacial lakes. Through this study, we analyzed the rate of area change of glacial lakes for different timeframes and observed that it is varying with time. The results signify the direct or indirect global climate variation impacts on glacial lakes. Therefore, analyzing the dynamics of the expansion and contraction of glacial lakes is greatly important to take necessary measures against glacial hazards such as physical mitigation, creating awareness, and installing a early warning system. These hazards are highly disastrous for the downstream communities and infrastructure. Hence, it is vital to investigate the influencing parameters that may help predict the possibilities of these disasters and the potentiality for devastation. Furthermore, this study may be helpful for volumetric trend analysis and peak discharge computation, which are the critical parameters for GLOF modeling and GLOF hazardous lake identification.

ACKNOWLEDGMENT

We downloaded the open access Landsat images (TM, ETM+ and OLI) available on the United States Geological Survey (USGS; http://earthexplorer.usgs.gov/). We would like to thank the anonymous reviewer and editors for the valuable comments to improve the manuscript.

AUTHORS' CONTRIBUTIONS

Deepali Gaikwad planned the methodology of the present work and carried out the extraction of the glacial lakes. Supratim Guha applied inferential statistics. The research work was supervised by Reet Kamal Tiwari. The original draft prepared by Deepali Gaikwad and reviewed by Supratim Guha and Reet Kamal Tiwari. All authors interpreted the results and contributed to the final manuscript.

REFERENCES

Aggarwal, S., Rai, S.C., Thakur, P.K., Emmer, A. 2017. "Inventory and Recently Increasing GLOF Susceptibility of Glacial Lakes in Sikkim, Eastern Himalaya." *Geomorphology* 295: 39–54. https://doi.org/10.1016/j.geomorph.2017.06.014.

Ahluwalia, R.S., Rai, S.P., Gupta, A.K., Dobhal, D.P., Tiwari, R.K., Garg, P.K., Kesharwani, K. 2016. "Towards the Understanding of the Flash Flood Through Isotope Approach in Kedarnath Valley in June 2013, Central Himalaya, India." *Natural Hazards: Journal of the International Society for the Prevention and Mitigation of Natural Hazards* 82: 321–332.

Ashraf, A., Naz, R., Iqbal, M.B. 2015. "Heterogeneous Expansion of End-Moraine Dammed Lakes in the Hindukush-Karakoram-Himalaya Ranges of Pakistan during 2001–2013." *Journal of Mountain Science* 12: 1113–1124. https://doi.org/10.1007/s11629-014-3245-4

Ashraf, A., Naz, R., Iqbal, M.B. 2017. "Altitudinal Dynamics of Glacial Lakes under Changing Climate in the Hindu Kush, Karakoram, and Himalaya Ranges." *Geomorphology* 283: 72–79. ISSN 0169-555X. https://doi.org/10.1016/j.geomorph.2017.01.033

Ashraf, A., Iqbal, M.B., Mustafa, N., Naz, R., Ahmad, B. 2021. "Prevalent Risk of Glacial Lake Outburst Flood Hazard in the Hindu Kush–Karakoram–Himalaya Region of Pakistan." *Environmental Earth Sciences* 80: 451. https://doi.org/10.1007/s12665-021-09740-1.

Baig, S.U., Khan, H., Muneeb, F., Dad, K. 2020. "Formation of a Hazardous Ice-dammed Glacier Lake: A Case Study of Anomalous Behavior of Hassanabad Glacier System in the Karakoram." *SN Applied Sciences* 2: 1285. https://doi.org/10.1007/s42452-020-2989-4.

Basnett, S., Kulkarni, A.V., Bolch, T. 2013. "The Influence of Debris Cover and Glacial Lakes on the Recession of Glaciers in Sikkim Himalaya." *Indian Journal of Glaciology* 59: 1035–1046. https://doi.org/10.3189/2013JoG12J184.

Begam, S., Sen, D. 2019. "Mapping of Moraine Dammed Glacial Lakes and Assessment of Their Areal Changes in the Central and Eastern Himalayas Using Satellite Data." *Journal of Mountain Science* 16: 77–94. https://doi.org/10.1007/s11629-018-5023-1.

Belal, A.A., Moghanm, F.S. 2011. "Detecting Urban Growth Using Remote Sensing and GIS Techniques in Al Gharbiya Governorate, Egypt." *The Egyptian Journal of Remote Sensing and Space Science* 14: 73–79. https://doi.org/10.1016/j.ejrs.2011.09.001.

Bhattacharya, A., Ghosh, S., Mukherjee, K. 2018. "Multi-decadal Mass Budget and Area Change of Some Eastern Himalayan Glaciers (Nepal-Sikkim) Using Remote Sensing Techniques." *2018 4th International Conference on Recent Advances 458 in Information Technology (RAIT)*, Dhanbad, 1–6. https://doi.org/10.1109/RAIT.2018.8388976.

Brun, F., Berthier, E., Wagnon, P., Kääb, A., Treichler, D. 2017. "A Spatially Resolved Estimate of High Mountain Asia Glacier Mass Balances from 2000 to 2016." *Nature Geosci* 10: 668–673. https://doi.org/10.1038/ngeo2999.

Chen, F., Zhang, M., Guo, H., Allen, S., Kargel, J.S., Haritashya, U.K., Watson, C.S. 2021. "Annual 30 m Dataset for Glacial Lakes in High Mountain Asia from 2008 to 2017." *Earth System Science Data* 13(2): 741–766.

Cheng, J., Song, C., Liu, K., Ke, L., Chen, T., Fan, C. 2021. "Regional Assessment of the Potential Risks of Rapid Lake Expansion Impacting on the Tibetan Human Living Environment." *Environmental Earth Sciences* 80: 166. https://doi.org/10.1007/s12665-021-09470-4.

Cook, S.J., Kougkoulos, I., Edwards, L.A., Dortch, J., Hoffmann, D. 2016. "Glacier Change and Glacial Lake Outburst Flood Risk in the Bolivian Andes." *The Cryosphere* 10: 2399–2413. https://doi.org/10.5194/tc-10-2399-2016.

Debnath, M., Syiemlieh, H.J., Sharma, M.C., Kumar, R., Chowdhury, A., Lal, U. 2018. "Glacial Lake Dynamics and Lake Surface Temperature Assessment Along the Kangchengayo-Pauhunri Massif, Sikkim Himalaya, 1988–2014." *Remote Sensing Applications: Society and Environment* 9: 26–41. https://doi.org/10.1016/j.rsase.2017.11.002.

Duan, H., Yao, X., Zhang, D., Qi, M., Liu, J. 2020. "Glacial Lake Changes and Identification of Potentially Dangerous Glacial Lakes in the Yi'ong Zangbo River Basin." *Water* 12: 538. https://doi.org/10.3390/w12020538.

Eamer, J., Ahlenius, H., Prestrud, P. 2007. "United Nations Environment Programme." In: *Global Outlook for Ice & Snow. Division of Early Warning and Assessment (DEWA).* Nairobi, Kenya: United Nations Environmental Programme.

Eisinga, R., Heskes, T., Pelzer, B., Te Grotenhuis, M. 2017. "Exact P-values for Pairwise Comparison of Friedman Rank Sums, with Application to Comparing Classifiers." *BMC Bioinformatics* 18: 68. https://doi.org/10.1186/s12859-017-1486-2.

Emmer, A., Loarte, E.C., Klimeš, J., Vilímek, V. 2015. "Recent Evolution and Degradation of the Bent Jatunraju Glacier (Cordillera Blanca, Peru)." *Geomorphology* 228: 345–355. https://doi.org/10.1016/j.geomorph.2014.09.018.

Gardelle, J., Arnaud, Y., Berthier, E. 2011. "Contrasted Evolution of Glacial Lakes Along the Hindu Kush Himalaya Mountain Range between 1990 and 2009." *Global and Planetary Change* 75: 47–55. https://doi.org/10.1016/j.gloplacha.2010.10.003.

Garg, P.K., Shukla, A., Jasrotia, A.S. 2019. "On the Strongly Imbalanced State of Glaciers in the Sikkim, Eastern Himalaya, India." *Science of The Total Environment* 691: 16–35. https://doi.org/10.1016/j.scitotenv.2019.07.086.

Garg, P.K., Shukla, A., Tiwari, R.K., Jasrotia, A S. 2017. "Assessing the Status of Glaciers in Part of the Chandra Basin, Himachal Himalaya: A Multiparametric Approach." *Geomorphology* 284: 99–114.

Govindha Raj, B.K., Kumar, V.K., Remya, S.N. 2013. "Remote Sensing-based Inventory of Glacial Lakes in Sikkim Himalaya: Semi-automated Approach Using Satellite Data." *Geomatics Natural Hazards and Risk* 4: 241–253. https://doi.org/10.1080/19475705.2012.707153.

GSI. 2012. *Geology and Mineral Resources of the States of India.* Geological Survey of India.

Guha, S., Barik, D.K., Mandla, V.R. 2019. "Landscape Changes and Sustainable Development Policy in a Developing Area: A Case Study in Chirrakunta Rurban Cluster." In: Singh, H., Garg, P., Kaur, I. (Eds.), *Proceedings of the 1st International Conference on Sustainable Waste Management through Design, Lecture Notes in Civil Engineering.* Cham: Springer International Publishing, 68–77. https://doi.org/10.1007/978-3-030-02707-0_10.

Guha, S., Mandla, V.R., Barik, D.K., Das, P., Rao, V.M., Pal, T., Rao, P.K. 2018. "Analysis of Sustainable Livelihood Security: A Case Study of Allapur s Rurban Cluster." *Journal of Rural Development* 37: 365–382. https://doi.org/10.25175/jrd/2018/v37/i2/129703.

How, P., Messerli, A., Mätzler, E., Santoro, M., Wiesmann, A., Caduff, R., Langley, K., Bojesen, M.H., Paul, F., Kääb, A., Carrivick, J.L. 2021. "Greenland-wide Inventory of Ice Marginal Lakes Using a Multi-method Approach." *Scientific Reports* 11: 4481. https://doi.org/10.1038/s41598-021-83509-1.

Huggel, C., Kääb, A., Haeberli, W., Teysseire, P., Paul, F. 2002. "Remote Sensing Based Assessment of Hazards from Glacier Lake Outbursts: A Case Study in the Swiss Alps." *Canadian Geotechnical Journal* 39: 316–330. https://doi.org/10.1139/t01-099.

Jain, S.K., Mir, R.A. 2019. "Glacier and Glacial Lake Classification for Change Detection Studies Using Satellite Data: A Case Study from Baspa Basin, Western Himalaya." *Geocarto International* 34: 391–414. https://doi.org/10.1080/10106049.2017.1404145.

Janský, B., Šobr, M., Engel, Z. 2010. "Outburst Flood Hazard: Case Studies from the Tien-Shan Mountains, Kyrgyzstan." *Limnologica* 40: 358–364. https://doi.org/10.1016/j.limno.2009.11.013.

Jha, L.K., Khare, D. 2017. "Detection and Delineation of Glacial Lakes and Identification of Potentially Dangerous Lakes of Dhauliganga Basin in the Himalaya by Remote Sensing Techniques." *Nat Hazards* 85: 301–327. https://doi.org/10.1007/s11069-016-2565-9.

Khadka, N., Chen, X., Nie, Y., Thakuri, S., Zheng, G., Zhang, G. 2021. "Evaluation of Glacial Lake Outburst Flood susceptibility using multi-criteria assessment framework in Mahalangur Himalaya." *Frontiers in Earth Science* 8: 601288.

Kougkoulos, I., Cook, S.J., Edwards, L.A., Clarke, L.J., Symeonakis, E., Dortch, J.M., Nesbitt, K. 2018. "Modelling Glacial Lake Outburst Flood Impacts in the Bolivian Andes." *Nat Hazards* 94: 1415–1438. https://doi.org/10.1007/s11069-018-3486-6.

Kulkarni, A.V., Karyakarte, Y. 2014. "Observed Changes in Himalayan Glaciers." *Current Science* 106: 237–244.

Kumar, R., Bahuguna, I.M., Ali, S.N., Singh, R. 2020. "Lake Inventory and Evolution of Glacial Lakes in the Nubra-Shyok Basin of Karakoram Range." *Earth Syst Environ* 4: 57–70. https://doi.org/10.1007/s41748-019-00129-6.

Liu, W., Chen, X., Ran, J., Liu, L., Wang, Q., Xin, L., Li, G. 2020. "LaeNet: A Novel Lightweight Multitask CNN for Automatically Extracting Lake Area and Shoreline from Remote Sensing Images." *Remote Sensing* 13: 56. https://doi.org/10.3390/rs13010056.

Livingstone, S.J., Clark, C.D., Woodward, J., Kingslake, J. 2013. "Potential Subglacial Lake Locations and Meltwater Drainage Pathways Beneath the Antarctic and Greenland Ice Sheets." *The Cryosphere* 7: 1721–1740. https://doi.org/10.5194/tc-7-1721-2013.

Maanya, U.S., Kulkarni, A.V., Tiwari, A., Bhar, E.D., Srinivasan, J. 2016. "Identification of Potential Glacial Lake Sites and Mapping Maximum Extent of Existing Glacier Lakes in Drang Drung and Samudra Tapu Glaciers, Indian Himalaya." *Current Science* 111: 553–560.

Majeed, U., Rashid, I., Sattar, A., Allen, S., Stoffel, M., Nüsser, M., Schmidt, S. 2021. "Recession of Gya Glacier and the 2014 Glacial Lake Outburst Flood in the Trans-Himalayan Region of Ladakh, India." *Science of The Total Environment* 756: 144008. https://doi.org/10.1016/j.scitotenv.2020.144008.

Mason, K. 1929. "Indus Floods and Shyok Glaciers." *Himalayan Journal* 1: 10–29.

Mergili, M., Müller, J.P., Schneider, J.F. 2013. "Spatio-temporal Development of High-Mountain Lakes in the Headwaters of the Amu Darya River (Central Asia)." *Global and Planetary Change* 107: 13–24. https://doi.org/10.1016/j.gloplacha.2013.04.001.

Mitkari, K.V., Arora, M.K., Tiwari, R.K. 2017. "Extraction of Glacial Lakes in Gangotri Glacier Using Object-based Image Analysis." *IEEE Journal of Selected Topics in Applied Earth Observations and Remote Sensing* 10: 5275–5283. https://doi.org/10.1109/JSTARS.2017.2727506.

Muneeb, F., Baig, S.U., Khan, J.A., Khokhar, M.F. 2021. "Inventory and GLOF Susceptibility of Glacial Lakes in Hunza River Basin, Western Karakorum." *Remote Sensing* 13: 1794. https://doi.org/10.3390/rs13091794.

Murty, C.V.R. 2004. "Learning Earthquake Design and Construction 7. How Buildings Twist During Earthquakeseason*son* 9: 86–89. https://doi.org/10.1007/BF02834977.

Pandey, P., Ali, S. N., Champati Ray, P. K. 2021. "Glacier-Glacial Lake Interactions and Glacial Lake Development in The Central Himalaya, India (1994–2017)." *Journal of Earth Science* 32(6): 1563–1574.

Patel, L.K., Sharma, P., Laluraj, C.M., Thamban, M., Singh, A., Ravindra, R. 2017. "A Geospatial Analysis of Samudra Tapu and Gepang Gath Glacial Lakes in the Chandra Basin, Western Himalaya." *Nat Hazards* 86: 1275–1290. https://doi.org/10.1007/s11069-017-2743-4.

Philip, G., Sah, M.P. 2004. "Mapping Repeated Surges and Retread of Glaciers using IRS-1C/1D Data: A Case Study of Shaune Garang Glacier, Northwestern Himalaya." *International Journal of Applied Earth Observation and Geoinformation* 6: 127–141. https://doi.org/10.1016/j.jag.2004.09.002.

Raina, V.K., Srivastava, D. 2008. "Glacier Atlas of India." *Geological Society of India*: 315.

Remya, S.N., Kulkarni, A.V., Pradeep, S., Shrestha, D.G. 2019. "Volume Estimation of Existing and Potential Glacier Lakes, Sikkim Himalaya, India." *Current Science* 116(4).

Salerno, F., Thakuri, S., D'Agata, C., Smiraglia, C., Manfredi, E.C., Viviano, G., Tartari, G. 2012. "Glacial Lake Distribution in the Mount Everest Region: Uncertainty of Measurement and Conditions of Formation." *Global and Planetary Change* 92–93: 30–39. https://doi.org/10.1016/j.gloplacha.2012.04.001.

Sattar, A., Goswami, A., Kulkarni, A.V., Emmer, A. 2020. "Lake Evolution, Hydrodynamic Outburst Flood Modeling and Sensitivity Analysis in the Central Himalaya: A Case Study." *Water* 12: 237. https://doi.org/10.3390/w12010237.

Sharma, M.L., Maheshwari, B.K., Singh, Y., Sinvhal, A. 2012. "Damage Pattern during Sikkim. India Earthquake of September 18, 2011." *Indian Institute of Technology Roorkee*, Uttarakhand, India: 10.

Shrestha, F., Gao, X., Khanal, N.R., Maharjan, S.B., Shrestha, R.B., Wu, L., Mool, P.K., Bajracharya, S.R. 2017. "Decadal Glacial Lake Changes in the Koshi Basin, Central Himalaya, from 1977 to 2010, Derived from Landsat Satellite Images." *Journal of Mountain Science* 14: 1969–1984. https://doi.org/10.1007/s11629-016-4230-x.

Shukla, A., Garg, P.K., Srivastava, S. 2018. "Evolution of Glacial and High-Altitude Lakes in the Sikkim, Eastern Himalaya Over the Past Four Decades (1975–2017)." *Frontiers in Environmental Science* 6: 81. https://doi.org/10.3389/fenvs.2018.00081.

Smalheiser, N.R. 2017. "Nonparametric Tests, in: Data Literacy." *Elsevier*: 157–167. https://doi.org/10.1016/B978-0-12-811306-6.00012-9.

Sun, J., Zhou, T., Liu, M., Chen, Y., Shang, H., Zhu, L., Shedayi, A.A., Yu, H., Cheng, G., Liu, G., Xu, M., Deng, W., Fan, J., Lu, X., Sha, Y. 2018. "Linkages of the Dynamics of Glaciers and Lakes with the Climate Elements Over the Tibetan Plateau." *Earth-Science Reviews* 185: 308–324. https://doi.org/10.1016/j.earscirev.2018.06.012.

Tomczyk, A.M., Ewertowski, M.W. 2020. "UAV-Based Remote Sensing of Immediate Changes in Geomorphology Following a Glacial Lake Outburst Flood at the Zackenberg River, Northeast Greenland." *Journal of Maps* 16: 86–100. https://doi.org/10.1080/17445647.2020.1749146.

Vilca, O., Mergili, M., Emmer, A., Frey, H., Huggel, C. 2021. "The 2020 Glacial Lake Outburst Flood Process Chain at Lake Salkantaycocha (Cordillera Vilcabamba, Peru)." *Landslides* 18: 2211–2223. https://doi.org/10.1007/s10346-021-01670-0.

Wang, X., Guo, X., Yang, C., Liu, Q., Wei, J., Zhang, Y., . . . Tang, Z. 2020. "Glacial Lake Inventory of High-Mountain Asia in 1990 and 2018 Derived From Landsat Images." *Earth System Science Data* 12(3): 2169–2182. https://doi.org/10.12072/casnw.064.2019.db

Watson, C.S., Quincey, D.J., Carrivick, J.L., Smith, M.W. 2016. "The Dynamics of Supraglacial Ponds in the Everest Region, Central Himalaya." *Global and Planetary Change* 142: 14–27. https://doi.org/10.1016/j.gloplacha.2016.04.008.

Wilson, R., Glasser, N.F., Reynolds, J.M., Harrison, S., Anacona, P.I., Schaefer, M., Shannon, S. 2018. "Glacial Lakes of the Central and Patagonian Andes." *Global and Planetary Change* 162: 275–291. https://doi.org/10.1016/j.gloplacha.2018.01.004.

Yao, X., Liu, S., Sun, M., Wei, J., Guo, W. 2012. "Volume Calculation and Analysis of the Changes in Moraine-dammed Lakes in the North Himalaya: A Case Study of Longbasaba Lake." *Journal of Glaciology* 58: 753–760. https://doi.org/10.3189/2012JoG11J048.

Zhang, G., Bolch, T., Chen, W., Crétaux, J.-F. 2021. "Comprehensive Estimation of Lake Volume Changes on the Tibetan Plateau during 1976–2019 and Basin-wide Glacier Contribution." *Science of The Total Environment* 772: 145463. https://doi.org/10.1016/j. scitotenv.2021.145463.

Zhang, G., Yao, T., Xie, H., Wang, W., Yang, W. 2015. "An Inventory of Glacial Lakes in the Third Pole Region and Their Changes in Response to Global Warming." *Global and Planetary Change* 131: 148–157.https://doi.org/10.1016/j.gloplacha.2015.05.013.

4 Modeling Avalanche Susceptible Zones across the Indo-China Border around the Galwan Valley, Ladakh (India)

Geoinformatics-based Perspective in a High-Altitude Battlefield

Arvind Chandra Pandey*, Shubham Bhattacharjee, Munizzah Salim, and Chandra Shekhar Dwivedi

Department of Geoinformatics, School of Natural Resource Management, Central University of Jharkhand, Ranchi- 835222, Jharkhand, India

CONTENTS

4.1 Introduction 54
4.2 Study Area 55
4.3 Data and Software Used 56
4.4 Methodology 57
    4.4.1 Data Processing, Feature Extraction and Preparation
    of Thematic Layers 57
    4.4.2 Implementation of the AHP Model for
    Susceptibility Analysis 58
    4.4.3 Accuracy and Validation of Results 61
4.5 Results and Discussion 61
    4.5.1 Demarcation of Avalanche Impact Landforms and
    Geomorphological Features 61
    4.5.2 Avalanche Susceptibility Mapping 63
    4.5.3 Accuracy and Validation of Results 64

DOI: 10.1201/9781003265160-5

          4.5.3.1 ROC–AUC Curve Analysis 65
          4.5.3.2 Frequency Ratio Analysis 65
4.6   Conclusion 66
Acknowledgment 66
Conflicts of Interest/Competing Interests 66
Authors' Contributions 66
References 66

4.1 INTRODUCTION

Sliding down of accumulation of fresh snow layer over the dense layer down the slope results in an avalanche (Jacob et al., 2011). Both natural factors (earthquake, heavy precipitation) and anthropogenic factors (controlled explosion, skiing activities) act as the triggering cause of an avalanche event. European countries, such as France, Switzerland, and Austria, and some North American countries, such as Canada and United States, were deeply affected by avalanches as governed by the deaths of more than 2000 people in the past 15 years (Daoikb, 2000). Snow avalanches in European Alps are a prominent cause of severe injuries and deaths of more than dozens of people every winter (Techel et al., 2016).

The Indian mountainous states falling alongside the Greater Himalayan range are more susceptible to avalanche risk, rendering 91 villages in Jammu and Kashmir (JK), 104 villages in Himachal Pradesh, and 16 villages in Uttarakhand under more risk during the winter season (Ganju and Dimri, 2004). The Nubra-Shyok basin, JK, India, witnessed nearly 75% of deaths due to snow avalanches, rugged terrain, and harsh climatic conditions (Pandit and Mishra, 2014). During the Sino-India conflict in 1962 military troops were highly affected due to avalanche events and subsequently Snow and Avalanche Study Establishment (SASE) under Defense Research and Development Organisation (DRDO) was set up in 1969 to overcome avalanche-induced hazards (Podolskiy et al., 2009).

Avalanche hazard risk has increased due to urban expansion and road construction activities in mountainous areas (Beven and Kirkby, 1979). Residential areas in avalanche-prone zones, as well as public infrastructure, are also impacted by these events (Blahut et al., 2017). Snow at speeds of 200 km/hr running through valley bottoms for hundreds of meters portrays a large avalanche capable of destroying concrete structures and forests (Jamieson and Stethem, 2002).

The prediction of potential hazard zones and susceptibility analysis play a vital role in risk reduction and mitigation of avalanche hazards (Bhargavi and Jyothi, 2009; Bohner and Antonic, 2009) in addition to scientific knowledge of triggering factors and patterns of avalanche activity are very imperative for hazard zonation analysis and forecasting avalanche events (McClung, 2002). The inclusion of dynamic metrological parameters, that is, temperature, precipitation, snowfall, and wind speed, along with static terrain parameters, that is, slope, aspect, relief, curvature, and groundcover, can be used as key inputs for predicting potential avalanche hazard zones. From a geomorphological perspective different types of erosional and depositional landforms (snow avalanche impact landforms) are produced by avalanches (Barbolini et al., 2011; Bebi et al., 2009), but prediction cannot be done directly or manually for a particular time frame and is a complicated process (McClung and Schaerer, 1993).

Due to short-observation cycled snow-cover dynamics, inaccessible places of snow cover, harsh climate and rugged terrain, ground-based in situ measurements of avalanche are difficult to obtain. Hence, remote-sensing technology proved to be an effective tool to acquire the latest terrain features along with snow-cover conditions with efficient spatiotemporal coverage (Singh, 2001). A combination of remote sensing and Geographic Information Systems (GIS) witnessed to be promising to map and classify avalanche susceptibility and hazard zones (Salzmann et al., 2004).

Preparation of avalanche susceptibility maps has been accomplished in many studies using different approaches including methods like multicriteria decision analysis (MCDA) and frequency ratio methods (Bui et al., 2019; Bunn et al., 2011). Some researchers used dynamic numerical models to carry out such studies (Cama et al., 2017; Chen et al., 2017). A fuzzy logic rule-based expert system model has been described to study the behavior and release of wet snow avalanches in the Italian Ortler Alps (Zischg et al., 2005). The expert system model has been used by Maggioni and Gruber (2005) to identify potential release areas (PRAs) of avalanche using terrain features, that is, slope, aspect, relief, and curvature. The multicriteria decision-making (MCDM) method with usability of remote sensing has been applied over the Gangotri Glacier to demarcate avalanche hazards (Snehmani et al., 2014).

Analytical Hierarchy Process (AHP)–based models proved to be a promising platform for complex decision problems and have been efficiently used for studying hazard risks by classifying them into different zones based on proneness (Malczewski, 2006; Nefeslioglu et al., 2013). Feizizadeh et al. (2014) has combined the AHP and the fuzzy logic method to procure landslide susceptible zones with satisfactory accuracy estimates. The AHP was considered a reasonable approach by Kayastha et al. (2013). After comparing the AHP, logistic regression and Bayesian models, Althuwaynee et al. (2014) found AHP as the most reliable method for criteria rating. Avalanche hazard zones for 10 major river basins of Uttarakhand have been mapped using the AHP method (Kishor, 2017). Snow avalanche susceptibility of road network in Alaknanda basin was carried out with usability of the AHP method and remote sensing, which revealed roads in lower elevations are less susceptible, and susceptibility increases with an increase in elevation (Singh et al., 2018).

The present study focused on mapping the avalanche susceptible zones in the area extended between the Galwan Valley and Pangong-Tso Lake in Eastern Ladakh by incorporating the AHP method using various terrain and meteorological parameters. Avalanche impact landforms were mapped to identify potential release areas of the avalanche as the demarcation of the exact location of the avalanche is quite difficult through satellite imagery. The geomorphological setting of the area has been studied to identify the landforms that are more prone to avalanches. Accuracy estimation and validation of the work were done by using receiver operating characteristics (ROC) curve estimation and frequency ratio analysis.

4.2   STUDY AREA

The study incorporated the area extended between the Galwan Valley and Pangong-Tso Lake in the vicinity of the disputed border region between Ladakh (India) and Aksai Chin (an area administered by China) having an area of about 10,656 km² between the coordinates 37.36° to 38.63° N latitude and 79.25° to 81.29° E longitude

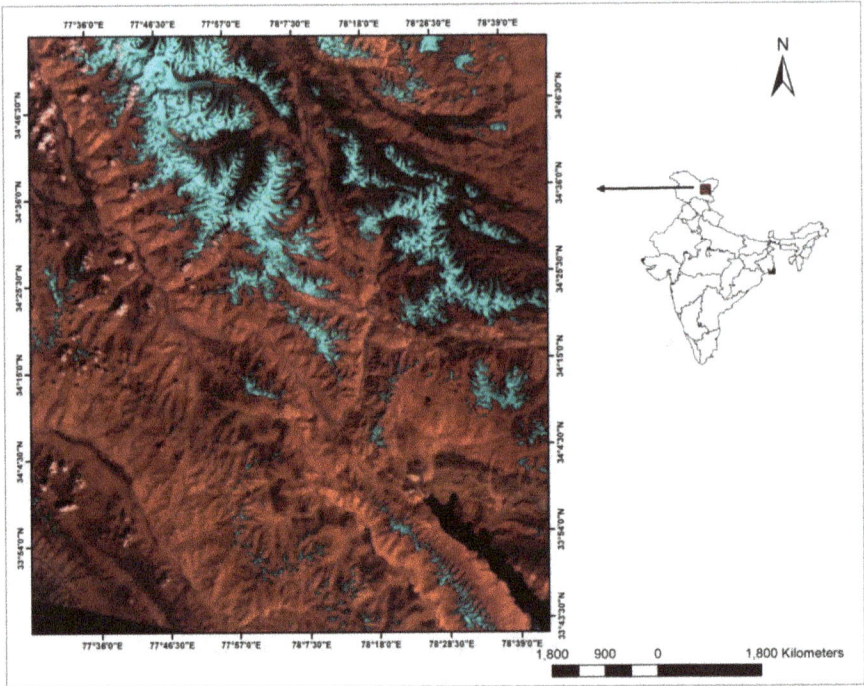

FIGURE 4.1 Location map of the study area with satellite image (FCC) showing barren slopes in red color, snow/ice cover in white/blue, and Pangong-Tso Lake (southeast portion) in black.

(Figure 4.1). The area is characterized by a semiarid climate with an average annual precipitation of less than 200 mm. The temperature ranges between 10–20°C in the peak summer season and dips to −40°C in winter. The Galwan River flows from Samzungling, Aksai Chin, to join the Shyok River, Ladakh, India. Pangong-Tso Lake (elevation: 4225 m, length: 134 km, average width: 5 km) is located on the southeastern flank of the study area line over which the line of actual control (LAC) passes. During the winter season, the majority of the area got snow-covered and Pangong-Tso Lake freezes completely. The study area is selected due to its strategic significance. The present study could be utilized for planning infrastructure development and safe movement of military troops across the Galwan Valley.

4.3   DATA AND SOFTWARE USED

Details of various data sets used in the present study are listed in Table 4.1. Satellite data processing and thematic map generation were performed using ERDAS IMAGINE and Arc Map software, respectively. Microsoft Excel has been utilized for the preparation of the ROC curve used in the validation section. A detailed flowchart is shown in Figure 4.2.

TABLE.4.1

Data Sets Used in the Present Study

Satellite Data	Resolutions	Time Period	Purpose	Source
Sentinel-2	10 m	12-August 2019 (Pre-snowfall);24-March 2020 (Post- snowfall)	Groundcover mapping; delineation of road networks; demarcation of avalanche impact landforms; geomorphologicalsetting	Copernicus
ASTER-GDEM	30 m	2012	Thematic layers preparation	NASA Earthdata
Precipitation		November2019-March 2020	Precipitation distribution	Power NASA

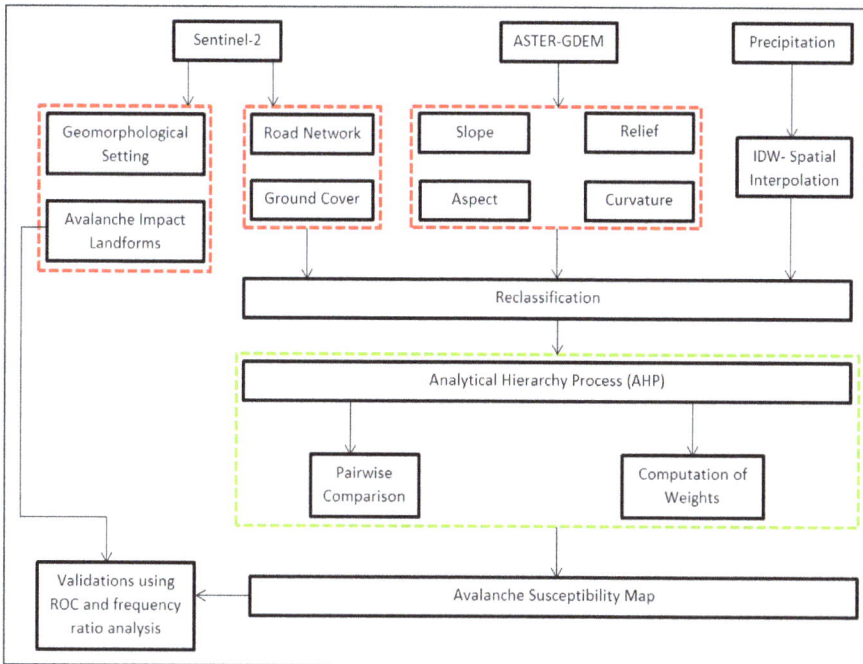

FIGURE 4.2   Methodology flowchart for the present study.

4.4   METHODOLOGY

4.4.1   DATA PROCESSING, FEATURE EXTRACTION AND PREPARATION OF THEMATIC LAYERS

Sentinel-2 satellite images were acquired for pre-snowfall and post-snowfall periods. Pre-snowfall imagery was used to extract major roads present in the study area. Snow and non-snow-covered areas were discerned using post-snowfall

imagery with the usability of an unsupervised classification method. Avalanche impact landforms and geomorphological settings have been mapped using the combination of both pre- and post-snowfall satellite images. Important erosional and depositional landforms were identified based on their image characteristics and locations. Erosional landforms, such as avalanche furrows, generally appear in groups of two or four and look like a needle-scratched pattern in satellite imagery. Avalanche gullies are the result of successive avalanche boulder impacts and can be identified as striae in gully sidewalls in satellite imagery. Depositional landforms, like avalanche boulder tongue, are tongue-shaped structures produced by avalanche debris accumulations and can be marked under successive avalanche gullies in the imagery. Talus cones are cone-shaped debris accumulation along the flanks of river valleys formed due to many small rock falls and may be seen under snow-covered peaks and talus slopes. Geomorphological landforms like glacier accumulation/ablation zones, deglaciated valleys, active floodplain, and major rivers have been marked manually in order to identify their proneness to avalanche events in that area.

Errors related to digital elevation models (DEMs) have been discarded very carefully in order to obtain accuracy in the DEM data set. Geolocation error was removed using the image-to-image co-registration with the satellite imagery and projecting the DEM into the same projection as satellite imagery using reprojection tool in Arc map 10.3. DEM data void errors, which represent negative relief features were removed by filling the voids using the fill tool in the hydrology section in Arc map 10.3. Slope, aspect, relief, and curvature layers have been prepared using respective tools to obtain slope inclination, slope direction, elevation, and concavity/convexity in the study area.

Precipitation point data (Nov 2019 to Mar 2020) collected from power NASA (https://power.larc.nasa.gov) was distributed spatially using inverse distance weightage interpolation tool to establish the distribution of precipitation over the area.

The derived slope map (Figure 4.3a) was classified into five classes from <5° to >50° using the reclassification tool. An aspect map (Figure 4.3b), which shows slope direction, was classified into 10 classes, including flat surface. The relief map (Figure 4.3c) was prepared by classifying DEM into six classes ranging from <4000 m to >6000 m. The curvature map (Figure 4.3d) was classified into three major categories, that is, concave, flat and convex. Five multiring buffer was formed on both sides of road, starting from 50 m to 250 m (Figure 4.3e) to access proximity to roads. The ground cover map (Figure 4.3f) was classified into snow/ice cover and non-snow areas. Cumulative winter precipitation distribution in mm (Figure 4.3g) was classified into five classes showing zones of higher and lower precipitation.

4.4.2 IMPLEMENTATION OF THE AHP MODEL FOR SUSCEPTIBILITY ANALYSIS

Ratings for all the classes of each factor map were given based on expert knowledge and judgments. The chances of occurrence of snow avalanche on slopes below 20° are very rare on account of stable slope conditions, and steeper slopes >40° are also less prone to avalanche as snow accumulation here is not supported, and a much less

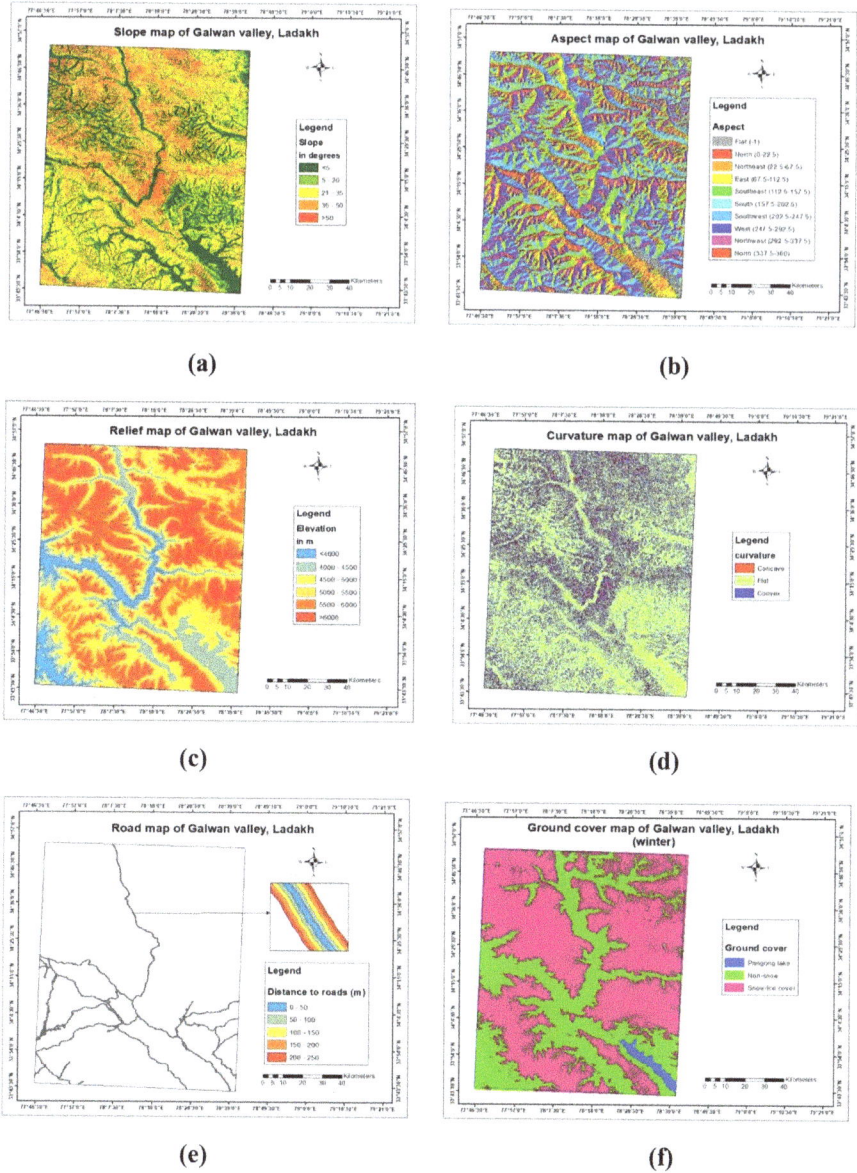

FIGURE 4.3 Thematic layers showing (a) slope, (b) aspect, (c) relief, (d) curvature, (e) proximity to roads, (f) groundcover, and (g) cumulative winter precipitation distribution in the study area.

amount of snow is available for sliding down the slope whereas slopes ranging from 20°–40° witnessed more hazardous avalanche due to accumulation of more snow and sufficient amount of shear stress to start avalanche event (Ancey, 2009). Therefore, the class of slope falling between 21°–35° was given more priority followed by class

from 36°–50°. Snowpack characteristics and stability are directly influenced by the direction of slopes and the sun. Leeward slopes are more prone to avalanche events compared to windward slopes due to the accumulation of drifted snow on the leeward direction from the windward slope by the wind. A northern aspect, having most of the avalanche slopes, is more susceptible to these events in the western Himalayas (Sharma and Ganju, 2000). Thus north and northeast aspects were given more priority as compared to other classes. The risk of avalanche hazards increases with increasing elevation as air temperature decreases and because of longer duration of snow availability for avalanching. However, at lower elevations, melting is prominent due to higher insolation. In the western part of Indian Himalayas, the elevation ranging from 2700–6000 m is prone to avalanching activities, and it becomes more dangerous between 5000–5600 m (Sharma and Ganju, 2000; Ganju et al., 2002). Thus, elevation class ranging between 5000–5500 m was given more priority. A concave surface is more stable as compared to a convex surface due to its holding capacity of a large amount of snow (Nagarajan et al., 2014). Thus, more priority was given to the convex curvature class. Construction activities, like roads, keep triggering avalanche events. So a buffer class nearby the road, that is, 50 m, was given more priority due to its more vulnerability. Snow/ice-covered zones were given more priority for avalanching activity as compared to non-snow-covered zones. The study area has very few instances of precipitation events. However, more precipitation contributes more to avalanching activities; thus, the class >200 mm was given more priority than others. After a pairwise comparison of each factor, the weight value was obtained, and the weighted sum was applied to receive the susceptibility layer. All the factors, along with their classes and class priority, ratings are given in Table 4.2.

TABLE 4.2
Ratings Based on the Priority of Respective Classes

Factors	Classes	Ratings
Slope	<5°	1
	5°–20°	4
	21°–35°	9
	36°–50°	5
	>50°	3
Aspect	Flat	1
	North	9
	Northeast	8
	East	4
	Southeast	5
	South	3
	Southwest	1
	West	2
	Northwest	7
Relief	<4000 m	1
	4000–4500 m	3
	4500–5000 m	4

Factors	Classes	Ratings
	5000–5500 m	9
	5500–6000m	5
	>6000m	2
Curvature	Concave	1
	Flat	3
	Convex	9
Distance to Roads	0–50 m	8
	50–100 m	6
	100–150 m	4
	150–200 m	2
	200–250 m	1
Groundcover	Snow/Ice cover	9
	Non-snow	4
Precipitation	<50 mm	1
	50–100 mm	3
	101–150 mm	4
	151–200 mm	6
	>200 mm	8

4.4.3 ACCURACY AND VALIDATION OF RESULTS

The accuracy and validation of obtained susceptibility map were done using receiver operating characteristics (ROC) curve and frequency ratio analysis. Instances of avalanche impact erosional landforms, that is, avalanche furrows and gullies, were used to create training data sets. For frequency ratio analysis, different frequency ratios of each susceptible zone have been obtained by viewing raster training data sets over the susceptibility map.

4.5 RESULTS AND DISCUSSION

4.5.1 DEMARCATION OF AVALANCHE IMPACT LANDFORMS AND GEOMORPHOLOGICAL FEATURES

Avalanche impact erosional (i.e., avalanche furrows and gullies) and depositional (i.e. avalanche boulder tongues, talus cones) landforms (Figure 4.4) have been marked based on their image characteristics following visual image interpretation. Talus cones were generally seen along the river valley flanking the river. Avalanche boulder tongues were seen just below avalanche gullies, where successive avalanche boulder impact was witnessed alongside the major roads.

However, avalanche gullies in continuation were observed along Pangong lake, which may be due to the formation of small water-filled depressions formed by repetitive avalanching. Three-dimensional (3-D) views (Figure 4.5a–4.5d) have also been acquired to witness and validate the landforms in a perspective view. Talus slopes under snow-cover peaks clearly indicated the presence of talus cones. Avalanche furrows appeared as a needle-scratched pattern near the peaks.

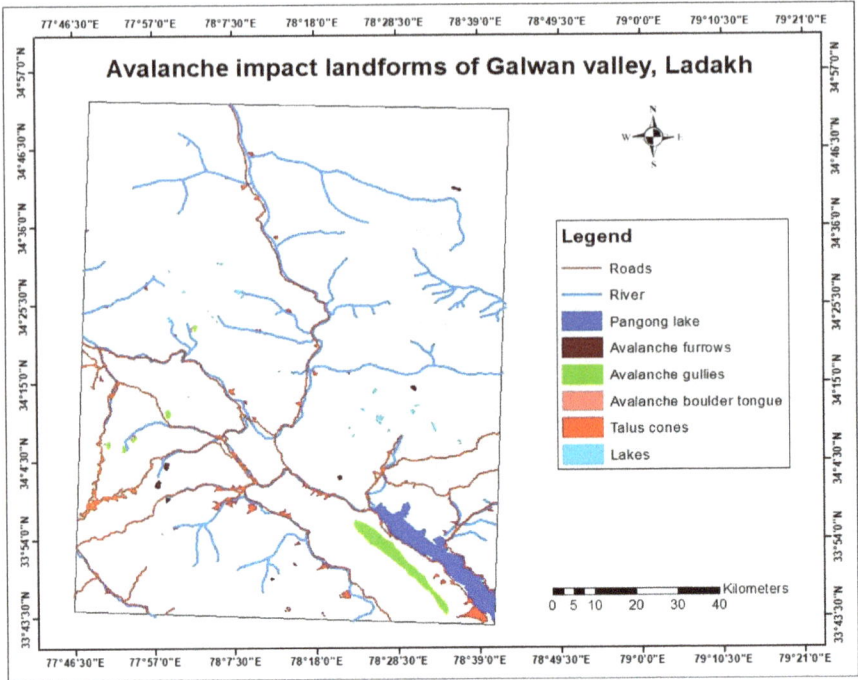

FIGURE 4.4 Avalanche impact erosional and depositional landforms in the Galwan Valley, Ladakh, deduced from satellite image.

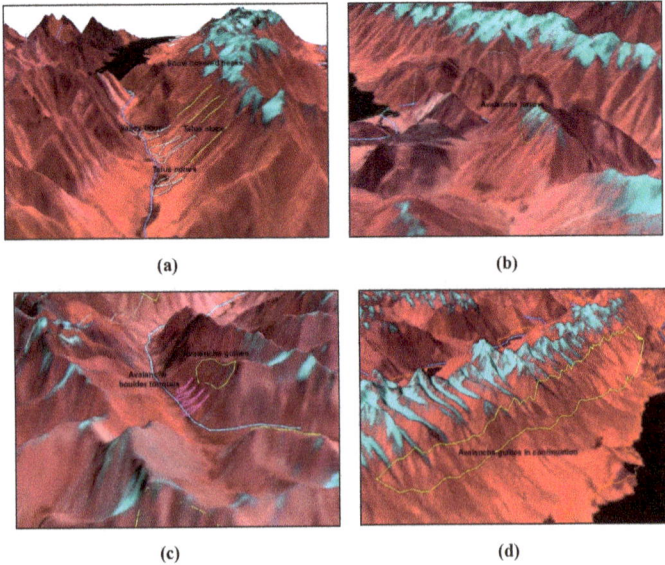

FIGURE 4.5 3-D views showing (a) talus slopes and cones, (b) avalanche furrows, (c) avalanche gullies and boulder tongues, and (d) gullies in continuation.

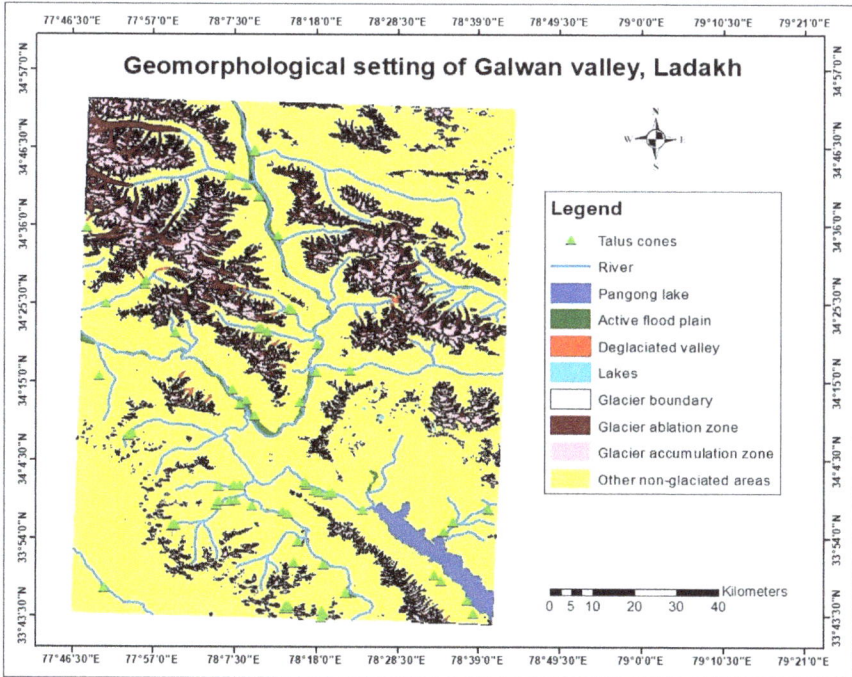

FIGURE 4.6 Avalanche impact erosional and depositional landforms mapped using Sentinel-2 satellite image.

Geomorphological factors (Figure 4.6) present in the study area were mapped to identify the most vulnerable landform in case of avalanching activity, which revealed more avalanche impact landforms were noticed over the valley slopes in the vicinity of active flood plains, deglaciated valleys, and glacier ablation zones.

4.5.2 AVALANCHE SUSCEPTIBILITY MAPPING

An avalanche susceptibility map (Figure 4.7) was prepared by incorporating the thematic layers with usability of the AHP model and GIS. A pairwise comparison between selected factors was done, and a matrix (Table 4.3) was prepared to observe the priority of one factor over another. The weight value of each factor was obtained.

The slope factor weight value was estimated to be 0.15, followed by relief (0.08), groundcover (0.05), aspect (0.03), curvature (0.029), distance to roads (0.026) and precipitation (0.016). The distribution of weights clearly indicated more effect of terrain factors such as slope and relief. The susceptibility map was classified into five zones, namely, very low (14.87%), low (25.3%), moderate (24.06%), high (23.8%) and very high (11.97%) to understand the level of impacts. The southwestern and northeastern slopes of the study area are more susceptible to avalanching. Slopes near river

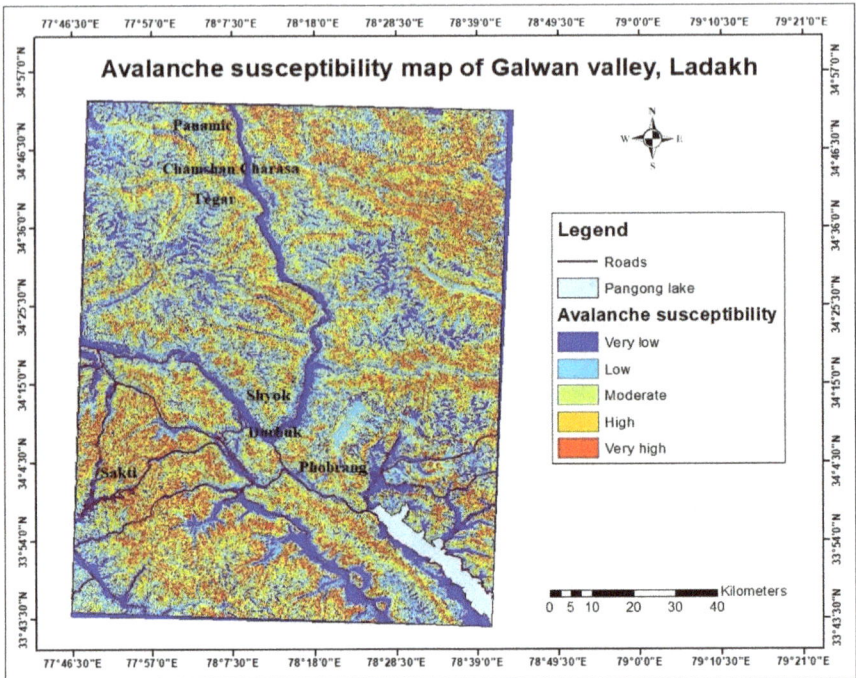

FIGURE 4.7   Avalanche susceptibility map of the Galwan Valley, Ladakh.

TABLE 4.3
Pairwise Comparison of Factors Associated with Avalanching Activities

Layers	Slope	Aspect	Relief	Curvature	Groundcover	Precipitation	Distance to Roads
Slope	1	1/5	1/2	1/4	1/6	1/6	1/6
Aspect	5	1	2	1	1/2	1/5	1/4
Relief	2	1/2	1	1/2	1/5	1/6	1/5
Curvature	4	1	2	1	1/2	1/6	1/4
Groundcover	6	2	5	2	1	1/6	1/5
Precipitation	6	5	6	6	6	1	1/4
Distance to roads	6	4	5	4	5	4	1

valleys and Pangong Lake are also highly susceptible due to the presence of both landforms created by avalanche.

4.5.3   ACCURACY AND VALIDATION OF RESULTS

The results produced in the study were validated using two different methods, namely, ROC curves and frequency ratio analysis.

4.5.3.1 ROC–AUC Curve Analysis

ROC–AUC (area under curve; Figure 4.8) has been plotted to observe the possibility of true-positive and false-positive estimations. Values ranging from 0.5–1 (50–100%) can be considered good with a high possibility of acceptance, whereas values below 50% can be considered random acceptance (Hanley and McNeil, 1983). The selected training data sets (avalanche impact erosive landform polygons) were used in accordance with the classified susceptibility map to gain true positive and false positive rates.

AUC value of 0.734 was obtained in this study, which showed 73.4% of avalanche-impacted landform pixels were properly accommodated in the avalanche susceptibility map.

4.5.3.2 Frequency Ratio Analysis

The avalanche impact landform pixels were viewed over the susceptibility map to gain frequency ratios (Table 4.4) of all five susceptible zones. High-frequency ratio values in more susceptible zones carry out good results (Jaccard, 1990).

A high-frequency ratio, namely, 0.01437, has been obtained for the very high susceptible zone, followed by the high zone, with a value of 0.00846, which support the accuracy of the result obtained in the study.

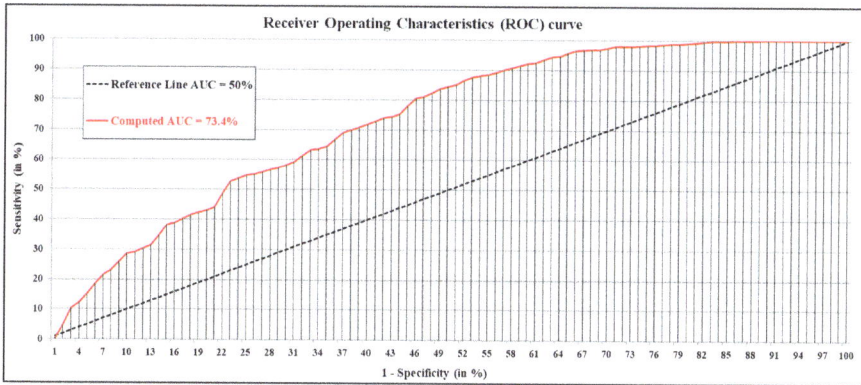

FIGURE 4.8 ROC–AUC curve estimation for the avalanche susceptibility map.

TABLE 4.4
Frequency Ratios of Different Susceptible Classes

Classes	No. of Class Pixels	No. of Training Pixels	Frequency Ratio (FR)
Very low	1,758,343	826	0.00047
Low	2,991,543	6676	0.00223
Moderate	2,845,867	16,398	0.00576
High	2,814,914	23,827	0.00846
Very high	1,414,065	20,323	0.01437

4.6 CONCLUSION

Highly rugged terrain and harsh climatic conditions in border areas around the Galwan Valley–Pangong Lake area during the winter season hinder in situ estimations and monitoring of avalanche risk zones through field investigations, rendering remote sensing and GIS as the best techniques for temporal monitoring of avalanche impact zones in a near-real-time frame. Demarcation of avalanche impact landforms through satellite images helped to identify the possible avalanche release areas. Geomorphological setting showed traces of avalanching activities in the vicinity of river valley floors followed by deglaciated valley slopes and glacier ablation zones. The avalanche susceptibility map demonstrated 11.97% and 23.8% of the total study area falling under the very high and high zones. The AUC value of 73.4% and a higher frequency ratio value in the very high zone depicted the adequate accuracy of the results, proving AHP is a good model for carrying out avalanche-prone zone mapping in data-scarce regions of strategic importance. The satellite image–based landform mapping and avalanche hazard zones delineated could be utilized for planning infrastructure development and safe movement of military troops along the border regions, especially in the situation of border area dispute between Indian and China during avalanche-prone winter seasons.

ACKNOWLEDGMENT

Authors thanks to United State Geological Survey and NASA Earthdata, and Power NASA for providing Sentinel satellite data, ASTER GDEM, and precipitation point locations.

CONFLICTS OF INTEREST/COMPETING INTERESTS

The authors declare no conflicts of interest.

AUTHORS' CONTRIBUTIONS

A.C.P.: conceptualization, methodology, analysis, visualization, validation, writing—original draft, review and editing. **S.B. and M.S.:** methodology, software, analysis, visualization, validation, writing- original draft, review and editing. **C.S.D.:** analysis, validation, writing- review and editing.

REFERENCES

Althuwaynee FO, Pradhan B, Park H, Lee JH. 2014. A novel ensemble bivariate statistical evidential belief function with knowledge-based analytical hierarchy process and multivariate statistical logistic regression for landslide susceptibility mapping. *Catena*, 114, 21–36.

Ancey C. 2009. *Snow Avalanches*. New York: Wiley & Sons, PMID: 18936996

Barbolini M, Pagliardi M, Ferro F, Corradeghini P. 2011. Avalanche hazard mapping over large undocumented areas. *Natural Hazards*, 56, 451–464.

Bebi P, Kulakowski D, Rixen C. 2009. Snow avalanche disturbances in forest ecosystems-State of research and implications for management. *Forest Ecology and Management*, 257, 1883–1892.

Beven KJ, Kirkby MJ. 1979. A physically based, variable contributing area model of basin hydrology/Un modele a base physique de zone d'appel variable de l'hydrologie du basin versant. *Hydrological Sciences Journal*, 24, 43–69.

Bhargavi P, Jyothi S. 2009. Applying native bayes data mining technique for classification of agricultural land soils. *International Journal of Computer Science and Network Security*, 9, 117–122.

Blahut J, Klimes J, Balek J, Hajek P, Cervena L, Lysak J. 2017. Snow avalanche hazard of the Krkonose National Park, Czech Republic. *Journal of Maps*, 13, 86–90.

Bohner J, Antonic O. 2009. Land-surface parameters specific to topo-climatology. *Developments in Soil Science*, 33, 195–226.

Bui DT, Ngo P-TT, Pham TD, Jaafari A, Minh NQ, Hoa PV, Samui PA. 2019. A novel hybrid approach based on a swarm intelligence optimized extreme learning machine for flash flood susceptibility mapping. *Catena*, 179, 184–196.

Bunn AG, Hughes MK, Salzer MW. 2011. Topographically modified tree-ring chronologies as a potential means to improve paleoclimate inference. *Climatic Change*, 105, 627–634.

Cama M, Lombardo L, Conoscenti C, Rotigliano E. 2017. Improving transferability strategies for debris flow susceptibility assessment: Application to the Saponara and Itla catchments (Messina, Italy). *Geomorphology*, 288, 52–65.

Chen W, Pourghasemi HR, Kornejady A, Zhang N. 2017. Landslide spatial modelling Introducing new enembles of ANN, MaxEnt, and SVM machine learning techniques. *Geoderma*, 305, 314–327.

Daoikb. 2000. *Meterolojik Kaynakli DogalAfetler Alt Komisyonu*. Ankara: Basilmamis Rapor.

Feizizadeh B, Shadman RM, Jankowaski P, Blaschke T. 2014. A GIS-based extended fuzzy multi-criteria evaluation for landslide susceptibility mapping. *Computers & Geosciences*, 73, 208–221.

Ganju A, Dimri A. 2004. Prevention and mitigation of avalanche disasters in Western Himalayan region. *Natural Hazards*, 31, 357–371.

Ganju A, Thakur NK, Rana V. 2002. Characteristics of avalanche accidents in Western Himalayan region. *Proceedings of International Snow Science Workshop*, Canada, pp. 200–207.

Hanley JA, McNeil BJ. 1983. A method of comparing the areas under receiver operating characteristic curves derived from the same cases. *Radiology*, 148(3), 839–843.

Jaccard C. 1990. Fuzzy factorial analysis of snow avalanches. *Natural Hazards*, 3(4), 329–340.

Jacob N, Srinivasulu J, Varadan G, Reddy R, Kumari A, Jyothi V. 2011. Emerging trends in Snow Avalanche Modelling. *Diamensions and Direction of Geospatial Industry*, 18–21 January, Hyderabad, India.

Jamieson B, Stethem C. 2002. Snow avalanche hazards and management in Canada: challenges and progress. *Natural Hazard*, 26(1), 35–53.

Kayastha P, Dhital MR, De Smedt F. 2013. Application of the analytical hierarchy process (AHP) for landslide susceptibility mapping: A case study from the Tinau watershed, west Nepal. *Computers & Geosciences*, 52, 398–408.

Kishor SK. 2017. *Mapping and Modelling of Potential Avalanche Zones in Parts of Uttarakhand: An Earth Observation Initiative*. Mtech, Indian Institute of Remote Sensing, Andhra University, India.

Maggioni M, Gruber U. 2005. The influence of topographic parameters on avalanche release dimension and frequency. *Cold Regions Science and Technology*, 37, 407–419.

Malczewski J. 2006. GIS-based multicriteria decision analysis a survey of the literature. *International Journal of Geographical Information Sciences*, 20(7), 703–726.

McClung DM. 2002. The elements of applied avalanche forecasting, part i: The human issues. *Natural Hazards*, 25, 111–129.

McClung DM, Schaerer P. 1993. *The Avalanche Handbook*. Seattle, WA: The Mountaineers Books.

Nagarajan R, Venkataraman G, Snehmani. 2014. Rule based classification of potential snow avalanche areas. *Natural Resources and Conservation*, 2(2), 11–24.

Nefeslioglu HA, Sezer EA, Gokceoglu C, Ayas Z. 2013. A modified analytical hierarchy process (M-AHP) approach for decision support systems in natural hazard assessments. *Computers and Geosciences*, 59, 1–8.

Pandit R, Mishra A. 2014. 18 years after he went missing armyman's body found in Siachen. *The Time of Inida*, August 20.

Podolskiy EA, Sato A, Komori J. 2009. Avalanche issue in Western Himalaya, India. *International Symposium on Snow and Avalanche*, 498–502.

Salzmann N, Allg B, Huggel A, Huggel C, Wer & Wilfried, B & Haeberli, Salzmann H. 2004. Assessment of the hazard potential of ice avalanches using remote sensing and GIS modelling. *Norwegian Journal of Geography*, 58(2), 74–84.

Sharma SS, Ganju A. 2000. Complexities of avalanche forecasting in Western Himalaya–An overview. *Cold Regions Science and Technology*, 31(2), 95–102.

Singh P. 2001. Snow and Glacier Hydrology. *Springer Science & Business Media (Netherland)*, ISBN-978-0-7923-6767-3.

Singh V, Thakur PK, Garg V, Aggarwal SP. 2018. Assessment of snow avalanche susceptibility of road network—a case study of Alaknanda Basin. *International Archives of the Photogrammetry, Remote Sensing and Spatial Information Sciences*, 5, 20–23.

Snehmani, BA, Pandit A, Ganju A. 2014. Demarcation of Potential avalanche sites using remote sensing and ground observations: A case study of Gangotri Glacier. *Geocarto International*, 29(5), 520–535.

Techel F, Jarry F, Kronthaler G, Mitterer S, Nairz P, Pavsek M, Valt M, Darms G. 2016. Avalanche fatalities in the European alps: Long-term trends and statistics. *Geographica Helvetica*, 71, 147–159.

Zischg A, Fuchs S, Keiler M. 2005. Modelling the system behaviour of wet snow avalanches using an expert system approach for risk management on high alpine traffic roads. *Natural Hazards Earth System Sciences*, 5(6), 821–832.

5 Flood Monitoring and Assessment over the Himalayan River Catchment

Kavita Kaushik[1], Arvind Chandra Pandey[1,],*
Bikash Ranjan Parida[1], and Navneet Kumar[2]

1 Department of Geoinformatics, School of Natural Resource Management, Central University of Jharkhand; Ranchi-835222, India

2 Department of Ecology and Natural Resources Management, Center for Development Research (ZEF), University of Bonn, Germany

CONTENTS

5.1 Introduction 69
5.2 Study Area 71
5.3 Materials and Method 73
    5.3.1 MODIS NRT Data 73
    5.3.2 CHIRPS Rainfall Data 73
    5.3.3 Copernicus LULC 74
    5.3.4 Methodology 74
5.4 Results 75
    5.4.1 Rainfall Distribution 75
    5.4.2 Flood Extent over Bihar 76
    5.4.3 Flood Stagnancy in Bihar 79
    5.4.4 Stagnant Water–Affected Cropland and Built-Up Area 80
5.5 Discussion 81
5.6 Conclusion 83
References 84

5.1 INTRODUCTION

In recent times, climate change is mostly an anthropogenic phenomenon with enough historical evidence to support it with the variability in rainfall, cyclonic patterns, and temperature changes. The South Asian climate is dominated by the southwest monsoon, with

DOI: 10.1201/9781003265160-6

spatial and temporal variation throughout the region (Shrestha 2008). The shifting of the monsoon season influences the variability of rainfall with serious consequences on the environment. A delay of 15 days was predicted in the onset of monsoons in Southeast Asia in near future (Loo, Billa, and Singh 2015). Asian continent can be considered the most vulnerable to climate change because of the higher population density and higher exposure to floods, which is severely felt in agriculture (Dhanya and Ramachandran 2016). In the Indian subcontinent, floods have been affecting the economy, environment, and population on a very large scale, with different reasons associated with their occurrence. Some regions are largely affected by the rainfall condition due to excessive rainfall generated runoff from the higher elevation areas toward the lower elevation areas (Parida et al. 2017). There is a strong association between large flood events and the variability in monsoonal rainfall intensity that has been recognized for a long time (Kale 2012). Geomorphological settings, such as the inflow and outflow of the sewerage limits, also contribute to flooding (Ray et al. 2019). The rise in temperature caused by global warming results in an increment in the ocean heat content, causing mean sea-level rise because of the thermal expansion of the seawater (Mimura 2013) increasing the vulnerability of the coastal cities. The natural factors that enhance the risk of flooding in the coastal region are low relief and topography, storm surges, cyclones, and heavy rainfall (Dhiman et al. 2018).

The availability of advanced satellite technology in remote-sensing and Geographic Information System (GIS) processing systems can be effectively used for mapping flood disasters and major risk zones. It is highly beneficial in predicting the future scenarios of the flood events and for planning mitigation and management plans for pre- and post-disaster conditions (Sundaram and Yarrakula 2017). The impact of the flood is a function of frequency, the magnitude and power of flood, the duration of the excessive flow of water, the changes of the channel planform and the cross-section geometry of the river, and several others (Das 2019). In the flood-prone Himalayas, determining the magnitude and frequency of floods is important for estimating the likelihood of future floods (Wasson et al. 2013). Riverine flooding is a serious issue in July and August in Nepal. The temporal changes in the rainfall extremes were analyzed using long-term continuous records for the 1970–2012 period by applying the Mann–Kendall test for the identification of trend and, furthermore, the Sen's slope to calculate the magnitude of the trend (Karki et al. 2017). The results concluded an increment in the pre-monsoonal precipitation extremes over the lowland and central hills of Nepal. While the monsoon precipitation is increasing in western mountains and central hills, decreasing precipitation was observed from the central mountains, central lowlands, and eastern lowlands (Karki et al. 2017). Remote-sensing-derived Tropical Rainfall Measuring Mission (TRMM) based rainfall products have been widely used for many hydrological applications (Parida et al. 2018; Shukla et al. 2019). TRMM Precipitation Radar (PR) version 6 product "2A25 near-surface rainfall" of 11 years (1998–2008) was used to analyze the rainfall–elevation relationship in the central Himalayan region for the monsoon and pre-monsoon season, which revealed a positive and strong relationship between rainfall and elevation in both the seasons (Shrestha, Singh, and Nakamura 2012). There was a large amount of rainfall observed over higher elevations during the pre-monsoon seasons (Shrestha, Singh, and Nakamura 2012). The temporal rainfall variation in the Bhagmati River basin based on the data available at the Department of Hydrology and Meteorology, Government of Nepal, revealed a significant rising trend in water levels in the summer season (Shrestha and Sthapit 2016).

There are significant alternations in morphometric parameters due to climatic changes. In a study over the Gandak River Basin (GRB), drainage morphometric characteristics through geologic, topographic, and hydrological information are identified to assess the extreme weather events. The Standardized Precipitation Index (SPI) and Rainfall Anomaly Index (RAI) were used to derive the extreme rainfall events from TRMM precipitation products, which concluded that 1998, 2007, 2011, 2013, and 2017 experienced extreme flood events leading to channel shifting (Chaubey et al. 2021). The three biggest river systems of Kosi, Gandaki, and Karnali originate in the high mountain glaciers of Nepal and flow through the country to enter India through the state of Bihar. These rivers have been classified with heavy rains, leading to landslides in Nepal during the monsoon season and creating alarming situations in Bihar (Deoras 2021).

Bihar state is highly vulnerable to floods due to its recurrent behavior in the northern Bihar region with a strong spatial pattern among the vulnerable districts of the state (Jha and Gundimeda 2019). The northern and Central Bihar districts along the river Ganga are susceptible to the severe impacts of flooding with 50.95% of the total area under high and very high risk (Pandey, Singh, and Nathawat 2010). During the floods of 2017, 13% of the total area of the Darbhanga district of North Bihar was submerged and various flood patches were observed in central, northern, and western districts because of the presence of various water channels (Tripathi et al. 2020). The floodwater duration is very high in terms of these districts for which disaster preventive and mitigation measures need to be adopted along with risk reduction measures to minimize the disastrous impact of flooding. Hence, the main objective of this study is to analyze rainfall and flooding patterns in the Himalayan catchment region. Specifically, to characterize 2021 flood events with respect to inundation and stagnancy period, we have considered only the state of Bihar where large-scale flooding and adverse impacts on agriculture were observed.

5.2   STUDY AREA

The study area comprises the Himalayan catchment region mainly including three states of India (Uttarakhand, Uttar Pradesh, Bihar) and Nepal (Figure 5.1). The region is drained by the Ganga River, which is the longest river in India. The mighty river begins at the confluence of the Alaknanda and Bhagirathi Rivers, which meet at Dev Prayag in Uttarakhand. The Bhagirathi River rises at the foot of Gangotri Glacier and has been considered the true source of Ganga. The Ganga River Basin is divided into Upper Ganga Basin, Middle Ganga Basin, and Lower Ganga Basin (Shukla et al. 2019). The major right-bank tributaries of Ganga River include Yamuna, Chambal, Banas, Sind, Betwa, Ken, Son, and Damodar. The Ramganga, Gomti, Ghaghra, Kali, Gandak, Burhi, and Kosi are the left-bank tributaries of Ganga.

Kali, Karnali, Kosi, and Gandak are transboundary rivers flowing through Nepal and India. Kosi is widely known as the sorrow of Bihar, attributed to its unstable nature for carrying heavy silt during the monsoon season. Ghaghara River, alternatively known as Karnali, originates in the Mapchachungo Glacier, is a transboundary perennial river originating from the Tibetan plateau near Lake Mansarovar, and joins Ganga at Chhapra in Bihar. The rivers originating in Nepal and draining in Bihar create havoc conditions during the monsoon season.

The land cover of the study area is dominated by cropland, because of the presence of an agricultural belt over the Uttar Pradesh and Bihar region. The Himalayan

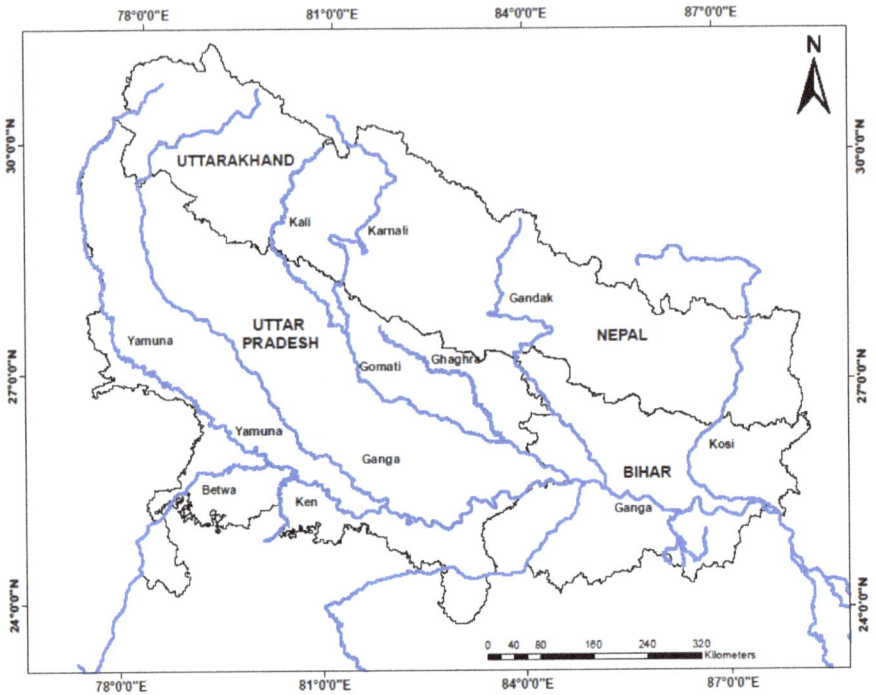

FIGURE 5.1   Map of the study area with major rivers and tributaries.

mountain ranges are confined to the upper limits of the study area including northern Nepal and Uttarakhand. The upper Himalayan region in Nepal occupies 15% of the total area of the country; it is extremely cold and inhospitable. Most of the country is formed of middle hills and the lower Himalayas and is far more fertile than the upper Himalayas, owing to which it holds the largest population of the country in the middle hills. The southernmost part of the country is the Tarai Region, exceedingly fertile with a subtropical climate. Nepal receives rainfall ranging from 50–5000 mm in different seasons. Altitude and topography play a vital role in rainfall distribution; because of this variation, three areas have been identified as the highest rainfall regions, namely, the southern slopes of the Makalu range in the eastern region, the Jugal range in central region, and south of the Annapurna range in the western region.

The geography of Uttarakhand offers a wide range of landforms, ranging from the snow region, Jugal range in central region, and south of the Annapurna range in the western region. It covers mountains to river valleys, hills, and densely forested areas. The Ganga and Yamuna originate in the glaciers of Uttarakhand. The topography of Uttar Pradesh can be divided into Himalayan foothills in the north, the Gangetic plains, and the Vindhyan hills. Uttar Pradesh is a highly populated region because of the availability of fresh water from rivers and its flat topography, which makes it ideal for agriculture. In terms of rainfall, the northeastern districts of Uttar Pradesh receive a higher amount of rainfall as compared to southwest Uttar Pradesh. Bihar state (covering parts of the study area) has most of its geographical area under

cultivation and has one of the highest population densities in the country. The study area mostly receives a maximum of precipitation during the monsoon season from July to September and is highly prone to flooding in the low-lying region.

5.3 MATERIALS AND METHOD

In the present study, the flood condition delineation during the monsoon season (June–October) 2021 of Bihar was based on 14-day composite of the Moderate Resolution Imaging Spectroradiometer (MODIS) near-real-time (NRT) product acquired from https://floodmap.modaps.eosdis.nasa.gov. To observe the rainfall distribution across the Himalayan catchment, the Climate Hazards Centre InfraRed Precipitation with Station data (CHIRPS) daily precipitation data were acquired from https://data.chc. ucsb.edu/products/CHIRPS-2.0/. The details of the data are given in Table 5.1.

5.3.1 MODIS NRT DATA

MODIS operates on 36 spectral bands with a spatial resolution of 250 m to 1000 m. MODIS NRT flood product is provided by Dartmouth Flood Observatory. The flood product delineates water from other areas by using its infrared bands toward different objects based on their reflectance or absorption. Flooded pixels are detected as a composite product that is available at 3-day and 14-day composites. Two or more observations are utilized as cloud shadow appears similar to water. Often, time-series images are employed to remove cloud shadow noise from the NRT flood product because the cloud moves on and its shadow is rarely found in the same locations.

5.3.2 CHIRPS RAINFALL DATA

CHIRPS is a quasi-global rainfall data set that is available from 1981 to the near present. CHIRPS incorporates in-situ station data with 0.05° resolution satellite imagery to create a gridded rainfall time series for trend analysis. CHIRPS provides

TABLE 5.1
Data Set Used in the Present Study

Data Set Used	Spatial and Temporal Resolution	Acquisition Date	Purpose	Source
CHIRPS rainfall data	0.05*0.05 degree, daily	May–October (2021)	Rainfall distribution	("Climate Hazards Center, UC Santa Barbara")
MODIS NRT	250m, daily	May–October (2021)	Flood inundation extent	("NASA GSFC NRT Global Flood Mapping")
Copernicus Global Land Cover	100m	2019	Affected LULC extraction	(Buchhorn et al. 2020)

blended gauge-satellite precipitation estimates covering most parts of the global land regions that have low latency, a long period of record, high resolution, and low bias (Funk et al. 2015).

5.3.3 COPERNICUS LULC

The Copernicus Global Land Service (CGLS) operates on a multipurpose service component that provides biophysical products based on the status of the land surface at a global scale. Copernicus land cover maps are available at a moderate spatial resolution of 100 m that primarily targets the land cover and its changes (Buchhorn et al. 2020).

5.3.4 METHODOLOGY

CHIRPS daily precipitation data were used for the observation of rainfall distribution from May to October 2021 over the Himalayan catchment. The daily precipitation data were firstly merged into a 14-days composite to keep in accordance with the MODIS NRT flood data. ArcGIS software was used for the processing of data, where firstly a composite was created. From the acquired composite of the images, the spatial analyst tool was used for generating a sum total output of the data using the cell statistics tool. The workflow has been shown in Figure 5.2.

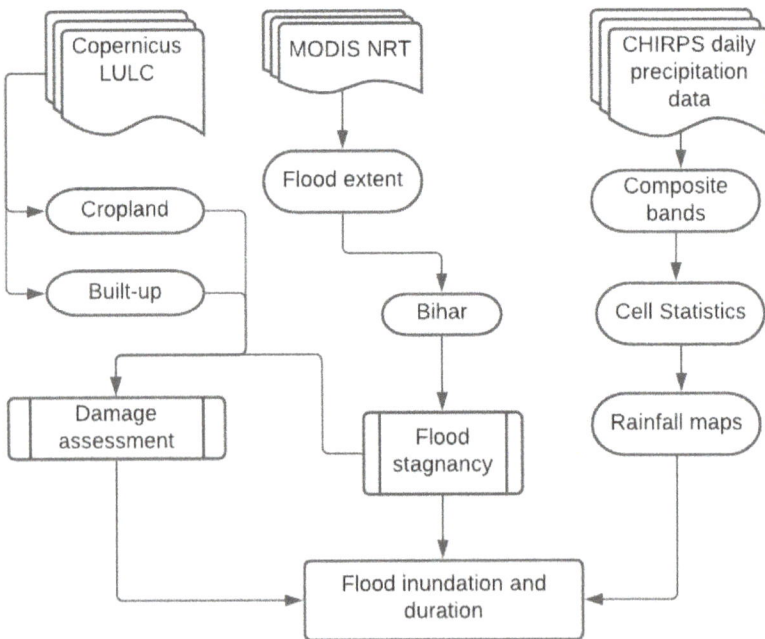

FIGURE 5.2 Flowchart of the study.

The flood extent and the stagnant water mapping were done for the Himalayan catchment area and Bihar, respectively. The flood extent was observed using the MODIS NRT data with a 14-days composite period for May to October in 2021. The raster data were downloaded from the NASA site, and the extent of flooding in the Himalayan catchment was extracted. The permanent water bodies have been subtracted using the Copernicus flood water layer from area calculation. After the flood maps were prepared for the Himalayan catchment area, it was observed that Bihar was affected to a major extent during the monsoon season and heavy rainfall in the Himalayan River catchments all led to disastrous impacts in Bihar. The data for the flood-affected area of Bihar were then used for determining the stagnancy period of water. The period for stagnancy estimation was divided into four classes, namely, the inundated area for less than 15 days, 15–30 days, 30–45 days and more than 45 days. For the division of the area into the specified categories of stagnancy, the MODIS NRT flood product for 15 days composite was used to identify the area under flood for less than 15 days. For estimating the area affected between 15–30 days, an intersect tool in ArcGIS was used by which the common area between the two files (a composite of 15 days and the other proceeding 15 days) was derived. The same processing was followed for 30–45 days, by intersecting three composites, and more than 45 by intersecting four composites of NRT flood product. The composite flood inundated area of more than 30 days was used to acquire the damage to cropland and built-up using the Copernicus LULC. The damage assessment was done using the clipping tool in ArcGIS and then the area was calculated.

5.4   RESULTS

5.4.1   RAINFALL DISTRIBUTION

The rainfall maps were used to observe the rainfall pattern from May to October for the year 2021 over the Himalayan region, including Nepal and three states of India, namely, Bihar, Uttar Pradesh, and Uttarakhand (Figure 5.3). In June, the rainfall intensity of 0–100 mm was observed in Uttarakhand and the western part of Uttar Pradesh, and Nepal recorded an intensity of 100–200 mm, which increased to 100–300 mm in parts of Uttar Pradesh, Bihar, and Nepal and 300–500 mm in central Nepal in mid-June. By the end of the month, the intensity remained higher at the Nepal–Bihar border. The start of July was marked by lower rainfall intensity of 0–100 mm in Uttarakhand and western Uttar Pradesh, whereas the intensity of 100–300 mm was observed in the eastern part of Uttar Pradesh, Bihar, with Central Nepal recording an increased intensity of 300–500 mm, which greatly decreased by mid-July with Uttarakhand, Uttar Pradesh, Bihar, and Nepal receiving rainfall intensity between 0–200 mm in some scattered parts of the study area. The end of the month witnessed an increase by 100–200 mm rainfall intensity covering major parts of Uttarakhand, Uttar Pradesh, Bihar, and Nepal, and 200–300 mm was observed along with the southern districts of Uttar Pradesh, Uttarakhand, and Bihar.

In August, the intensity remained higher, with a maximum rainfall intensity of 600–700 mm observed in the elevated parts of Nepal. Rainfall intensity of 100–300 mm was observed in Uttarakhand, the western part of Uttar Pradesh, and southern

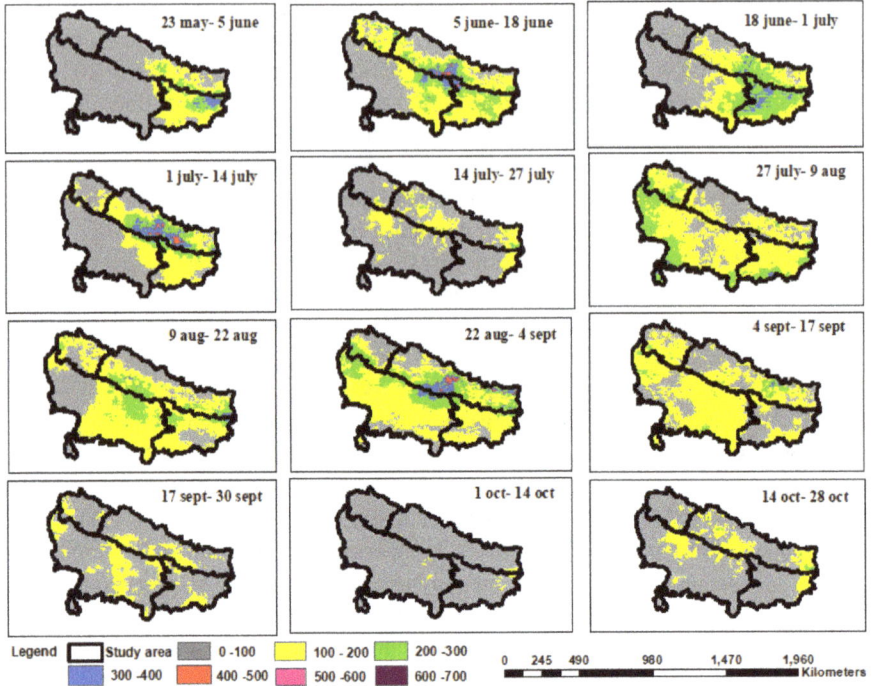

FIGURE 5.3 Rainfall intensity in mm over the Himalayan catchment during May 23–28 October 2021.

Bihar (Figure 5.3). The higher reaches of Uttar Pradesh, Bihar, and Central Nepal received higher rainfall intensity of 300–600 mm. By the start of September, the rainfall intensity decreased considerably with rainfall intensity of 200–300 mm observed in Uttar Pradesh and Nepal, whereas Bihar and Uttarakhand received rainfall intensity lower than 100 mm. The rainfall intensity indicated a decreasing pattern throughout September with a similar pattern by the start of October. The end of the month was remarked with rainfall intensity of 100–300 mm in the northern limits of Uttar Pradesh, Uttarakhand, Bihar, and Nepal.

5.4.2 FLOOD EXTENT OVER BIHAR

The flood extent was acquired using MODIS NRT flood product over the Himalayan catchment, which led to the observation that Nepal, Uttar Pradesh and Uttarakhand have shown almost negligible flooding from May to August. By the end of August to October, some parts of Uttar Pradesh were flood-affected in minor terms along the Ganga River, whereas Bihar has witnessed intense flooding for a longer period compared to other states of the study area. Due to the excessive runoff from Nepal, and subsequently increased flow from Uttarakhand and Uttar Pradesh, it leads to large-scale flooding in Bihar. Therefore, the flood progression at 14 days interval for

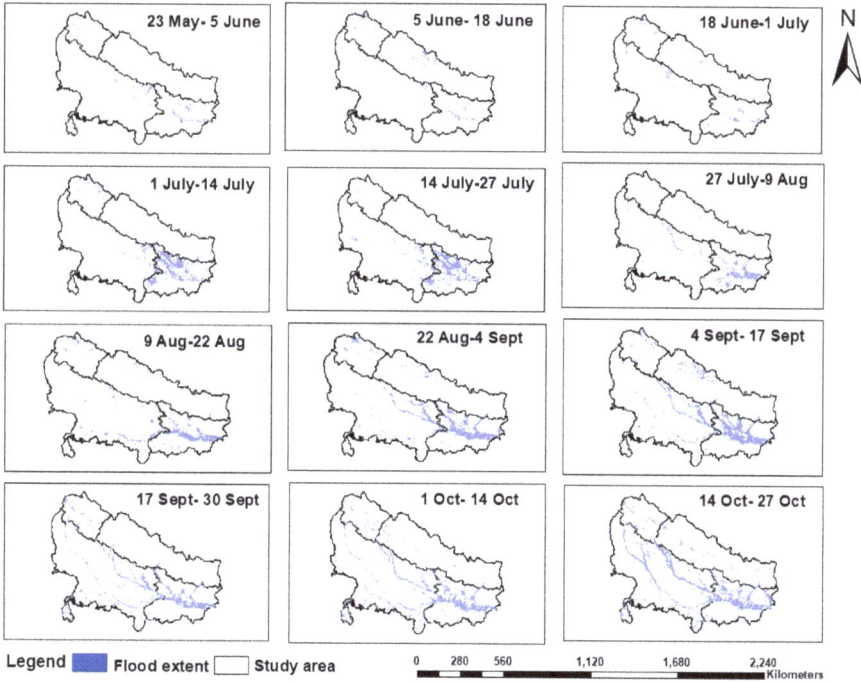

FIGURE 5.4 Flood progression at 14 days intervals based on MODIS NRT data during the monsoon season in 2021.

the Himalayan catchment is presented in Figure 5.4, and the corresponding areal statistics for flood inundated area of Bihar are given in Figure 5.6. From the end of May to the start of September, the flood shows changes in location in different periods. In June, flooding in the Ganga River is confined to Patna or Bhagalpur. By the start of July, the Himalayan River catchment leads to flooding in the upper Bihar region, leading to an increase in the water level of the Ganga River, resulting in flooding in Patna and Ganga Plains, which reduces by mid-August. In mid-August, the rainfall has affected mostly the area surrounding the Ganga River, although the rainfall in the Himalayan catchment continues to flood the upper Bihar region by the end of August. By the start of September, heavy rainfall in the upper Ganga catchment has widely affected the upper Bihar region and the area affected was estimated to be the highest for the month of September. The higher elevated glacier areas of Uttarakhand and Nepal have been misclassified as flood pixels.

This flood pattern shows that the July flood was more intensive in higher elevation parts of Bihar, with rivers coming from Nepal creating flooding, whereas in August and September, the lower elevation zones, mainly along the Ganga River with its catchment in Uttaranchal, created the flood. A cumulative flood-affected area of Bihar was estimated to be 9911.64 km^2 (or 10.5% of the total geographical area; Figure 5.5).

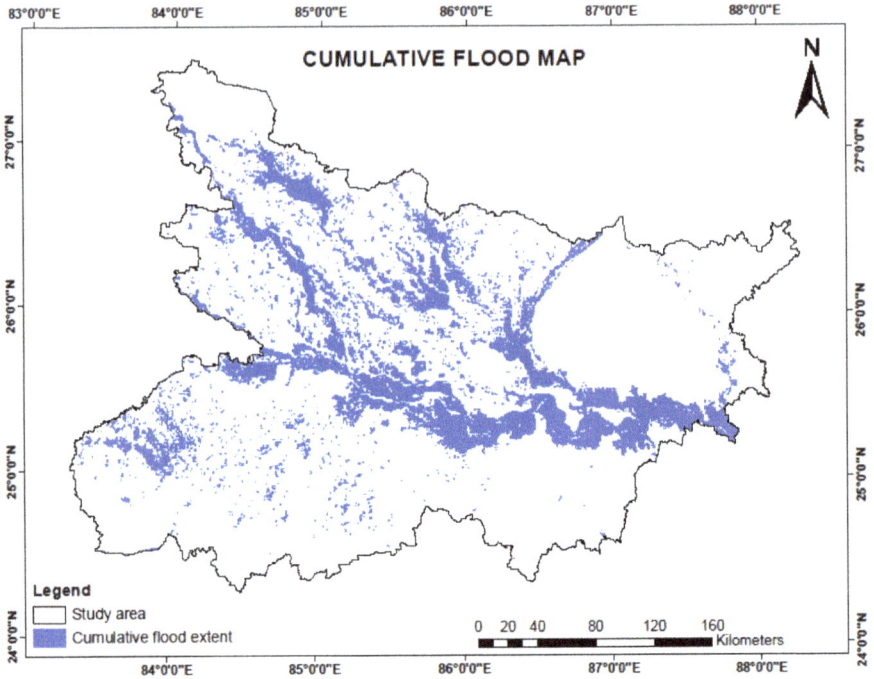

FIGURE 5.5   Cumulative flood-affected area of Bihar during the 2021 flood event.

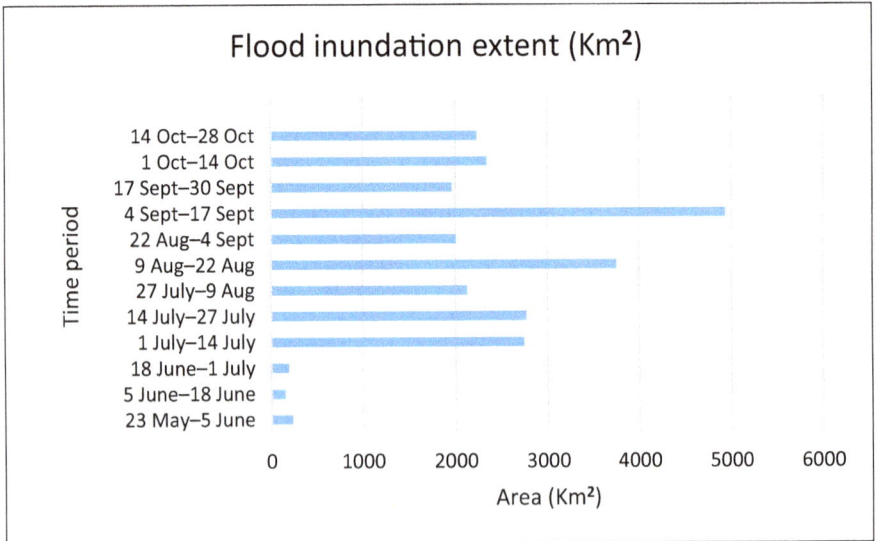

FIGURE 5.6   Flood inundation extent statistics during monsoon season in 2021. The total geographical area of Bihar is 94234.6 km².

5.4.3 Flood Stagnancy in Bihar

The flood stagnancy has been derived according to flood progression during the monsoon season in 2021 and is presented in Figure 5.7. The corresponding area statistics have been given in Table 5.2. The flood duration that has been estimated by listing the area affected by flood for the period of 28 days, 136.94 km² has been affected during the starting of flood season and has stayed from May 23–June 18, mostly affecting the Darbhanga, Patna, and Vaishali districts (Figure 5.7). From June 5–July 1, a 124.73 km² area has been continuously submerged by river water, affecting Patna and Darbhanga districts. During the period of 18th June 18–July 14, only 25.7 km² was found to be submerged, mostly in Patna district. Patna shows extreme flooding over the months, the water stayed for more than a month in Patna followed by Darbhanga district. From the start of the monsoon period, floods intensified in Bihar, and consequently, the affected area increased by more than ten times as recorded in the previous month. The statistics for the period of July 1–July 27 mark 2048.59 km² of area submerged over the 28-day period affected various districts (Darbhanga, Gopalganj, Madhubani, Muzaffarpur, Pashchim Champaran, Patna, Purba Champaran, Samastipur, Saran, Sitamarhi, Vaishali), which decreased to 632.28 km² between July 14–August 9, affecting Patna, Samastipur, Madhepura, Khagaria, Darbhanga, and Bhagalpur. Seven districts (Samastipur, Patna, Munger, Khagaria, Katihar, Bhagalpur, Begusarai) had 1136.93 km² area confined with floodwater from July 27–August 22, whereas an area of 1009.91 km² was submerged with floodwater from August 9–September 4, mainly affecting Saran, Samastipur, Patna, Munger, Lakhisarai, Katihar, Bhagalpur, and Begusarai. Between August 22–September 17, an estimated area of 1462.94 km² was flood affected, majorly confined to the areas of Begusarai, Bhagalpur, Darbhanga, Katihar, Khagaria, Lakhisarai, and Munger. This was followed by an increment during September 4–September 30 of a 2334.87-km² area affecting a large number of districts (Begusarai, Bhagalpur, Darbhanga, Gopalganj, Katihar, Khagaria, Madhepura, Muzaffarpur, Patna, Purba Champaran, Vaishali).

By the start of October, the area under inundation increased even more as compared to the other months considered. Many districts were found affected by floodwater with a total area of 2232.2 km² from September 17–October 14. Furthermore, the inundated area decreased to 2229.08 km² by end of the month (October 14–27), which affected the districts Begusarai, Bhagalpur, Darbhanga, Katihar, Khagaria, Lakhisarai, Madhepura, Munger, Patna, Purnia, Samastipur, Saran, Sheikhpura, and Vaishali.

The area observed under the flood stagnancy period of 30–45 days during the months from June to September was estimated to be 3190.15 km², affecting many districts, including Bhagalpur, Darbhanga, Katihar, Khagaria, Lakhisarai, Madhepura, Munger, Muzaffarpur, Patna, Samastipur, Saran, and Vaishali. Out of the 13 districts listed, all were found to be affected by stagnant water for more than 45 days or between 40–60 days with an area of 2640.42 km², all of which is confined to the areas near the Ganga River channel.

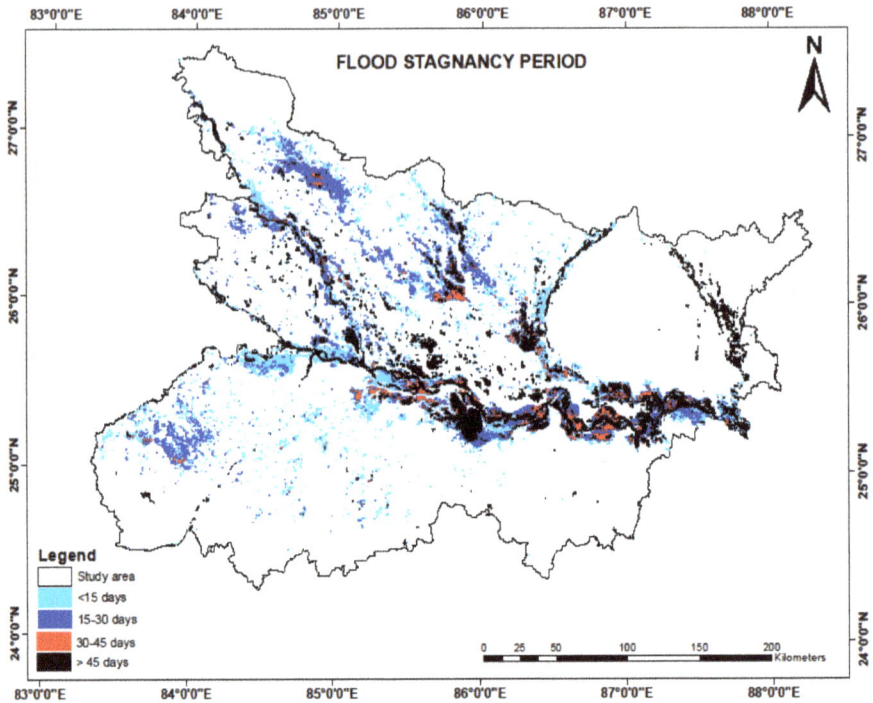

FIGURE 5.7   Stagnant period of water in Bihar during the 2021 flood event.

TABLE 5.2

Inundated Area (km²) as per the Stagnancy Periods

Stagnated Period (days)	Area Inundated (km²)
<15 days	9313.23
15–30 days	5962.48
30–45 days	3190.15
>45 days	2640.42
Composite (1–60 days)	9911.64

5.4.4   STAGNANT WATER–AFFECTED CROPLAND AND BUILT-UP AREA

The stagnant water for a longer duration of time adversely affects any area in terms of crop yields and the spread of various waterborne diseases and ultimately affects a large part of the population. For the estimation of affected cropland due to the stagnancy of water for more than 30 days, the Copernicus LULC data were used that concluded that out of the composite area (for more than 30 days) of 5962.48 km², an extensive cropland area of about 5428.73 km² (91%) and a built-up area of 138.19 km² (2%) were affected (Figure 5.8). The districts with cropland affected by floodwater

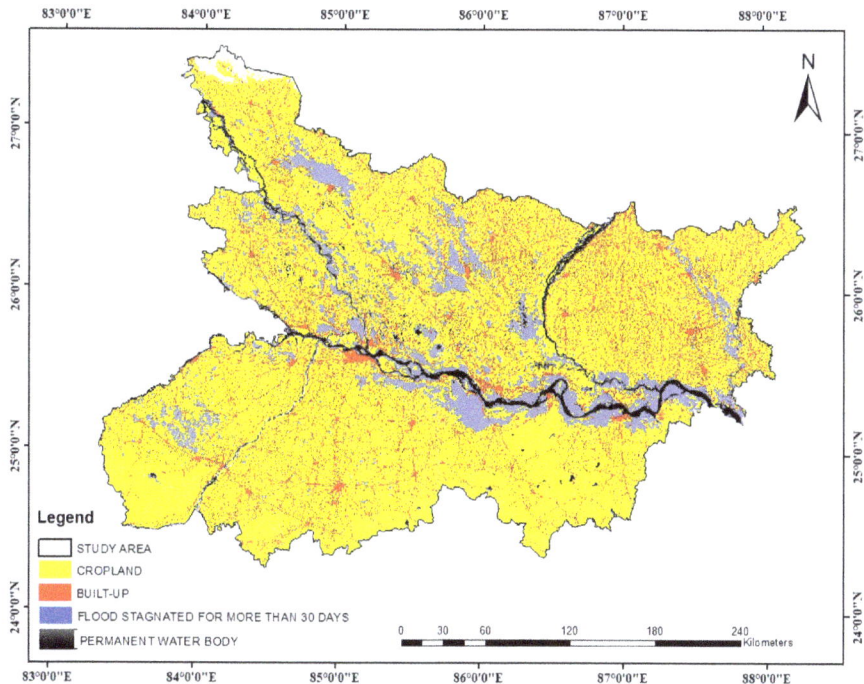

FIGURE 5.8   Affected cropland and buildup due to stagnant water for more than 30 days in Bihar during the 2021 flood event.

were Darbhanga, Gopalganj, Khagaria, Lakhisarai, Muzaffarpur, Patna, Purba Champaran, Samastipur, and Saran. It was observed that the districts along the Ganga River channel or various tributaries, including Kosi, Gandak, and Kali, were affected throughout the monsoon period and were submerged for more than 30 days at least. Although Bihar is highly dependent on monsoon rains for the flourishing growth of Kharif crops, the early onset of monsoon adversely affects the crops and results in a high decline in crop production.

5.5   DISCUSSION

Floods in Bihar are mostly attributed to hydrological, topographical, and fluvial reasons. Floods also occur due to breach of artificial embankments made by the state government, which was constructed as part of flood control measures. For instance, in August 2008, a major flood in North Bihar occurred due to the failure of embankment on River Kosi that caused significant loss of life, crops, and livestock (Kafle, Khanal, and Dahal 2017). Several past studies also indicated that river water from upstream catchment causes floods in North Bihar (Kafle, Khanal, and Dahal 2017; Tripathi, Parida, and Pandey 2019). As per the estimates using satellite data, about 20.88% (4108 km^2) of areas have been inundated during flood events in 2017 in North

Bihar with a floodwater depth of 0.1–1 m as estimated from SAR-based data and the Digital Elevation Model (Parida et al. 2021). It found that effectively, floodwater stagnated for a period of 12–24 days, affecting croplands and urban areas. Rättich et al. (2020) reported that an area of about 4,600 km^2 was inundated based on the multitemporal satellite data during July–September 2017, and effectively, the water-stagnation period varies from 30–40 days (mean of 26 days). The latest study found that flood causes inundation up to 34% based on MODIS NRT historical flood product from 2000 to 2020 (Tripathi, Pandey, and Parida 2022). Another study revealed that an area of 7682.38 km^2 (or 8.15% of the geographical area of Bihar) was affected in Bihar from June 30–July 18, 2020 (Lal, Prakash, and Kumar 2020). The floods have affected almost 19.48% (77,493.1 km^2) of areas in the Ganga-Brahmaputra Basin and caused significant impacts on agriculture (23.68–28.47% of cropland area) and buildup (5.66–9.15% of the built-up area). Moreover, they found millions of population were as affected by floods, and the highest impact was reported from Bihar (11.8 million) as most of people reside in the lower valley that is also near the river (Pandey, Kaushik, and Parida 2022).

As per our study, we found that the maximum inundated area was 9911.64 km^2 (or 10.5% out of the total geographical area) in Bihar, where the flood water stagnated for a period of 1–60 days during the 2021 flood events. The estimated stagnated water was much higher than previous reported (Parida et al. 2021; Rättich et al. 2020), possibly attributed to different flood events (i.e., 2017) and the use of a shorter temporal scale based on SAR data (e.g., July–August). The other possibility is that the rainfall intensity was much higher in the 2021 flood event than in 2017. The outcomes of the present study are very useful for flood management and preparedness to mitigate risk associated with society. However, the present approaches have some limitations like the rainfall data and MODIS NRT product, which were very coarse in spatial resolution. Therefore, alternative data sets such as SAR-based inundation mapping would be more crucial for the accurate estimation of flood areas as well as stagnant water.

The situation of floods in India is really complex and its management is a tedious process. After Independence, the Government of India has worked on taking up different measures to limit the consequence of flooding and reducing the chances of its occurrence by setting up various committees, forces, and other working groups. They have formulated various policies for helping in water resource management and guiding recommendations. Most of these strategies were focused on minimizing the impact of floods by focusing more on structural measures and emergency responses (Mohanty, Mudgil, and Karmakar 2020). One such strategy was Dam Safety Organization, which was set up to provide procedures of dam safety for all dams across the country (Pillai and Giraud 2014), which will assist state governments in identifying the potential distress causes and recommending the remedial measures in any event of dam failure. To reduce the damage during the flood, a set of guidelines of emergency action plans and operating manuals were prepared by every state. However, there were only 7% (349) of emergency action plans and 5% (231) of operating manuals out of the total of 4862 large dams were prepared as of March 2016 (Mohanty, Mudgil, and Karmakar 2020).

After the flood in 1954 in Bihar, the seriousness was acknowledged for flood pre-vention and to protect the floodplains. Since then, efforts began for flood protection with controlling the riverine flooding, and there was a massive investment toward the construction of the structural measures. The structural measures are the physi-cal constructions such as embankments that are used to block the flow of floodwater from rivers to the floodplain, reservoirs for detention, retention pond to lower the flow of flooding, and improvement in the drainage situations of the river basins that help in reducing the possible impact of flooding, but in the long run, these methods have not been helpful.

A shift is recommended to India's flood management policy from "passive response or the structural measures" to "progressive response" focusing more on the nonstructural measures (Padma 2020). The nonstructural measures that are men-tioned in flood-related policies include regulations, flood forecasting and zoning, land-use planning, and others. However, they are hardly practiced in a few parts of the country and are yet to be enacted extensively by the states (Prasad and Mukherjee 2014). Various other nonstructural measures include changes in cropping patterns, training and public awareness, flood warning system, flood resistance infrastructure, institutional arrangements, and local disaster action plans.

The rivers originating in Nepal, including the Kosi, the Gandak, and the Kali, all bring heavy water discharge to the lower plains of Bihar, leading to a disastrous situation. To resolve the recurrent flooding in the Ganga-Brahmaputra Basin, inter-linking river projects are also being considered. The main purpose of interlinking the rivers is to join the major rivers through reservoirs and canals from a surplus area to a deficit one. Based on this, the river with extensive water surplus can provide relief to the area of the river with a shortage of water, which can provide aid during disas-ters like flood and drought (Bandyopadhyay and Perveen 2008; Parida and Oinam 2015). Another major focus has been towards silt management, as the Ganga carries heavy sediment loads that raise the riverbeds, leading to extensive flooding (WALMI 2021). The fast-flowing rivers such as the Gandak, the Bagmati, the Kosi, and the Mahananda are generally prone to silting up. The raised riverbed disturbs the natural course of the river, and the river starts flowing toward lateral paths, leading to the breaching of embankments (Kansal, Kishore, and Kumar 2016).

5.6 CONCLUSION

The changing pattern of climate highly influences the life form in a general sense. The variability in rainfall extremes has affected countries and states on a recur-rent basis. In the present study, the rainfall intensity was observed to determine the increased drainage of major rivers during the monsoon season resulting in flooding situations in Bihar. The pattern of rainfall observed through the months of May–October reveals that higher intensity rainfall over the Tibetan Plateau results in a heavy discharge of water into the catchment areas of Nepal, creating strong pres-sure on the Indian river embankments. Bihar is considered as India's highest flood-prone region, which is affected throughout the monsoon season due to increased discharge from the Himalayan River catchment. The months of July and August

were embarked with the highest intensity over the Nepal mountains, reaching up to 600–700 mm, whereas 100–300 mm of rainfall intensity was consistently observed over the Uttarakhand and Uttar Pradesh regions. Therefore, added drainage from the higher elevated areas, such as Uttarakhand and Uttar Pradesh to Bihar in the lower Ganga basin and from the three major river systems (Kosi-Gandak-Karnali) of Nepal, gives rise to alarming situations in Bihar.

The flood extent of Bihar revealed many submerged patches in various districts from May–October. The areas near the Ganga River channel are flooded every year during the monsoon. The Himalayan River catchment contributes highly toward the extent of flood inundation of Bihar. The stagnant period of water in Bihar revealed many districts being affected by floodwater for more than 30–45 days at times. The highly flood-prone districts were Bhagalpur, Darbhanga, Katihar, Khagaria, Lakhisarai, Madhepura, Munger, Muzaffarpur, Patna, Samastipur, Saran, and Vaishali, where the water was stagnant for over a month, resulting in damage to croplands and settlement areas. The identified districts need different flood protection measures and some of the vital strategies are structural (e.g., interlinking of rivers, wetlands development, dredging, etc.) and nonstructural (e.g., flood plain zoning, forecasting, disaster mitigation, preparedness, and response policy).

REFERENCES

Bandyopadhyay, Jayanta, and Shama Perveen. 2008. "The Interlinking of Indian Rivers: Questions on the Scientific, Economic and Environmental Dimensions of the Proposal." 53–76. Available online: https://www.jagranjosh.com/general-knowledge/advantages-and-disadvantages-of-interlinking-rivers-in-india-1506409679-1 (Accessed January 2, 2022).

Buchhorn, M., B. Smets, L. Bertels, B. De Roo, M. Lesiv, N.-E. Tsendbazar, M. Herold, and S. Fritz. 2020. "Copernicus Global Land Service: Land Cover 100m: Collection 3: Epoch 2019: Globe." https://doi.org/10.5281/zenodo.3939050. Available online: https://land.copernicus.eu/global/products/lc (Accessed December 15, 2021).

Chaubey, Pawan K., Prashant K. Srivastava, Akhilesh Gupta, and R. K. Mall. 2021. "Integrated Assessment of Extreme Events and Hydrological Responses of Indo-Nepal Gandak River Basin." *Environment, Development and Sustainability* 23 (6): 8643–8668. https://doi.org/10.1007/s10668-020-00986-6.

"Climate Hazards Center- UC Santa Barbara." Available online: www.chc.ucsb.edu/data (Accessed February 2, 2022).

Das, Sumit. 2019. "Geospatial Mapping of Flood Susceptibility and Hydro-Geomorphic Response to the Floods in Ulhas Basin, India." *Remote Sensing Applications: Society and Environment* 14: 60–74. https://doi.org/10.1016/j.rsase.2019.02.006.

Deoras, Akshay. 2021. "Heavy Rains in Nepal May Threaten Bihar Again." www.downtoearth.org.in/blog/heavy-rains-in-nepal-may-threaten-bihar-again-45768.

Dhanya, P., and A. Ramachandran. 2016. "Farmers' Perceptions of Climate Change and the Proposed Agriculture Adaptation Strategies in a Semiarid Region of South India." *Journal of Integrative Environmental Sciences* 13 (1): 1–18. https://doi.org/10.1080/1943815X.2015.1062031.

Dhiman, Ravinder, Renjith Vishnu Radhan, T. I. Eldho, and Arun Inamdar. 2018. "Flood Risk and Adaptation in Indian Coastal Cities: Recent Scenarios." *Applied Water Science* 9 (1): 5. https://doi.org/10.1007/s13201-018-0881-9.

Funk, Chris, Pete Peterson, Martin Landsfeld, Diego Pedreros, James Verdin, Shraddhanand Shukla, Gregory Husak, et al. 2015. "The Climate Hazards Infrared Precipitation with Stations—a New Environmental Record for Monitoring Extremes." *Scientific Data* 2 (1): 150066. https://doi.org/10.1038/sdata.2015.66.

Jha, Rupak Kumar, and Haripriya Gundimeda. 2019. "An Integrated Assessment of Vulnerability to Floods Using Composite Index—A District Level Analysis for Bihar, India." *International Journal of Disaster Risk Reduction* 35: 101074. https://doi.org/10.1016/j.ijdrr.2019.101074.

Kafle, K. R., S. N. Khanal, and R. K. Dahal. 2017. "Consequences of Koshi Flood 2008 in Terms of Sedimentation Characteristics and Agricultural Practices." *Geoenvironmental Disasters* 4 (1): 4. https://doi.org/10.1186/s40677-017-0069-x.

Kale, Vishwas. 2012. "On the Link between Extreme Floods and Excess Monsoon Epochs in South Asia." *Climate Dynamics* 39 (5): 1107–1122. https://doi.org/10.1007/s00382-011-1251-6.

Kansal, M. L., Kumar Abhishek Kishore, and Prashant Kumar. 2016. "Sedimentation: It's Impact and Management." *Conference: 31st Indian Engineering Congress*, The Institution of Engineers in Kolkata, India.

Karki, Ramchandra, Shabeh ul Hasson, Udo Schickhoff, Thomas Scholten, and Jürgen Böhner. 2017. "Rising Precipitation Extremes across Nepal." *Climate* 5 (1): 4. https://doi.org/10.3390/cli5010004.

Lal, Preet, Aniket Prakash, and Amit Kumar. 2020. "Google Earth Engine for Concurrent Flood Monitoring in the Lower Basin of Indo-Gangetic-Brahmaputra Plains." *Natural Hazards* 104 (2): 1947–1952. https://doi.org/10.1007/s11069-020-04233-z.

Loo, Yen Yi, Lawal Billa, and Ajit Singh. 2015. "Effect of Climate Change on Seasonal Monsoon in Asia and Its Impact on the Variability of Monsoon Rainfall in Southeast Asia." *Geoscience Frontiers*, Special Issue: Geoinformation Techniques in Natural Hazard Modeling 6 (6): 817–823. https://doi.org/10.1016/j.gsf.2014.02.009.

Mimura, Nobuo. 2013. "Sea-Level Rise Caused by Climate Change and Its Implications for Society." *Proceedings of the Japan Academy, Series B* 89 (7): 281–301. https://doi.org/10.2183/pjab.89.281.

Mohanty, Mohit Prakash, Sahil Mudgil, and Subhankar Karmakar. 2020. "Flood Management in India: A Focussed Review on the Current Status and Future Challenges." *International Journal of Disaster Risk Reduction* 49: 101660. https://doi.org/10.1016/j.ijdrr.2020.101660.

"NASA GSFC NRT Global Flood Mapping." Available online: https://floodmap.modaps.eosdis.nasa.gov/ (Accessed December 15, 2021).

Padma, T. V. 2020. "New Recommendations for a Proactive Flood Policy in India." http://eos.org/articles/new-recommendations-for-a-proactive-flood-policy-in-india.

Pandey, Arvind Chandra, Kavita Kaushik and Bikash Ranjan Parida. 2022. "Google Earth Engine for Large-scale Flood Mapping Using SAR Data and Impact Assessment on Agriculture and Population of Ganga-Brahmaputra Basin." *Sustainability* 14 (7): 4210. https://doi.org/10.3390/su14074210.

Pandey, A. C., Suraj Kumar Singh, and M. S. Nathawat. 2010. "Waterlogging and Flood Hazards Vulnerability and Risk Assessment in Indo Gangetic Plain." *Natural Hazards* 55 (2): 273–289. https://doi.org/10.1007/s11069-010-9525-6.

Parida, B. R., Sailesh N. Behera, Bakimchandra Oinam, Arvind Pandey, and Nilendu Singh. 2017. "Evaluation of Satellite-Derived Rainfall Estimates for an Extreme Rainfall Event Over Uttarakhand, Western Himalayas." *Hydrology* 4 (2): 22. https://doi.org/10.3390/hydrology4020022.

Parida, B. R., Sailesh N. Behera, Bakimchandra Oinam, N. R. Patel, and R. N. Sahoo. 2018. "Investigating the Effects of Episodic Super-Cyclone 1999 and Phailin 2013 on Hydro-Meteorological Parameters and Agriculture: An Application of Remote Sensing." *Remote Sensing Applications: Society and Environment* 10: 128–137. https://doi.org/10.1016/j.rsase.2018.03.010

Parida, B. R., and B. Oinam. 2015. Unprecedented Drought in North East India Compared to Western India. *Current Science* 109 (11): 2121–2126.

Parida, B. R., Gaurav Tripathi, Arvind Chandra Pandey, and Amit Kumar. 2021. "Estimating Floodwater Depth Using SAR-Derived Flood Inundation Maps and Geomorphic Model in Kosi River Basin (India)." *Geocarto International*: 1–26. https://doi.org/10.1080/10106049.2021.1899298

Pillai, B. R. K., and S. Giraud. 2014. "Dam Rehabilitation and Improvement Project (DRIP- India) Dams to Be Rehabilitated." *Proceedings of First National Dam Safety Conference*, pp. 24–25. Available online: https://scholar.googleusercontent.com/scholar?q=cache:5NVD5Yy6GzoJ:scholar.google.com/+BRK+pillai&hl=en&as_sdt=0,5 (Accessed December 2, 2021).

Prasad, Eklavya, and Nandan Mukherjee. 2014. "Situation Analysis on Floods and Flood Management." *Report: Ecosystems for Life: A Bangladesh-India Initiative*: 124.

Rättich, Michaela, Sandro Martinis, and Marc Wieland. 2020. "Automatic Flood Duration Estimation Based on Multi-Sensor Satellite Data." *Remote Sensing* 12 (4): 643. https://doi.org/10.3390/rs12040643.

Ray, Kamaljit, Prabha Pandey, C. Pandey, A. P. Dimri, and K. Kishore. 2019. "On the Recent Floods in India." *Current Science* 117. https://doi.org/10.18520/cs/v117/i2/204-218.

Shrestha, Dibas, Prasamsa Singh, and Kenji Nakamura. 2012. "Spatiotemporal Variation of Rainfall over the Central Himalayan Region Revealed by TRMM Precipitation Radar." *Journal of Geophysical Research: Atmospheres* 117 (D22). https://doi.org/10.1029/2012JD018140.

Shrestha, Mandira. 2008. "Impacts of Flood in South Asia." *Journal of South Asia Disaster Studies* 1 (1): 85–106.

Shrestha, Rajendra, and Azaya Sthapit. 2016. "Temporal Variation of Rainfall in the Bagmati River Basin, Nepal." *Nepal Journal of Science and Technology* 16: 31. https://doi.org/10.3126/njst.v16i1.14355.

Shukla, Anoop Kumar, Chandra Shekhar Prasad Ojha, Rajendra Prasad Singh, Lalit Pal, and Dafang Fu. 2019. "Evaluation of TRMM Precipitation Dataset over Himalayan Catchment: The Upper Ganga Basin, India." *Water* 11 (3): 613. https://doi.org/10.3390/w11030613.

Sundaram, Sreechanth, and Kiran Yarrakula. 2017. "Multi-Temporal Analysis of Sentinel-1 SAR Data for Urban Flood Inundation Mapping—Case Study of Chennai Metropolitan City." *Indian Journal of Ecology* 44: 564–568.

Tripathi, Gaurav, Arvind Chandra Pandey, and Bikash Ranjan Parida. 2022. "Flood Hazard and Risk Zonation in North Bihar Using Satellite-Derived Historical Flood Events and Socio-Economic Data." *Sustainability* 14 (3): 1472. https://doi.org/10.3390/su14031472.

Tripathi, Gaurav, Arvind Chandra Pandey, Bikash Ranjan Parida, and Amit Kumar. 2020. "Flood Inundation Mapping and Impact Assessment Using Multi-Temporal Optical and SAR Satellite Data: A Case Study of 2017 Flood in Darbhanga District, Bihar, India." *Water Resources Management* 34. https://doi.org/10.1007/s11269-020-02534-3.

Tripathi, Gaurav, Bikash Ranjan Parida, and Arvind Chandra Pandey. 2019. "Spatio-Temporal Rainfall Variability and Flood Prognosis Analysis Using Satellite Data over North Bihar during the August 2017 Flood Event." *Hydrology* 6 (2): 38. https://doi.org/10.3390/hydrology6020038.

WALMI. 2021. "Expert Committee Sheds Light on Bihar's Mounting Silt Crisis." Available online: www.downtoearth.org.in/news/governance/expert-committee-sheds-light-on-bihar-s-mounting-silt-crisis-61668.

Wasson, R. J., Y. P. Sundriyal, Shipra Chaudhary, Manoj K. Jaiswal, P. Morthekai, S. P. Sati, and Navin Juyal. 2013. "A 1000-Year History of Large Floods in the Upper Ganga Catchment, Central Himalaya, India." *Quaternary Science Reviews* 77: 156–166. https://doi.org/10.1016/j.quascirev.2013.07.022.

6 Flood Inundation and Floodwater Depth Mapping Using Synthetic Aperture Radar Data in the Gandak River Basin

Gaurav Tripathi[1], *Bhawani Shankar Phulwari*[1],
Bikash Ranjan Parida[1,*], *Arvind Chandra Pandey*[1],
and Mukunda Dev Behera[2]

1 Department of Geoinformatics, School of
  Natural Resources Management, Central
  University Jharkhand, Ranchi-835222, India

2 Centre for Oceans, Rivers, Atmosphere and
  Land Sciences, IIT Kharagpur, India-721302

CONTENTS

6.1 Introduction 90
6.2 Study Area 92
6.3 Materials and Methods 93
    6.3.1 Sentinel-1 SAR Data 93
    6.3.2 Copernicus Land-Use Product (2019) 94
    6.3.3 CHIRPS Data 94
    6.3.4 ALOS-Based DEM Data 94
    6.3.5 Methods 95
        6.3.5.1 Flood Extent and Duration Mapping 96
        6.3.5.2 Floodwater Depth Estimation 96
6.4 Results 97
    6.4.1 Multitemporal Precipitation Based on CHIRPS 97
    6.4.2 Spatiotemporal Flood Inundation Extent in August 2017 97
    6.4.3 Flood Inundation Impact on Varied LULC 100
    6.4.4 Floodwater Depth on 13 August 2017 101
6.5 Discussion 102
6.6 Conclusion 104
Acknowledgment 104

DOI: 10.1201/9781003265160-7

Conflicts of Interest 104
References 104

6.1 INTRODUCTION

Flooding is the most recurring natural disaster that affects most of the population worldwide. As compared to the previous two decades (2000–2019), higher impacts of economic losses (US$151.6 billion) were reported in 2020, with an increasing number of flood events (EMDAT 2021). In 2020, nearly 201 out of 389 natural disasters were reported, and floods caused the death of 15,080 people and affected 98.4 million population across the world (EMDAT 2021). The impact of disaster events was felt heavily throughout Asia as it shares 41% of the total events and 64% of the total people affected (EMDAT 2021). In India, Bihar is one of the most vulnerable states to flooding (Sinha et al. 2008; Tripathi et al. 2019), followed by other disasters, namely, droughts, heat/cold waves, river erosions. Flood is the most prevalent natural disaster in Bihar, and it occurs every year, causing massive loss of life and property (Sinha et al. 2008). Increased population pressure at sensitive locations, excessive building density, low construction quality, and improper management in preparedness and mitigation measures have all contributed to increasing the vulnerability to natural hazards (Alahacoon et al. 2018; Blöschl et al. 2019; Kumar and Sahoo 2021). In recent years, floods are most concerning natural disasters occurring very frequently, so remote-sensing-based techniques are preferred over the traditional methods of flood assessment. Furthermore, the current procedures of flood mitigation measures are not adequate where we need some innovative approaches and satellite-based inputs (Clement et al. 2018; Parida et al. 2017) to minimize the risk factors associated with floods.

Remote-sensing techniques were widely used for environmental monitoring and applications, such as drought and flood at higher spatial resolution (Parida 2006). The utility of remote sensing for rapid, accurate flood mapping, monitoring, and management has been reported by several studies (Agnihotri et al. 2019; Tripathi et al. 2020). To extract flood inundation, various types of remote-sensing data sources (i.e., optical and microwave) have been applied. Importantly, microwave radar is sensitive in discriminating water and land surfaces, which allows retrieving flood pixels over large-scale areas (Bates et al. 2006; Pandey et al. 2022; Sv et al. 2017). Most of the space-borne satellite data have detected floodwater from a medium to high spatial resolution (i.e., 5–25 m Zhou and Zhang 2017; Shakya et al. 2021; Tripathi et al. 2022). Many studies have been also conducted to monitor flood extent by employing multi-source satellite data by integrating both optical and Synthetic Aperture Radar (SAR) sensors. Integrated time-series river flow data and Moderate Resolution Imaging Spectroradiometer (MODIS) images were also integrated to map inundation dynamics at the Chao Phraya River basin in Thailand (Zhou and Zhang 2017). Powell et al. (2014) evaluated the application of remote sensing-based indices like Normalized Difference Vegetation Index (NDVI) based on the time time-series series satellite data set from Landsat, MODIS, and AVHRR (Advanced Very High-Resolution Radiometer) and further utilized to characterize flood extent in floodplain wetlands in Australia. Optical sensors derived spectral indices, such as the Normalized Difference Water Index (NDWI), the modified NDWI, and the Water Ratio Index (WRI), were

employed to assess flooded regions, including the soil moisture zoning (Singh et al. 2015; Patel et al. 2019). Particularly, both large- and small-scale flood mapping and flood dynamics were also studied widely based on the SAR data (Schumann et al. 2007; Matgen et al. 2011; Tripathi et al. 2020). Satellite-based information, namely, pre/post-flood information and hydrographs, were coupled in the Geographic Information System (GIS) domain for the flood-induced damage assessment (Tripathi et al. 2022). The digital elevation model (DEM) data and flood inundation maps were integrated for floodwater depth estimations (Manfreda and Samela 2019; Parida et al. 2021). Floodwater depth estimation is another vital aspect in determining the severity of the flood and assessing the damages. Nakmuenwai et al. (2017) utilized high-resolution optical sensors data from the GeoEye-1 and ThaiChote-1 to validate flood extent derived using multitemporal dual-polarization RADARSAT-2 data during the 2011 flooding event in Central Thailand. Hydrographs were also used from different gauging stations to validate the flood progression and regression. Floodwater depth is another key component in determining flood magnitude and assessing flood damage that can be used as a potential flood management component.

Traditionally, floodwater depth estimation is done using sound and positioning equipment like echo sounders and multibeam sonar by installing them physically on ships (Özgen et al. 2016). Active (e.g., microwave radar and airborne laser radar) and passive (e.g., visible light remote sensing) types of remote-sensing data sets were used to estimate floodwater depth. Chang et al. (2010) proposed an algorithm to derive floodwater depth across the inundated region with the help of DEM. Several algorithms are being utilized to analyze floodwater depth across the inundated region extracted using satellite data (Scorzini et al. 2018). Flooded regions were combined with the DEM to acquire continuous elevation information distributed in the borderlines of flood patches after correcting the errors. The floodwater depth distribution within the boundary is estimated using bilinear interpolation after assuming the water surface in the inundated area is an inclining plane. Several researchers have also estimated floodwater depth by implying remote-sensing satellite data with the DEM, where river cross-section profiles along with the river were taken (Matgen et al. 2007; Zwenzner and Voigt 2009). There is a variety of one-, two-, and three-dimensional hydrological models in order to determine floodplain elevations and floodway encroachments, whereas the Hydrologic Engineering Center-River Analysis System model (i.e., open channel flow) was utilized as the primary model.

In particular, over the Gandak and Kosi River Basin in North Bihar, several studies have attempted to map flood inundation, but none of them have addressed floodwater depth and duration. For instance, RADARSAT–1 satellite data was used to map spatiotemporal flood dynamics in 2004 and 2008 over North Bihar. It employed image segmentation methods to extract flood pixels (Manjusree et al. 2012). The Otsu's threshold technique was applied on RADARSAT-2 images to map floodwater extent across Darbhanga district for flood events in 2011 (Manjusree et al. 2012). The Analytical Hierarchy Process (AHP) has been also used within the GIS domain to map floodwater extent in 2003 and 2006, resulting in the identification of flood risk zones inside the Kosi River basin (Sinha et al. 2008). To demarcate flood inundated regions, self-organizing Kohonen's maps were combined with an Artificial Neural Network (Kussul et al. 2011). No framework exists especially for characterizing

flood at a large scale (floodwater depth and duration), which generally needs high computational cost. Therefore, this study was carried out to evaluate the advantages of using Sentinel-1A-based SAR data in extracting, analyzing, and mapping floodwater propagation across the study region. The specific objectives are (1) to map spatiotemporal precipitation during August 2017, (2) to map flood inundation extent using temporal Sentinel-1A SAR data, (3) to calculate floodwater depth during peak flood, and (4) to assess the impact of flood over varied land use land cover (LULC).

6.2 STUDY AREA

The northern Bihar plains are India's most vulnerable region that is susceptible to flooding (Kumar and Sahoo 2021), resulting in recurrent flooding events in the last three decades (Alahacoon et al. 2018). Climatic-induced extreme events have led to more areas being affected by floods and so the flooded areas are increasing rapidly. There are two major rivers in northern Bihar, the Kosi and Gandak, and several smaller seasonal water channels such as Burhi Gandak, Baghmati, and Kamla-Balan, which results in frequent inundation, loss of life, and loss of property mainly over the lower catchment areas of North Bihar (Sinha et al. 2008).

The Gandak River (also known as the Narayani in Nepal) is the major river of Nepal, spanning more than 46,300 km^2 in the country (Figure 6.1). It starts from Tibetan Plateau, travels through the deep gorge of the great Himalayas, and meets the Ganga River near Patna, Bihar. The Gandak basin consists of 3, Dhaulagiri, Manaslu, and Annapurna, among the world's 14 highest peaks over 8,000 meters. However, the Dhaulagiri is highest among all. It lies between Kosi and Karnali (Ghaghara) from east to west. The region of study is composed of four districts (Paschim Champaran, Purba Champaran, Gopalganj in Bihar and Kushinagar in Uttar Pradesh) along the Gandak River and spatially located between 83° 52' E to 85° 27' E and 26° 20' N to 27° 53' N, spanning 14,164.75 km^2 of geographical region. Floods have been occurring in North Bihar for a long time as a consequence of the complex setting of seasonal/permanent water channels, unplanned built infrastructure, and poor flood management strategies, as well as the contribution in the form of silt from the greater Himalayas (Acharya and Prakash 2019).

The Gandak River is a major tributary of the Ganga River, which often gets flooded in large-scale areas. The study exhibited that in the last 30 years, the lower catchment areas have experienced the highest number of flood occurrences (Tripathi et al. 2022), causing significant loss of life and property. Short-term fluctuations in sediment load and water volume, as well as long-term climate changes, are the main factors behind flooding. Several remote-sensing and GIS-based approaches were considered to assess the associated flood risk that occurred due to riverine floods in the study area where it integrates hydro geomorphological, meteorological, and socioeconomical parameters.

Based on the Copernicus LULC data set (Figure 6.1), it was found that agriculture is the dominant class across the study region followed by vegetation, waterbody, built-up areas, and others, which accounts for 13,416.1 km^2 (94.71%), 336.5 km^2 (2.38%), 257.21 km^2 (1.82%), 87.7 km2 (0.62%), and 67.24 km^2 (0.47%), respectively.

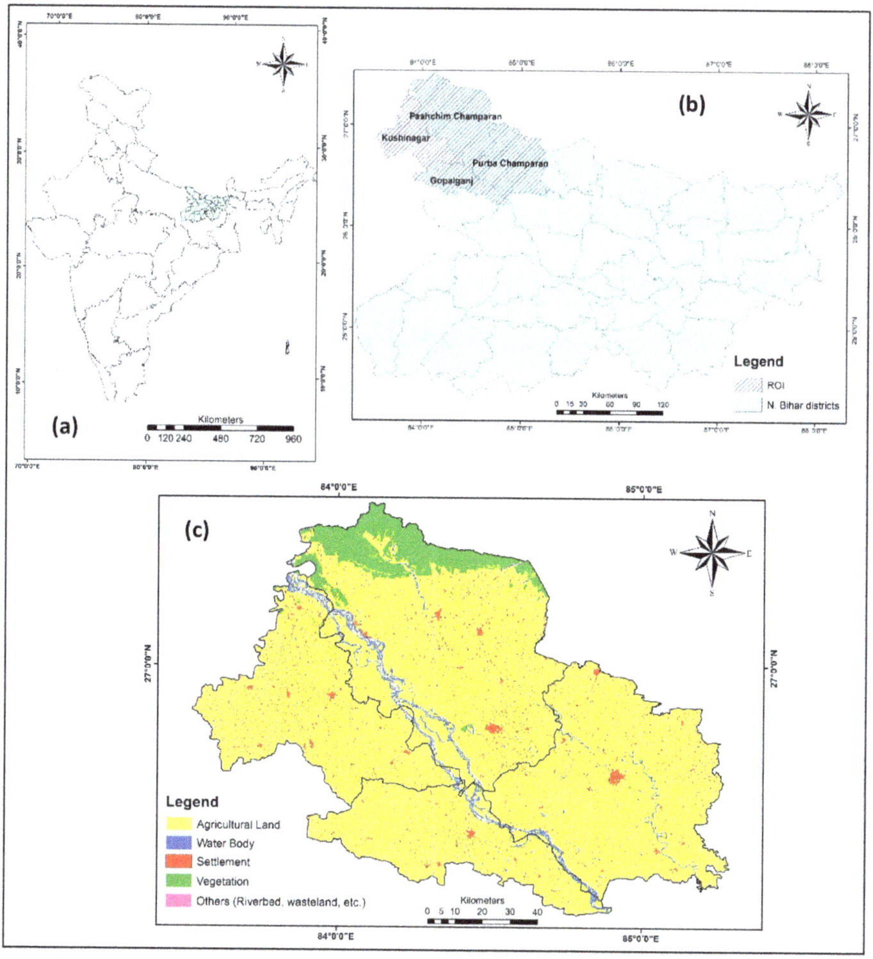

FIGURE 6.1 Location of the study area: The Gandak River Basin and the corresponding Copernicus-based LULC map.

6.3 MATERIALS AND METHODS

The data sets used in this study were Sentinel-1A-based SAR satellite data, rainfall data (from Climate Hazards Centre InfraRed Precipitation with Station [CHIRPS]), the Copernicus-based LULC data set, and Advanced Land Observation Satellite (ALOS)-based DEM data. Data used in this study are presented in Table 6.1.

6.3.1 SENTINEL-1 SAR DATA

The Sentinel-1 mission comprises a C-band SAR that enables acquiring imagery regardless of the weather. The SAR sensor provides microwave data in dual-polarization (HH + HV, VV + VH) mode (where HH stands for horizontally

TABLE 6.1

Data Used in This Study and Their Characteristics

Name of Data Set	Temporal Resolution	Spatial Resolution	Acquisition Date	Purpose	Source
Sentinel–1A/B	11 days	10 × 10 m	August 2017	Floodwater delineation, floodwater depth estimation	LP DAAC
Copernicus Global Land Cover	Yearly	100 × 100 m	–	LULC map	Copernicus (ESA)
CHIRPS	Daily	0.05° × 0.05°	August 2017	Rainfall maps	Climate Hazards Center
ALOS PALSAR DEM	–	12.5 × 12.5 m	–	Floodwater depth calculation	ASF

transmitted and horizontally received, HV for horizontally transmitted and vertically received, VV for vertically transmitted and vertically received, and VH for vertically transmitted and horizontally received). In this study, dual-polarization data have been used to map flood inundation and flood duration. Furthermore, the flood pixels were also used to estimate floodwater depth during August 2017.

6.3.2 COPERNICUS LAND-USE PRODUCT (2019)

The Copernicus Land Service provides global land cover products at 100-m spatial resolution. We have used the recent collection (version 3.0.1) acquired during 2015–2019 using PROBA-V satellite data. In this study, the land-cover data sets were used to assess flood impact over various LULC classes present in the study region.

6.3.3 CHIRPS DATA

CHIRPS is a global high-resolution satellite-derived precipitation data available at 0.05° * 0.05° (~5 km) spatial resolution between 50° S and 50° N at daily, monthly, and pentadal levels, composed of satellite as well as ground-based observations data followed by the bias correction. This data set allows the assessment of climatic variability and rainfall-induced events in a historical context (+35 years' rainfall data). Precipitation outputs are publicly available in different geospatial formats used in a variety of application areas and purposes. The rainfall data set for August and September 2017 was used in this study to link flood extent followed by rainfall over the region.

6.3.4 ALOS-BASED DEM DATA

The ALOS-PALSAR-based DEM has 12.5-m spatial resolution provided by Japan Aerospace Agency (JAXA). The data set was used to estimate floodwater depth.

6.3.5 Methods

Based on the flood progression and regression across the study region, a flood duration map was also prepared in August 2017. However, for linking rainfall patterns with flood inundation patterns, we used daily CHIRPS satellite-derived rainfall data of the same duration. Furthermore, flood inundation extent was coupled with the ALOS-based DEM to estimate floodwater depth. Flood damage was assessed based on the Copernicus LULC product. The detailed methodology flow is presented in Figure 6.2.

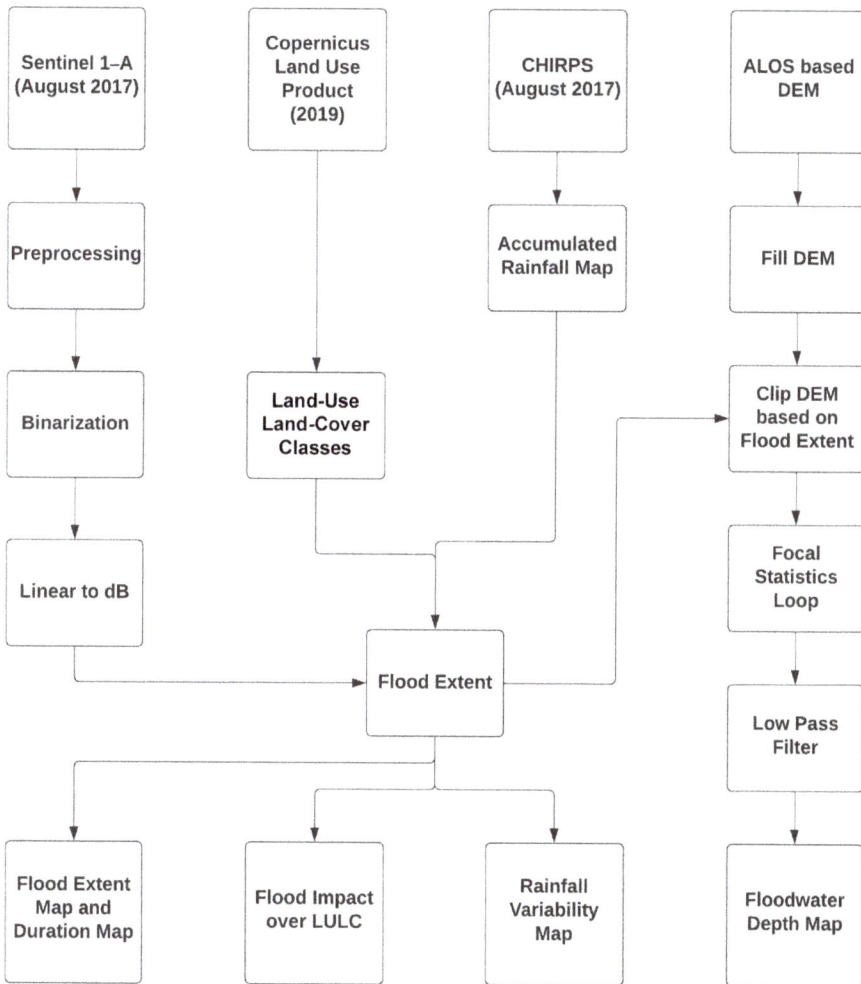

FIGURE 6.2 Methodology flowchart adopted for this study.

6.3.5.1 Flood Extent and Duration Mapping

The flood inundation extent during August 2017 was derived based on multitemporal Sentinel 1-A-based SAR images. However, a flood duration map was prepared based on overlay analysis. The workflow involves image preprocessing, floodwater extraction, and the creation of flood maps to analyze floodwater propagation and flood duration. The SAR sensor is an active microwave system that records objects from the surface in the form of a backscatter coefficient. A nearly flat surface with less or no obstruction reflects the majority of the radiation away from the sensor, which leads to a darker tone of pixels. However, rough surfaces might reflect more radiation toward the sensor, which can be seen in a brighter tone (Matgen et al. 2007).

The European Space Agency (ESA)-based open-source software SNAP has been used for preprocessing the Sentinel-1 data. To extract floodwater, SAR satellite images were calibrated first to relate pixel DN values with the radar backscatter value applied over the whole image. SAR is an active sensor system that led to producing speckle noises in satellite images. To remove this error, a refined lee filter is applied using 3 × 3 window size. Terrain corrections were performed by using Shuttle Radar Topographic Mission (SRTM)-based DEM data of 3 arcsec. In this process, the image and the DEM were resampled using the bilinear interpolation method. After correcting the image geometrically, Sentinel-1A images were classified into two classes, water and non-water. Microwave signals are very sensitive to surface roughness so that inundated regions will appear in a dark tone, whereas non-water areas can be clearly delineated as it appears in a brighter tone. To separate flooded regions from non-flooded areas, backscatter values were converted into dB values with 8-bit radiometric resolution (0–255) using the threshold method and the random forest classifier (Tripathi et al., 2020). The density-slicing technique was also applied later on using a threshold value to clearly differentiate water from non-water areas. To get the actual flood inundated area, we subtracted the area of permanent water bodies (PWBs) from flood inundation. Flood progression and regression were analyzed by super-imposing multiple flood inundation layers to show how floodwater traveled through time. The ERDAS IMAGINE and ArcGIS were the software used for image processing, analyzing, and mapping the flood-affected regions within the basin. Multitemporal flood inundation extents were coupled to generate a flood duration map and represented in four classes as follows: 0 days, 9 days, 12 days, and 24 days.

6.3.5.2 Floodwater Depth Estimation

Floodwater depth was computed by subtracting the elevation of any pixel of a particular cross section with the highest elevation value using DEM data. A one-dimensional-based geomorphic method was used here to calculate floodwater depth over the study region. It is based on FwDET geomorphic model in which we incorporated floodwater extent in the polygon and DEM data (Cohen et al. 2018). At first, the model assigned elevation values to different flood patches starting from the boundary cell toward the center. The water depth within the flooded domain is calculated using the local maximum floodwater height. To do so, a cell within the flooded domain is given the elevation of its nearest boundary cell, which is then subtracted from the cell's surface

elevation to calculate the local water depth (derived from a DEM). A python script was used to automate the process, which included several ArcGIS tools (within the ArcPy Python library). These methods can also be implemented using other geospatial analysis packages (GDALs; e.g., www.gdal.org). The geomorphic method includes the following steps, and more details can be seen in Cohen et al. (2018).

Step 1—To identify the boundary cell of flood patches

Step 2—Extraction of the elevation values of particular boundary cells, elevation of boundary cells

Step 3—In this step, elevation values of the boundary cells will be assigned to domain cells.

Step 4—Estimation of floodwater depth

Step 5—Revising floodwater depth area after smoothing the result

6.4 RESULTS

6.4.1 MULTITEMPORAL PRECIPITATION BASED ON CHIRPS

The accumulated spatiotemporal rainfall variability maps based on CHIRPS data during August 2017 flooding event across the Gandak River Basin are shown in Figure 6.3. The accumulated rainfall maps were shown that indicate areas of intense rainfall. Rainfall maps exhibited that the accumulated rainfall during the period of 1–4 August was up to 130 mm. During 1–4 August 2017, intense rainfall occurred in Purbi and Paschhim Champaran districts. All these regions are associated with the Gandak River, which creates flooding-like conditions toward downstream areas. The precipitated water traveled through the Tibetan plateau, through the gorge of the Himalayas in Nepal afterward and finally through a gentle slope in the North Bihar region to the lower catchment areas of North Bihar. The analysis exhibited that during 5–13 August, the accumulated rainfall received was more intense than previous dates during this period; the maximum rainfall received was observed to be about 200 mm. Later on, during the period of 14–16 August, the lower amount of rainfall over the catchment was associated with fewer days considered, and the maximum rainfall was about 90 mm. During 17–28 August, rainfall increased relative to the previous dates, and the maximum rainfall was up to 200 mm.

6.4.2 SPATIOTEMPORAL FLOOD INUNDATION EXTENT IN AUGUST 2017

The flood dynamics derived using SAR satellite data during the August 2017 flood event across the study region were shown in Figure 6.4. The flood inundation map exhibited that on 4 August 2017, a total of 112 km^2 of the geographical area was inundated (Table 6.2). This further inundates the eastern part of the study region and region along the river as well. The situation further deteriorated owing to torrential rainfall (Figure 6.3), and the flood-inundated area increased to 730.57 km^2 and 1331 km^2 during 13 and 16 August 2017, respectively. Consequently, the northern, eastern, and western parts of the study region were heavily affected. On 28 August 2017, the flooding area that was inundated receded to 340 km^2 following the rainfall

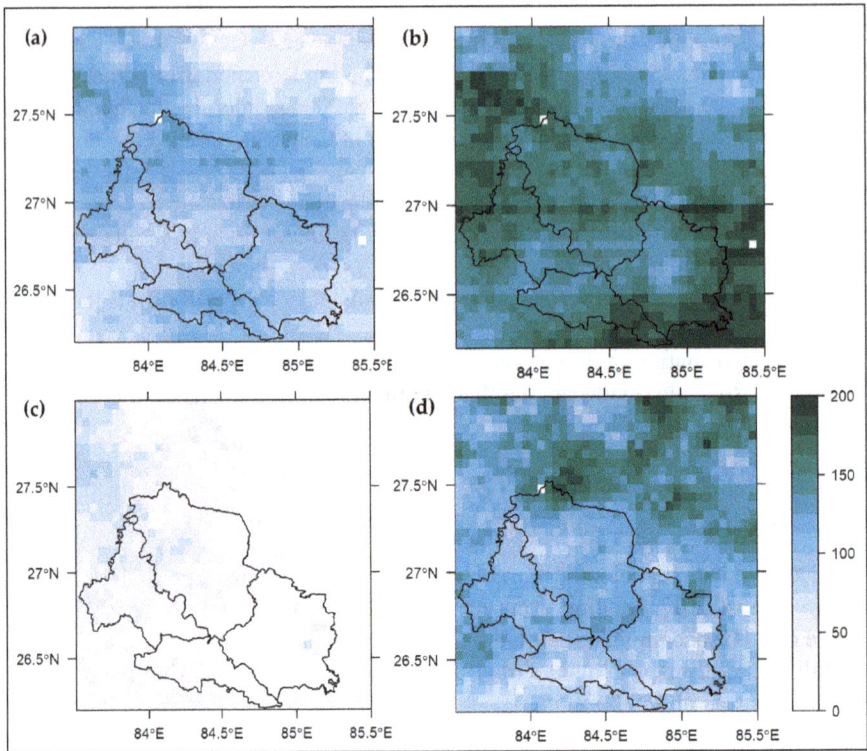

FIGURE 6.3 The accumulated spatiotemporal rainfall variability in mm for (a) 1–4 August, (b) 5–13 August, (c) 14–16 August, and (d) 17–28 August 2017, over the study region. The Environmental Systems Research Institute shape file of the Gandak River Basin was overlaid to identify the districts that received rainfall.

TABLE 6.2

Area under Flood during August 2017

Date of Flood	Inundated Area (km²)
04 Aug 2017	112.4 (0.8%)
13 Aug 2017	730.57 (5.15%)
16 Aug 2017	1331 (9.4%)
28 Aug 2017	340.09 (2.4%)

(Figure 6.3). After 16 August, flood regression took place. It was exhibited that flood inundation during 16 August 2017 was categorized as very high areal flood inundation extent, accounting for 9.4% of the area (Table 6.2).

Flood inundation extents on 4, 13, 16, and 28 August were super-imposed and analyzed to find out the nature of the spatial extent and the duration of flood through time. The result exhibited that flooding water stayed for a maximum of 24 days across the study region (Figure 6.5). The study also exhibited that flooding water

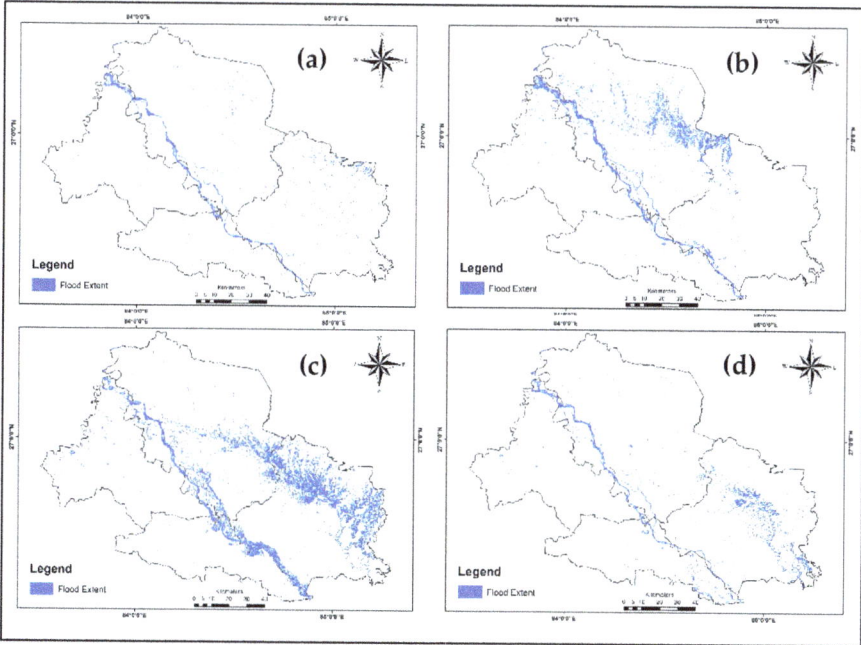

FIGURE 6.4 Flood water extent over the Gandak River's catchment during (a) 4 August, (b) 13 August, (c) 16 August, and (d) 28 August 2017.

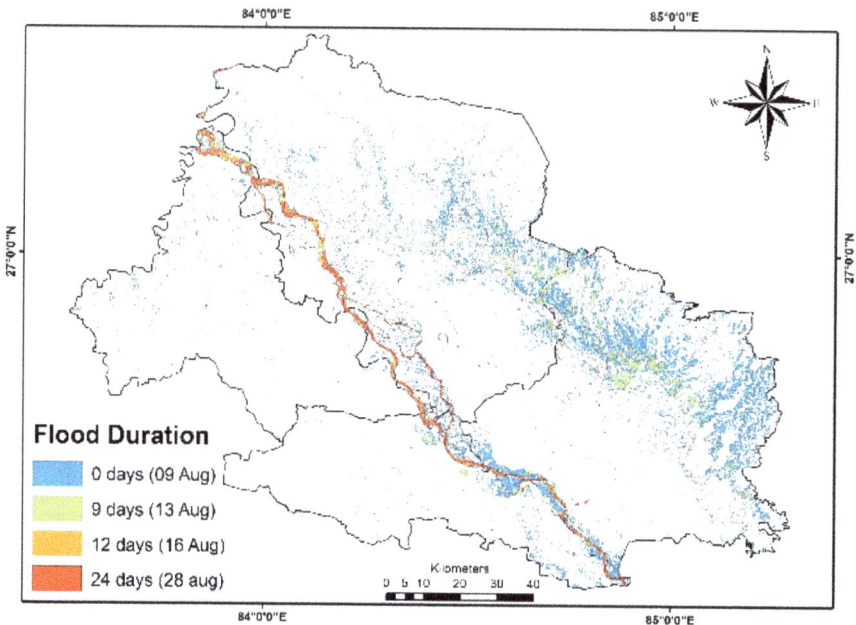

FIGURE 6.5 Flood duration map over the Gandak River's catchment during August 2017.

stayed for 12 and 24 days along the river channel mostly, which was categorized as the vulnerable zone to flooding. Flooding water stayed for 9 days were the extents over the eastern, northern, and central parts of the study region.

6.4.3 FLOOD INUNDATION IMPACT ON VARIED LULC

The spatial LULC map showed agriculture was a predominant class in the Gandak River Basin (Figure 6.6). Flood impacts were assessed on varied LULC classes, and the respective statistics are shown in Table 6.3. The statistics exhibited that agricultural land (81.84%) was highly affected due to flood, followed by waterbody (14.86%), vegetation/forest (1.49%), built-up areas (1.08%), and others (0.73%).

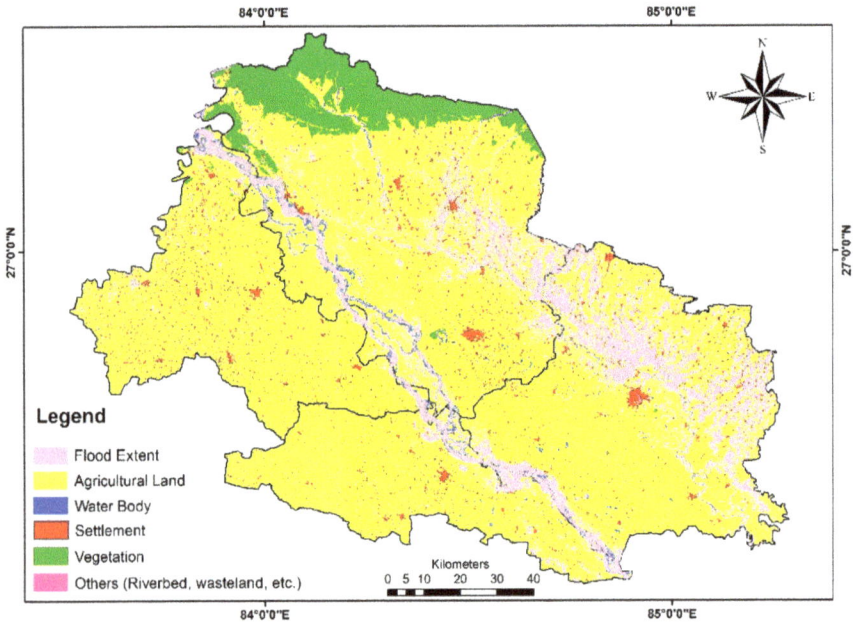

FIGURE 6.6   Composite flood map overlaid on Copernicus-based LULC map.

TABLE 6.3
Affected LULC Classes by Flood on 16 August. The percentage was calculated with respect to the total affected area (1331 km²)

LULC Class	Inundated LULC Classes (km²)
Built up	14.39 (1.08%)
Agricultural land	1089.33 (81.84%)
Water body	197.78 (14.86%)
Vegetation	19.8 (1.49%)
Others	9.7 (0.73%)

6.4.4 FLOODWATER DEPTH ON 13 AUGUST 2017

The floodwater depth map was derived based on the methodology shown in Figure 6.2. The floodwater depth map illustrates the sensitivity associated with the variation of water depth to identify and assess their relative impacts (Figure 6.7). The floodwater

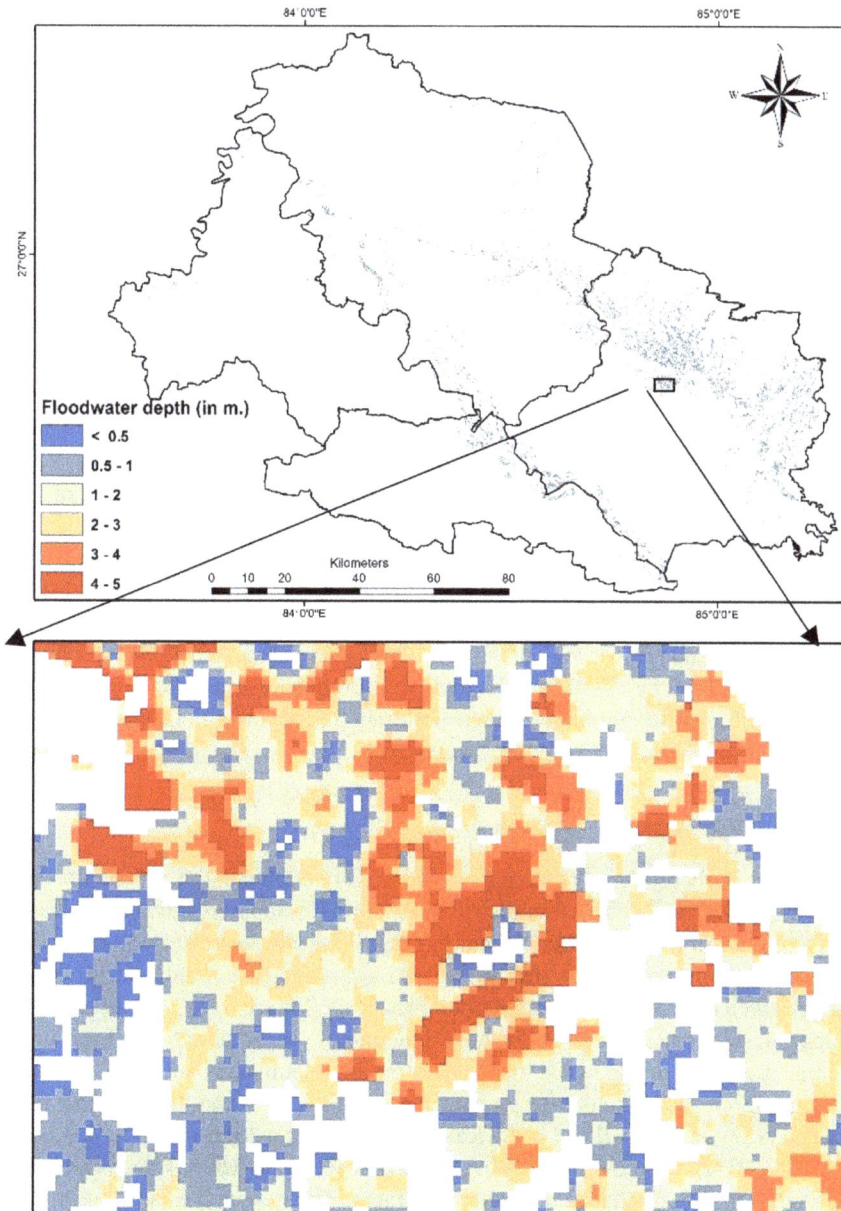

FIGURE 6.7   Floodwater depth over study region on 13 August 2017.

TABLE 6.4

District-Wise Floodwater Depth and Its Corresponding Areal Extent

Depth Range (m)	Gopalganj (km²)	Kushinagar (km²)	Purba Champaran (km²)	Paschim Champaran (km²)	Total
0–0.5	32.84	5.99	172.80	50.56	262.2
0.5–1	21.91	5.14	142.55	50.60	220.2
1–2	15.72	3.303	128.72	41.02	188.76
2–3	3.43	0.77	31.87	10.98	47.05
3–4	0.87	0.22	5.66	2.92	9.68
4–5	0.28	0.08	1.38	0.94	2.68
Total	75.08	15.52	483.01	157.04	730.57

depth was estimated and classified into six different classes (<0.5 m, 0.5–1 m, 1–2 m, 2–3 m, 3–4 m, 4–5 m) to observe floodwater depth across three districts of north Bihar (i.e., Purba Champaran, Paschhim Champaran, Gopalganj) and one from Uttar Pradesh (Kushinagar). The floodwater depth map has shown that among the four districts, Purba Champaran was highly affected, with 483.01 km² (66.11% with respect to total flood water extent on 13 August 2017) of areal inundation, followed by Paschhim Champaran (21.5%), Gopalganj (10.28%), and Kushinagar (2.12%), respectively (Table 6.4). The results also exhibited that a total of 730.57 km² of geographical land was inundated on 13 August 2017.

The floodwater depth map revealed that Purba Champaran district has an area of almost 173 km² under floodwater 0.5 m deep, followed by Goplagnaj district, with 32.84 km². An approximate total of 36% and about 30% of the land area (with respect to total inundation on 13 August 2017) was submerged under floodwater depths that were in the less-than-0.5-m and 0.5–1.0-m categories. However, 25.83%, 6.44%, 1.32%, and 0.37% of the area were estimated to be under floodwater depths of 1–2 m, 2–3 m, 3–4 m, and 4–5 m, respectively.

6.5   DISCUSSION

The present study proposed to examine and assess flood inundation extent, floodwater depth, and its impact over four districts (Paschim Champaran, Purba Champaran, and Gopalganj in Bihar and Kushinagar in Uttar Pradesh) situated along the Gandak River that occurred in August 2017. Here, we have proposed a feasible methodology that has given clear insights into the flood prognosis (Figure 6.2). Torrential rainfall in upper catchment areas result in huge amounts of water flow toward the lower catchment areas, causing havoc and inundating the landmasses near rivers and the surrounding areas. Therefore, to prepare a flood mitigation and management plan, it is necessary to have all the historical information about the intensity and duration of floods. In this case, the remote sensing-based approach has been proved as a key input for flood-induced damage assessment. This research establishes a good relationship between the emergency and response systems that could provide useful

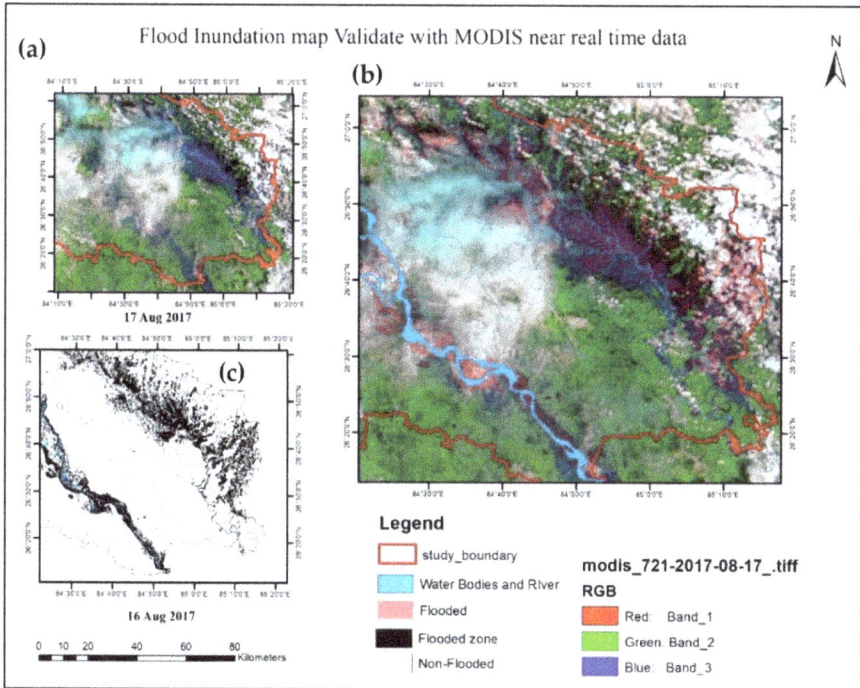

FIGURE 6.8 Validation of flood extent based on MODIS data (a–b) and SAR data (c) over the study region on 16 and 17 August 2017.

information for reducing risk. The flood inundation validation is shown in Figure 6.8, where we show the applicability of SAR data over optical during disasters.

The flood duration map shows the region near the riverbed remains inundated for a longer time than any other region, this is due to the fact, that is, the most low-lying area within the region of study. It is evident that the small villages along the river are at high risk of getting inundated by flood (Tripathi et al. 2022) and these households might get affected by diseases and other post-flood hazard calamities due to prolonged stagnant water. Agricultural land gets highly affected due to prolonged floodwater inundation (Pandey et al. 2022) as it uprooted planted crops. On the other hand, agricultural land cannot survive under flooded conditions and, thus, results in socioeconomic loss and might result in a scarcity of food. The northeastern region of the basin experiences more inundation in compressed to the western region. A major portion of the settlements, approximately 14.39 km^2, and cropland, 1089.33 km^2, were affected by prolonged inundation. The study region, just adjacent to the riverbed is most affected by the intensity of stagnancy of accumulation or inundation of floodwater. Similar findings are also reported by Pandey et al. (2022) over the North Bihar region, which revealed nearly 23.29 million of the population was affected by floods in the Ganga-Brahmaputra basin. The highest impacts of floods are from the Bihar state as people live in the lower valley as well as near to the riverbank owing to their dependency on river water.

6.6 CONCLUSION

An overview of the importance of microwave remote sensing data for flood inundation mapping is provided including the major processing elements. Important methodologies were also discussed along with the flood water distribution, duration, and dynamics. This study analyzed flood water extent to monitor the floodwater propagation dynamics (progression and regression) in the part of Gandak River Basin (Bihar), India, using multitemporal Sentinel-1-based SAR data. In the present study, both the spatial and temporal dynamics of the flooding were analyzed as identifying flood-damaged zones is highly essential for effective flood response. The highest inundation was found over 1331 km^2 (9.4%) of area of the basin with a maximum stagnancy of floodwater by 12–24 days. A floodwater depth of up to 2 m was mostly prevalent over the Gandak River Basin. The croplands are adversely affected by floods, covering 82% of the agricultural area with the stagnancy of water by 2–3 weeks.

Based on the findings, it is obvious that the use of SAR data has proved utility, potentiality, and capability in monitoring flood events, precisely identifying the flooded area, and determining the duration of flooding using multitemporal SAR-based satellite data. This study also demonstrates the locations of flood-prone areas and their characteristics, which are the key inputs for flood inundation modeling and forecasting. The derived information will be extremely beneficial for planning and management purposes to reduce the impact of flooding.

ACKNOWLEDGMENT

Authors thanks to Copernicus ESA and Japan Aerospace Exploration Agency (JAXA) for providing Sentinel satellite data and ALOS-based DEM data set.

CONFLICTS OF INTEREST

The authors declare no conflicts of interest.

REFERENCES

Acharya, A., and Prakash, A. 2019. "When the River Talks to Its People: Local Knowledge-based Flood Forecasting in Gandak River Basin, India." *Environmental Development* 31: 55–67. https://doi.org/10.1016/j.envdev.2018.12.003

Agnihotri, A. K., Ohri, A., Gaur, S., Shivam, Das, N., and Mishra, S. 2019. "Flood Inundation Mapping and Monitoring Using SAR Data and Its Impact on Ramganga River in Ganga Basin." *Environmental Monitoring and Assessment* 191(12): 760. https://doi.org/10.1007/s10661-019-7903-4

Alahacoon, N., Matheswaran, K., Pani, P., and Amarnath, G. 2018. "A Decadal Historical Satellite Data and Rainfall Trend Analysis (2001–2016) for Flood Hazard Mapping in Sri Lanka." *Remote Sensing* 10(3): 448. https://doi.org/10.3390/rs10030448

Bates, P. D., Wilson, M. D., Horritt, M. S., Mason, D. C., Holden, N., and Currie, A. 2006. "Reach Scale Floodplain Inundation Dynamics Observed Using Airborne Synthetic Aperture Radar Imagery: Data Analysis and Modelling." *Journal of Hydrology* 328(1–2): 306–318. https://doi.org/10.1016/j.jhydrol.2005.12.028

Blöschl, G., Hall, J., Viglione, A., Perdigão, R. A. P., Parajka, J., Merz, B., Lun, D., Arheimer, B., Aronica, G. T., Bilibashi, A., et al. 2019. "Changing Climate Both Increases and Decreases European River Floods." *Nature* 573(7772): 108–111. https://doi.org/10.1038/s41586-019-1495-6

Chang, L.-C., Shen, H.-Y., Wang, Y.-F., Huang, J.-Y., and Lin, Y.-T. 2010. "Clustering-Based Hybrid Inundation Model for Forecasting Flood Inundation Depths." *Journal of Hydrology* 385(1–4): 257–268. https://doi.org/10.1016/j.jhydrol.2010.02.028

Clement, M. A., Kilsby, C. G., and Moore, P. 2018. "Multi-Temporal Synthetic Aperture Radar Flood Mapping Using Change Detection: Multi-Temporal SAR Flood Mapping Using Change Detection." *Journal of Flood Risk Management* 11(2): 152–168. https://doi.org/10.1111/jfr3.12303

Cohen, S., Brakenridge, G. R., Kettner, A., Bates, B., Nelson, J., McDonald, R., Huang, Y.-F., Munasinghe, D., and Zhang, J. 2018. "Estimating Floodwater Depths from Flood Inundation Maps and Topography." *Journal of the American Water Resources Association* 54(4): 847–858. https://doi.org/10.1111/1752-1688.12609

EMDAT 2021. "The Non-COVID Year in Disasters. Brussels: CRED; 2021." https://emdat.be/sites/default/files/adsr_2020.pdf. Accessed on February 02, 2022.

Kumar, J., and Sahoo, S. 2021. "Monitoring North Bihar Flood of 2020 Using Geospatial Technologies." In: Rai, P. K., Singh, P., Mishra, V. N., editors. *Recent Technologies for Disaster Management and Risk Reduction.* Cham: Springer International Publishing, p. 135–155. https://doi.org/10.1007/978-3-030-76116-5_9

Kussul, N., Shelestov, A., and Skakun, S. 2011. "Flood Monitoring from SAR Data." In: Kogan, F., Powell, A., and Fedorov, O., editors. *Use of Satellite and In-Situ Data to Improve Sustainability.* Dordrecht: Springer Netherlands, pp. 19–29. https://doi.org/10.1007/978-90-481-9618-0_3

Manfreda, S., and Samela, C. 2019. "A Digital Elevation Model Based Method for a Rapid Estimation of Flood Inundation Depth." *Journal of Flood Risk Management* 12(S1:e12541): 1–10. https://doi.org/10.1111/jfr3.12541

Manjusree, P., Prasanna Kumar, L., Bhatt, M. C., Rao, G. S., and Bhanumurthy, V. 2012. "Optimization of Threshold Ranges for Rapid Flood Inundation Mapping by Evaluating Backscatter Profiles of High Incidence Angle SAR Images." *International Journal of Disaster Risk Science* 3(2): 113–122. https://doi.org/10.1007/s13753-012-0011-5

Matgen, P., Hostache, R., Schumann, G., Pfister, L., Hoffmann, L., and Savenije, H. H. G. 2011. "Towards an Automated SAR-Based Flood Monitoring System: Lessons Learned from Two Case Studies." *Physics and Chemistry of the Earth, Parts A/B/C* 36(7–8): 241–252. https://doi.org/10.1016/j.pce.2010.12.009

Matgen, P., Schumann, G., Henry, J.-B., Hoffmann, L., and Pfister, L. 2007. "Integration of SAR-Derived River Inundation Areas, High-Precision Topographic Data and a River Flow Model Toward Near Real-time Flood Management." *International Journal of Applied Earth Observation and Geoinformation* 9(3): 247–263. https://doi.org/10.1016/j.jag.2006.03.003

Nakmuenwai, P., Yamazaki, F., and Liu, W. 2017. "Automated Extraction of Inundated Areas from Multi-Temporal Dual-Polarization RADARSAT-2 Images of the 2011 Central Thailand Flood." *Remote Sensing* 9(1): 78. https://doi.org/10.3390/rs9010078

Özgen, I., Zhao, J., Liang, D., and Hinkelmann, R. 2016. "Urban Flood Modeling Using Shallow Water Equations with Depth-dependent Anisotropic Porosity." *Journal of Hydrology* 541: 1165–1184. https://doi.org/10.1016/j.jhydrol.2016.08.025

Pandey, A. C., Kaushik, K., and Parida, B. R. (2022). "Google Earth Engine for Large-scale Flood Mapping Using SAR Data and Impact Assessment on Agriculture and Population of Ganga-Brahmaputra Basin." *Sustainability* 14(7): 4210. https://doi.org/10.3390/su14074210

Parida, B. R. 2006. "Analysing the Effect of Severity and Duration of Agricultural Drought on Crop Performance Using Terra-MODIS Satellite Data and Meteorological Data." M.Sc. Dissertation, The International Institute for Geo-Information Science and Earth Observation, The Netherlands.

Parida, B. R., Behera, S., Bakimchandra, O., Pandey, A., and Singh, N. 2017. "Evaluation of Satellite-Derived Rainfall Estimates for an Extreme Rainfall Event over Uttarakhand, Western Himalayas." *Hydrology* 4(2): 22. https://doi.org/10.3390/hydrology4020022

Parida, B. R., Tripathi, G., Pandey, A. C., and Kumar, A. 2021. "Estimating Floodwater Depth Using SAR-Derived Flood Inundation Maps and Geomorphic Model in Kosi River Basin (India)." *Geocarto International*: 1–26. https://doi.org/10.1080/10106049.2021.1899298

Patel, N. R., Mukund, A., and Parida, B. R. 2022. "Satellite-Derived Vegetation Temperature Condition Index to Infer Root Zone Soil Moisture in Semi-Arid Province of Rajasthan, India." *Geocarto International* 37(1): 179–195. https://doi.org/10.1080/10106049.2019.1704074

Powell, S. J., Jakeman, A., and Croke, B. 2014. "Can NDVI Response Indicate the Effective Flood Extent in Macrophyte Dominated Floodplain Wetlands?" *Ecological Indicators* 45: 486–493. https://doi.org/10.1016/j.ecolind.2014.05.009

Schumann, G., Matgen, P., Hoffmann, L., Hostache, R., Pappenberger, F., and Pfister, L. 2007. "Deriving Distributed Roughness Values from Satellite Radar Data for Flood Inundation Modelling." *Journal of Hydrology* 344(1–2): 96–111. https://doi.org/10.1016/j.jhydrol.2007.06.024

Scorzini, A., Radice, A., and Molinari, D. 2018. "A New Tool to Estimate Inundation Depths by Spatial Interpolation (RAPIDE): Design, Application and Impact on Quantitative Assessment of Flood Damages." *Water* 10(12): 1805. https://doi.org/10.3390/w10121805

Shakya, A., Biswas, M., and Pal, M. 2021. "Parametric Study of Convolutional Neural Network Based Remote Sensing Image Classification." *International Journal of Remote Sensing* 42(7): 2663–2685. https://doi.org/10.1080/01431161.2020.1857877

Singh, K. V., Setia, R., Sahoo, S., Prasad, A., and Pateriya, B. 2015. "Evaluation of NDWI and MNDWI for Assessment of Waterlogging by Integrating Digital Elevation Model and Groundwater Level." *Geocarto International* 30: 650–661. https://doi.org/10.1080/10106049.2014.965757

Sinha, R., Bapalu, G. V., Singh, L. K., and Rath, B. 2008. "Flood Risk Analysis in the Kosi River Basin, North Bihar Using Multi-parametric Approach of Analytical Hierarchy Process (AHP)." *Journal of the Indian Society of Remote Sensing* 36(4): 335–349. https://doi.org/10.1007/s12524-008-0034-y

Sv, S. S., Roy, P. S., Chakravarthi, V., Srinivasarao, G., and Bhanumurthy, V. 2017. "Extraction of Detailed Level Flood Hazard Zones Using Multi-Temporal Historical Satellite Data-Sets—a Case Study of Kopili River Basin, Assam, India." *Geomatics, Natural Hazards and Risk* 8(2): 792–802. https://doi.org/10.1080/19475705.2016.1265014

Tripathi, G., Pandey, A. C., and Parida, B. R. 2022. "Flood Hazard and Risk Zonation in North Bihar Using Satellite-Derived Historical Flood Events and Socio-Economic Data." *Sustainability* 14(3): 1472. https://doi.org/10.3390/su14031472

Tripathi, G., Pandey, A. C., Parida, B. R., and Kumar, A. 2020. "Flood Inundation Mapping and Impact Assessment Using Multi-Temporal Optical and SAR Satellite Data: A Case Study of 2017 Flood in Darbhanga District, Bihar, India." *Water Resources Management* 34(6): 1871–1892. https://doi.org/10.1007/s11269-020-02534-3

Tripathi, G., Parida, B. R., and Pandey, A. C. 2019. "Spatio-Temporal Rainfall Variability and Flood Prognosis Analysis Using Satellite Data over North Bihar during the August 2017 Flood Event." *Hydrology* 6(2): 38. https://doi.org/10.3390/hydrology6020038

Zhou, S. L., and Zhang, W. C. 2017. "Flood Monitoring and Damage Assessment in Thailand Using Multi-temporal HJ-1A/1B and MODIS Images." *IOP Conference Series: Earth and Environmental Science* 57: 012016. https://doi.org/10.1088/1755-1315/57/1/012016

Zwenzner, H., and Voigt, S. 2009. "Improved Estimation of Flood Parameters by Combining Space Based SAR Data with Very High Resolution Digital Elevation Data." *Hydrology and Earth System Sciences* 13(5): 567–576. https://doi.org/10.5194/hess-13-567-2009

Section II

Geospatial Approaches for Monitoring Water Resources and Climate Change

7 Climate Change Impact on the Hydrological Extremes of a River Basin in the Hindu Kush Himalayan Region
A Case Study of the Marsyangdi River Basin, Nepal

Prajwal Neupane[1], Sangam Shrestha[1,], and Suwas Ghimire[1]*

1 Water Engineering and Management, School of Engineering and Technology, Asian Institute of Technology, P.O. Box 4 Klong Luang, Pathum Thani 12120, Thailand

CONTENTS

7.1 Introduction 112
7.2 Study Area and Data.. 113
    7.2.1 Study Area 113
    7.2.2 Data.. 115
7.3 Methodology 116
    7.3.1 Future Climate Projection 116
    7.3.2 Future Streamflow Projection... 117
    7.3.3 SWAT Model Calibration and Validation................................ 118
    7.3.4 Hydrological Extremes Analysis 119
7.4 Results and Discussion 119
    7.4.1 Projected Future Temperature 120
    7.4.2 Projected Future Rainfall 120
    7.4.3 SWAT Model Performance Evaluation 123
    7.4.4 Climate-Change Impact on Water Balance Components................ 124
    7.4.5 Future Hydrology under Climate Change 126
    7.4.6 Climate-Change Impact on Extreme Events 126

DOI: 10.1201/9781003265160-9

7.5   Conclusion 130
Acknowledgments...131
References..134

7.1   INTRODUCTION

The prime reason for rising global surface temperature is the well-mixing of green-house gases (GHGs) in the atmosphere due to various human activities other than natural drivers and the earth's internal variability. This is also the cause of inevitable global climate change, driving the global citizens to face extreme hydro-climatic situations, depletion of resources with time, and other climate-related hazards (IPCC 2021; Chen et al. 2020; Tabari 2021; Baig et al. 2021). The Hindu Kush Himalayan (HKH) region possesses highly diverse climatic conditions varying within its pecu-liar topographical variety and is highly vulnerable to climate change. (IPCC 2007; Kang et al. 2010; Sarıkaya et al. 2013; IPCC 2021). The Himalayas stretching from east to west is the home of more than 100,000 km^2 of glacier and snow cover, initiat-ing numerous snow-fed rivers and draining about 60% of the annual discharge as the water melts (Yao et al. 2012; Group et al. 2015; Adnan et al. 2017), which is the key source of fresh water for the downstream ecosystem and humankind. Such water bodies are expected to face the serious impact of climate change due to alteration of the annual hydrological cycle, in terms of both magnitude and frequency. The study area Marsyangdi River Basin (MRB) in Nepal is the river basin in the HKH region originating from Himalayan meltwater. MoHA/GoN (2017) revealed Nepal as one of the most disaster-susceptible countries in the world ranking (4th in climate-related hazards; 30th in flood hazard). So extreme hydroclimatic events are the triggering factors of water-induced disasters and associated damages, lost lives, and properties.

Several studies have been carried out to date in terms of climate change in such mountainous river basins of the region indicating varying (mostly increasing) cli-mate variables (temperature and precipitation). The rapid warming and increment of rainfall in highlands trigger the consequences like speedy melting and shrinkage of freshwater storage (glacier and snow cover) in the third pole (Azmat et al. 2018; Bajracharya et al. 2018; Budhathoki et al. 2021). As a result, the possibility rises for upsurging water levels in rivers, frequent flash flooding, erosion and landslides, and similar water-induced hazards. Most of the previously mentioned studies are based on data sets from Coupled Model Intercomparison Project Phase 5 (CMIP5) and are mostly concerned to investigate changes in average hydrology and water availability. Most of the studies in these hydroclimatic regions have limited exploration in the area like comparing and estimating the scale of changes in water balance and hydro-logical extremes. Also, very few studies have utilized the latest climate data set from Coupled Model Intercomparison Project Phase 6 (CMIP6).

This study focuses on the assessment and analysis of the hydrological extremes induced under the future climatic conditions at the basin outlet of MRB. The study is expected to explore the revised climate condition in the future as per the CMIP6 project General Circulation Models (GCMs) data set. Subsequently, it uses a semi-distributed physical hydrological model to generate daily discharge (average and extremes) from the snow-fed river to assess and analyze the hydrological extremes between baseline and

future periods under climate change in the mountainous river basin. The research also explores basin-scale changes in future climate and hydrology (average and extremes). Climate variables, namely, temperature (maximum and minimum) and precipitation, are projected for the future periods, Near Future (2015–2044), Mid Future (2045–2074), and Far Future (2075–2100), using the data set from CMIP6 project GCMs (ACCESS-CM2, BCC-CSM2-MR, and EC-Earth3) under three Shared Socioeconomic Pathways (SSPs) climate scenarios SSP1-2.6, SSP2-4.5, and SSP5-8.5. Projected future climate is fed into a hydrological model, Soil and Water Assessment Tool (SWAT) to simulate future daily streamflow at the basin outlet. Additionally, a hydrological tool, Indicators of Hydrologic Alteration (IHA), was used for the extreme event analysis.

The outcome of this study will aid to develop a better understanding of basin-scale future climatic conditions, changes in hydrology in the future under changing climate, and analysis of hydrological extremes at the basin outlet. This study was executed by adopting efficient and highly adopted tools and techniques for the reliability of its outcome. This is believed to become an asset and assist the scientific community, policymakers, water management practitioners, and other stakeholders to foresee possible hydro-climatic hazards, develop proper plans and policies to address such undesirable situations, and manage water resources of the study area or similar basins.

7.2 STUDY AREA AND DATA

7.2.1 STUDY AREA

The study was carried out in the MRB, a mountainous river basin in Nepal from the HKH region. Geographically, it extends from 27°50′42″ N to 28°54′11″ N latitude and 83°47′24″ E to 84°48′04″ E longitude, draining water from a total area of 4,056 km² at Bimalnagar station (Hydrological station no. 439). It is one of the tributaries of the Sapta-Gandaki river basin system, the second-largest river basin in Nepal, which drains water from Central and Western Nepal to the Ganges River basin before meeting the ocean at the Bay of Bengal. The river basin is home to a diverse range of flora and fauna comprising the significant area of the Annapurna Conservation Area Project (ACAP) boasting many snow-clad mountains above 7000 m above mean sea level (AMSL). The river also has some cultural value to the locals/residents and serves as a source of drinking water and irrigation. Most importantly, it is well known for hydroelectricity production due to its high current and considerable amount of flow throughout the year. Currently, there are three major hydropower projects along the mainstream with two of them within the study area as shown in Figure 7.1.

Physio graphically, the Marsyangdi river basin extends from High Himalayas (7,800 m AMSL) in the North to the Siwalik region (about 200 m AMSL) in the South with an approximate river length of 150 km from the headwaters to the outlet and an average basin slope of 29.42°. The climate in the basin is determined by its topography ranging from the chilling cold at high mountains upstream to hot and humid subtropical at Siwalik downstream. The Marsyangdi River catchment has a significantly snow-covered and glaciated area contributing snowmelt as a significant part of streamflow. More than 20 glaciers exist in the basin (WACREP 2019), and more than 12% of the basin area remains under snow cover throughout the year.

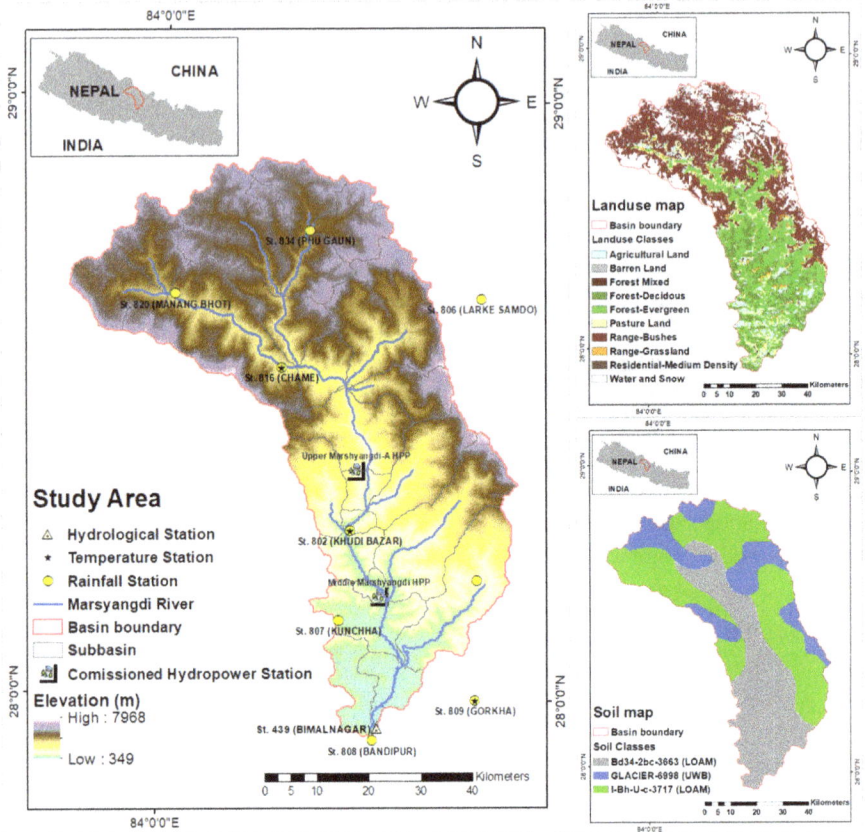

FIGURE 7.1   MRB and spatial distribution of hydro-meteorological stations (left panel). The right panels correspond to land-use (top) and soil maps (bottom) of the study basin.

The climate in the study area is predominately influenced by the summer monsoon, starting in June and extending till September. Due to the high gradient of the basin, generally, climate diverges from subtropical in the lower belt to the arctic in the higher altitudes. The average annual temperature for the lower areas of catchment like Bimalnagar, Tanahu is about 17°C (minimum) and 30°C (maximum). For higher settlements like Chame and Manang, the maximum and minimum temperatures are 17°C and 4°C, respectively, Department of Hydrology and Meteorology (DHM), Nepal. For places in higher altitudes than Chame, temperature declines with an increase in altitude. After analyzing observed temperature data from 1982 to 2011 for daily maximum and minimum temperature, the basin average maximum temperature (T_{max}) is found to be 23.3 ± 0.9°C and minimum temperature (T_{min}) was found to be 11.7 ± 0.6°C.

Precipitation pattern over the basin mostly varies with conditions like altitude and landforms. The average annual rainfall in the basin is about 1666.59 ± 170 mm.

This figure was obtained from daily rainfall data analysis for nine rainfall stations from 1982 to 2011. The average annual is similar to the one presented in the study by Parajuli et al. (2016), and about 75% of this rainfall occurs during the monsoon (June to September) below snowline altitude. Seasons within Nepal are categorized as Pre-Monsoon/Spring (March, April, May) Monsoon/Summer (June, July, August, September), Post-Monsoon/Autumn (October, November), and Winter (December, January, February; DHM 2015). Although the winter season receives minimal rainfall in the lower part of the basin, snowfall is received in higher altitudes.

7.2.2 DATA

A short description of the data used is presented in Table 7.1. The present study has used mainly geospatial data, hydro-meteorological data, and the result of the climate model (GCM's data). The source of geospatial data like the digital elevation model (DEM) named AW3D30 (ALOS World 3D-30m) was from the Japanese Aerospace Space Agency (JAXA), land use from the European Space Agency (ESA), and soil map from FAO. All the hydro-meteorological data were collected from the DHM, Nepal, and further processed for the study. Initially, those data were recorded and acquired by the DHM, Nepal from its meteorological (3 temperature and 9 rainfall)

TABLE 7.1
Summary of Data and Their Corresponding Sources

SN	Data	Spatiotemp Oral Resolution	Duration	File Type	Sources
Geospatial data					
1	Topography (Digital Elevation Model, DEM)	30m × 30m		Raster	World Elevation Data, JAXA, www.eorc.jaxa.jp/ALOS/en/aw3d30/data/index.htm
2	Land-use map	300m × 300m	2015	Raster	ESA, www.esa-landcover-cci.org/?q=node/164
3	Soil map	1:5,000,000	2011	Vector	FAO, www.fao.org/geonetwork
Hydro-Meteorological data					
1	Temperature (°C)	Point/Daily	1982–2011	Text	DHM, Nepal
2	Rainfall (mm)	Point/Daily	1982–2011	Text	
3	Discharge (m³/s)	Point/Daily	1988–2011	Text	
Global Circulation Models SSP1-2.6, SSP2-4.5, SSP5-8.5					
1	ACCESS-CM2	1.25° × 1.25°	1982–2100	NetCDF	CSIRO and ACCESS, Australia
2	BCC.BCC-CSM2-MR	1.12° × 1.12°	1982–2100	NetCDF	BCC, China
3	EC-Earth3	0.7° × 0.7°	1982–2100	NetCDF	EC-Earth Consortium

Note: ALOS: Advanced Land Observing Satellite; JAXA: Japan Aerospace Exploration Agency; ESA: European Space Agency; FAO: Food and Agriculture Organization; DHM: Department of Hydrology and Meteorology; CSIRO: Commonwealth Scientific and Industrial Research Organization; ACCESS: Australian Research Council Centre of Excellence for Climate System Science; BCC: Beijing Climate Center

and one hydrological station. Similarly, results of climate models were downloaded from the World Climate Research Programme (WRCP) internet portal from the list of CMIP6 project. The selection of 3 GCMs (ACCESS-CM2, BCC-CSM2-MR, and EC-Earth3) in this study was based on the performance of climate models from CMIP6 projects from previous research/studies within a similar region (Mishra, Bhatia, and Tiwari 2020; Mondal et al. 2021).

7.3 METHODOLOGY

The overall methodological approach is illustrated in Figure 7.2. This chart provides insight into the steps carried out during the research and gives the flow direction of the study from beginning to completion.

7.3.1 FUTURE CLIMATE PROJECTION

Among the many bias-correction techniques, linear scaling was used in this study as its ease of use was suggested by many other studies (Shrestha, Acharya, and Shrestha 2017; Shrestha, Shrestha, and Shrestha 2017). This technique works based on monthly correction factors evaluated from observed data from ground stations and historical data from GCMs. A multiplicative factor is applied for precipitation whereas additive factors are utilized for temperature as shown in Equations 7.1, 7.2, 7.3, and 7.4. In this study three climate variables, namely, temperature maximum (T_{max}), temperature minimum (T_{min}), and precipitation (Ppt), from the GCMs went through bias correction.

$$P_{his}^* = P_{his}(d) \cdot \frac{\mu_m(P_{obs}(d))}{\mu_m(P_{his}(d))}, \tag{7.1}$$

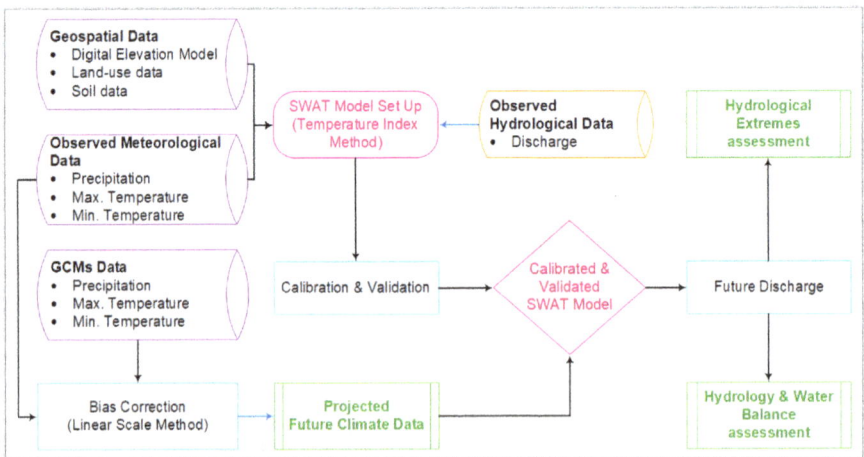

FIGURE 7.2 Methodological framework for climate-change projection and hydrological extremes assessment.

$$P^*_{fut} = P_{fut}(d) \cdot \frac{\mu_m\left(P_{obs}(d)\right)}{\mu_m\left(P_{his}(d)\right)}, \qquad (7.2)$$

$$T^*_{his} = T_{his}(d) + \mu_m\left(T_{obs}(d)\right) \pm \mu_m\left(T_{his}(d)\right), \qquad (7.3)$$

$$T^*_{fut} = T_{fut}(d) + \mu_m\left(T_{obs}(d)\right) \pm \mu_m\left(T_{his}(d)\right), \qquad (7.4)$$

where P = precipitation (mm), T = temperature (°C), μ_m = average of data set, his = historical data, fut = future data, obs = observed data, m = monthly data set, d = daily data set and (*) represents bias-corrected data.

Future climate projection for the required climate change scenario was carried out by using Atmospheric Oceanic Coupled Global Circulation Models (AOGCMs). These models are considered as most sophisticated tools for the assessment of climate change in the future. The uncertainties associated with climate response and future emissions are likely to exist in projected future climate. To reduce such uncertainties, simultaneously use of multiple GCM and emission scenarios were practiced. For this study, the data for future projection were used from the CMIP6 project, which possesses output from recent and updated climate models. The result after bias correction from selected multiple GCMs was then ensembled for both temperature and precipitation by using a simple average technique before projecting the future for each SSP scenario.

7.3.2 FUTURE STREAMFLOW PROJECTION

SWAT is a physical-based, spatially distributed (semi-distributed), and time-continuous hydrological model. This model is developed by the U.S. Department of Agriculture. SWAT as a feasible tool is extensively accepted as it can consider heterogeneous characteristics of the watershed to project the impacts of future hydro-climatic changes and address other environmental issues (Arnold et al. 1998; Bieger et al. 2017; Bajracharya et al. 2018). This model divides the basin into sub-basins based on topography and river network. Again, those sub-basins are divided into the number of hydrologic response units (HRUs). The whole model is composed of such HRUs, for the SWAT model HRU is the smallest unit and is generated by the combination of unique topography, soil category, and slope arrangement within a sub-basin based on user-defined thresholds. A land-use map, soil map, and slope classes are taken as input in the SWAT model to create an HRU. Water balance is the basic principle for simulation in the SWAT model, which is controlled by climate inputs like maximum and minimum surface air temperature and daily precipitation. The new area of studies covered by SWAT is eco-hydrological modeling, ecosystem services, sub-daily simulations, and pesticide and sediment load transport simulations (Tan et al. 2020). The water balance of a watershed is represented by Equation 7.5 (Arnold et al. 1998).

$$SW_t = SW_{orginal} + \sum_{i=1}^{t}\left(R_{daily} - Q_{surf} - E_a - W_{seep} - Q_{gw}\right) \qquad (7.5)$$

To find the final water content at the required time t (SW_t) in mm, we require initial/ original water content ($SW_{orginal}$) in mm, required time in days t, daily precipitation (R_{daily}) in mm, surface runoff (Q_{surf}), evapotranspiration (E_a) in mm, percolation toward the ground in mm, and the amount of base flow in the river (Q_{gw}) in mm.

Within SWAT, the snow module was also used to incorporate snowfall and snow-melt processes as a major hydrological component for such a mountainous basin. Snowmelt simulations in SWAT can only be carried out if snow exists in the sub-basin area. The categorized subbasins in SWAT can be further divided into several elevation bands that incorporate temperature and precipitation variation concerning changing altitude (Hartman et al. 1999). SWAT model has a snowmelt module majorly composed of snowpack and snowmelt.

Relative humidity, solar radiation, and wind speed are other data required by the SWAT hydrological model to estimate the potential evapotranspiration process, using Penman-Monteith (PM) equation in the general approach. Due to the scarcity of those data the Hargreaves equation was considered for the evapotranspiration process in this study. That is based on temperature rather than other parameters, unlike the PM equation. This equation is often adopted in the data-scarce condition.

7.3.3 SWAT MODEL CALIBRATION AND VALIDATION

Calibration is an integral part while setting up the model precisely according to a given set of local conditions so that model could reduce uncertainty while performing and predicting required results efficiently. Initially, the most sensitive parameters among all influencing parameters are estimated for the first step of calibration and validation of the SWAT model. The finalization of sensitive parameters is related to sensitivity analysis, in which the rate of change in output to changes in input for the model is observed and a relation is established (Arnold et al. 2012). Here, the calibration process of the SWAT model and sensitivity analysis will be executed by using an external tool SWAT-CUP by following the Sequential Uncertainty Fitting (SUFI-2) algorithm among available other options. It was selected since it came with parallel processing capabilities and does the work automatically (Abbaspour, Johnson, and van Genuchten 2004). This tool/software assists SWAT users to perform calibration of parameters with ease, efficiently, and automatically. The calibration process is performed by comparing a set of observed data with output from a model generated by injecting selected values for model input parameters in a user-created replication of physical conditions. Validation is the following and final step for model setup after calibration, where a model is run with a new set of data with the same parameters defined during the calibration process.

The result is compared with the observed set of data for an efficient prediction form model (Arnold et al. 2012). Further, during the calibration and validation process, the statistical performance of the model was evaluated based on Nash-Sutcliffe efficiency (NSE; Nash and Sutcliffe 1970), percentage bias (PBIAS; Vijai, Soroosh, and Ogou 1999), the coefficient of determination (R^2; Santhi et al. 2001), and root mean square error observations standard deviation ratio (RSR; Moriasi et al. 2007).

$$NSE = 1 - \frac{\sum_{i=1}^{n}\left(Q_i^o - Q_i^s\right)^2}{\sum_{i=1}^{n}\left(Q_i^o - \bar{Q}^o\right)^2} \, , \tag{7.6}$$

$$PBIAS = \frac{\sum_{i=1}^{n}\left(Q_i^o - Q_i^s\right)}{\sum_{i=1}^{n}\left(Q_i^o\right)} \times 100 \, , \tag{7.7}$$

$$R^2 = \frac{\sum_{i=1}^{n}\left(Q_i^o - \bar{Q}^o\right)\left(Q_i^s - \bar{Q}^s\right)}{\sqrt{\sum_{i=1}^{n}\left(Q_i^o - \bar{Q}^o\right)^2}\sqrt{\sum_{i=1}^{n}\left(Q_i^s - \bar{Q}^s\right)^2}} \, , \tag{7.8}$$

$$RSR = \frac{RMSE}{STDEV_o} = \frac{\left[\sqrt{\sum_{i=1}^{n}\left(Q_i^o - Q_i^s\right)^2}\right]}{\left[\sqrt{\sum_{i=1}^{n}\left(Q_i^o - \bar{Q}^o\right)^2}\right]} \, , \tag{7.9}$$

where, Q_i^o is observed discharge, $\bar{Q}^o$ is mean of observed discharge, Q_i^s is simulated discharge, $\bar{Q}^s$ is mean of simulated discharge, i is the counter for individual observed and simulated discharge, and n is the number of total samples.

7.3.4 HYDROLOGICAL EXTREMES ANALYSIS

Extreme event analysis was carried out at the basin outlet of the study area, for which the IHA (Richter et al. 1996; RICHTER et al. 1997) tool was considered. This tool comprises 33 parameters that represent hydrological variability and represents several aspects of hydrological extremes (Bharati et al. 2016). Moreover, in this study, the change of the future maximum and minimum flow parameters, parameter group 2 (1-day, 3-day, 7-day, 30-day, 90-day), regarding the baseline period were computed and compared. For the maximum and the minimum flow values for each duration, the IHA calculates the running average which gives a better estimation of the highest and lowest river discharge in the required duration of time. Similarly, for the seasonal extreme events analysis, the estimated future magnitude of maximum and minimum flow at the basin outlet was compared with that of the baseline period. The highest and lowest flow for each season under all projected future scenarios was computed and compared with the maximum and minimum river discharge from the baseline period.

7.4 RESULTS AND DISCUSSION

The basin is expected to be warmer and receive more rainfall under all future climatic scenarios in all future periods. Both maximum and minimum temperatures are estimated to increase by about 5°C by the end of 2100 under the SSP5-8.5 scenario. Similarly, a significant change in rainfall is also projected, in which an additional 731

mm of rainfall is expected than that of the baseline period (1666.59 ± 170 mm) by end of the century under the highest radiative forcing climate scenarios. In addition to that, this topic also discusses future climate trends and spatiotemporal changes in the climate components.

7.4.1 PROJECTED FUTURE TEMPERATURE

The ensemble results from selected GCMs projected maximum temperature (T_{max}) for the basin, which indicates that under almost all climate scenarios the rising trend is expected to occur in future periods. Among the scenarios, the rise in the SSP5-8.5 scenario is more significant in comparison to other climate scenarios as per Figure 7.3 (top). The trend analysis also suggests that the average rise of T_{max} ranges from 1°C to 5.2°C under all scenarios in different future periods. In an additional analysis of spatiotemporal changes of T_{max} throughout the basin (Supplementary Figure S7.1), the maximum temperature is projected to increase all over the basin from snow-capped mountains in the north to plain valleys in the south. A significant rise of T_{max} is expected to occur in the southern part of the study area due to which that region of the basin is anticipated to be warmer than the baseline period. Similarly, an increase of T_{max} in the northern region is also projected which signifies a warmer climate in the mountains.

The minimum temperature (T_{min}) of the study area was projected and the annual trend was plotted for the ensemble result from GCMs in Figure 7.3 (bottom). The graph signifies the increasing trend of the T_{min} under almost all climate scenarios in future periods. Similar to the T_{max} under the SSP5-8.5 scenario, the temperature rise is noteworthy in future periods. The average rise of T_{min} in future periods ranges from 1.2–5.3°C under all climate scenarios. In addition, spatiotemporal analysis (Supplementary Figure S7.2) shows that the hike of T_{min} is anticipated to be all over the basin. Also, with the rise of T_{min} the northern part of the study area—a mountainous region—is expected to experience warmer winter. Such conditions could be the cause of the loss of snow masses in that region.

Contrasting to the other two scenarios SSP1-2.6, one of the optimistic scenarios projects the rise of both maximum and minimum temperature till the end of mid-future and then shows the decreasing trend. Under this scenario, global CO_2 emissions are anticipated to be severely cut at a mild pace to achieve net-zero emissions after 2050 along with substantial land-use change (significant increment in global forest cover).

7.4.2 PROJECTED FUTURE RAINFALL

According to the ensemble result of projected rainfall in the MRB, the continuous rise in annual rainfall is expected under all climate scenarios and in all future periods (Supplementary Figure S7.3). The increase in the rainfall is estimated in the range between 144 mm to 731 mm for the least to high radiative forcing scenarios in different future periods. The overall rainfall distribution in the basin and expected change with time for projected future rainfall are presented in Figure 7.4. As the most rainfall is concentrated in the southern part of the basin in the baseline period, the

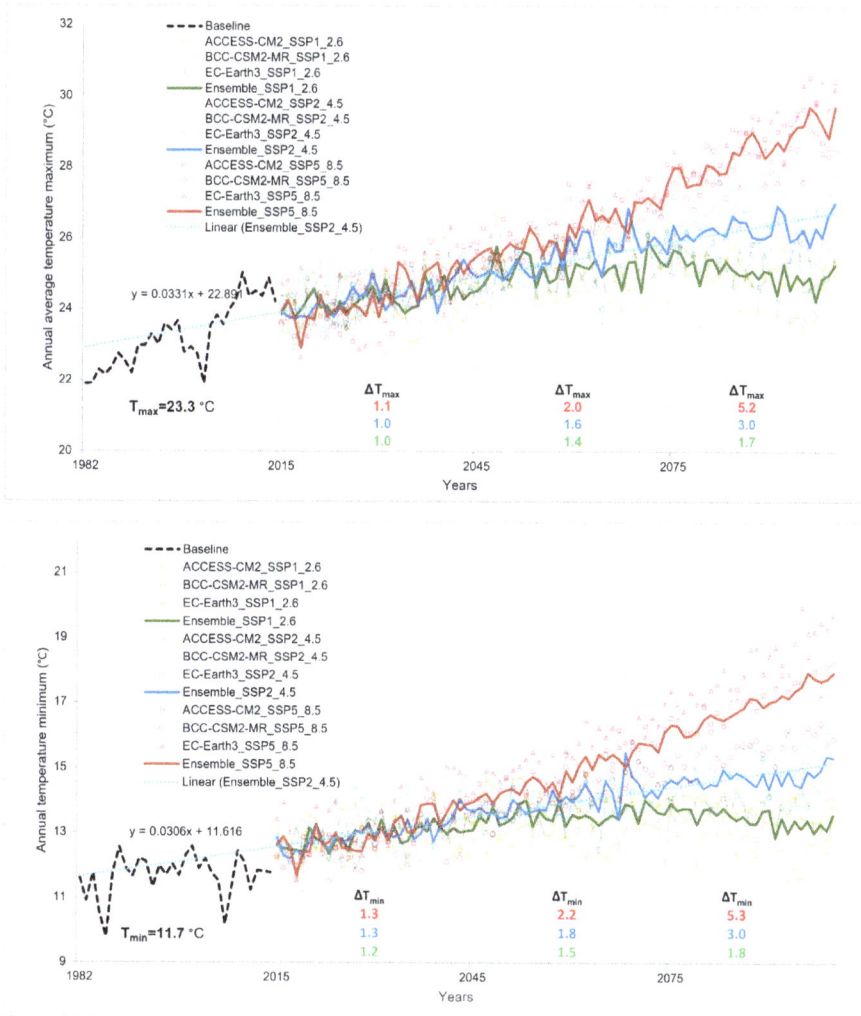

Top chart axis labels:
Annual average temperature maximum (°C)

Legend:
- - - Baseline
ACCESS-CM2_SSP1_2.6
BCC-CSM2-MR_SSP1_2.6
EC-Earth3_SSP1_2.6
—— Ensemble_SSP1_2.6
ACCESS-CM2_SSP2_4.5
BCC-CSM2-MR_SSP2_4.5
EC-Earth3_SSP2_4.5
—— Ensemble_SSP2_4.5
ACCESS-CM2_SSP5_8.5
BCC-CSM2-MR_SSP5_8.5
EC-Earth3_SSP5_8.5
—— Ensemble_SSP5_8.5
······ Linear (Ensemble_SSP2_4.5)

$y = 0.0331x + 22.89$

$T_{max} = 23.3\ °C$

ΔT_{max}	ΔT_{max}	ΔT_{max}
1.1	2.0	5.2
1.0	1.6	3.0
1.0	1.4	1.7

Years: 1982, 2015, 2045, 2075

Bottom chart axis labels:
Annual temperature minimum (°C)

Legend:
- - - Baseline
ACCESS-CM2_SSP1_2.6
BCC-CSM2-MR_SSP1_2.6
EC-Earth3_SSP1_2.6
—— Ensemble_SSP1_2.6
ACCESS-CM2_SSP2_4.5
BCC-CSM2-MR_SSP2_4.5
EC-Earth3_SSP2_4.5
—— Ensemble_SSP2_4.5
ACCESS-CM2_SSP5_8.5
BCC-CSM2-MR_SSP5_8.5
EC-Earth3_SSP5_8.5
—— Ensemble_SSP5_8.5
······ Linear (Ensemble_SSP2_4.5)

$y = 0.0306x + 11.616$

$T_{min} = 11.7\ °C$

ΔT_{min}	ΔT_{min}	ΔT_{min}
1.3	2.2	5.3
1.3	1.8	3.0
1.2	1.5	1.8

Years: 1982, 2015, 2045, 2075

FIGURE 7.3 Projected future maximum (top) and minimum (bottom) temperature in the MRB under SSP1-2.6, SSP2-4.5, and SSP5-8.5 scenarios compared to the baseline period (1982–2011).

continuous increase of rainfall is also anticipated to occur in the southern region of the basin in most of the climate scenarios and future periods. But in the far future under the SSP5-8.5 scenario, the rainfall is likely to extend significantly toward the northern territory of the basin. The escalated magnitude of rainfall in the southern region signifies that during the wet season this part of the basin gives a major contribution to the discharge at the outlet point of the basin.

Seasonal percentage change of future rainfall has been presented in Figure 7.5 as derived from the ensemble result of three GCMs. The graph suggests that projected rainfall in all seasons except winter is estimated to increase. A significant rise

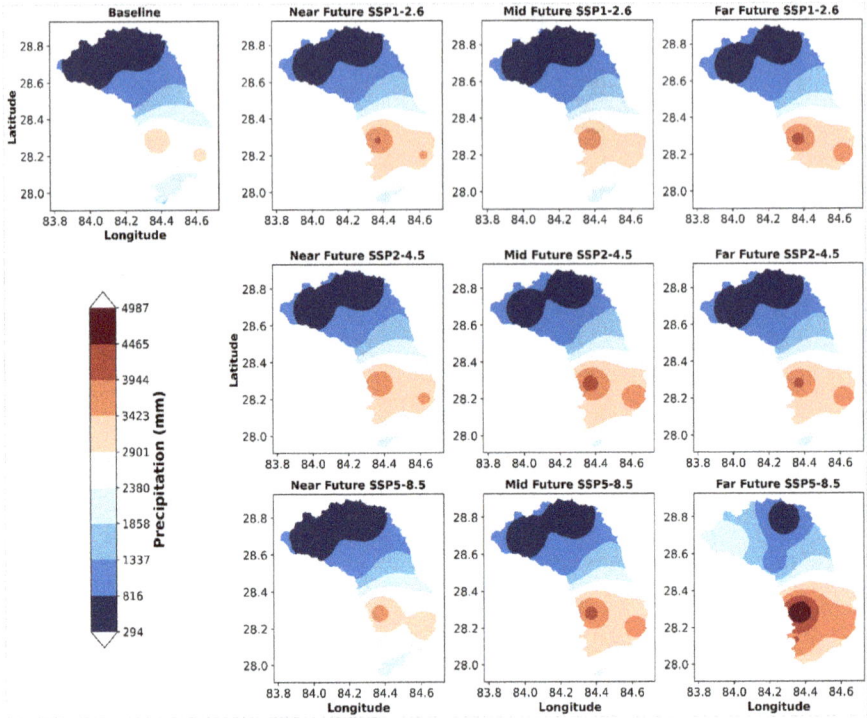

FIGURE 7.4 Spatial distribution of future rainfall in the MRB under SSP1-2.6, SSP2-4.5, and SSP5-8.5 scenarios compared to the baseline period (1982–2011).

FIGURE 7.5 Projected future seasonal rainfall (%) in the MRB under SSP1-2.6, SSP2-4.5, and SSP5-8.5 scenarios compared to the baseline period (1982–2011).

in percentage change is foreseen for the pre-monsoon season; however, in terms of volume, future monsoon is likely to receive more downpour. Also, this study area receives very little rain in winter. Even with a 24% decrease in winter rainfall in the future, the rainfall would not alter the total volume of annual rainfall by a significant amount.

Climate change is evident from this research that is coherent to IPCC (2021) in regional climate projection for South Asia. Similar results for South Asia and the HKH region can be observed from some recent studies (Almazroui et al. 2020; Mishra, Bhatia, and Tiwari 2020) working with future climate data sets from the recent CMIP6 project. Those studies suggest the possibility of temperature rise by up to 6°C under the SSP5-8.5 extreme climate scenario by 2100. In terms of rainfall, the result from this study coincides with the aforementioned studies, which advocate an increment of average annual rainfall by about 8% to 44%.

7.4.3 SWAT MODEL PERFORMANCE EVALUATION

The SWAT hydrological model was used in this study and the basin outlet was considered for the calibration and validation point. The calibration and validation were carried out using the SWAT-CUP program and the period of 16 years (1988-2003) was regarded for calibration and 7 years (2004–2011) for validation. Here, the year 2005 was not considered for validation due to missing rainfall and discharge data. Missing rainfall data were replaced by −999 as recommended by the SWAT manual for continuous discharge simulation. Fontaine et al. (2002) highlight that the warmup period plays a significant role to maintain required hydrological conditions such as initial snow content in the basin, groundwater conditions, and storage of water in unsaturated soil. In this study, 6 years (1982–1987) of warmup period was considered.

After running the sensitivity test and sorting 18 parameters as sensitive as well as major influencing parameters to the hydrological process of the basin, those parameters were considered, and their values were altered to best fit the model as per the observed daily hydrology. Among which the top 6 parameters are the most sensitive parameters based on the P-value from global sensitivity analysis in SWAT CUP that highly influences the discharge simulation. Table 7.2 shows the list of those parameters, their sensitivity rank, and best-fitted values and the detailed explanation of all parameters are presented in Supplementary Table S7.1. The accuracy of the model was tested by checking the fitness of observed and simulated flow (Figure 7.6, top). Besides this, NSE, R^2, PBIAS, and RSR were also calculated and found to be 0.78, 0.80, −14.97, and 0.47, respectively, in calibration and 0.80, 0.82, 8.19, and 0.45, respectively, in the validation process. Figure 7.6 (bottom) shows the observed and simulated regression plot for the calibration and validation period for the hydrological station. The value of the coefficient of determination (R^2) shows a satisfactory result for the model acceptance in both calibration and validation processes. From the scatterplot, we can observe the scattering of observed versus simulated points from the mean, where higher values seem to be overpredicted and lowest values seem to be under predicated than that of observed data. Overall, both plots show a positive correlation between observed and simulated data.

TABLE 7.2

SWAT Model Parameters with Their Sensitivity Rank and Fitted Values

Sensitivity Rank	Parameters	Parameter Range		Default Value	Fitted Value	Unit
		Min	Max			
1	r__CN2.mgt	−0.15	0.15	hru	−0.149	% Change
2	v__SMTMP.bsn	2.57	2.61	0.5	2.55	°C
3	v__ALPHA_BF.gw	0.25	0.42	0.048	0.367	days
4	v__REVAPMN.gw	460	465	750	463.93	mm
5	v__GW_REVAP.gw	0.05	0.15	0.02	0.77	
6	v__TLAPS.sub	−5.89	−5.92	−6	−5.89	°C/km
7	v__SFTMP.bsn	0.269	0.363	1	0.33	°C
8	v__SNOCOVMX.bsn	22.67	26.87	1	25.66	mm
9	v__GW_DELAY.gw	45.73	75	31	75	days
10	v__ESCO.hru	0.95	1	1	0.967	factor
11	v__PLAPS.sub	0.085	0.094	0	0.094	mm/km
12	v__SMFMX.bsn	6.56	6.56	4.5	6.62	mm/°C-day
13	v__SMFMN.bsn	2.55	3.16	4.5	2.35	mm/°C-day
14	v__TIMP.bsn	0.77	0.79	1	0.77	
15	v__LAT_TTIME.hru	16.65	52.52	0	17	days
16	v__EPCO.hru	0.95	0.97	1	0.95	factor
17	v__SNO50COV.bsn	0.515	0.538	0.5	0.52	
18	v__GWQMN.gw	108.83	121.7	1000	113.03	mm

Note: "Factor" represents any ratio without any unit; "hru" represents the value for the parameter that differs from each hydrological response unit (HRU) along with its individual property of the HRU.

7.4.4 CLIMATE-CHANGE IMPACT ON WATER BALANCE COMPONENTS

The major water balance component here as inflow into the MRB is precipitation which further occurs into two forms rainfall (about 81%) and snow (about 19%). One of the major outflow components of the water balance is water yield (about 64%) which is composed of snowmelt, surface runoff, lateral flow, and groundwater flow for MRB. Second, evapotranspiration (about 20%) consists of sublimation from snow and evaporation from land and water bodies. The remaining quantity of water is changed in the storage in the basin, losses occurred by water trapped in ponds and deep percolation and unaccounted process due to lack of data in the model like relative humidity, solar radiation, wind speed, and other required data for basin characteristics representation by model. With the rise of the total volume in the basin precipitation, an increase in the inflow of the basin and other water balance components related to outflow are also expected to rise accordingly as shown in Figure 7.7. Significant rise is estimated for water yield in all future periods under selected future scenarios from the basin outlet. Evapotranspiration in the future period compared to the baseline period (1982–2011) doesn't show a noticeable difference under all SSP scenarios.

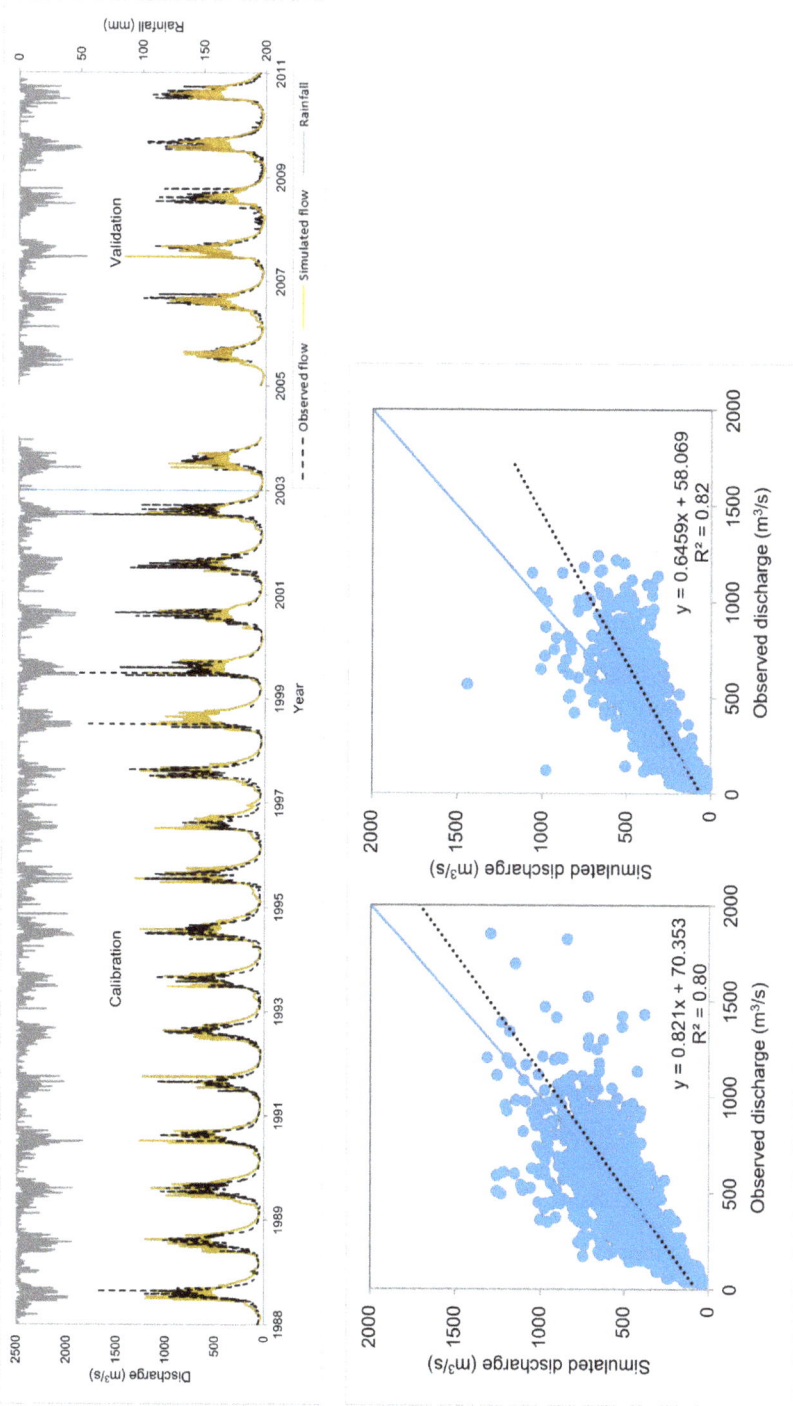

FIGURE 7.6 Simulated and observed hydrographs at basin outlet (Bimalnagar Station) during calibration (1988–2003) and validation (2004–2011) periods (top) along with their respective regression plots (Calibration: bottom left, Validation: bottom right). The blue line in regression plots represents a perfect coefficient of determination (COD) while the dotted line represents the COD from the model.

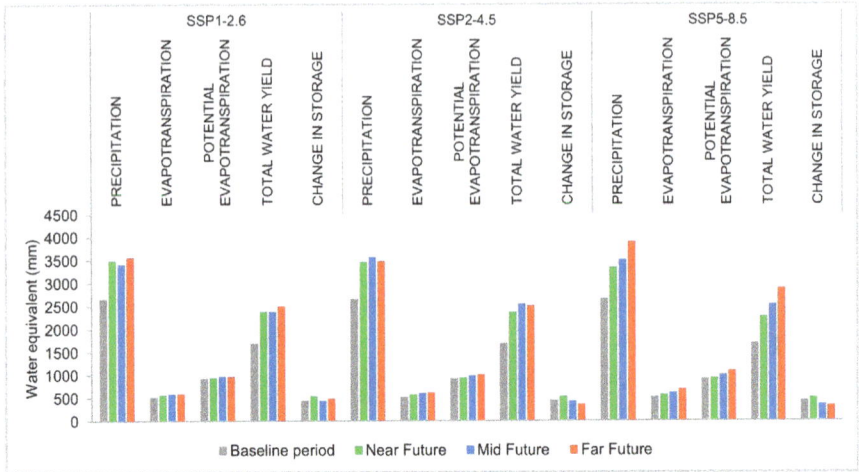

FIGURE 7.7   Future water balance components in the MRB under SSP1-2.6, SSP2-4.5, and SSP5-8.5 scenarios compared to the baseline period (1982–2011).

7.4.5   FUTURE HYDROLOGY UNDER CLIMATE CHANGE

The simulated future discharge under the projected climatic condition of MRB indicates that the future discharge is likely to increase in all future periods under all emission scenarios. For the near future, the maximum change is predicted under the SSP1-2.6 scenario, for the mid-future period under the SSP2-4.5 scenario highest rise in discharge is estimated, but in the far future, the highest change (increment) of discharge is expected to occur under SSP5-8.5 scenario, which concludes that overall discharge at basin outlet is increasing as presented in Supplementary Figure S7. 4. The seasonal analysis as illustrated in Figure 7.8 and the percentage change of seasonal discharge showed that in all future periods, the discharge is expected to rise. A significant rise in discharge in the post-monsoon season can be experienced in the future as the result of an increase in snowmelt due to rising temperature in the basin. But in volumetric terms, the highest increment in discharge at the basin outlet is estimated to occur in the monsoon season, which is related to the increase in rainfall in the southern part of the basin in the wet season.

7.4.6   CLIMATE-CHANGE IMPACT ON EXTREME EVENTS

The hydrological extremes from the projected hydrology of MRB were finally analyzed using the IHA software/tool. This analysis focuses on the magnitude of the highest and lowest discharge at the basin outlet from observed and projected future daily streamflow. Thus, the result from parameter group 2 was primarily considered for this analysis. The percentage change of maximum and minimum flow at basin outlet from parameter group 2 are plotted and illustrated in Figure 7.9 (top and bottom). In general, a substantial increment in the extent of

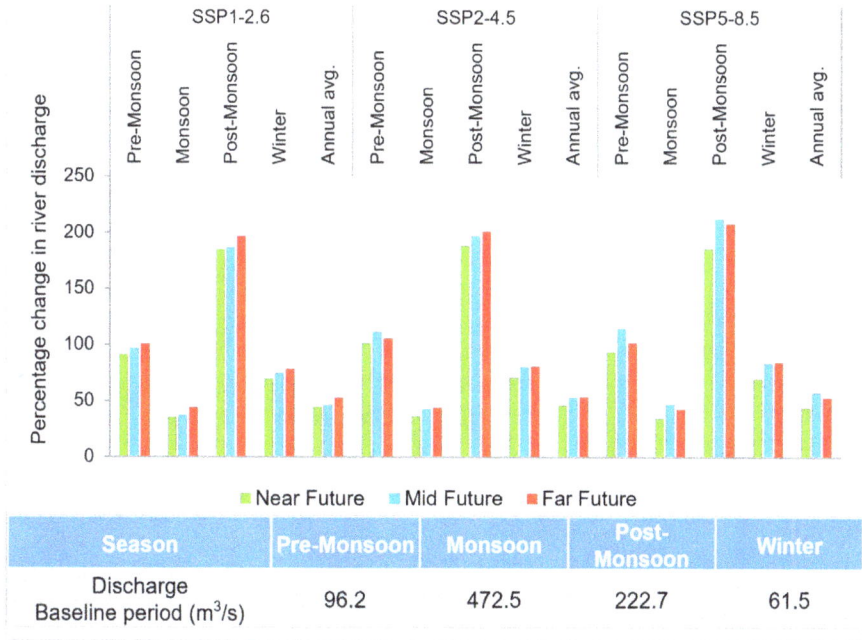

FIGURE 7.8 Projected future seasonal discharge (%) in the MRB under SSP1-2.6, SSP2-4.5, and SSP5-8.5 scenarios compared to the baseline period (1982–2011).

hydrological extremes is anticipated in the future period when compared to the baseline period.

The significant rise of 1-day maximum and minimum flows in the river is expected to occur, in all future periods under all three-climate scenarios. About 177% and 149% of the maximum increase is estimated in the far future under the SSP5-8.5 scenario for 1-day maximum and minimum flow, respectively. A significant rise in the 3-day, 7-day, 30-day, and 90-day average maximum flow is anticipated. Among which under the SSP5-8.5 climate scenario highest increment is estimated in the far future and a remarkable increase in the magnitude of 1-day and 3-day maximum flow indicates high flow condition is concentrated in shorter time duration in the future rather than events distributed in longer temporal scale. Meanwhile, for the future minimum flow parameters (1-day, 3-day, 7-day, 30-day, 90-day), the change (rise) ranges from 70% to 150%, mostly near to the double the volume of the baseline period. In contrast to the maximum flow, future low flow parameters at the basin outlet are not concentrated at a point in time. Overall, the increment of the minimum flow indicates an assurance of more water availability in the basin outlet than that of the baseline period in the dry season as well.

Besides that, the hydrological extremes were analyzed on a seasonal basis. This analysis compares maximum and minimum seasonal flow at the basin outlet between the baseline period and projected future discharge (Figure 7.10). Here,

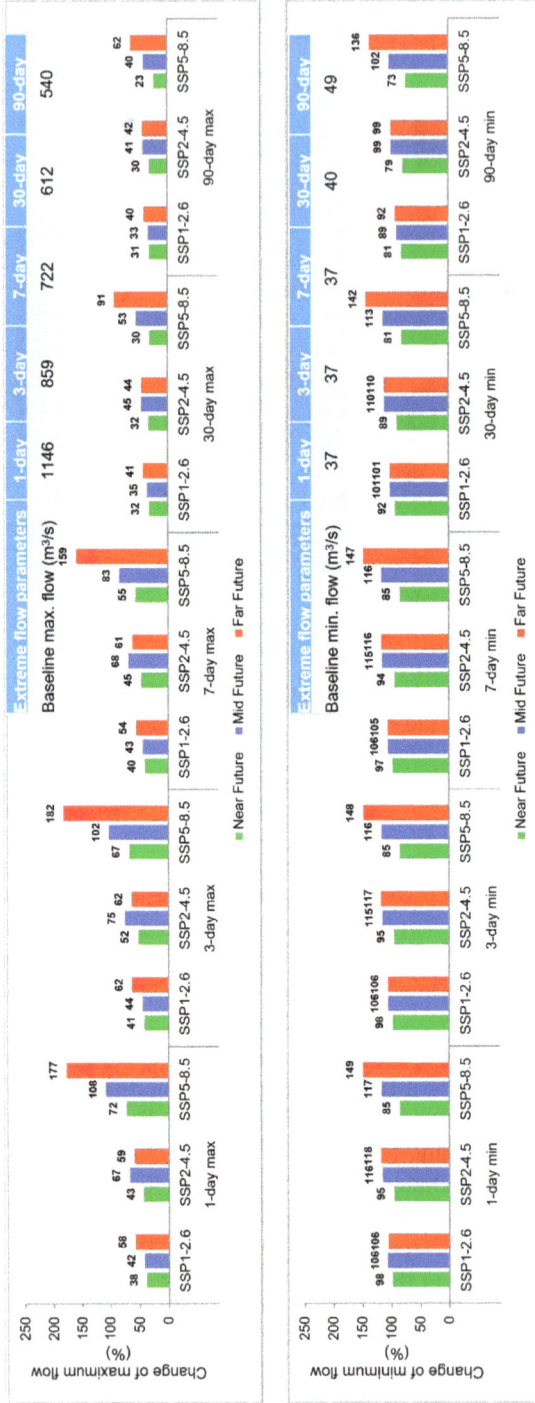

FIGURE 7.9   Projected future maximum (top) and minimum (bottom) flow (%) in the MRB under SSP1-2.6, SSP2-4.5, and SSP5-8.5 scenarios compared to the baseline period (1982–2011).

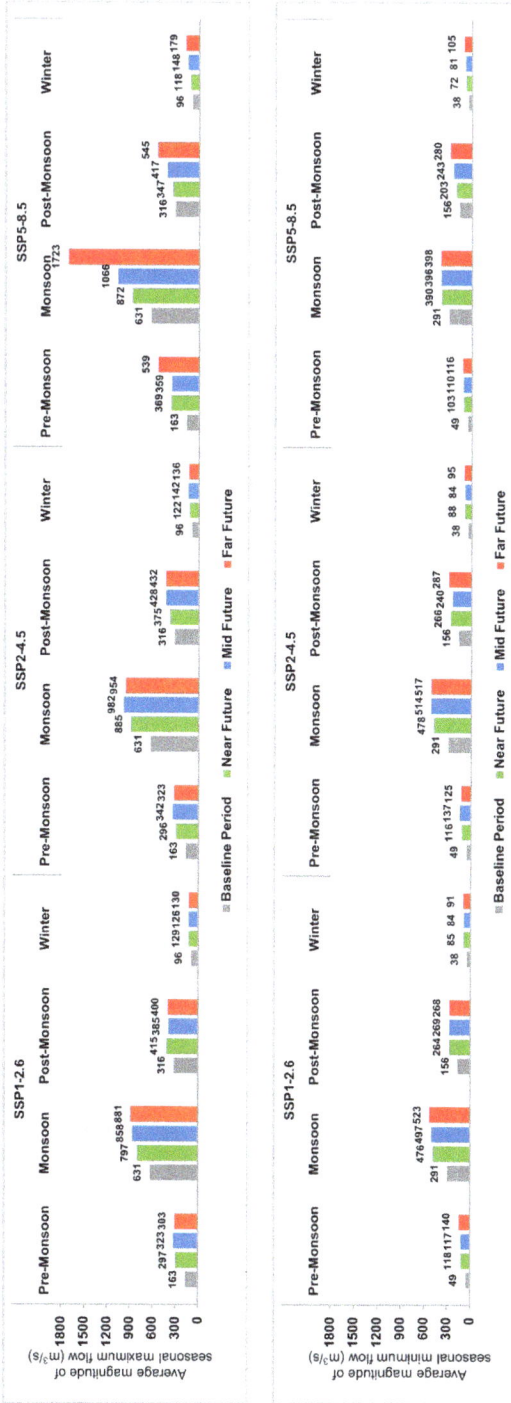

FIGURE 7.10 Projected future seasonal maximum (top) and minimum (bottom) flow (%) in the MRB under SSP1-2.6, SSP2-4.5, and SSP5-8.5 scenarios compared to the baseline period (1982–2011).

the average magnitude of both hydrological extremes, in terms of seasonal river discharge is expected to rise for all seasons under each future climate scenario and future period. In the case of maximum seasonal discharge, the highest increment by around 1092 m³/s is anticipated for the monsoon season under SSP5-8.5 in the far future, whereas the winter season is expected to experience the least augmentation of around 34 m³/s under SSP1-2.6 in near future. Similarly, the minimum average magnitude of streamflow at basin outlet on the seasonal basis implies a significant increment of flow for all seasons with the utmost increment in monsoon season while the least increment in winter is projected under all future climate scenarios. The average value of the extreme hydrology at the basin outlet is estimated to be around twice for each projected climate scenario than that of the baseline period.

7.5 CONCLUSION

This study investigates the evolution of hydrological extremes in the future under changed climatic conditions accounting for three climate variables: daily maximum temperature, minimum temperature, and rainfall obtained from ground stations and CMIP6 GCMs: ACCESS-CM2, BCC-CSM2-MR, and EC-Earth3. The climate projection was considered for three future periods: NF (2015-2044), MF (2045-2074), and FF (2076-2100) under three different SSPs: SSP1-2.6, SSP2-4.5, and SSP5-8.5. The MRB is expected to be warmer and wetter in the future under climate change. Both of the temperature variables are anticipated to rise by about 5°C under high–radiative forcing scenarios by 2100. As analogous to temperature, the rainfall is also estimated to gain a constant increment and hike up by about 731 mm, which is half the average annual rainfall during the baseline period (1982–2011). In terms of seasonal evaluation, all seasons excluding winter are expected to receive more rainfall. Changes in both the temperatures and rainfall are varying spatially and temporally throughout the basin where the southern part of the basin is expected to grow warmer and wetter than the rest. Also, future river discharge (average and extremes) is expected to increase at the basin outlet due to changes in climatic conditions. The average annual discharge is estimated to rise by 45% to 53%, whereas the increment in the magnitude of both maximum (20–182%) and minimum (73–149%) flow at the basin outlet is also anticipated under projected future climate.

Such a change in the spatiotemporal distribution of temperature and rainfall is likely to influence the water balance within the study basin, hence increasing challenges for water resources management. The escalating temperature and changes in precipitation patterns directly impact the rate of snow- and glacier melt in the basin. This triggers a change in the overall hydrology of the basin and alters the spatiotemporal water availability within the basin. Also, changes in hydro-climatic conditions are likely to affect the local residents, riverine ecosystem, and planned and commissioned water resources projects, along with other interconnected stakeholders. The

anticipated increase in magnitude of extreme hydrology, the maximum and minimum flow is going to increase the probability of floods in the wet season. In contrast, it also anticipates the availability of more water in the river as streamflow even the future rainfall is expected to decrease in the dry period in the future climatic scenario. These results from this study can be useful for researchers, policymakers, and stakeholders to explore hydro-climatic prospects in the HKH region, develop water resource management policies, disaster mitigation and preparedness plans, and water resource allocation policies for the study basin or other similar basins in the HKH region.

ACKNOWLEDGMENTS

The authors would like to express gratitude to Asian Institute of Technology (AIT), Thailand for the financial support and the Department of Hydrology and Meteorology (DHM), Nepal for providing several data sets.

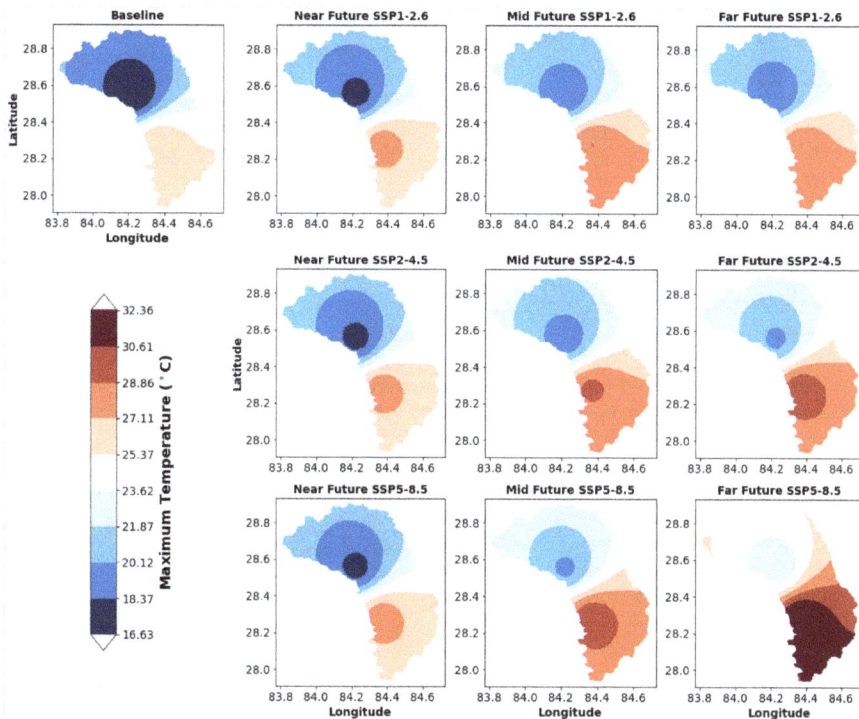

SUPPLEMENTARY FIGURE S7.1   Spatial distribution of future maximum temperature in the MRB under SSP1-2.6, SSP2-4.5, and SSP5-8.5 scenarios compared to the baseline period (1982–2011).

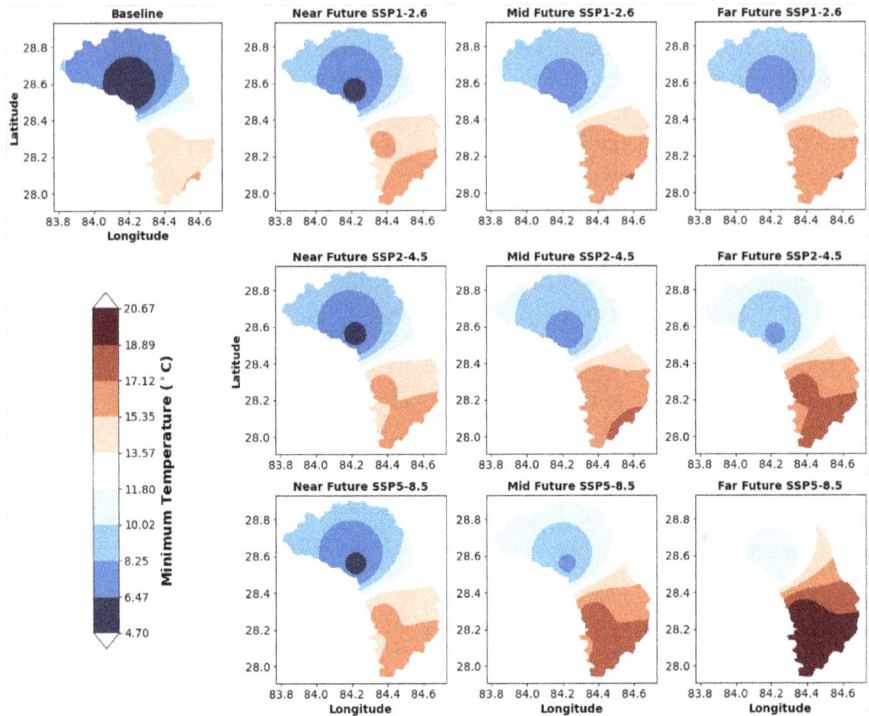

SUPPLEMENTARY FIGURE S7.2   Spatial distribution of future minimum temperature in the MRB under SSP1-2.6, SSP2-4.5, and SSP5-8.5 scenarios compared to the baseline period (1982–2011).

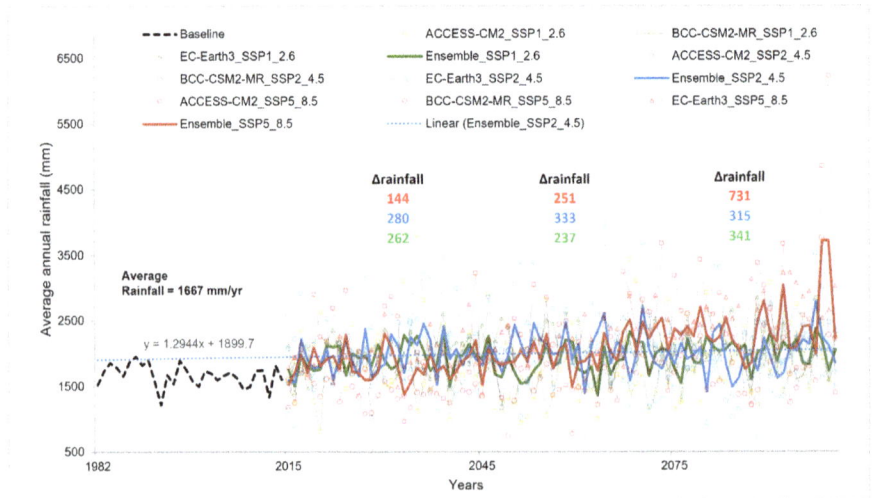

SUPPLEMENTARY FIGURE S7.3   Projected future average annual rainfall in the MRB under SSP1-2.6, SSP2-4.5, and SSP5-8.5 scenarios compared to the baseline period (1982–2011).

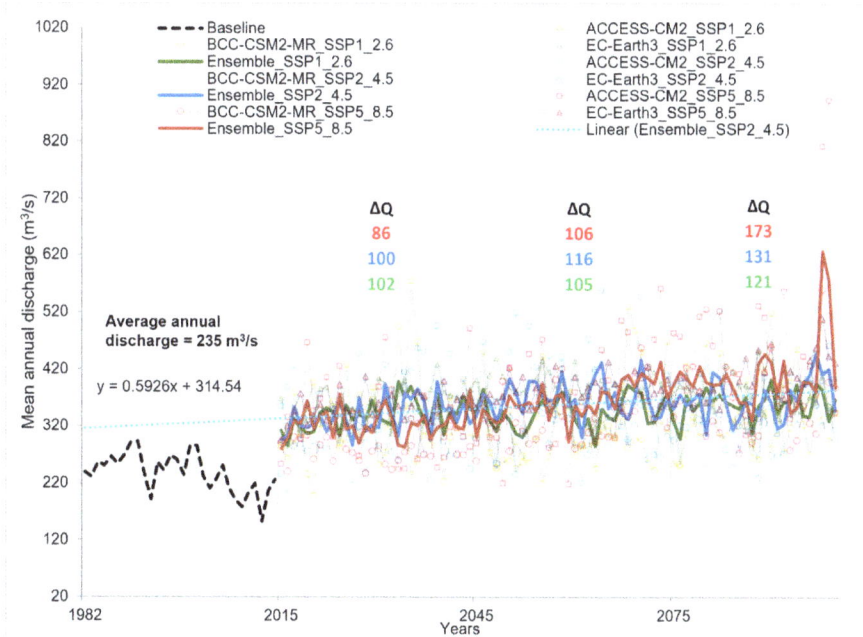

SUPPLEMENTARY FIGURE S7.4 Projected future average annual discharge at the MRB outlet under SSP1-2.6, SSP2-4.5, and SSP5-8.5 scenarios compared to the baseline period (1982–2011).

SUPPLEMENTARY TABLE S7.1

Description of the SWAT Model Parameters Used for the Study

S.N.	Parameters	Description
1	r__CN2.mgt	SCS runoff curve number
2	v__SMTMP.bsn	Snowmelt base temperature
3	v__ALPHA_BF.gw	Base flow alpha factor
4	v__REVAPMN.gw	Threshold depth of water in the shallow aquifer for "revap" to occur
5	v__GW_REVAP.gw	Groundwater "revap" coefficient
6	v__TLAPS.sub	Temperature-lapse rate
7	v__SFTMP.bsn	Snowfall temperature
8	v__SNOCOVMX.bsn	Snow water content that corresponds to 100% snow cover
9	v__GW_DELAY.gw	Groundwater delay
10	v__ESCO.hru	Soil evaporation compensation factor
11	v__PLAPS.sub	Precipitation lapse rate
12	v__SMFMX.bsn	Maximum melt rate for snow during year (summer solstice)
13	v__SMFMN.bsn	Minimum melt rate for snow during year (winter solstice)
14	v__TIMP.bsn	Snowpack temperature lag factor
15	v__LAT_TIME.hru	Lateral flow travel time
16	v__EPCO.hru	Plant uptake compensation factor
17	v__SNO50COV.bsn	Snow water equivalent that corresponds to 50% snow cover
18	v__GWQMN.gw	Threshold depth of water in the shallow aquifer required for return flow to occur

REFERENCES

Abbaspour, K C, C A Johnson, and M Th. van Genuchten. 2004. "Estimating Uncertain Flow and Transport Parameters Using a Sequential Uncertainty Fitting Procedure." *Vadose Zone Journal* 3 (4): 1340–1352. https://doi.org/https://doi.org/10.2136/vzj2004.1340.

Adnan, Muhammad, Ghulam Nabi, Muhammad Saleem Poomee, and Arshad Ashraf. 2017. "Snowmelt Runoff Prediction under Changing Climate in the Himalayan Cryosphere: A Case of Gilgit River Basin." *Geoscience Frontiers* 8 (5): 941–949. https://doi.org/10.1016/j.gsf.2016.08.008.

Almazroui, Mansour, Sajjad Saeed, Fahad Saeed, M Nazrul Islam, and Muhammad Ismail. 2020. "Projections of Precipitation and Temperature over the South Asian Countries in CMIP6." *Earth Systems and Environment* 4 (2): 297–320. https://doi.org/10.1007/s41748-020-00157-7.

Arnold, J G, D N Moriasi, P W Gassman, K C Abbaspour, M J White, R Srinivasan, C Santhi, et al. 2012. "SWAT: Model Use, Calibration, and Validation." *Transactions of the ASABE* 55 (4): 1491–1508. https://doi.org/https://doi.org/10.13031/2013.42256.

Arnold, J G, R Srinivasan, R S Muttiah, and J R Williams. 1998. "Large Area Hydrologic Modeling and Assessment Part I: Model Development." *JAWRA Journal of the American Water Resources Association* 34 (1): 73–89. https://doi.org/10.1111/j.1752-1688.1998.tb05961.x.

Azmat, Muhammad, Muhammad Uzair Qamar, Christian Huggel, and Ejaz Hussain. 2018. "Future Climate and Cryosphere Impacts on the Hydrology of a Scarcely Gauged Catchment on the Jhelum River Basin, Northern Pakistan." *Science of The Total Environment* 639: 961–976. https://doi.org/https://doi.org/10.1016/j.scitotenv.2018.05.206.

Baig, Muzaffar Ali, Qamruz Zaman, Shams Ali Baig, Muhammad Qasim, Umair Khalil, Sajjad Ahmad Khan, Muhammad Ismail, Sher Muhammad, and Shaukat Ali. 2021. "Regression Analysis of Hydro-Meteorological Variables for Climate Change Prediction: A Case Study of Chitral Basin, Hindukush Region." *Science of The Total Environment* 793: 148595. https://doi.org/https://doi.org/10.1016/j.scitotenv.2021.148595.

Bajracharya, Ajay Ratna, Sagar Ratna Bajracharya, Arun Bhakta Shrestha, and Sudan Bikash Maharjan. 2018. "Climate Change Impact Assessment on the Hydrological Regime of the Kaligandaki Basin, Nepal." *Science of the Total Environment* 625: 837–848. https://doi.org/10.1016/j.scitotenv.2017.12.332.

Bharati, L, P Gurung, L Maharjan, and U Bhattarai. 2016. "Past and Future Variability in the Hydrological Regime of the Koshi Basin, Nepal." *Hydrological Sciences Journal* 61 (1): 79–93. https://doi.org/10.1080/02626667.2014.952639.

Bieger, Katrin, Jeffrey G Arnold, Hendrik Rathjens, Michael J White, David D Bosch, Peter M Allen, Martin Volk, and Raghavan Srinivasan. 2017. "Introduction to SWAT+, A Completely Restructured Version of the Soil and Water Assessment Tool." *JAWRA Journal of the American Water Resources Association* 53 (1): 115–130. https://doi.org/https://doi.org/10.1111/1752-1688.12482.

Budhathoki, Aakanchya, Mukand S Babel, Sangam Shrestha, Gunter Meon, and Ambili G Kamalamma. 2021. "Climate Change Impact on Water Balance and Hydrological Extremes in Different Physiographic Regions of the West Seti River Basin, Nepal." *Ecohydrology & Hydrobiology* 21 (1): 79–95. https://doi.org/https://doi.org/10.1016/j.ecohyd.2020.07.001.

Chen, Hui, Hailong Liu, Xi Chen, and Yina Qiao. 2020. "Analysis on Impacts of Hydro-Climatic Changes and Human Activities on Available Water Changes in Central Asia." *Science of the Total Environment* 737: 139779. https://doi.org/10.1016/j.scitotenv.2020.139779.

DHM. 2015. "Study of Climate and Climatic Variation over Nepal, Technical Report," no. June. www.dhm.gov.np/uploads/climatic/1407411953ClimateandClimaticvariability.pdf.

Fontaine, T A, T S Cruickshank, J G Arnold, and R H Hotchkiss. 2002. "Development of a Snowfall–Snowmelt Routine for Mountainous Terrain for the Soil Water Assessment Tool (SWAT)." *Journal of Hydrology* 262 (1): 209–223. https://doi.org/https://doi.org/10.1016/S0022-1694(02)00029-X.

Group, Mountain Research Initiative Edw Working, N Pepin, R S Bradley, H F Diaz, M Baraer, E B Caceres, N Forsythe, et al. 2015. "Elevation-Dependent Warming in Mountain Regions of the World." *Nature Climate Change* 5 (May): 424–430. https://doi.org/10.1038/nclimate2563.

Hartman, Melannie D, Jill S Baron, Richard B Lammers, Donald W Cline, Larry E Band, Glen E Liston, and Christina Tague. 1999. "Simulations of Snow Distribution and Hydrology in a Mountain Basin." *Water Resources Research* 35 (5): 1587–1603. https://doi.org/10.1029/1998WR900096.

IPCC. 2007. *Climate Change 2007—The Physical Science Basis: Working Group I Contribution to the Fourth Assessment Report of the IPCC.* Assessment Report (Intergovernmental Panel on Climate Change).: Working Group. Cambridge University Press. https://books.google.co.th/books?id=8-m8nXB8GB4C.

IPCC. 2021. "Climate Change 2021: The Physical Science Basis. Contribution of Working Group I to the Sixth Assessment Report of the Intergovernmental Panel on Climate Change [Masson-Delmotte, V, P Zhai, A Pirani, S L Connors, C Péan, S Berger, N Caud, Y Chen]." Cambridge University Press, no. 3949, In Press. www.ipcc.ch/report/ar6/wg1/downloads/report/IPCC_AR6_WGI_Full_Report.pdf.

Kang, Shichang, Yanwei Xu, Qinglong You, Wolfgang-Albert Flügel, Nick Pepin, and Tandong Yao. 2010. "Review of Climate and Cryospheric Change in the Tibetan Plateau." *Environmental Research Letters* 5 (1): 15101. https://doi.org/10.1088/1748-9326/5/1/015101.

Mishra, Vimal, Udit Bhatia, and Amar Deep Tiwari. 2020. "Bias-Corrected Climate Projections for South Asia from Coupled Model Intercomparison Project-6." *Scientific Data* 7 (1): 1–13. https://doi.org/10.1038/s41597-020-00681-1.

MoHA/GoN. 2017. "Disaster Risk Reduction in Nepal: Status, Achievements, Challenges and Ways Forward." *National Position Paper for the Global Platform on Disaster Risk Reduction 22–26 May 2017*, Cancun, Mexico, no. May: 1–9. http://drrportal.gov.np/uploads/document/892.pdf.

Mondal, Sanjit Kumar, Jinlong Huang, Yanjun Wang, Buda Su, Jianqing Zhai, Hui Tao, Guojie Wang, Thomas Fischer, Shanshan Wen, and Tong Jiang. 2021. "Doubling of the Population Exposed to Drought over South Asia: CMIP6 Multi-Model-Based Analysis." *Science of The Total Environment* 771: 145186. https://doi.org/10.1016/j.scitotenv.2021.145186.

Moriasi, D N, J G Arnold, M W Van Liew, R L Bingner, R D Harmel, and T L Veith. 2007. "Model Evaluation Guidelines for Systematic Quantification of Accuracy in Watershed Simulations." *Transactions of the ASABE* 50 (3): 885–900. www.scopus.com/inward/record.uri?eid=2-s2.0-34447500396&partnerID=40&md5=50b5724614f28257edef46d43db96018.

Nash, J E, and J V Sutcliffe. 1970. "River Flow Forecasting through Conceptual Models Part I – A Discussion of Principles." *Journal of Hydrology* 10 (3): 282–290. https://doi.org/https://doi.org/10.1016/0022-1694(70)90255-6.

Parajuli, Achut, Lochan Prasad Devkota, Thirtha Raj Adhikari, Susmita Dhakal, and Rijan Bhakta Kayastha. 2016. "Impact of Climate Change on River Discharge and Rainfall Pattern: A Case Study from Marshyangdi River Basin, Nepal." *Journal of Hydrology and Meteorology* 9 (1): 60–73. https://doi.org/10.3126/jhm.v9i1.15582.

Richter, Brian D, Jeffrey V Baumgartner, Jennifer Powell, and David P Braun. 1996. "A Method for Assessing Hydrologic Alteration within Ecosystems." *Conservation Biology* 10 (4): 1163–1174. https://doi.org/https://doi.org/10.1046/j.1523-1739.1996.10041163.x.

Richter, Brian D, Jeffrey V Baumgartner, Robert Wigington, and David P Braun. 1997. "How Much Water Does a River Need?" *Freshwater Biology* 37 (1): 231–249. https://doi.org/https://doi.org/10.1046/j.1365-2427.1997.00153.x.

Santhi, C, J G Arnold, J R Williams, W A Dugas, R Srinivasan, and L M Hauck. 2001. "Validation of the SWAT Model on a Large Rwer Basin with Point and Nonpoint Sources." *JAWRA Journal of the American Water Resources Association* 37 (5): 1169–1188. https://doi.org/10.1111/j.1752-1688.2001.tb03630.x.

Sarıkaya, Mehmet A, Michael P Bishop, John F Shroder, and Ghazanfar Ali. 2013. "Remote-Sensing Assessment of Glacier Fluctuations in the Hindu Raj, Pakistan." *International Journal of Remote Sensing* 34 (11): 3968–3985. https://doi.org/10.1080/01431161.2013. 770580.

Shrestha, Manish, Suwash Chandra Acharya, and Pallav Kumar Shrestha. 2017. "Bias Correction of Climate Models for Hydrological Modelling—Are Simple Methods Still Useful?" *Meteorological Applications* 24 (3): 531–539. https://doi.org/10.1002/met.1655.

Shrestha, Sangam, Manish Shrestha, and Pallav Kumar Shrestha. 2017. "Evaluation of the SWAT Model Performance for Simulating River Discharge in the Himalayan and Tropical Basins of Asia." *Hydrology Research* 49 (3): 846–860. https://doi.org/10.2166/ nh.2017.189.

Tabari, Hossein. 2021. "Extreme Value Analysis Dilemma for Climate Change Impact Assessment on Global Flood and Extreme Precipitation." *Journal of Hydrology* 593: 125932. https://doi.org/https://doi.org/10.1016/j.jhydrol.2020.125932.

Tan, Mou Leong, Philip W Gassman, Xiaoying Yang, and James Haywood. 2020. "A Review of SWAT Applications, Performance and Future Needs for Simulation of Hydro-Climatic Extremes." *Advances in Water Resources* 143 (June): 103662. https://doi. org/10.1016/j.advwatres.2020.103662.

Vijai, Gupta Hoshin, Sorooshian Soroosh, and Yapo Patrice Ogou. 1999. "Status of Automatic Calibration for Hydrologic Models: Comparison with Multilevel Expert Calibration." *Journal of Hydrologic Engineering* 4 (2): 135–143. https://doi.org/10.1061/ (ASCE)1084-0699(1999)4:2(135).

WACREP, Water and Climate Resilience Program. 2019. "State of Conflict on Water Resources and Benefit Sharing in Marsyangdi River Basin." Kathmandu.

Yao, Tandong, Lonnie Thompson, Wei Yang, Wusheng Yu, Yang Gao, Xuejun Guo, Xiaoxin Yang, et al. 2012. "Different Glacier Status with Atmospheric Circulations in Tibetan Plateau and Surroundings." *Nature Climate Change* 2 (9): 663–667. https://doi. org/10.1038/nclimate1580.

8 Climate-Change Projections in the Himalayan River Basin Using CMIP6 GCMs
A Case Study in the Koshi River Basin, Nepal

Pragya Pradhan* and Sangam Shrestha
Water Engineering and Management, School of
Engineering and Technology, Asian Institute of
Technology, Pathum Thani 12120, Thailand

CONTENTS

8.1 Introduction 138
8.2 Study Area: The KRB 139
8.3 Materials and Methods 139
    8.3.1 Data Description 139
    8.3.2 SSP Scenarios and GCMs 141
    8.3.3 Methods 142
        8.3.3.1 Bias Correction of GCM Data 142
8.4 Results...143
    8.4.1 Performance of Bias Correction for GCM Data.............................143
    8.4.2 Future Rainfall Projection 146
    8.4.3 Future Temperature Projection 150
8.5 Discussions 157
    8.5.1 Climate-Change Impact on Rainfall and Temperature of
        Different Regions of the KRB 157
    8.5.2 Policy Implications for Water Resources Management....................158
8.6 Conclusion 158
Acknowledgment 161
References...161

DOI: 10.1201/9781003265160-10

8.1 INTRODUCTION

The Himalayan mountains are storage of freshwater resources. Monsoon rainfall is the major source of water for people living in the region for their agricultural production, hydropower generation, and daily household activities (Buytaert et al., 2010; Rajbhandari et al., 2016). The sustainability of the water resources in the Himalayas region is affected by climate change. Climate change has become a key factor in the ongoing environmental changes experienced by the High Mountain region (Wester et al., 2019). According to the sixth assessment report (AR6), of Intergovernmental Panel on Climate Change (IPCC), warming has occurred in the Himalayas, the Swiss Alps, and the Central Andes and has risen with altitude. Climate change influences the climate extremes, and it is expected to affect the mountainous river basin's water resources and biodiversity, especially at higher elevations with higher global warming (Thapa et al., 2020). It was reported that the warming over High Mountain Asia is higher by 11% compared to the northern hemisphere continental surfaces (Lalande et al., 2021).

Climate change and its impact on climate variables such as rainfall, maximum temperature (T_{max}) and minimum temperature (T_{min}) of the Himalayan region are prominent and visible. Some studies estimated that the warming rate of 0.16°C/decade had increased to 0.32°C/decade in winter season, which causes the snow, glacier, and ice to melt, leading to an increase in meltwater (Wang et al., 2008). Snow-cover areas will decrease during the 21st century in most of the region ranging from the European Alps, the Rocky Mountains, the Scandinavian Mountains to High Mountain Asia (IPCC, 2021). Mishra (2020) reported that future climate of South Asia in the 21st century will be wetter by 13–30%. Karim (2020) recently found that in Pakistan, the minimum temperatures in winters are rising rapidly by 0.17–0.37°C/decade compared to summer temperature. The extreme rainfall events increased by 25–80% is predicted all over the India in coming future periods (Gupta et al., 2020). Talchabhadel and Karki (2019) estimated the heavy rainfall extremes all over Nepal, which can instigate different natural calamities and can affect agricultural productivity. In the Himalayan region, there is a greater likelihood of more frequent natural calamities, such as flash floods, landslides, and cattle diseases (You et al., 2017). The central Himalayan region is not only expected to experience severe drought conditions but also floods in early future periods (Sharma et al., 2021). The climate change impact studies mainly use different climate models such as regional climate models (RCMs) and general circulation models (GCMs). Different GCMs and RCMs project the future climate of Himalayan River basins (Lutz et al., 2016; Kaini et al., 2019; Chettri et al., 2021). Guo et al. (2018) used the multiple RCMs to simulate the climate over Tibetan Plateau and projected that the warming in the region will increase by 1.38°C. Similarly, the future rainfall (up to 2100) of Upper Indus Basin was projected using GCMs and future rainfall will increase by 2.2% in the dry season and 15.9% in the wet season (Khan and Koch, 2018). However, selecting suitable GCMs/RCMs among many GCMs/RCMs is very challenging. Kaini (2019) evaluates Coupled Model Intercomparison Project Phase 5 (CMIP5) based GCMs for Koshi River Basin (KRB) Nepal using an advanced envelope-based selection approach and projected the variability of rainfall and temperatures under the two future scenarios representative concentration pathways (RCPs) RCP 4.5 and RCP 8.5. Similarly, Pradhan (2021) evaluated the 12 GCMs from CMIP5 for the Koshi River Basin, Nepal, based on historical observation and compromise programming.

The previous studies mainly used CMIP5 models in the KRB (Khadka et al., 2014; Agarwal et al., 2016; Rajbhandari et al., 2016). Recently, the CMIP6 model was released, which is a revised and improved climate model. There are major bio-geochemical and physical climate systems in CMIP6 with various parameterization schemes (Eyring et al., 2016). As new sets of data have been released recently, only a few studies have used the CMIP6 model outputs for the projection of the future climate of Nepal and very few in the Himalayan River Basin of Nepal. There is a broad range of improvements and changes from the CMIP5 ensemble to CMIP6 (Wu et al., 2019; Gusain et al., 2020). One of the main improvements is using a new set of scenarios based on socioeconomic assumptions and radiative forcing pathways, which are considered more realistic future scenarios (Cook et al., 2020; Narsey et al., 2020; Shrestha et al., 2020). Climate is changing rapidly and experiencing the extremes frequently; there should be the use of a new CMIP6 model, which consist of updated emission pathways in regional and global scales that discover a more comprehensive range of possible future outcomes (Almazroui et al., 2020; Alaminie et al., 2021).

This study focuses on the climate variability over Koshi River Basin, simulated with CMIP6 models under different Shared Socioeconomic Pathways (SSPs). The study's main objective is to conduct the climate change analysis of the climate variables, that is, rainfall and temperature, to analyze the past and future climate trends. This study also describes the future projection and its changes in seasonal and annual mean temperature and rainfall over KRB, Nepal. Moreover, the impact of climate change on the low-lying plains area, the middle mountains, and high Himalayas region are discussed in the study. Based on the results of this study obtained from the CMIP6 climate models in the Hindu Kush Himalaya region, some reforms are required for the existing policies.

8.2 STUDY AREA: THE KRB

The KRB is a transboundary river basin that passes through China (43%), Nepal (42%), and India (15%) with a total catchment area of 74,030 km² (FMIS, 2012) (Figure 8.1). The basin ranges from 85°22′–88°21′ E and 26°47′–29°12′ N, and it has wide elevation, ranging from 113 meter above sea level (m.a.s.l) in the south to 8848 m.a.s.l in the north (Khadka et al., 2020). The KRB is a physio-graphically dominated river basin, and it is divided into three different regions from the high Himalayas to the middle mountains to the low-lying plains area (Rajbhandari et al., 2016). This basin experiences four different kinds of climatic seasons like pre-monsoon, monsoon, post-monsoon, and winter (Bharati et al., 2014). The annual rainfall of this basin ranges from 32.5 mm to 925.25 mm, of which 80% is monsoon rainfall. The rainfall in monsoon is high-intensity rainfall, which causes different extreme hydrological events like floods and landslides (Agarwal et al., 2016). The basin experiences a wide annual temperature range from −3°C to 28°C. Additionally, the basin is dominated by snow and glaciers that contribute to the runoff of the major rivers (Immerzeel et al., 2009).

8.3 MATERIALS AND METHODS

8.3.1 Data Description

The Department of Hydrology and Meteorology (DHM), Nepal, provided the meteorological data for this study. The study was carried out using the data acquired from

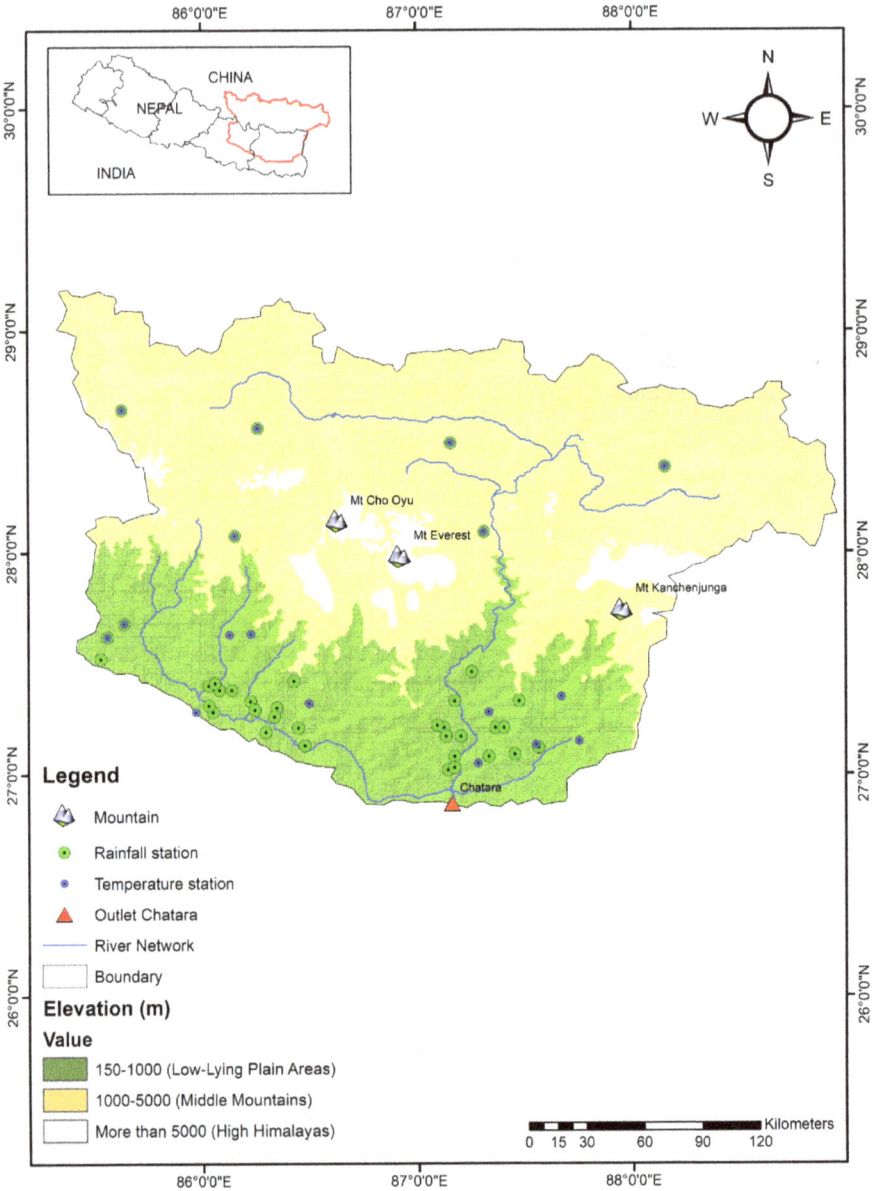

FIGURE 8.1 Location map and the hydro-meteorological stations in the KRB, Nepal (Pradhan et al., 2021).

the 17 temperature stations and 38 rainfall stations located in the middle mountains and low-lying plains area of the KRB. Observed data from 1980 to 2015 is taken up for climate impact studies, but there is always a scarcity of data in the Himalayan region (Agarwal et al., 2016). To address this scarcity in this study, we used gridded data

TABLE 8.1

Summary of Data and Corresponding Source

Data	Duration	Resolution (temporal/spatial)	Source
Meteorological Data (rainfall and temperature)	1980–2015	Point/Daily	(DHM), Nepal
APHRODITE	1980–2015	0.25° × 0.25°/Daily	Asian Precipitation—Highly Resolved Observational Data Integration Toward Evaluation (http://aphrodite.st.hirosaki-u.ac.jp/download/)
CPC_NOAA	1979–2015	0.5° × 0.5°/Daily	Climate Prediction Centre (CPC) (/www.cpc.ncep.noaa.gov/)

sets such as Climate Prediction Centre (CPC) (CPC_NOAA) temperature data (0.5° × 0.5°) and Asian Precipitation—Highly Resolved Observational Data Integration Towards Evaluation (APHRODITE) precipitation data (0.25° × 0.25°) to create six rainfall and temperature station for the ungauged area of the Himalayan region of KRB and used as observational data (Table 8.1). These gridded data sets are mainly chosen for this study because these gridded data sets mainly used the observed station data from all over Nepal and generated reliable rainfall and temperature data for all over Nepal and the Himalayan region (Duncan et al., 2012).

8.3.2 SSP SCENARIOS AND GCMS

The IPCC released the latest set of scenarios known as SSPs, which guide global socioeconomic development trends. It provides a worldwide perspective for regional and global climate change studies. Five distinct shared socioeconomic pathways represent possible future states of society and the environment. SSPs are built on five narratives: sustainable development, inequality, fossil-fueled development, regional rivalry, and middle-of-the-road development (Rihai et al., 2017). Among five SSPs, in this study, two SSPs, that is, SSP2, known as the middle of the road, and SSP5, known as fossil-fueled development, were used. The two SSPs-based scenarios used in this study for the climate projection are SSP2-4.5 and SSP5-8.5, representing medium and high emissions of GHGs, respectively. These two scenarios are chosen because it represents the medium and high range of future forcing pathways with a radiative forcing of 4.5 w/m² and 8.5 w/m² in 2100, respectively (O'Neill et al., 2016).

In this study, four different CMIP6 GCMs are used from four different institutes. The GCMs models were selected because most of the studies carried out in the region have utilized and reported better than others (Mishra et al., 2020; Chhetri et al., 2021; Pradhan et al., 2021). The summary of the models, the resolution, and their home institution are described in Table 8.2. Additional details can be found on the website of CMIP6 (https://esgf-index1.ceda.ac.uk/search/cmip6-ceda/).

TABLE 8.2

List of Climate Models Selected from CMIP6

Model	Resolution (lat × lon)	Institute
BCC-CSM2-MR	1.1° × 1.1°	Beijing Climate Center (BCC)
CNRM-CM6	1.4° × 1.4°	Centre National de Recherches Meteorologiques (France)
CanESM5	2.8° × 2.8°	Canadian Centre for Climate Modelling and Analysis, Canada
IPSL-CM6A-LR	1.27° × 2.5°	Institute Pierre-Simon Laplace (France)

8.3.3 METHODS

8.3.3.1 Bias Correction of GCM Data

Bias is the alteration between the estimated and the actual parameter values, and the change made to the predicted values to reflect the observed distribution and statistics is known as bias correction (Trzaska and Schnarr, 2014). Bias-correction methods are based on gamma-gamma transformation, which improves the frequency and amount of raw GCM precipitation at the grid nodes. There are different bias correction methods, namely, linear bias correction, quantile mapping, local intensity scaling, variance scaling, and delta-change method, among many more (Teutschbein and Seibert, 2012).

In this study, quantile mapping was used for bias correction of GCMs model because this method corrects the bias in climate data of selected GCMs since it decreases the bias in daily precipitation and temperature by roughly one order of magnitude (Jakob et al., 2011). This method is executed in the R languageusing the "qmap" package (Gudmundsson, 2014). For quantile mapping the following equations are applied:

$$P_{his}(d)^* = F_{obs,m}^{-1}\left\lfloor F_{his,m}\left(P_{his,m}\right)\right\rfloor$$

$$P_{sim}(d)^* = F_{sim,m}^{-1}\left\lfloor F_{sim,m}\left(P_{sim,m}\right)\right\rfloor$$

$$T_{his}(d)^* = F_{obs,m}^{-1}\left\lfloor F_{his,m}\left(T_{his,m}\right)\right\rfloor$$ (8.1)

$$T_{sim}(d)^* = F_{sim,m}^{-1}\left\lfloor F_{sim,m}\left(T_{sim,m}\right)\right\rfloor,$$

where T = temperature, P = precipitation, d = daily, m = monthly* bias corrected, *his* = Raw GCM data, *obs* = observed data, *sim* = raw GCM future data, F = cumulative distribution function (CDF), F^{-1} = inverse of the CDF.

For bias correction of the GCMs with observation data from ground stations and gridded observation data for ungauged stations, daily rainfall, and temperature data from 1980–2014 were used (Figure 8.2). With the climate-change scenarios, the projection of the future climate uses four CMIP6 GCMs: BCC-CSM2-MR, CanESM5, CNRM-CM6, and IPSL-CM6A-LR.

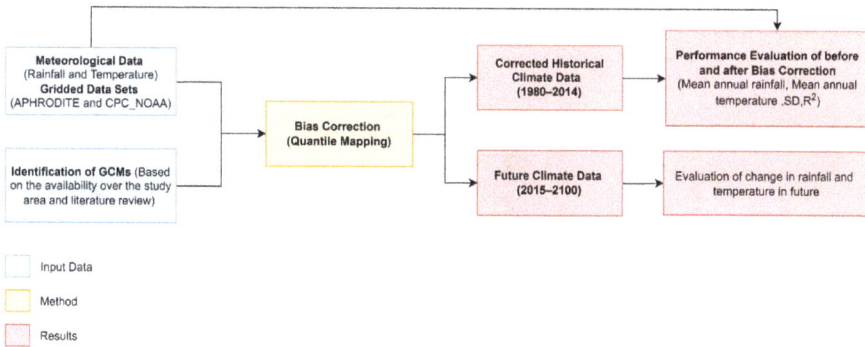

FIGURE 8.2   Research framework.

After bias correction, we obtained corrected historical data from 1980–2014 and future time-series climate data from 2015–2100, which is analyzed and compared with the observed climate data regarding different future periods, seasonal change, annual average, and SSP-RCP scenario change. The future time-series climate data is divided into three future period, namely, the near-future (2021–2045), mid-future (2046–2076), and far-future (2077–2100) periods. The standard deviation (σ), the coefficient of determination (R^2), the mean temperature, and the average annual rainfall are applied to evaluate corrected historical data from CMIP6 GCM compared to observed climate variables.

8.4   RESULTS

8.4.1   PERFORMANCE OF BIAS CORRECTION FOR GCM DATA

The four statistical analysis methods were used to evaluate bias correction performance: the standard deviation (σ), the coefficient of determination (R^2), the mean temperature, and the average annual rainfall. The performance was evaluated using daily data from 1980–2014. The results show the model performance has improved after the bias correction. The comparison of monthly observed and GCM data for rainfall, T_{max} and T_{min}, for the baseline period (Figure 8.3). The figure shows that the bias-corrected rainfall values from GCMs follow a similar monthly rainfall pattern for the low-lying plains region, but the CanESM5 model overestimated rainfall values. The maximum temperature and minimum temperature values of GCMs after bias correction perfectly correlated with observed data for low-lying plains regions. Similarly, after bias correction, the monthly rainfall values for the middle mountains and the high Himalayas are perfectly correlated with the observed data, whereas the minimum and maximum temperature values of the middle mountains are overestimated after bias correction. For the high Himalayas, temperature values are estimated with the same values as observed data (Supplementary Figures S8.1 and S8.2).

   The performance of the models has been improved after the bias correction for all three climate variables. The average annual rainfall and standard deviation of

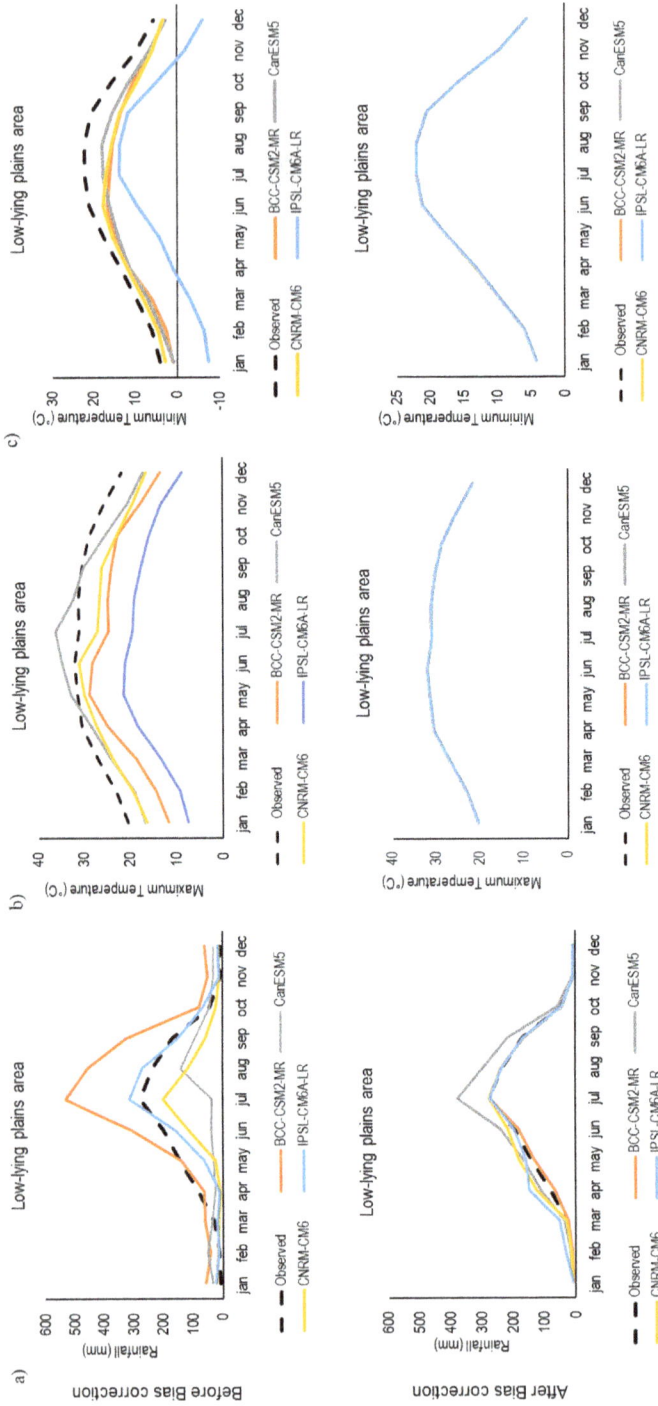

FIGURE 8.3   Performance of bias correction before and after of (a) rainfall, (b) maximum temperature, and (c) minimum temperature for low-lying plains area of the KRB, Nepal.

corrected GCMs are similar to the observed rainfall data for all gauging stations from 1980–2014 (Table 8.3). Similarly, the average maximum and minimum temperature of corrected GCMs are close to the observed minimum and maximum temperature of all temperature stations (Tables 8.4 and 8.5).

TABLE 8.3
Performance for Evaluation Bias Correction for Rainfall

Regions	GCMs	Before Bias Correction			After Bias Correction		
		R²	σ	Avg annual rainfall (mm)	R²	σ	Avg annual rainfall (mm)
Low-Lying Plains Area (1209 mm)	BCC-CSM2-MR	0.03	11.52	2157	0.04	7.59	1167
	CNRM-CM6	0.004	5.72	592	0.06	8.79	1550
	CanESM5	0.03	5.21	624	0.05	6.93	1315
	IPSL-CM6A-LR	0.08	5.72	1123	0.05	6.78	1343
Middle Mountains (1998 mm)	BCC-CSM2-MR	0.065	12.25	2167	0.14	8.82	2016
	CNRM-CM6	0.007	6.88	576	0.15	8.88	2145
	CanESM5	0.005	6.55	616	0.16	8.20	2180
	IPSL-CM6A-LR	0.12	7.75	1906	0.19	7.62	2225
High Himalayas (413 mm)	BCC-CSM2-MR	0.03	8.03	1740	0.06	2.11	417
	CNRM-CM6	0.001	2.46	355	0.06	2.10	418
	CanESM5	0.07	4.66	1573	0.07	2.07	415
	IPSL-CM6A-LR	0.05	6.23	1889	0.09	2.00	414

TABLE 8.4
Performance for Evaluation Bias Correction for Maximum Temperature

Regions	GCMs	Before Bias Correction			After Bias Correction		
		R²	σ	Avg Tmax (°C)	R²	σ	Avg Tmax (°C)
Low-Lying Plains Area (27.7°C)	BCC-CSM2-MR	0.47	6.75	21.1	0.49	4.81	27.7
	CNRM-CM6	0.50	7.52	26.5	0.50	4.82	27.8
	CanESM5	0.46	5.66	23.8	0.49	4.82	27.7
	IPSL-CM6A-LR	0.48	5.63	15.4	0.49	4.83	27.7
Middle Mountains (22.5°C)	BCC-CSM2-MR	0.52	6.58	18.1	0.49	4.51	25.0
	CNRM-CM6	0.52	6.61	22.8	0.49	4.54	24.5
	CanESM5	0.51	5.28	20.0	0.52	4.74	24.7
	IPSL-CM6A-LR	0.49	5.28	23.0	0.52	4.46	24.7
High Himalayas (7.5°C)	BCC-CSM2-MR	0.67	7.97	5.6	0.69	6.37	7.4
	CNRM-CM6	0.62	10.35	4.2	0.59	6.35	7.5
	CanESM5	0.69	8.70	−4.4	0.65	6.32	7.5
	IPSL-CM6A-LR	0.64	5.31	−5.3	0.66	6.27	7.4

TABLE 8.5

Performance for evaluation bias correction for minimum temperature

Regions	GCMs	Before Bias Correction			After Bias Correction		
		R²	σ	Avg Tmin (°C)	R²	σ	Avg Tmin (°C)
Low-Lying	BCC-CSM2-MR	0.74	5.90	9.8	0.80	6.93	13.9
Plains reas	CNRM-CM6	0.71	6.49	10.4	0.81	6.94	13.8
(14°C)	CanESM5	0.73	5.57	10.5	0.82	6.95	13.7
	IPSL-CM6A-LR	0.77	8.10	2.9	0.81	6.91	14
Middle	BCC-CSM2-MR	0.77	5.76	6.8	0.76	5.69	14.6
Mountains	CNRM-CM6	0.74	6.62	8.4	0.74	5.83	14.0
(12.4°C)	CanESM5	0.77	5.52	7.1	0.76	5.81	14.2
	IPSL-CM6A-LR	0.80	7.62	10.4	0.76	5.80	14.2
High Himalayas	BCC-CSM2-MR	0.73	9.58	−5.2	0.87	8.25	−6.7
(−6.7°C)	CNRM-CM6	0.69	12.98	−18.4	0.85	8.29	−6.8
	CanESM5	0.75	12.75	−17.4	0.86	8.28	−6.6
	IPSL-CM6A-LR	0.74	12.17	−23.1	0.87	8.22	−6.6

8.4.2 FUTURE RAINFALL PROJECTION

The predicted future rainfall in the KRB for all three regions, that is, low-lying plains area, the middle mountains, and high Himalayas under the SSP2-4.5 and SSP5-8.5 scenarios, is continually increasing (Figures 8.4, 8.5, and 8.6). The average rainfall for the low-lying plains area was 1210 ± 5.5mm; for the middle mountains, it was 1998 ± 7.54 mm; and for the high Himalayas, it was 413 ± 2.03mm from 1980–2014. The observed average annual rainfall was compared with projected rainfall from four CMIP6 GCMs, namely, the BCC-CSM2-MR, the CanESM5, the CNRM-CM6, and the IPSL-CM6A-LR, for the near future (2015–2045), the mid-future (2046–2076), and the far future (2077–2100) under the SSP2-4.5 and SSP5-8.5 scenarios. The projected future rainfall is increasing as expected. The CNRM-CM6 and the IPSL-CM6A-LR are the models with the highly distributed rainfall increasing annual average rainfall for all future periods in the SSP2-4.5 and in the SSP5-8.5 scenarios. The average annual rainfall is higher in the SSP5-8.5 than in the SSP2-4.5 because the average annual rainfall is expected to rise by more than 50% in all future periods. Previous studies on climate projection using RCP scenarios predicted that the average annual rainfall would increase for the KRB for all future periods (Rajbhandari et al., 2018; Kaini et al., 2019). Similarly, recent studies that used SSP scenarios also predicted that average annual rainfall would increase predominantly (Mishra et al., 2020; Chhetri et al., 2021).

Additionally, this study examined the changes in seasonal rainfall of the KRB. There are four seasons in Nepal as categorized by the DHM: pre-monsoon (March–May), monsoon (June–September), post-monsoon (October–November), and winter season (December–February). Past studies report that the future rainfall may increase or decrease depending on the season of the KRB (Rajbhandari et al., 2016; Rajbhandari et al., 2018; Kaini et al., 2019).

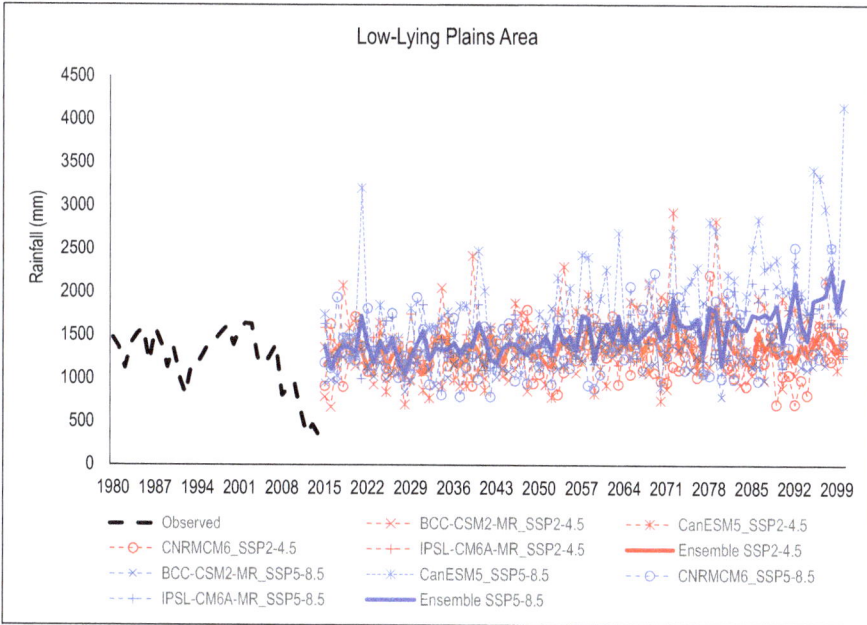

FIGURE 8.4 Projected annual rainfall under the SSP2-4.5 and SSP5-8.5 scenarios for the low-lying plains area of the KRB, Nepal.

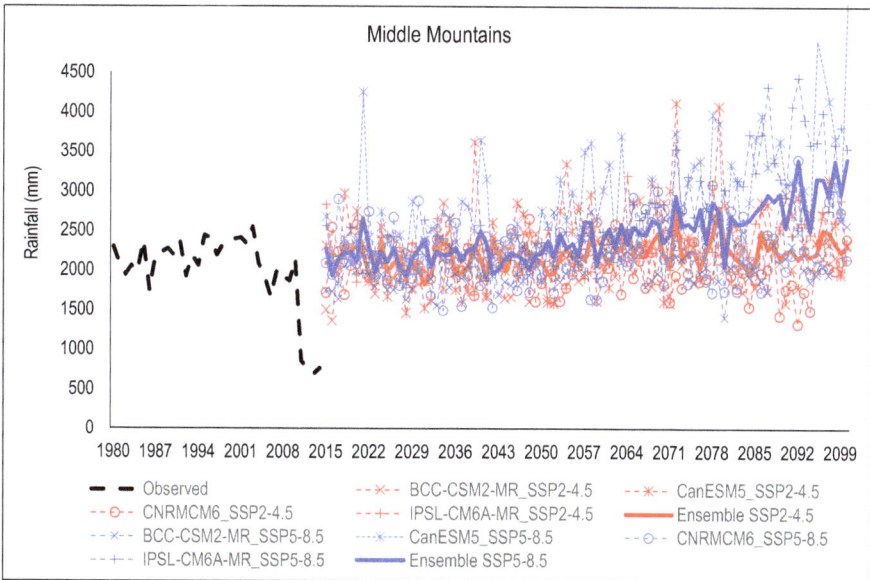

FIGURE 8.5 Projected annual rainfall under the SSP2-4.5 and SSP5-8.5 scenarios for the middle mountains of the KRB, Nepal.

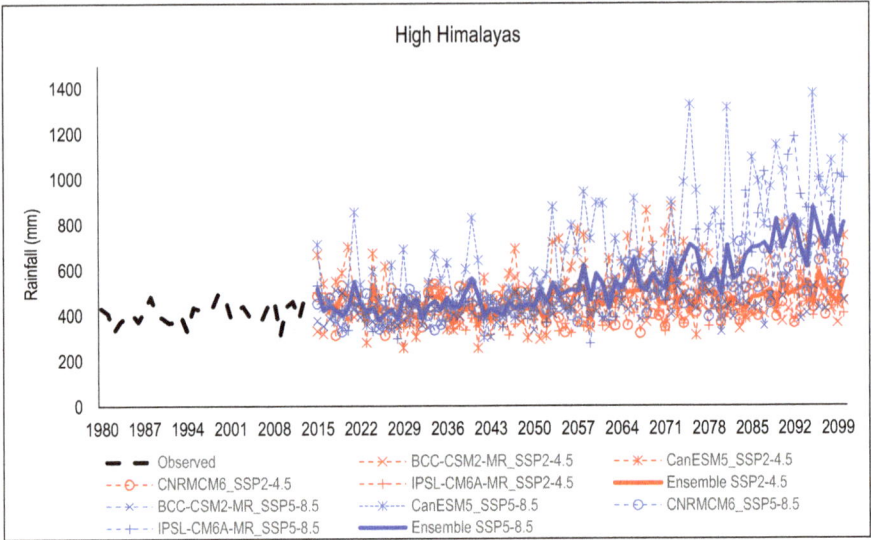

FIGURE 8.6   Projected annual rainfall under the SSP2-4.5 and SSP5-8.5 scenarios for the high Himalayas of the KRB, Nepal.

TABLE 8.6
Rainfall Projection under the SSP2-4.5 and SSP5-8.5 Scenarios for Three Future Periods

Region	Period	Rainfall (mm) SSP2-4.5	SSP5-8.5
Low-Lying Plains Area	Baseline (1980–2014)	1210±5.5	
	2020s (2015–2045)	1318 ± 4.89	1386 ± 5.03
	2050s (2046–2076)	1415 ± 5.08	1559 ± 5.72
	2080s (2077–2100)	1451 ± 5.14	1797 ± 6.65
Middle Mountains	Baseline (1980–2014)	1998 ± 7.54	
	2020s (2015–2045)	2189 ± 6.63	2265 ± 6.82
	2050s (2046–2076)	2299 ± 6.84	2494 ± 7.75
	2080s (2077–2100)	2379 ± 7.00	2955 ± 9.37
High Himalayas	Baseline (1980–2014)	413 ± 2.03	
	2020s (2015–2045)	444 ± 1.50	455 ± 1.52
	2050s (2046–2076)	485 ± 1.67	543 ± 1.84
	2080s (2077–2100)	506 ± 1.69	706 ± 2.42

The seasonal percentage change between observed rainfall and future rainfall under SSP2-4.5 and SSP5-8.5 scenarios are quite visible (Table 8.7). During the pre-monsoon season, the CNRMCM6 and IPSL-CM6A-MR models show that rainfall increases by more than 100% for the far-future period under the SSP2-4.5 SSP5-8.5 scenarios for both the low-lying plains area and the middle mountains. In contrast,

TABLE 8.7

Seasonal Percentage Change of Rainfall under the SSP2-4.5 and SSP5-8.5 Scenarios for Three Future Periods

Region	GCMs	Pre-Monsoon (%)						Monsoon (%)						Post-Monsoon (%)						Winter (%)					
		SSP2-4.5			SSP5-8.5			SSP2-4.5			SSP5-8.5			SSP2-4.5			SSP5-8.5			SSP2-4.5			SSP5-8.5		
		NF	MF	FF	NF	MF	FF	NF	MF	FF	NF	MF	FF	NF	MF	FF	NF	MF	FF	NF	MF	FF	NF	MF	FF
Low-Lying Plains Area	BCC-CSM2-MR	-1	17	13	-1	11	36	4	12	10	-5	18	19	-4	31	-3	25	28	54	-16	55	16	20	32	12
	CANESM5	-5	3	6	19	-1	24	12	39	47	39	65	112	43	28	123	46	106	282	-2	4	18	-15	-29	-49
	CNRMCM6	46	57	35	50	76	106	-1	-7	-1	4	5	10	-16	-8	-28	-6	-33	31	-5	-36	-38	3	3	-41
	IPSL-CM6A-MR	69	71	95	68	88	118	1	1	1	-3	13	23	3	9	16	0	-1	36	38	54	92	80	100	45
	Ensemble	**27**	**37**	**37**	**34**	**43**	**71**	**4**	**11**	**14**	**9**	**25**	**41**	**6**	**15**	**27**	**16**	**25**	**101**	**4**	**19**	**22**	**22**	**27**	**-8**
Middle Mountains	BCC-CSM2-MR	2	19	20	2	11	32	3	7	7	-4	12	13	1	10	-11	20	24	47	-15	38	8	9	20	6
	CANESM5	17	26	27	36	21	44	10	29	35	29	48	79	35	25	80	31	71	198	-2	0	15	-13	-21	-34
	CNRMCM6	48	52	37	50	67	86	1	-4	-1	4	5	6	-10	-7	-23	-6	-27	18	26	3	-7	28	30	-1
	IPSL-CM6A-MR	43	41	61	53	53	77	2	5	15	-2	20	76	41	55	52	38	34	99	124	122	160	138	149	119
	Ensemble	**28**	**35**	**36**	**35**	**38**	**60**	**4**	**9**	**14**	**7**	**21**	**44**	**17**	**21**	**25**	**20**	**25**	**90**	**33**	**41**	**44**	**41**	**44**	**22**
High Himalayas	BCC-CSM2-MR	-1	9	-7	4	4	32	4	12	14	-5	14	18	-11	50	4	7	-9	20	-10	35	0	11	24	11
	CANESM5	-7	20	35	11	33	67	14	44	39	37	77	107	43	77	92	1	153	219	30	37	67	2	33	38
	CNRMCM6	9	3	15	5	12	24	12	10	24	7	24	57	0	-17	-11	1	-22	36	32	12	13	25	63	23
	IPSL-CM6A-MR	0	-1	1	19	8	37	-1	1	13	-3	17	115	47	78	58	37	57	177	2	5	57	33	43	19
	Ensemble	**0**	**8**	**11**	**10**	**14**	**40**	**7**	**17**	**23**	**9**	**33**	**74**	**20**	**47**	**36**	**11**	**45**	**113**	**14**	**22**	**34**	**18**	**41**	**23**

for the far-future period under the SSP5-8.5 scenarios, the pre-monsoon rainfall of the high Himalayas region will increase by 20–60%.

In the monsoon season, the CanESM5 and IPSL-CM6A-MR are the models which show rainfall will increase by 20–40% for the far-future period under SSP2-4.5 and more than 100% for the far-future period under SSP5-8.5 scenarios. The monsoon rainfall will predominantly increase in the far-future period under the SSP5-8.5 scenario than in the near-future period for all three regions of the KRB. The monsoon rainfall will highly increase in the high Himalayas region as compared to other regions. In the post-monsoon season, the CanESM5 model projected the future rainfall would increase predominantly in the mid-future and far-future periods under the SSP5-8.5 scenario. At the same time, the BCC-CSM2-MR and CNRMCM6 models projected that post-monsoon rainfall would decrease by 10–25% for future periods under the SSP2-4.5 and SSP5-8.5 scenarios.

In the winter season, the IPSL-CM6A-MR is the model that shows that the rainfall will increase in the winter season by more than 80% for all future periods under the SSP2-4.5 and SSP5-8.5 scenarios for both the low-lying plains area and the middle mountains. For the high Himalayas region, the winter rainfall will increase by 20–50% for all future SSP2-4.5 and SSP5-8.5 scenarios.

Overall, ensemble results show that the pre-monsoon, monsoon, and winter seasons' rainfall is gradually increasing in all future periods for all three regions under both scenarios. The future rainfall is increasing with high percentage in far-future periods as compared to near and mid-future periods. The post-monsoon season is expecting heavy rainfall in all three regions during far-future period. The high Himalayas region is expecting heavy rainfall in both the monsoon season and post-monsoon season, which will increase the snowmelt and glacier melt rates, affecting the people living in the region.

8.4.3 FUTURE TEMPERATURE PROJECTION

The annual average baseline (1980–2014) maximum temperature (T_{max}) of KRB for three different regions was observed to be 27.7 ± 4.8°C for the low-lying plains area, 22.4 ± 4.2°C for the middle mountains, and 7.4 ± 6.06°C for the high Himalayas, and the minimum temperature (T_{min}) was observed to be 13.9 ± 6.9°C for the low-lying plains area, 12.2 ± 5.5°C for the middle mountains, and −6.7 ± 8.2°C for the high Himalayas, respectively. The projected future T_{max} and T_{min} are distributed to three different future periods, namely, the near future (2015–2045), the mid-future (2046–2076), and the far future (2077–2100) for all four CMIP6 GCMs under the SSP2-4.5 and SSP5-8.5 scenarios. Generally, both T_{max} and T_{min} are expected to increase throughout the KRB for all three future periods. The high Himalayas region is projected to go through more warming than other regions. The previous studies show that the basin will experience the elevation-dependent warming as warming will be higher with higher elevation (Gobiet et al., 2014; Pepin et al., 2015).

The future maximum temperature (T_{max}) is expected to increase by 1–5°C for both thlow-lying plains area and the middle mountains for the future periods under the SSP2-4.5 SSP5-8.5 scenarios. Similarly, for the Himalayan region, the T_{max} is expected to increase by 2–7°C. The trend of T_{max} under the SSP2-4.5 and SSP5-8.5

scenarios for all three regions are presented in Figures 8.7, 8.8, and 8.9. The magnitude of increase in temperature under SSP5-8.5 is greater than that of the SSP2-4.5 scenario. Almazroui et al. (2020) projected the annual mean temperature of the South Asian region using CMIP6 models and found out that temperatures will be higher for SSP5-8.5 scenarios than other SSP scenarios.

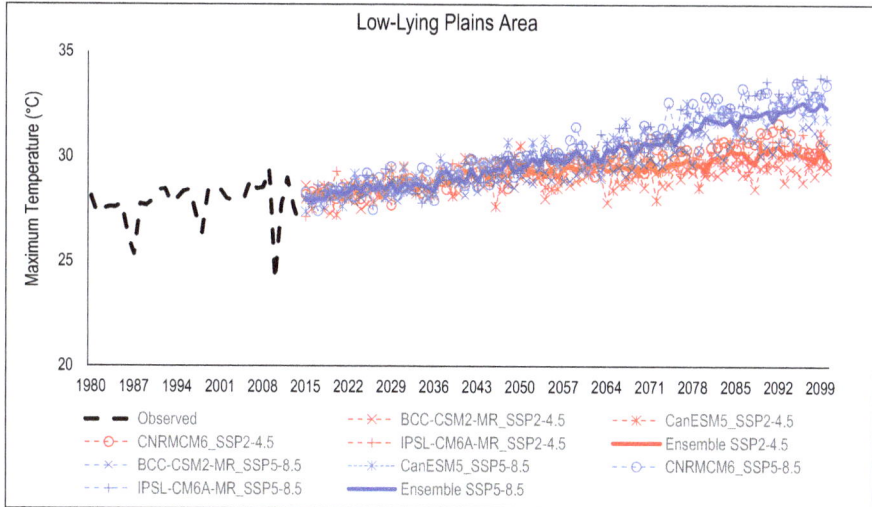

FIGURE 8.7 Projected future maximum temperature under the SSP2-4.5 and SSP5-8.5 scenarios for the low-lying plains area of the KRB.

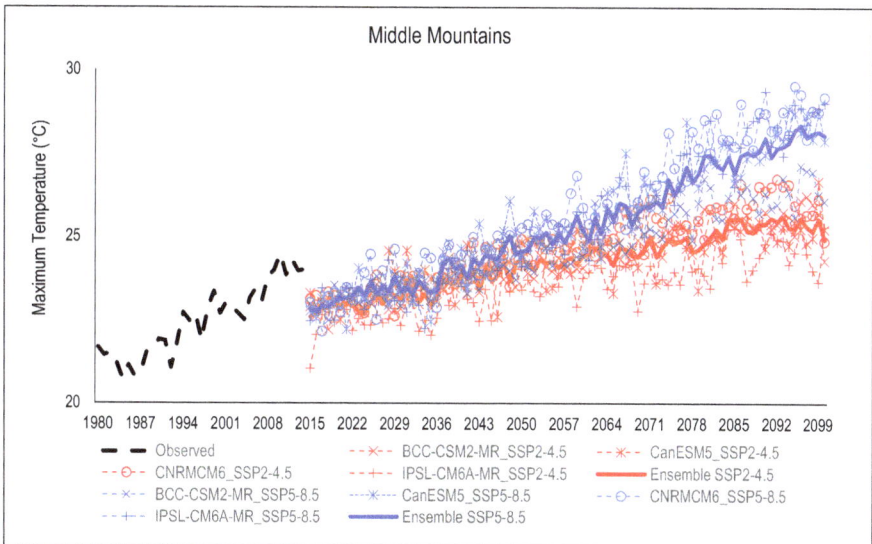

FIGURE 8.8 Projected future maximum temperature under the SSP2-4.5 and SSP5-8.5 scenarios for the middle mountains of the KRB.

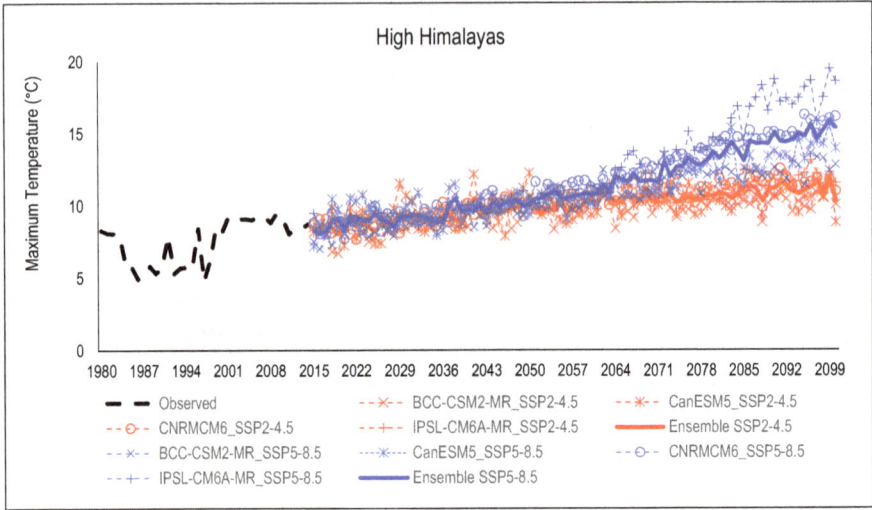

FIGURE 8.9 Projected future maximum temperature under the SSP2-4.5 and SSP5-8.5 scenarios for the high Himalayas of the KRB.

TABLE 8.8

Maximum Temperature Projections under the SSP2-4.5 and SSP5-8.5 Scenarios for the Three Future Periods

Region	Period	Maximum Temperature (°C) SSP2-4.5	SSP5-8.5
Low-Lying Plains Area	Baseline (1980–2014)	27.7 ± 4.8°C	
	2020s (2015–2045)	28.5 ± 4.2°C	28.5 ± 4.2°C
	2050s (2046–2076)	29.4 ± 4.2°C	30.0 ± 4.2°C
	2080s (2077–2100)	30.0 ± 4.2°C	31.9 ± 4.2°C
Middle Mountains	Baseline (1980–2014)	22.49 ± 4.2°C	
	2020s (2015–2045)	23.2 ± 3.79°C	23.5 ± 3.75°C
	2050s (2046–2076)	24.3 ± 3.75°C	25.3 ± 3.73°C
	2080s (2077–2100)	25.2 ± 3.67°C	27.5 ± 3.41°C
High Himalayas	Baseline (1980–2014)	7.47 ± 6.06°C	
	2020s (2015–2045)	8.9 ± 5.57°C	9.2 ± 5.38°C
	2050s (2046–2076)	10.2 ± 5.37°C	11.2 ± 5.36°C
	2080s (2077–2100)	10.3 ± 5.45°C	14.2 ± 5.52°C

The future minimum temperature (T_{min}) is expected to increase by 2–7°C in all three regions of the KRB for the future periods under the SSP2-4.5 and SSP5-8.5 scenarios. In near-future period, T_{min} will increase by 1 or 2°C. Still, the KRB will face severe warming for the far-future period, with temperatures increasing by 7°C and beyond. The trend of minimum temperature under the SSP2-4.5 and SSP5-8.5

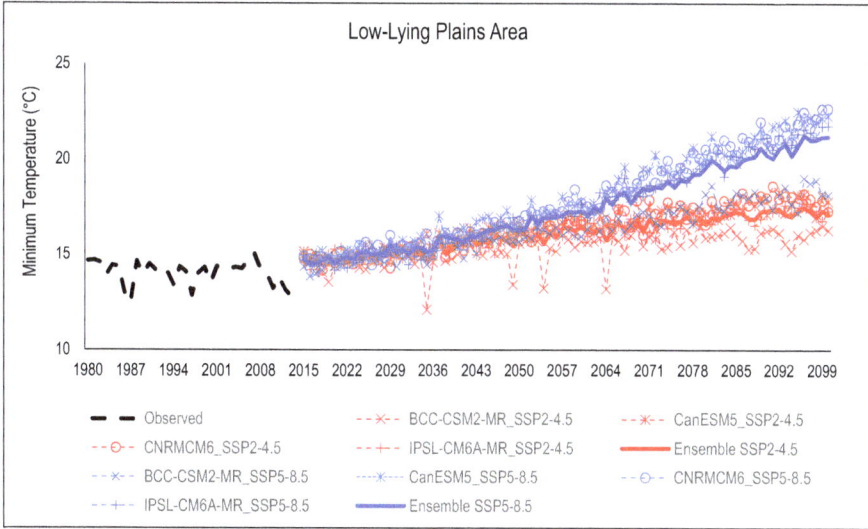

FIGURE 8.10 Projected future minimum temperature under the SSP2-4.5 and SSP5-8.5 scenarios for low-lying plains area of the KRB.

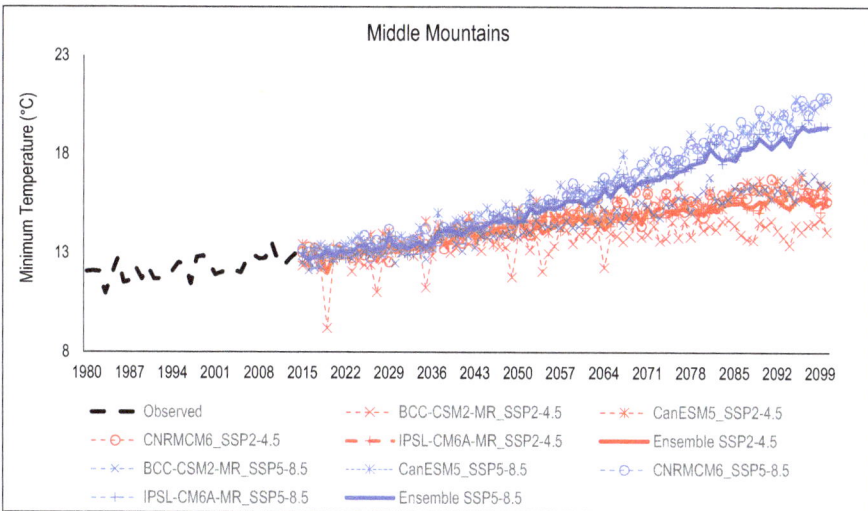

FIGURE 8.11 Projected future minimum temperature under the SSP2-4.5 and SSP5-8.5 scenarios for the middle mountains of the KRB.

scenarios is presented in Figures 8.10, 8.11, and 8.12. The magnitude of increment under the SSP5-8.5 is greater than that of the SSP2-4.5 scenario (Table 8.9).

The future maximum temperature and minimum temperatures are also analyzed seasonally, and Tables 8.10 and 8.11 shows the absolute seasonal change between

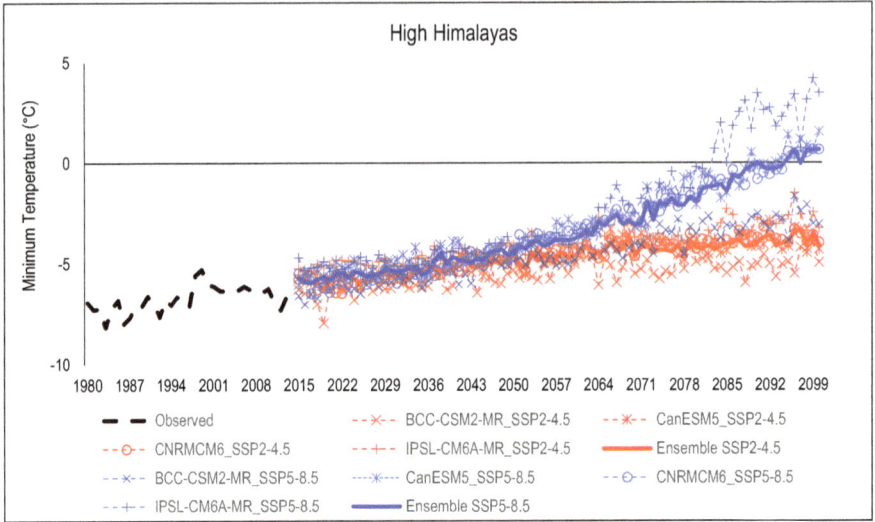

FIGURE 8.12 Projected future maximum temperature under the SSP2-4.5 and SSP5-8.5 scenarios for the high Himalayas of the KRB.

TABLE 8.9

Minimum Temperature Projections under the SSP2-4.5 and SSP5-8.5 Scenarios for the Three Future Periods

Region	Period	Minimum Temperature (°C)	
		SSP2-4.5	SSP5-8.5
Low-Lying Plain Areas	Baseline (1980–2014)	13.9 ± 6.9°C	
	2020s (2015–2045)	15.0 ± 6.8°C	15.2 ± 6.6°C
	2050s (2046–2076)	16.2 ± 6.8°C	17.4 ± 6.7°C
	2080s (2077–2100)	17.0 ± 6.7°C	20.1 ± 6.6°C
Middle Mountains	Baseline (1980–2014)	12.2 ± 5.5°C	
	2020s (2015–2045)	13.2 ± 5.6°C	13.4 ± 5.3°C
	2050s (2046–2076)	14.5 ± 5.7°C	15.7 ± 5.5°C
	2080s (2077–2100)	15.4 ± 5.6°C	18.3 ± 5.3°C
High Himalayas	Baseline (1980–2014)	−6.7 ± 8.2°C	
	2020s (2015–2045)	−5.5 ± 8.2°C	−5.3 ± 8.1°C
	2050s (2046–2076)	−4.5 ± 8.3°C	−3.4 ± 8.5°C
	2080s (2077–2100)	−3.9 ± 8.4°C	−0.5 ± 8.9°C

observed data and future temperature data. During the pre-monsoon season, all four GCMs show that the future T_{max} and T_{min} will increase by 5°C for the low-lying plains area and the middle mountains, whereas, for the high Himalayas, the CNRMCM6 and IPSL-CM6A-MR models show both minimum and maximum temperatures to increase by more than 5°C for the SSP5-8.5 scenario.

TABLE 8.10

Absolute Seasonal Change of Maximum Temperatures under the SSP2-4.5 and SSP5-8.5 Scenarios for the Three Future Periods

Region	GCMs	Pre-Monsoon (°C)						Monsoon (°C)						Post-Monsoon (°C)						Winter (°C)					
		SSP2-4.5			SSP5-8.5			SSP2-4.5			SSP5-8.5			SSP2-4.5			SSP5-8.5			SSP2-4.5			SSP5-8.5		
		NF	MF	FF	NF	MF	FF	NF	MF	FF	NF	MF	FF	NF	MF	FF	NF	MF	FF	NF	MF	FF	NF	MF	FF
Low-Lying Plains Area	BCC-CSM2-MR	0.7	1.6	2.3	0.9	2.3	3.4	0.4	0.9	1.1	0.5	1.1	2.3	0.7	1	1.3	0.4	1.4	2.4	0.7	1.3	1.6	0.7	1.8	3.5
	CANESM5	1.4	1.8	2.7	1.3	3.7	5.9	0.6	1.7	2.3	0.7	2.3	3.8	0.5	1	1.7	0.7	1.7	2.9	0.6	1.6	1.7	1	2.1	4.2
	CNRMCM6	1	2.3	3.3	1.1	3.1	5.9	0.8	1.7	2.3	0.6	2	3.6	1	2.1	3	1.3	3.2	4.8	0.6	2.2	3.1	1	2.7	5.3
	IPSL-CM6A-MR	1.1	2.2	2.6	0.8	2.8	4.8	0.9	2	2.7	1	2.6	4.8	0.7	1.8	2.7	0.9	2.9	5.1	0.9	1.7	2.2	0.7	2.5	5.2
	Ensemble	1.1	2.0	2.7	1.0	3.0	5.0	0.7	1.6	2.1	0.7	2.0	3.6	0.7	1.5	2.2	0.8	2.3	3.8	0.7	1.7	2.2	0.9	2.3	4.6
Middle Mountains	BCC-CSM2-MR	0.8	2.2	3.1	1.2	2.7	3.8	0.6	1.3	2.3	0.7	1.6	2.9	0.8	2.1	3.2	0.9	2.5	3.9	0.9	2.1	3.6	1.1	2.5	4.5
	CANESM5	1.4	2.2	2.8	1.3	3.7	6.1	0.7	1.9	2.4	0.4	2.5	3.9	0.7	1.8	2.4	1.2	3	5.3	1.2	2.3	2.5	1.6	3.3	5.9
	CNRMCM6	1	2.4	3.4	1.1	3.2	6	1	2.1	2.7	0.9	2.6	4.4	1.3	2.8	4.1	1.7	4.6	7.1	0.8	2.6	3.7	1.2	3.3	6.6
	IPSL-CM6A-MR	0	1.3	1.7	0.6	2.7	4.8	0.3	1.2	1.9	1	2.5	4.7	0.1	1.5	2.9	1.2	4	7.1	0.3	1.2	1.9	0.9	2.9	6.3
	Ensemble	0.8	2.0	2.8	1.1	3.1	5.2	0.7	1.6	2.3	0.8	2.3	4.0	0.7	2.1	3.2	1.3	3.5	5.9	0.8	2.1	2.9	1.2	3.0	5.8
High Himalayas	BCC-CSM2-MR	1	2.3	3.3	1.1	3.3	4.8	0.9	1.6	2	0.8	2.4	4	2.1	2.8	3.5	1.8	4.1	5.7	1.2	2.8	3.1	1.6	3.6	5.7
	CANESM5	2.3	2.4	2.5	2.3	3.7	6.5	1	1.6	2.4	0.8	2.1	3.7	1.8	3.5	3.8	2.8	3.7	5.7	1.8	3.6	3.5	3	4.4	8.1
	CNRMCM6	1.6	3	3.9	1.7	4.1	8	1.5	3.2	3.9	1.8	4	6.2	1.6	3.2	4.5	2	5.1	8.4	1.3	3.1	3.9	1.7	3.9	7.4
	IPSL-CM6A-MR	1.7	2.9	3.8	1.8	3.8	7.3	1.6	2.8	4.4	1.5	4.3	11.9	1.8	3.2	4.2	1.7	4.7	8.9	1.9	3	4	2.2	4.3	7
	Ensemble	1.7	2.7	3.4	1.7	3.7	6.7	1.3	2.3	3.2	1.3	3.2	6.5	1.8	3.2	4.0	2.1	4.4	7.2	1.6	3.1	3.6	2.1	4.1	7.1

TABLE 8.11

Absolute Seasonal Change of Minimum Temperatures under the SSP2-4.5 and SSP5-8.5 Scenarios for the Three Future Periods

Region	GCMs	Pre-Monsoon (°C) SSP2-4.5			Pre-Monsoon SSP5-8.5			Monsoon (°C) SSP2-4.5			Monsoon SSP5-8.5			Post-Monsoon (°C) SSP2-4.5			Post-Monsoon SSP5-8.5			Winter (°C) SSP2-4.5			Winter SSP5-8.5		
		NF	MF	FF	NF	MF	FF	NF	MF	FF	NF	MF	FF	NF	MF	FF	NF	MF	FF	NF	MF	FF	NF	MF	FF
Low-Lying Plains Area	BCC-CSM2-MR	0.7	1.3	2.2	1	2.6	4	0.5	1.2	1.6	0.8	1.9	3	1.2	2	2.7	1.2	3.3	5.3	0.7	1.3	1.8	0.9	2.3	4.2
	CANESM5	1.3	2.4	3.1	1.4	4.1	7.6	1.2	2.7	3.8	1.6	4.4	7.6	1.6	2.5	4.1	1.8	4.4	7.9	1.2	2.5	3.1	1.4	3	5.3
	CNRMCM6	1.4	3.2	4.4	1.5	4.7	9	1	2.2	2.9	1.1	3.1	5.3	1.1	2.9	4	1.8	4.4	6.9	0.9	2.8	3.9	1.2	3.8	7.7
	IPSL-CM6A-MR	1.1	2.4	3.3	1.2	3.6	6.9	1	2.2	3.3	1	3.5	6.4	1.3	3.1	3.8	1.3	4.5	7.4	0.8	1.8	2.4	0.9	2.9	5.8
	Ensemble	1.1	2.3	3.3	1.3	3.8	6.9	0.9	2.1	2.9	1.1	3.2	5.6	1.3	2.6	3.7	1.5	4.2	6.9	0.9	2.1	2.8	1.1	3.0	5.8
Middle Mountains	BCC-CSM2-MR	−0.1	1.1	1.9	0.8	2.4	3.6	0.6	1.3	1.9	0.8	2.2	3.5	0.9	2	2.7	1.1	3.3	5.4	0.6	1.4	1.8	0.8	2.2	3.8
	CANESM5	1.5	3	4	1.2	3.5	6.5	1.2	2.8	4	1.6	4.7	7.8	1.5	2.8	4.1	1.9	4.7	8.7	1.2	2.4	3	1.4	3.3	5.8
	CNRMCM6	1.2	3	4.1	1.4	4.2	7.8	1.1	2.4	3.3	1.2	3.5	6.2	1.1	2.8	3.9	1.7	4.3	7.5	0.9	2.7	3.7	1.2	3.7	7.6
	IPSL-CM6A-MR	1	2.3	3.2	1.1	3.3	5.9	1.2	2.6	3.6	1.3	3.8	5.9	1.2	3.1	4	1.3	4.8	7.9	0.9	2	2.7	1.1	3.2	6.5
	Ensemble	0.9	2.4	3.3	1.1	3.4	6.0	1.0	2.3	3.2	1.2	3.6	5.9	1.2	2.7	3.7	1.5	4.3	7.4	0.9	2.1	2.8	1.1	3.1	5.9
High Himalayas	BCC-CSM2-MR	0.3	1.4	1.9	0.9	2.7	4.1	0.6	1.2	1.5	0.6	1.8	2.8	1.2	1.6	2.1	1.2	0.6	4.1	0.7	1.7	2	1.1	2.5	4.4
	CANESM5	1.5	2.2	2.4	1.4	3.2	5.7	1.4	2.6	3	1.9	4.2	7	1.6	1	4	2.3	2.8	8.4	0.9	1.8	1.8	1.2	2.2	5.2
	CNRMCM6	0.9	1.8	2.3	0.9	2.4	5.2	1.5	3.1	4	1.7	4.8	7.7	1.3	0.5	3.4	1.4	1.8	7.3	0.7	1.7	2	0.9	2.5	4.2
	IPSL-CM6A-MR	2	2.8	3.3	2	3.6	7.3	1.8	2.9	4.3	1.8	4.4	10.6	2	0.8	3.8	2	2	8.1	1.7	2.3	2.9	1.7	3.1	6.4
	Ensemble	1.2	2.1	2.5	1.3	3.0	5.6	1.3	2.5	3.2	1.5	3.8	7.0	1.5	1.0	3.3	1.7	1.8	7.0	1.0	1.9	2.2	1.2	2.6	5.1

During the monsoon season, the maximum temperature is projected to rise by an average of 2–3°C for all future periods under the SSP2-4.5 and SSP5-8.5 scenarios in all three regions. The minimum temperature is projected to rise by an average of 4–5°C for all future periods under SSP2-4.5 and SSP5-8.5 scenarios. The temperature is significantly increasing under SSP5-8.5 scenarios than SSP2-4.5 scenarios.

During the post-monsoon season, T_{max} is expected to rise by more than 6°C in the Himalayan region. In the low-lying plains area and the middle mountains, the maximum temperature will rise by an average of 4–5°C. Similarly, T_{min} is expected to rise by more than 5°C under the SSP5-8.5 scenario for all three regions.

During the winter season, the high Himalaya's maximum temperature is expected to experience more warming, with an increase of temperature by more than 6°C in mid-future and far-future periods. Similarly, the maximum temperatures will increase with an average of 2–3°C for the near-future and mid-future periods in the low-lying plains area and the middle mountains region. The minimum temperatures of the low-lying plains area and the middle mountains region are expected to increase more than the high Himalayas regions.

Overall, the ensemble results show that both the maximum temperature and minimum temperature will increase with an absolute change of 5–7°C in the middle mountains and the high Himalayas region. The maximum and minimum temperature in all three regions is increasing by 1–3°C in the coming near-future and mid-future periods. The absolute seasonal change of both temperatures shows that the high Himalayas region will go through temperature variation and will experience more warming compared to the other regions.

8.5   DISCUSSIONS

8.5.1   CLIMATE-CHANGE IMPACT ON RAINFALL AND TEMPERATURE OF DIFFERENT REGIONS OF THE KRB

Climate change has significantly impacted the KRB and its regions mainly through variation in rainfall, T_{max} and T_{min}. The future rainfall in all three regions is expected to increase by more than 50%. According to the results, climate change will severely impact all three regions, and they are quite vulnerable because of extreme rainfall events in the pre-monsoon, monsoon, and post-monsoon seasons. From the preceding ensemble results, all three regions are vulnerable. Climate change can have severe impact on the high Himalayas region because the region will experience heavy rainfall and absolute temperature change in the future periods. The high Himalayas region is the snow-dominated region of the KRB, and snowmelt has an equal contribution to runoff generation in the basin (Siderius et al., 2013). The result shows that rainfall can increase by more than 50–60% in the Himalayan region, which would have severe consequences on snow, glaciers, and the livelihoods of those living in the region. Recent studies reveal that snow accumulation is decreasing, and if global warming increases, then the Himalayan glaciers will melt. The snowmelt rate will rise in the near future, and summer streamflow will be high in the Himalayas (Sharma et al., 2021). Similarly, Chettri et al. (2021) projected the future rainfall in the mid-hills of Nepal and predicted that future rainfall would increase predominantly during

all four seasons. The impact of increase in rainfall and temperature affects the runoff generation in the basin, which directly affects the livelihoods of those living in all three regions of the basin. Natural hazards, like landslides, floods, and soil erosion, are likely to occur in future due to extreme climatic events that will affect the basin economy, agriculture, and people's livelihoods.

The KRB will experience more warming in the future, as the future maximum temperature (T_{max}) and minimum temperature (T_{min}) will increase. Mishra (2020) projected the future temperature of South Asia and found out that the temperature will get warmer by 3–5°C. Similarly, the future T_{max} and T_{min} of the high Himalayas region of the KRB will increase, and the region will get warmer by 3–7°C, which will directly affect the snow dynamics of the basin. Higher temperatures can cause a reduction in snowfall and snow-cover duration (Magnusson et al., 2010). The low-lying plains area and the middle mountains will also experience hotter days in the coming future periods.

8.5.2 POLICY IMPLICATIONS FOR WATER RESOURCES MANAGEMENT

The Hindu Kush Himalayan region is the source of water resources for the people living in the Himalayan region, and the KRB is a tributary. Climate change has a significant impact on the availability of water resources. There are different policies and adaptation strategies planned and implemented to manage the water resources of the Hindu Kush Himalayan region from a regional to a basin scale. Our results can contribute to forming the new and revised policies and climate adaptation strategies for managing water resources in the KRB. According to the Climate Change Policy (2011), the climate change impact assessment was carried out in KRB using CORDEX RCMs of phase 5. So, our study will be beneficial for revising policy because the study used the latest CMIP6 models and SSPs, which guide different climate-change research and findings at a global and a regional scale.

The KRB is an agricultural basin that mainly depends on the seasonally rainfed irrigation through local streams, springs, and snowmelt. Our study can contribute to the revision of water management policies and agricultural policies for water availability and management. The CMIP6 GCMs projected that the future rainfall would increase and have high chances of extreme flood events. The KRB is affected by floods and droughts, and both of them are recurring in nature (Devkota and Gyawali, 2015; Wu et al., 2019; Sharma et al., 2021). So, our study will benefit the climate researcher and policymakers to predict future disasters and plan the strategies and measures against the disaster.

8.6 CONCLUSION

This study examines the potential impact of climate change on precipitation and temperature in different regions of the KRB, Nepal. The research used the four CMIP6 GCMs data sets of the BCC, CANESM-CCCma, CNRMCM-CERFACS, and IPSL to project the future climate for the near future (2015–2045), the mid-future (2046–2076), and the far future (2077–2100) under the SSP2-4.5 and SSP5-8.5 scenarios. The climate projection was carried out using Quantile Mapping to

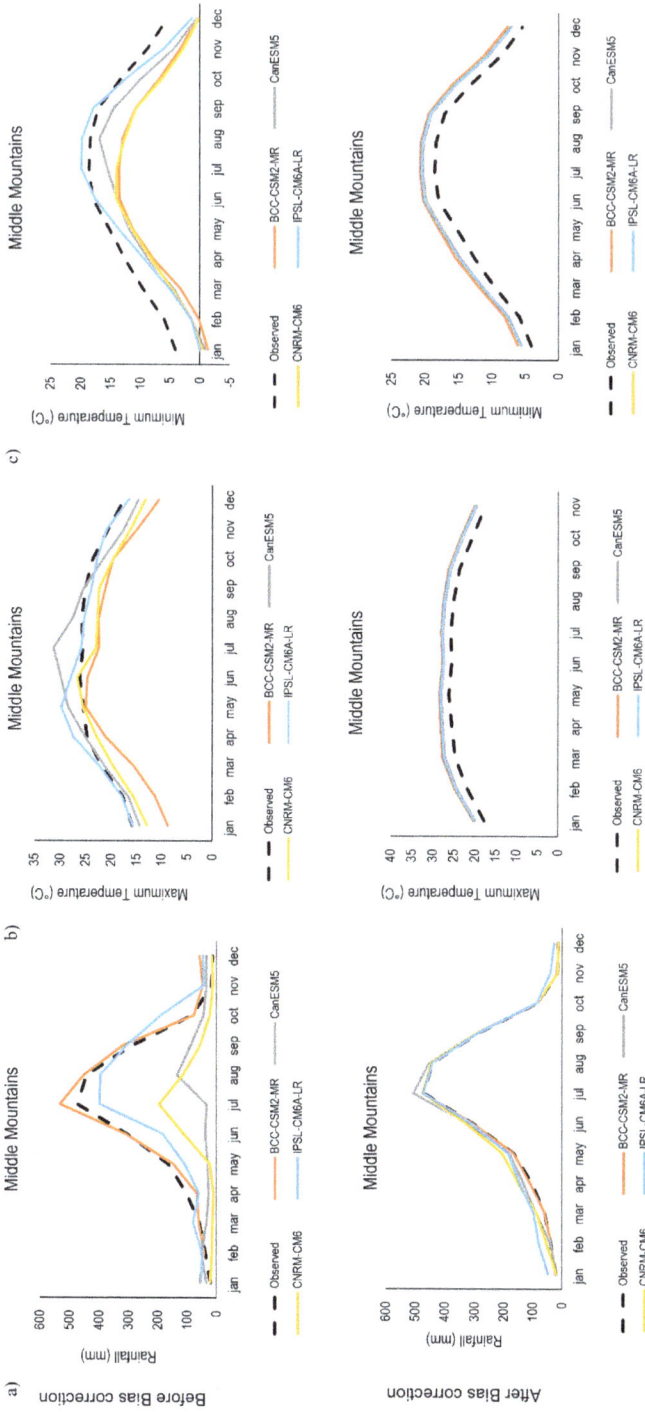

FIGURE SUPPLEMENTARY S8.1 Performance of bias correction before and after of (a) rainfall, (b) maximum temperature, and (c) minimum temperature for the middle mountains of the KRB, Nepal.

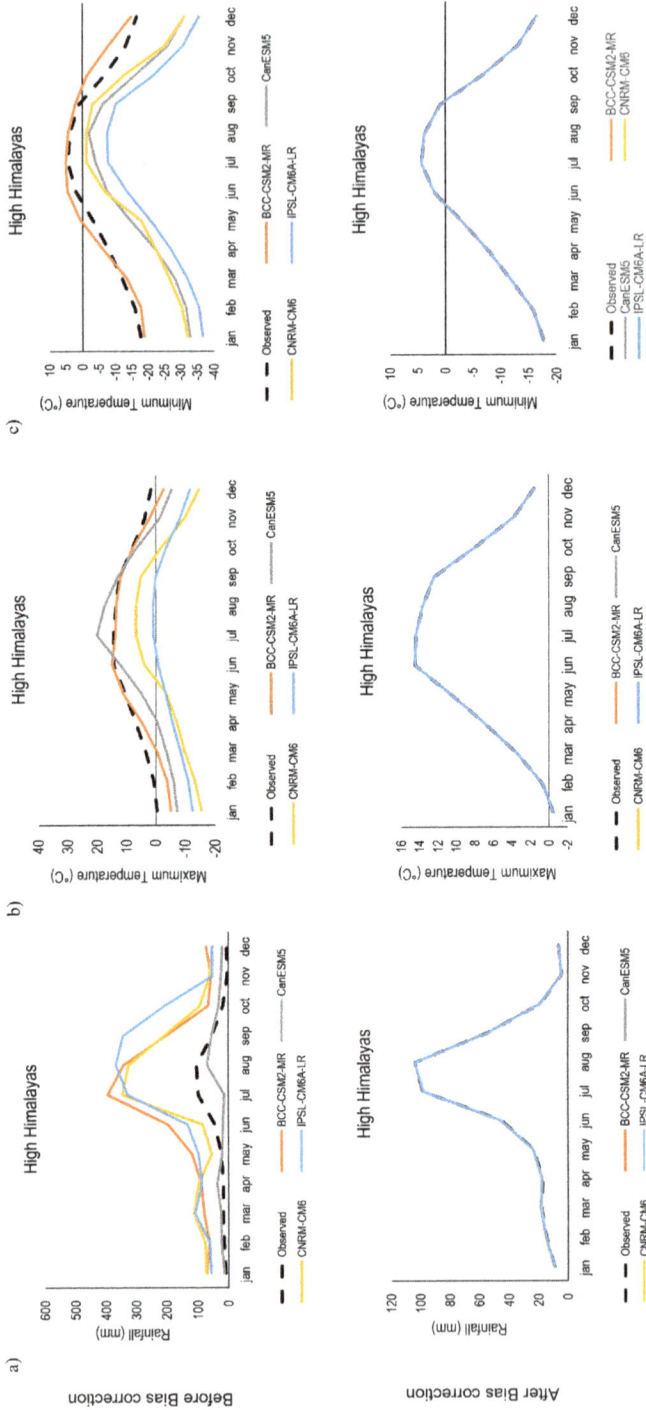

FIGURE SUPPLEMENTARY S8.2　Performance of bias correction before and after of (a) rainfall, (b) maximum temperature, and (c) minimum temperature for the high Himalayas of the **KRB**, Nepal

correct the bias in climate models. The observed climate data for 17 temperature stations and 38 rainfall stations for 1980–2014 were used. The evaluation of bias correction shows moderate performance of all four climate models with rainfall, T_{max} and T_{min}. The CNRMCM6 and IPSL-CM6A-MR are the models that projected that future rainfall and temperatures would increase in all three regions of KRB. The future climate values projected for rainfall and temperature highly vary against the baseline period by more than 10% for rainfall and an average of 3–4°C for temperature all over the KRB.

The results show that the different regions of the KRB will experience different magnitudes and intensities of rainfall seasonally in the future period. The future rainfall will increase in the monsoon season and non-monsoon season as well. Himalayan regions expect to experience an increase in rainfall by 50–60%, which will directly affect the rate of runoff of the river basin. The results also show that future temperature extremes are likely to rise approximately by 2–7°C for all three regions, which means that all three regions will experience more warming in coming future periods.

ACKNOWLEDGMENT

The authors would like to express their gratitude to the Department of Hydrology and Meteorology (DHM) of Nepal, for the data to facilitate this research work.

REFERENCES

Agarwal, A., Babel, M. S., Maskey, S., Shrestha, S., Kawasaki, A., & Tripathi, N. K. (2016). Analysis of Temperature Projections in the Koshi River Basin, Nepal. *International Journal of Climatology*, 36(1), 266–279. https://doi.org/10.1002/joc.4342

Alaminie, A. A., Tilahun, S. A., Legesse, S. A., Zimale, F. A., Tarkegn, G. B., & Jury, M. R. (2021). Scenarios for the UBNB (Abay). Ethiopia. *Water*, 13, 2110. https://doi.org/10.3390/w13152110

Almazroui, M., Saeed, S., Saeed, F., Islam, M. N., & Ismail, M. (2020). Projections of Precipitation and Temperature over the South Asian Countries in CMIP6. *Earth Systems and Environment*, 4(2), 297–320. https://doi.org/10.1007/s41748-020-00157-7

Bharati, L., Gurung, P., Jayakody, P., Smakhtin, V., & Bhattarai, U. (2014). The Projected Impact of Climate Change on Water Availability and Development in the Koshi Basin, Nepal. *Mountain Research and Development*, 34(2), 118–130. https://doi.org/10.1659/mrd-journal-d-13-00096.1

Buytaert, W., Vuille, M., Dewulf, A., Urrutia, R., Karmalkar, A., & Celleri, R. (2010). Uncertainties in Climate Change Projection and Regional Downscaling in the Tropical Andes: Implications for Water Resources Management. *Hydrology and Earth System Sciences*, 14, 1247–1258.

Chettri, R., Pandey, V. P., Talchabhadel, R., et al. (2021). How do CMIP6 Models Project Changes in Precipitation Extremes Over Seasons and Locations Across the Mid-hills of Nepal? *Theoretical and Applied Climatology*, 145, 1127–1144. https://doi.org/10.1007/s00704-021-03698-7

Cook, B. I., Mankin, J. S., Marvel, K., Williams, A. P., Smerdon, J. E., & Anchukaitis, K. J. (2020). Twenty-First Century Drought Projections in the CMIP6 Forcing Scenarios. *Earth's Future*, 8(6), 1–20. https://doi.org/10.1029/2019EF001461

Devkota, L. P., & Gyawali, D. R. (2015). Impacts of Climate Change on the Hydrological Regime and Water Resources Management of the Koshi River Basin, Nepal. *Journal of Hydrology: Regional Studies*, 4, 502–515.https://doi.org/10.1016/j.ejrh.2015.06.023

Duncan, J. M. A., & Biggs, E. M. (2012). Assessing the Accuracy and Applied Use of Satellite-derived Precipitation Estimates Over Nepal. *Applied Geography*, 34, 626–638.

Eyring, V. et al. (2016). Overview of the Coupled Model Intercomparison Project Phase 6 (CMIP6) Experimental Design and Organization. *Geoscientific Model Development*, 9, 1937–1958.

FMIS. (2012). *Flood Management Information System*, Water Resources Department, Bihar, http://fmis.bih.nic.in/Riverbasin.html#kosi (accessed, August 5, 2012).

Gobiet, A., Kotlarski, S., Beniston, M., Heinrich, G., Rajczak, J., & Stoffel, M. (2014). 21st-Century Climate Change in the European Alps-A Review. *Science of the Total Environment*, 493, 1138–1151. https://doi.org/10.1016/j.scitotenv.2013.07.050

Gudmundsson, L. (2014). *QMAP: Statistical Transformations for Postprocessing Climate Model Output*. R package version.

Guo, D. L., Sun, J. Q., & Yu, E. T. (2018). Evaluation of CORDEX Regional Climate Models in Simulating Temperature and Precipitation Over the Tibetan Plateau. *Atmospheric and Oceanic Science Letters*, 11(3), 219–227. https://doi.org/10.1080/16742834.2018.1451725

Gupta, V., Singh, V., & Jain, K. M. (2020). Assessment of Precipitation Extremes in India During the 21st Century Under SSP1-1.9 Mitigation Scenarios of CMIP6 GCMs. *Journal of Hydrology*, 590, 125422, ISSN 0022-1694, https://doi.org/10.1016/j.jhydrol.2020.125422

Gusain, A., Ghosh, S., & Karmakar, S. (2020). Added Value of CMIP6 Over CMIP5 Models in Simulating Indian Summer Monsoon Rainfall. *Atmospheric Research*, 232(September 2019), 104680. https://doi.org/10.1016/j.atmosres.2019.104680

Immerzeel, W. W., Droogers, P., de Jong, S. M., & Bierkens, M. F. P. (2009). Large-scale Monitoring of Snow Cover and Runoff Simulation in Himalayan River Basins Using Remote Sensing. *Remote Sensing of Environment*, 113(1), 40–49. https://doi.org/10.1016/j.rse.2008.08.010

IPCC. (2021). *Climate Change 2021: The Physical Science Basis. Contribution of Working Group I to the Sixth Assessment Report of the Intergovernmental Panel on Climate Change* [Masson-Delmotte, V., P. Zhai, A. Pirani, S. L. Connors, C. Péan, S. Berger, N. Caud, Y. Chen]. Cambridge University Press, In Press, 3949. www.ipcc.ch/report/ar6/wg1/downloads/report/IPCC_AR6_WGI_Full_Report.pdf

Jakob Themeßl, M., Gobiet, A., & Leuprecht, A. (2011). Empirical-statistical Downscaling and Error Correction of Daily Precipitation from Regional Climate Models. *International Journal of Climatology*, 31(10), 1530–1544. https://doi.org/10.1002/joc.2168

Kaini, S., Nepal, S., Pradhananga, S., Gardner, T., & Sharma, A. K. (2019). Representative General Circulation Models Selection and Downscaling Climate Data for the Transboundary Koshi River Basin in China and Nepal. *International Journal of Climatology*, 1–19. https://doi.org/10.1002/joc.6447

Karim, R., Tan, G., Ayugi, B., Babaousmail, H., & Liu, F. (2020). Evaluation of Historical CMIP6 Model Simulations of Seasonal Mean Temperature Over Pakistan During 1970–2014. *Atmosphere*, 11(9). https://doi.org/10.3390/atmos11091005

Khadka, D., Babel, M. S., Shrestha, S., & Tripathi, N. K. (2014). Climate Change Impact on Glacier and Snow Melt and Runoff in Tamakoshi Basin in the Hindu Kush Himalayan (HKH) Region. *Journal of Hydrology*, 511, 49–60. https://doi.org/10.1016/j.jhydrol.2014.01.005

Khadka, M., Kayastha, R. B., & Kayastha, R. (2020). Future Projection of Cryospheric and Hydrologic Regimes in Koshi River Basin, Central Himalaya, Using Coupled Glacier

Dynamics and Glacio-hydrological Models. *Journal of Glaciology*, 66(259), 831–845. https://doi.org/10.1017/jog.2020.51

Khan, A. J., & Koch, M. (2018). Selecting and Downscaling a Set of Climate Models for Projecting Climatic Change for Impact Assessment in the Upper Indus Basin (UIB). *Climate*, 6(4), 1–20. https://doi.org/10.3390/cli6040089

Lalande, M., Ménégoz, M., Krinner, G., Naegeli, K., & Wunderle, S. (2021). Climate Change in the High Mountain Asia in CMIP6. *Earth System Dynamics Discussions*, 1–56. https://doi.org/10.5194/esd-2021-43

Lutz, A. F., Immerzeel, W. W., Biemans, H., ter Maat, H., Veldore, V., & Shrestha, A. (2016). Selection of Climate Models for Developing Representative Climate Projections for the Hindu Kush Himalayan Region. *Hi-Aware*, 46. http://lib.icimod.org/record/31874/files/HI-AWARE-WP1.pdf

Magnusson, J., Jonas, T., López, M.I., & Lehning. M. (2010, June 1) Snow Cover Response to Climate Change in a High Alpine and Half-glacierized Basin in Switzerland. *Hydrology Research*, 41 (3–4), 230–240. https://doi.org/10.2166/nh.2010.115

Mishra, V., Bhatia, U., & Tiwari, A. D. (2020). Bias-corrected Climate Projections for South Asia from Coupled Model Intercomparison Project-6. *Scientific Data*, 7, 338. https://doi.org/10.1038/s41597-020-00681-1

Narsey, S. Y., Brown, J. R., Colman, R. A., Delage, F., Power, S. B., Moise, A. F., & Zhang, H. (2020). Climate Change Projections for the Australian Monsoon from CMIP6 Models. *Geophysical Research Letters*, 47(13), 1–9. https://doi.org/10.1029/2019GL086816

O'Neill, B. C., Tebaldi, C., Van Vuuren, D. P., Eyring, V., Friedlingstein, P., Hurtt, G., Knutti, R., Kriegler, E., Lamarque, J. F., Lowe, J., Meehl, G. A., Moss, R., Riahi, K., & Sanderson, B. M. (2016). The Scenario Model Intercomparison Project (ScenarioMIP) for CMIP6. *Geoscientific Model Development*, 9(9), 3461–3482. https://doi.org/10.5194/gmd-9-3461-2016

Pepin, N., Bradley, R. S., Diaz, H. F., Baraer, M., Caceres, E. B., Forsythe, N., Fowler, H., Greenwood, G., Hashmi, M. Z., Liu, X. D., Miller, J. R., Ning, L., Ohmura, A., Palazzi, E., Rangwala, I., Schöner, W., Severskiy, I., Shahgedanova, M., Wang, M. B., . . . Yang, D. Q. (2015). Elevation-dependent Warming in Mountain Regions of the World. *Nature Climate Change*, 5(5), 424–430. https://doi.org/10.1038/nclimate2563

Pradhan, P., Shrestha, S., Sundaram, S. M., & Virdis, S. G. P. (2021). Evaluation of the CMIP5 General Circulation Models for Simulating the Precipitation and Temperature of the Koshi River Basin in Nepal. Journal of Water and Climate Change, 1–15. https://doi.org/10.2166/wcc.2021.124

Rajbhandari, R., Shrestha, A. B., Nepal, S., & Wahid, S. (2016). Projection of Future Climate Over the Koshi River Basin Based on CMIP5 GCMs. *Atmospheric and Climate Sciences*, 6(2), 190–204. https://doi.org/10.4236/acs.2016.62017

Rajbhandari, R., Shrestha, A. B., Nepal, S., & Wahid, S. (2018). Projection of Future Precipitation and Temperature Change over the Transboundary Koshi River Basin Using Regional Climate Model PRECIS. *Atmospheric and Climate Sciences*, 8(2), 163–191. https://doi.org/10.4236/acs.2018.82012

Riahi, K., Van Vuuren, D. P., Kriegler, E., Edmonds, J., Oneill, B. C., Fujimori, S., Bauer, N., Calvin, K., Dellink, R., & Fricko, O. (2017). The Shared Socioeconomic Pathways and Their Energy, Land Use, and Greenhouse Gas Emissions Implications: An Overview. *Global Environ Change*, 42, 153–168.

Sharma, S., Hamal, K., Khadka, N., Ali, M., Subedi, M., Hussain, G., Ehsan, M. A., Saeed, S., & Dawadi, B. (2021). Projected Drought Conditions Over Southern Slope of the Central Himalaya Using CMIP6 Models. *Earth Systems and Environment*, 0123456789. https://doi.org/10.1007/s41748-021-00254-1

Shrestha, A., Rahaman, M. M., Kalra, A., Jogineedi, R., & Maheshwari, P. (2020). Climatological Drought Forecasting Using Bias Corrected CMIP6 Climate Data: A Case Study for India. *Forecasting*, 2(2), 59–84. https://doi.org/10.3390/forecast2020004

Siderius, C., Biemans, H., Wiltshire, A., Rao, S., Franssen, W. H. P., Kumar, P., Gosain, A. K., van Vliet, M. T. H., & Collins, D. N. (2013). Snowmelt Contributions to Discharge of the Ganges. *Science of the Total Environment*, 468–469, S93–S101. https://doi.org/10.1016/j.scitotenv.2013.05.084

Talchabhadel, R., & Karki, R. (2019). Assessing Climate Boundary Shifting Under Climate Change Scenarios Across Nepal. *Environment Monitoring and Assessment*, 191, 520. https://doi.org/10.1007/s10661-019-7644-4

Teutschbein, C., & Seibert, J. (2012). Bias Correction of Regional Climate Model Simulations for Hydrological Climate-change Impact Studies: Review and Evaluation of Different Methods. *Journal of Hydrology*, 456–457, 12–29. https://doi.org/10.1016/j.jhydrol.2012.05.052

Thapa, S., Li, B., Fu, D., Shi, X., Tang, B., Qi, H., & Wang, K. (2020). Trend Analysis of Climatic Variables and Their Relation to Snow Cover and Water Availability in the Central Himalayas: A Case Study of Langtang Basin, Nepal. *Theoretical and Applied Climatology*, 140(3–4), 891–903. https://doi.org/10.1007/s00704-020-03096-5

Trzaska, S., & Schnarr, E. (2014). A Review of Downscaling Methods for Climate Change Projections. *United States Agency for International Development by Tetra Tech ARD* (September), 1–42.

Wang, B., Bao, Q., Hoskins, B., Wu, G., & Liu, Y. (2008). Tibetan Plateau Warming and Precipitation Changes in East Asia. *Geophysical Research Letters*, 35, L14 702. https://doi.org/10.1029/2008GL034330

Wester, P., Mishra, A., Mukherji, A., & Shrestha, A. B. (2019) *The Hindu Kush Himalaya Assessment—Mountains, Climate Change, Sustainability, and People.* Cham: Springer Nature Switzerland AG.

Wu, T. W., & Coauthors (2019). The Beijing climate center climate system model (BCC-CSM): The main progress from CMIP5 to CMIP6. *Geoscientific Model Development*, 12, 1573–1600. https://doi.org/10.5194/gmd-12-1573-2019

You, Q. L., Ren, G. Y., Zhang, Y. Q., Ren, Y. Y., Sun, X. B., Zhan, Y. J., Shrestha, A. B., & Krishnan, R. (2017). An overview of studies of observed climate change in the Hindu Kush Himalayan (HKH) region. *Advances in Climate Change Research*, 8(3), 141–147. https://doi.org/10.1016/j.accre.2017.04.001

9 Impact of Structural Barriers on the Morphology and Ecology of the Himalayan Rivers

Gaurav Kailash Sonkar and Kumar Gaurav*
Indian Institute of Science Education and Research
Bhopal, Madhya Pradesh, India. 462066

CONTENTS

9.1 Introduction 165
    9.1.1 Biodiversity Adaptation in Changing Riverine Landscape 166
    9.1.2 Modification of River Systems 167
9.2 Material and Methods...167
    9.2.1 Hydrological Alteration 167
    9.2.2 Geomorphic Alteration..168
    9.2.3 Indicator Species in the Ganga Basin and Their Status 170
    9.2.4 Threats to Exploratory Indicator Species 171
        9.2.4.1 Human Conflict and Habitat Quality Degradation 171
        9.2.4.2 Habitat Fragmentation by Structural Barriers 172
9.3 Results and Discussion 173
    9.3.1 Channel Morphology and Sediment Dynamics 173
    9.3.2 Hydraulic Variability 175
    9.3.3 Indicator Species Habitat Degradation by
        Human Intervention 176
        9.3.3.1 Hydraulic Condition...176
        9.3.3.2 Hydrological and Geomorphic Impact 179
9.4 Conclusion 181
References...181

9.1 INTRODUCTION

The Ganga and Brahmaputra basins on the Himalayan foreland are among the largest alluvial plains in the world. They support diverse freshwater fauna of invertebrates, fishes, reptiles, and mammals. The different hydro-geomorphic features of the riverine landscape control the formation and sustenance of aquatic floral and faunal species. Various fluvial processes dominate the stages of riverine landscape evolution.

DOI: 10.1201/9781003265160-11

Functional processes, such as erosion, aggradation, and exchange of nutrients with different floodplain water bodies, allow diverse succession stages to occur and heterogeneous diversity to thrive. This picture of heterogeneous diversity is mostly prevalent in the middle reaches of a river system. As pointed out by Ward (1998), the river continuum model shows that the biological community structure differs along a longitudinal gradient. Similarly, the Ganga River has its unique biodiversity distribution in three main eco-regions and climate regimes: the torrential mountain headwaters of the upper Ganga, the Ganga plain, and the deltaic region. The geomorphic features and the climate regime make up various biodiversity hot spots. In the torrential mountain streams of Arunachal Pradesh, the rocky streambeds are the preferential habitat of endangered hill stream catfish (Kachari et al. 2014). The shallow and cold headwaters, enriched with nutrients, are suitable habitats for crustaceans, annelids, and mollusks, which serve as prey bases of the endangered hill catfish (Kachari et al. 2014). In the Ganga plain, there are about 207 common plant species recorded in the active channels of the Ghaghra and Gandak Rivers (Singh et al. 2017). This can be owed to the similar physical conditions in both river environments. The biodiversity of the Himalayan rivers is under constant threat from human interference. For example, the distribution of the Eurasian otter (*Lutra lutra*) and the smooth-coated otter (*Lutrogale perspicillata*) is shrinking due to human-induced stressors, such as sand and boulder mining, pollution, hydropower plants, poaching, and loss of wetland habitat to agriculture and urbanization (Gupta et al. 2020; De Silva et al. 2015). Their high occupancy in the Kosi, Khoh, and Ramganga Rivers is due to high forest and grassland landcover but is significantly low in Song River due to agriculture and settlement (Gupta et al. 2020). Flow regulation and habitat degradation are two major drivers of contemporary changes in Himalayan River landscape ecology (Srivastava 2007; M. Singh, Sinha, and Tandon 2020).

9.1.1   BIODIVERSITY ADAPTATION IN CHANGING RIVERINE LANDSCAPE

The upper Ganga (Gangotri to Haridwar) and its tributaries are characterized by cold water, high flow velocity, large substratum, and low nutrients. The culmination of such factors sets the tone of faunal species that can adapt to such conditions. As we transition from the headwaters to the plains, the food chain trophic level tends to change as well and become more complex. The river transitions from a heterotrophic state in the middle headwaters to an autotrophic state in the foothills with an increase in fish diversity (Nautiyal 2010). The transition of the river from a confined in the mountains to an unconfined valley setting in the plains provides favorable conditions for more complex fauna to thrive (Sinha et al. 2017). Change in the valley setting allows more flexibility in the channel pattern (Sinha et al. 2017). Floodplain pockets diversify the riverine habitat. Geomorphic processes and features in the floodplain create micro and macro-habitat that naturally add to the carrying capacity of the river system. The periodic flood pulses in the high flow season sustain the micro- and macro-habitat of the floodplain. Different floodplain water bodies, such as oxbow lakes, wetlands, and back swamps, allow succession stages to occur, allowing riparian growth and biomass increase in the floodplains. Often, the oligotrophic lotic river system replenishes its nutrients by exchange from the riparian zones in the floodplain. A higher degree of lateral connectivity in the plain than in the mountain

reaches is one of the major reasons for a greater assemblage of fish species. The physical variables in the river system play a key role in the assemblage and distribution of fish diversity (Arunachalam 2000; Bain and Finn 2013; Dubey et al. 2012). Fishes with different morphological adaptations and sizes with respect to life stages occupy regions of different flow conditions in rivers (Arunachalam 2000). This can be because of changes in the substrate type and physical variables, such as channel width, depth, velocity, discharge, temperature, and entropy gain.

9.1.2 MODIFICATION OF RIVER SYSTEMS

The channel pattern and hydraulics determine the habitat template, which in turn is controlled by the flow. Any change in the flow regime, be it natural or human-induced can significantly alter the channel pattern and eventually affect the functional processes. Channel modification is an inherent phenomenon with flood interval and tectonic activity as major drivers of the changes. The temporal scale strongly influences the extent of fluvial landscape evolution, ranging from thousand years to a multi-decadal scale (Pal and Pani 2019; Phillips 1995; Singh, Prakash, and Shukla 2018). Anthropogenic interventions have drastically changed the rate of fluvial evolution by embanking, channelization, urbanization, and damming of the rivers (Best 2019; Gregory 2006; Gregory, Davis, and Downs 1992). The response of a river to such human interventions depends on its characteristics, regulation extent, geomorphic activity, and land-use change (Gregory 2006). Geomorphic changes in the fluvial system are immediate and often severe, which have ensuing ecological fallout both in the active channel and the floodplain. The hydro-geomorphic ramification of anthropogenic modification has influenced multiple studies on the behavior of river systems. In the following sections, we review, and present evidence of river system change from hydrological and geomorphic perspectives and their ecological implications.

9.2 MATERIAL AND METHODS

9.2.1 HYDROLOGICAL ALTERATION

The natural flow regime drives the ecological processes and is important in sustaining ecosystem integrity. The magnitude, frequency, duration, and rate of change in associated hydrological conditions are the key components that describe the flow regime (Poff et al. 1997). It reflects the inter-annual variability of hydro-climatological conditions (Petts 2009). Modification of one or more components of the flow regime can have a critical impact on the river ecosystem. A combination of components set the template to which a certain species adapt over an evolutionary period, defining the community composition in the face of dominating regime conditions (Lytle and Poff 2004; Poff et al. 1997). The first and foremost change brought on by flow modification is in the water and sediment transfer. Human-made barriers in rivers initiate an instantaneous reduction in flow, which leads to a string of connected events such as a decrease in instream flow depth, reduced frequency of bank-full discharge and overbank flow, low floodplain inundation during the peak flow season, and low

baseflow. Poor sediment and water delivery go hand in hand, and together they bring visible changes in the channel morphology. The sediment retention by regulated water instigates channel scouring and bed lowering, giving rise to a narrow and stable morphology. This phenomenon is best illustrated at the Farakka barrage, in the lower Ganga, where the downstream reach of the barrage is narrower as compared to the pre-barrage period (Sonkar and Gaurav 2020). These events are prominent downstream to the point of intervention, whereas in the upstream impoundment of water and sediment creates its own sets of geomorphologic and ecological problems. Sedimentation, together with impounded water, increases the frequency of overbank flooding, inundating the floodplain habitat more often.

Deviation from the natural flow regime systematically inhibits the habitat-forming geomorphic process, downgrading the river's quality (Latrubesse et al. 2017; Kuiper et al. 2014). In the 1980s, there was a growing awareness among the developed nations on the impact of flow regulation on the habitat and distribution of freshwater species. Water resource extraction for human use while maintaining the instream flow to sustain the natural pattern of habitat connectivity became a necessity. To address this, the approach of an ecologically acceptable flow regime, also known as environmental flow or e-flow, was brought into practice (Petts 2009). Determination of the e-flow in reaches under the anthropogenic stress is feasible, only if the historical records are available for the periods before perturbation began. Determining the ecological flow baseline is important to mimic the natural water condition under which a complex biological community exists. The e-flow is usually classified based on the annual mean percentage and frequency of inter-annual flow attributes (high flow, low flow, peak flow). The e-flow management methods are developed on two major approaches: hydrological and habitat. These are further integrated with other aspects such as geomorphology, sociocultural impact, and water quality parameters for a more holistic e-flow assessment (O'Keeffe et al. 2012). Examples of holistic methods are building block methodologies (BBMs) and Downstream Response to Imposed Flow Transformation (DRIFT; Tharme 2003). The hydrological flow approach analyses the historical flow records to develop a method that sustains flow to mimic the natural, climate drive variability, whereas the habitat flow approach takes into consideration the temporal variability of the hydraulic features (Petts 2009). Mathur and Kapoor (2015) have highlighted the environmental flow requirement centered on the keystone fish species of the upper Ganga segment. Tare et al. (2017) modified the BBM method to incorporate longitudinal and lateral connectivity to estimate the e-flow of the lean, monsoon, and high-flow conditions.

9.2.2 GEOMORPHIC ALTERATION

The flow in rivers controls the physical features of a habitat, such as hydraulics, grain size distribution, and channel morphologies. The formation of pools and riffles, bars, and channel patterns are modulated by the flow. The anthropogenic control of the natural flow regime through the obstruction, regulation, and channelization of rivers disrupts the longitudinal and lateral connectivity and its natural nutrient replenishment capability. The geomorphic feedback to flow alteration is evident in reaches on both upstream and downstream of structural barriers. The feedback is

quite contrasting and differs for water and sediment flux and the resultant channel pattern change (Figure 9.1).

By monitoring and mapping changes in a river system, we can ascertain the impact on the ecology and formulate necessary restoration and management strategies.

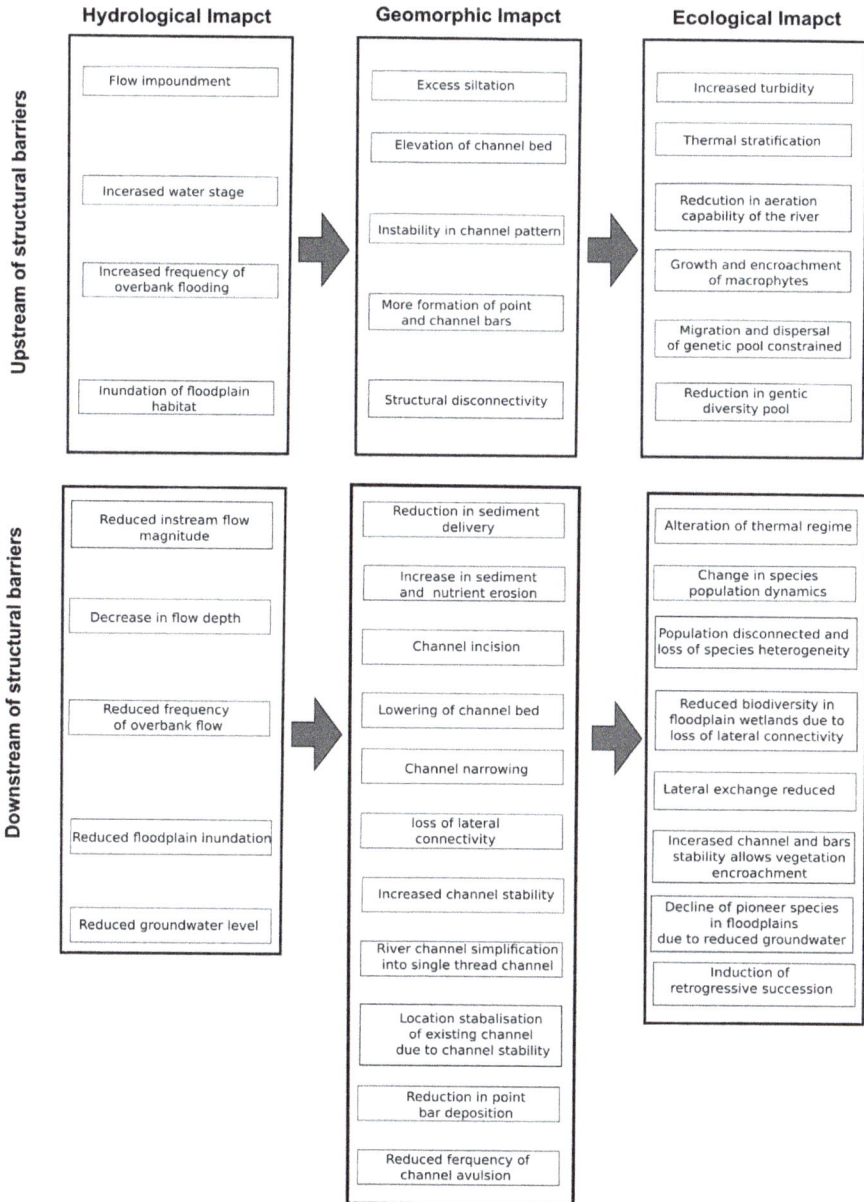

Hydrological Imapct

Geomorphic Imapct

Ecological Imapct

Upstream of structural barriers

Flow impoundment

Increased water stage

Increased frequency of overbank flooding

Inundation of floodplain habitat

Excess siltation

Elevation of channel bed

Instability in channel pattern

More formation of point and channel bars

Structural disconnectivity

Increased turbidity

Thermal stratification

Redcution in aeration capability of the river

Growth and encroachment of macrophytes

Migration and dispersal of genetic pool constrained

Reduction in gentic diversity pool

Downstream of structural barriers

Reduced instream flow magnitude

Decrease in flow depth

Reduced frequency of overbank flow

Reduced floodplain inundation

Reduced groundwater level

Reduction in sediment delivery

Increase in sediment and nutrient erosion

Channel incision

Lowering of channel bed

Channel narrowing

loss of lateral connectivity

Increased channel stability

River channel simplification into single thread channel

Location stabalisation of existing channel due to channel stability

Reduction in point bar deposition

Reduced ferquency of channel avulsion

Alteration of thermal regime

Change in species population dynamics

Population disconnected and loss of species heterogeneity

Reduced biodiversity in floodplain wetlands due to loss of lateral connectivity

Lateral exchange reduced

Increased channel and bars stability allows vegetation encroachment

Decline of pioneer species in floodplains due to reduced groundwater

Induction of retrogressive succession

FIGURE 9.1 Impact of structural barriers on geomorphic, hydraulic, and ecology of an alluvial river channel (modified after Sonkar and Gaurav 2020).

An economical and efficient way of monitoring the river ecosystem health is by the combined utilization of e-flow and river planform monitoring techniques. The reach-scale impact of structural barriers on the riverine system is reflected through the instream and floodplain geomorphic changes. To investigate temporal geomorphic and planform changes, remote sensing images can be used. Landsat images of the Earth Observation satellite missions provide a long time series (1972–present) of images that can be used to monitor the morphology of river channels. The Landsat images can be freely downloaded from the U.S. Geological Survey (USGS; https://earthexplorer.usgs.gov/). The wide temporal range of the Landsat mission is one of the best resources at our disposal for geomorphic mapping and change detection. The CORONA images are another important data set that is widely used to create the baseline map for morphological change analysis at a high spatial resolution (1.8–7.5 m). These images were acquired from 1960 to 1972 as a part of a spy mission by the United States. The CORONA images were declassified in 1995, now the images can be downloaded from the USGS web portal.

Similarly, ascertaining the width variability of an active channel together with the corresponding channel belt and floodplain during the monsoon and non-monsoon period on a multidecadal scale is one of the most effective ways to quantify the geomorphic changes in a river (Gupta, Atkinson, and Carling 2013; Sinha and Ghosh 2012). Flow variability and channel hydraulics configure the habitat template which determines the biological response (Jowett 1997). Assessing the habitat template under variable flow and their resultant hydraulics gives a measure of habitat suitability and the degree of habitat degradation. The use of an Acoustic Doppler Current Profiler (ADCP) provides a detailed analysis of flow characteristics at the cross-sectional level. By measuring the channel geometry (width and depth) and velocity in the upstream/downstream proximity of structural barriers, the impact of human interventions can be quantified. Analysis of the channel hydraulics and flow characteristics at variable flows will also help to predict the state of habitat under intervention.

9.2.3 INDICATOR SPECIES IN THE GANGA BASIN AND THEIR STATUS

Rivers are ecosystems with their unique community structure and function, but they are under severe pressure due to excessive water extraction and resource exploitation by humans. It is increasingly becoming difficult to find a balance between human needs and the minimum flow required to maintain ecosystem structure and function (Nale, Gosain, and Khosa 2017; Smakhtin and Anputhas 2006). Allowing an environmental flow that mimics the natural flow condition is being actively pursued under the Ganga River Basin Management Plan. The problem arises as to how can we accurately predict the minimum flow that would sustain the entire normal ecosystem functioning as well as meet the anthropogenic demands. Studies centered on the hydrological demands of Indicator species provide headway for sustainable water management (Mathur and Kapoor 2015). Indicator species are those organisms that act as a proxy for river ecosystem health. The term *indicator species* is sometimes interchangeably used with *keystone species*, with the difference being that keystone species have a large effect on their environment relative to their abundance

(Paine 1995). Adapting to hydro-geomorphic and structural changes in a river under the human-influenced flow regime has been challenging for the large aquatic vertebrates despite their wide acclimatization range. Some of the major Indicator species for an exploratory study on the health of the Ganga, Brahmaputra, and Meghna Rivers are the Ganga river dolphins (*Platanista gangetica*), gharials (*Gavialis gangeticus*), smooth-coated otter (*Lutrogale perspicillata*), monitor lizards (*Varanus bengalensis*), and freshwater turtles. From a conservation perspective, these are considered flagship species for a broader support campaign (Mathur and Kapoor 2015)

9.2.4 THREATS TO EXPLORATORY INDICATOR SPECIES

9.2.4.1 Human Conflict and Habitat Quality Degradation

The Ganga River dolphin and gharials, which were once widespread, are now endangered species (Figure 9.2) due to several factors, ranging from habitat fragmentation and degradation, human killing, and water pollution (Braulik and Smith 2017; Das, Das, and Dutta 2013). Human conflicts, such as accidental death by boats and deliberate killing of the river dolphins for their oil, have had a severe impact on their population in the past few decades (Sinha 1997, 2002). About 100 dolphins are killed annually by gillnets in the lower Ganga, and about 150 in the Brahmaputra, from Dubri to Saikowagaht (Mohan 1996). In 2004–2005, about 14 dolphins were killed by net catch and poaching, and 583 gill nets were recorded in the Brahmaputra River (Wakid 2009). Habitat utilization of the Ganga dolphin at the Vikramshila Gangetic Dolphin Sanctuary shows an estimated 85% spatial and 75% prey-resource overlap in fisheries and dolphins (Kelkar et al. 2010). In the Sapt Koshi and Karnali

FIGURE 9.2 Historical and present distribution of the Ganga river dolphins and gharials on the Himalayan foreland (image source: Google Earth).

Rivers of Nepal, there is an observed >60% exploitation rate of dolphin prey by humans (Paudel, Koprowski, and Cove 2020). Moreover, an increase of about 48% in behavioral change was observed in the freshwater cetacean exposed to fishing activity (Paudel, Koprowski, and Cove 2020). In the main channel and tributaries of the Barak River in Assam, the cetacean has extirpated, due to severe cases of poaching, net catch, prey-base depletion, and habitat destruction (Choudhury et al. 2019). This shows a high level of human–dolphin conflict in the Himalayan rivers. Gharials are in a much more dire state compared to other large aquatic vertebrates. They have a specific niche and geomorphic requirement of sandy river banks and large sand bars (Rao et al. 2013). In 1940, the estimated population of gharials was about 5,000–10,000, which was reduced to 200 in 1976 and remained the same in 2006 (Behera, Singh, and Sagar 2014). Furthermore, in 2008, nearly 111 gharials died in the Chambal Sanctuary, forcing authorities to reintroduce 494 gharials between 2009–2012 in another river habitat (Yadav, Nawab, and Khan 2013). In 2009–2012, the World Wildlife Fund and the Uttar Pradesh Forest Department selected a reach of the Ganga River near Hastinapur Wildlife Sanctuary. As this reach is free from anthropogenic pressure, and had the optimal gharial habitat such as large river width, shallow water, deep pools, moderate velocity, and sandy banks (Yadav, Nawab, and Khan 2013). Contaminants and pollutants are immediate threats to river health. Industrial and sewage waste in the middle Ganga alter the physiochemical properties of the river, making it highly undesirable as a habitat (Kumar, Jha, and Baier 2012; Santy, Mujumdar, and Bala 2020). High concentrations of heavy metals discharged from tanneries on the bank of river Buriganga, Bangladesh, were documented in the Ganga river dolphin (Alam and Sarker 2012). The river stretch between Kaachala Ghat and Kanpur in the middle Ganga is heavily polluted by industrial and terrestrial effluent, resulting in an absence of dolphins (Figure 9.2).

9.2.4.2 Habitat Fragmentation by Structural Barriers

The natural flow regime determines the hydrological and geomorphic characteristics to which dolphins have adapted by evolving their external morphology (Jensen et al. 2013; Poff et al. 1997). Radical changes in flow, channel hydraulics, sediment flux, and geomorphic characteristics of a river, prompted by the natural flow disruption, render the river unsuitable for habitation by the aquatic fauna. Paudel and Koprowski (2020) observed about an 18% decrease in the historical global distribution range of the Ganga River dolphin in the Ganga-Brahmaputra-Meghna and Karnaphuli River basins. A higher fragmentation rate is observed in the Ganga and its tributaries in Nepal and India than in rivers in Bangladesh (Sinha and Kannan 2014; Smith and Reeves 2000). The extirpated stretches of the Ganga, Brahmaputra, and Meghna River systems mostly include upstream and downstream reaches of certain large barrages in India and Nepal (Table 9.1). It is claimed by environmentalists, that about 940 dams, barrages, and weirs are present within the Ganga basin, obstructing and hampering the Ganga rejuvenation program (Press Trust of India 2018). The substantial number of structural barriers created to divert and store water in the Ganga basin has resulted in poor e-flow and low habitat quality.

TABLE 9.1

River Reaches Where the Population Extirpation of the Ganga River Dolphin Has Occurred in the Historical Range

River	River Stretch
Ganga	Haridwar barrage to Bijnor barrage
	Narora barrage to Kanpur barrage
Ramganga	Khoo barrage to Ganga River confluence
Yamuna	Okhla barrage to Chambal River confluence
Betwa	Rajghat to Chambal River confluence
Sone	Bandsagar dam to Ganga confluence
Kosi	Upstream of Sapt Kosi barrage
Mahakali	Upstream of Sharda barrage
Sharda	Sharda barrage to Lower Sharda barrage
Barak	Main channel of the Barak River

The diversion of water from barrages in the middle Ganga has reduced the overall flow by 10%, disrupting the lateral and longitudinal connectivity of the riverine habitat (WII-GACMC 2017). Apart from disconnectivity, structural barriers have a hydrological and geomorphic impact on the river system. Several authors have studied the impact of structural barriers on the riverine habitat morphology and ecology of the indicator species in the Ganga, Brahmaputra, and Meghna River systems, which we discuss in detail in a later section.

9.3 RESULTS AND DISCUSSION

9.3.1 Channel Morphology and Sediment Dynamics

In alluvial rivers, downstream sediment conveyance is necessary for the functional channel processes to occur. However, the reduced transportation of water and sediment interposes the dynamic equilibrium of the rivers. Flow regulation limits the sediment supply and channel capacity, which usually translates into a change in channel pattern. Sediment-free water (hungry waters) has the potential to retain the sediment load upstream that tends to erode the banks and incise the riverbed. Hungry water released during the monsoon erodes the channel bed and deposits them further downstream (Kondolf 1997). Generally, a greater degree of channel stability is observed due to the lack of sediment flux (Overeem, Kettner, and Syvitski 2013). This tends to alter multithread channels to a single thread (Ligon, Dietrich, and Trush 1995).

At the point of structural interventions (i.e., dams, barrages) in a river, impoundment and excess siltation are the major effects in the immediate upstream reach. This results in channel bed aggradation and eventually pattern instability. For example, channel bed aggradation of the Ganga River at Farakka (Figure 9.3b) has accelerated the channel mobility (Pal and Pani 2019; Sinha and Ghosh 2012; Thakur,

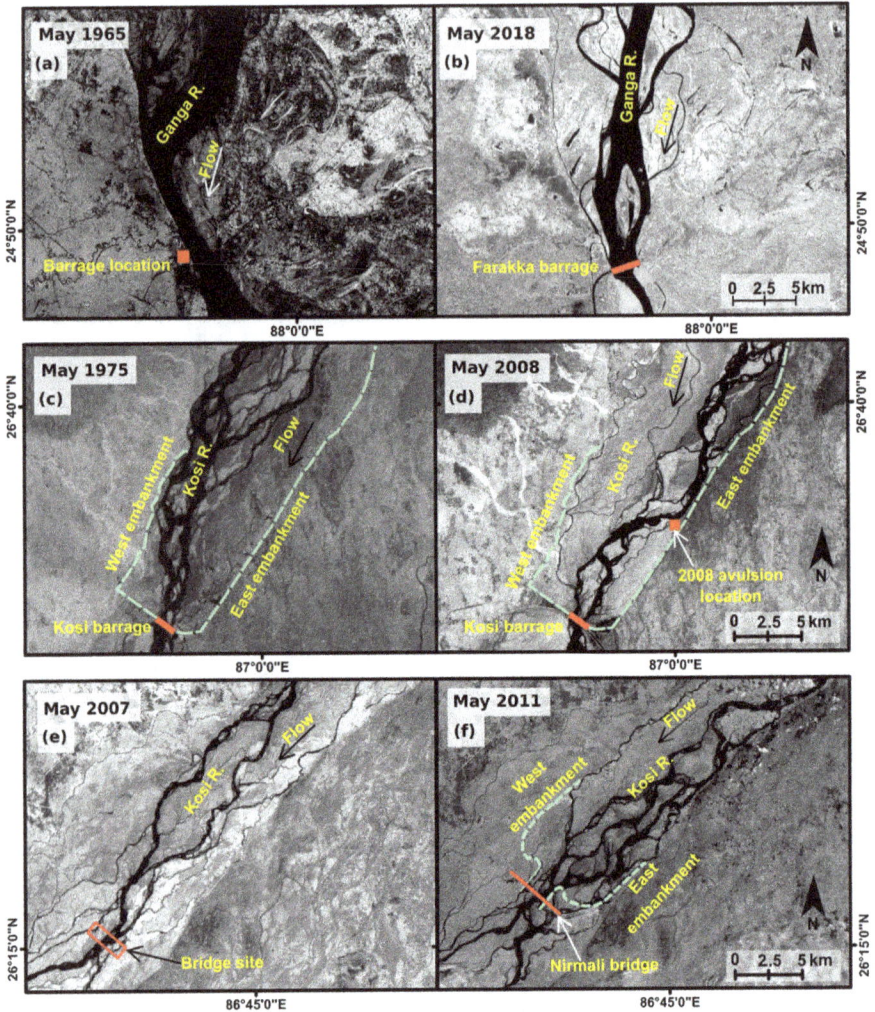

FIGURE 9.3 Structural intervention constructed to confine the flow pathways of a river has a major geomorphic consequence. The images on the left represent the pre-impact period and the images on the right show the post-impact period (image source: Corona, Landsat TM, & Landsat OLI).

Laha, and Aggarwal 2012). In some cases, the simultaneous formation of point and mid-channel bars with channel instability can cause the river channel to develop multithread channels (Thakur, Laha, and Aggarwal 2012). Furthermore, under high sediment aggradation rate, the formation of point and mid-channel bars becomes more frequent. In the Kosi River of northern Bihar, India, embankments constructed on either bank of the river have played a significant role in channel shifting and avulsion (Devkota, Crosato, and Giri 2012; Sinha et al. 2013). After the construction of

the Kosi barrage, the sedimentation rate upstream of the barrage has increased in the river at 0.12 m/yr (Devkota, Crosato, and Giri 2012). The high sedimentation rate and planform dynamics (Figures 9.3c and d) led to channel avulsion through an embankment breach and flooding in 2008 (Sinha et al. 2014). Another example of high sedimentation on planform is at the Nirmali bridge (also known as the Kosi Maha bridge) further downstream of the Kosi barrage. Two embankments were constructed about 10–15 km apart on either side of the channel belt. The embankment was designed as a bow-shaped funnel to confine the 11-km channel belt to pass through a 1.9-km passage at the bridge (Choubey 2012). As the flow is confined within the funnel passage, the river is forced to deposit sediment in reaches upstream of the bridge. Silt deposition has raised the level of the channel bed by about 4–5 m as compared to the associated floodplains. A large alluvial island can be seen in the immediate upstream of the Nirmali bridge (Figures 9.3e and f).

9.3.2 Hydraulic Variability

Discharge sets the first-order geometry of an alluvial river. Flow regulation at the barrages translates into the variability of channel depth and width. In the backwaters of Tilpara barrage over Mayurakasi River in West Bengal, India, there is an increase of about 30% and a 0.7-m reduction in the long-term trend of channel width and depth, respectively (Pal 2017). The channel bed aggradation was the highest near the obstruction point, thus elevating the water stage and directly contributing to the apparent lateral increase of water surface area (Figure 9.4). A similar phenomenon is observed upstream of the Farakka barrage (Thakur, Laha, and Aggarwal 2012). The water level at Farakka has increased from 15.24 m in the pre-barrage period to 21.95 m in the post-barrage period (Thakur, Laha, and Aggarwal 2012). On the other hand, downstream of the Farakka barrage, the post-barrage lean period (February–May)

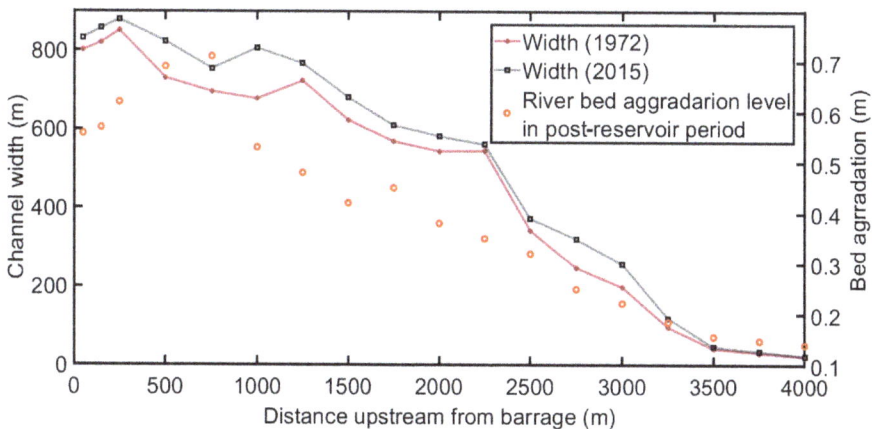

FIGURE 9.4 The pre- and post-impact period change the channel width upstream of the Tilpara barrage (data source: Pal 2017).

discharge has decreased by 1/3 (Sonkar and Gaurav 2020). This has reduced the wetted width by 3/4 (Sonkar and Gaurav 2020). This shows a contrasting impact on flow and sediment transport on either side of the interventions. Channel bed aggradation and water surface widening in backwaters contribute to an increase in the wetted perimeter and channel cross-sectional area. Pal (2017) and Thakur, Laha, and Aggarwal (2012) observed an increase in cross-sectional area upstream of Tiplara and Farakka barrages, respectively. Channel scouring and bank erosion are sighted as the major reason for channel shift of the Ganga River upstream of the Farakka barrage (Banarjee 1999; Rudra 2010; Sinha and Ghosh 2012; Thakur, Laha, and Aggarwal 2012). If we move further upstream from the point of intervention, the impact of excess sedimentation on the channel configuration becomes less prominent. One such example of channel geometry variation can be observed between the Bijnor and Narora barrages in the Ganga River (Figure 9.5). River reach with a deep cross section serves as a stable habitat for the Ganga river dolphins due to an adequate depth in the low flow season. Such deep cross sections are found upstream of the Narora barrage, where the channel has scoured near the banks to provide adequate forging depth (Figure 9.5b). On the other hand, at Garhmukteshwar, which is between the Bijnor (~85 km upstream) and the Narora barrage (~80 km downstream), the channel has a shallow cross section (Figure 9.5a). Reaches with such shallow cross sections are preferable only for longitudinal navigation, as depth acts as a limiting factor in niche selection. Another example is the Vikramshila Gangetic Dolphin Sanctuary, near Bhagalpur. The channel in the sanctuary is characterized by deep and wide cross sections (Figure 9.6). This sanctuary supports one of the highest populations of freshwater cetacean (Kelkar 2008). Altogether, channel geometry plays a vital role in habitat selection and navigation by the aquatic fauna.

9.3.3 INDICATOR SPECIES HABITAT DEGRADATION BY HUMAN INTERVENTION

9.3.3.1 Hydraulic Condition

The channel pattern and hydraulic geometry of regulated reaches serve as a sanctuary only if the habitat requirement conditions are met. Choudhary et al. (2012) conducted a study on the distribution and habitat use of the Ganga cetacean in a 332-km stretch from the Gandak barrage upstream to its confluence with the Ganga River downstream. They compared their finding with similar studies on rivers that host the cetacean population. They inferred that dolphin presence exhibit a positive correlation with reaches consisting of deep pools (5 m) and a meandering pattern. It can be argued that the dolphins may not utilize a certain portion of a river that does not accommodate the minimum required depth. The cetaceans have adapted to the varying habitat conditions, for example, the recorded depth of an actively used habitat ranges from 1–12 m (Table 9.2). Water diversion and excess siltation eliminate environmental variables that form suitable and adaptable habitat conditions. In the Barak River system of Assam, the cetacean population was recorded until 2008, but no sighting was confirmed in 2013 and 2014 (Choudhury et al. 2019). Furthermore, the land-use change and loss of riparian vegetation in the floodplain have amplified

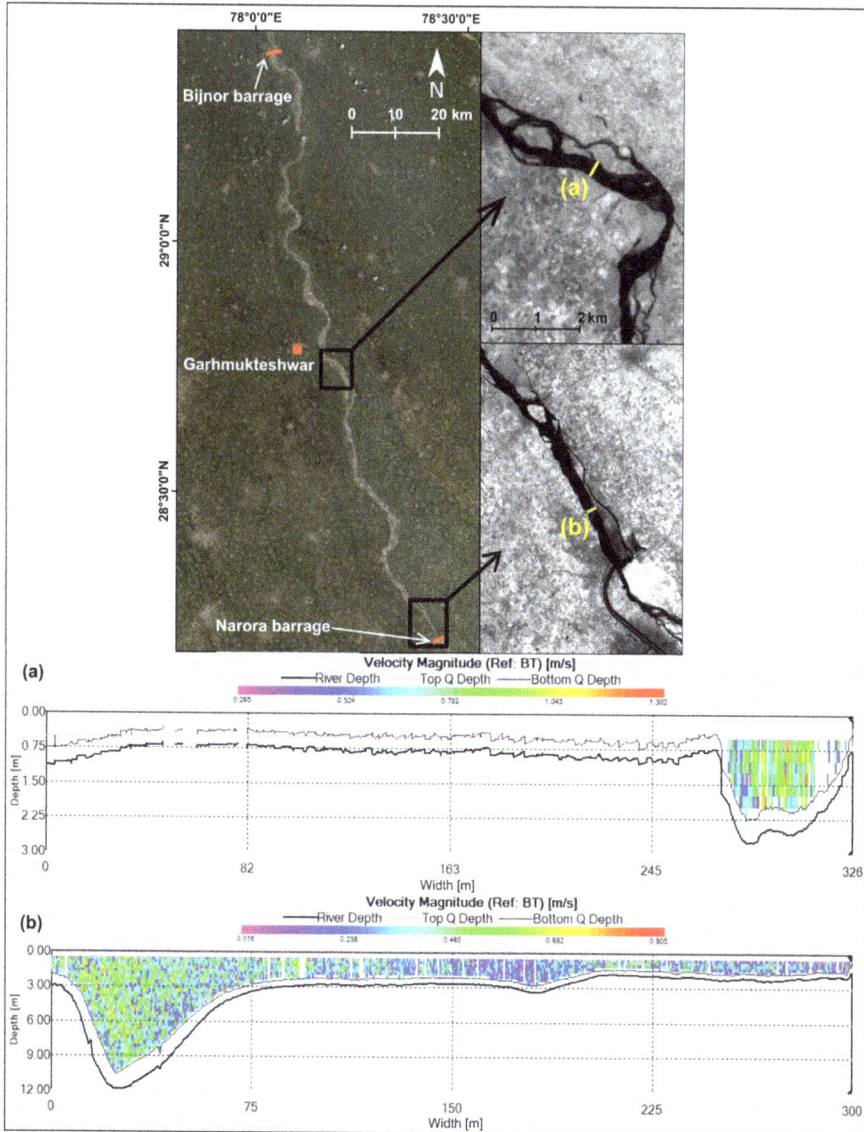

FIGURE 9.5   The cross-sectional profile of the Ganga River between Bijnor and Narora barrage, at Garhmukteshwar (a) and upstream of the Narora barrage (b) (image source: Landsat OLI).

erosion and landslide at banks, resulting in siltation of the channel bed (Mazumder 2014). This causes the aggradation of the channel bed to eventually reduce the depth (Mazumder et al. 2014). The prevalence of such conditions makes the river reach a highly unsuitable habitat for the freshwater cetacean (Kelkar 2008; Biswas and Boruah 2000).

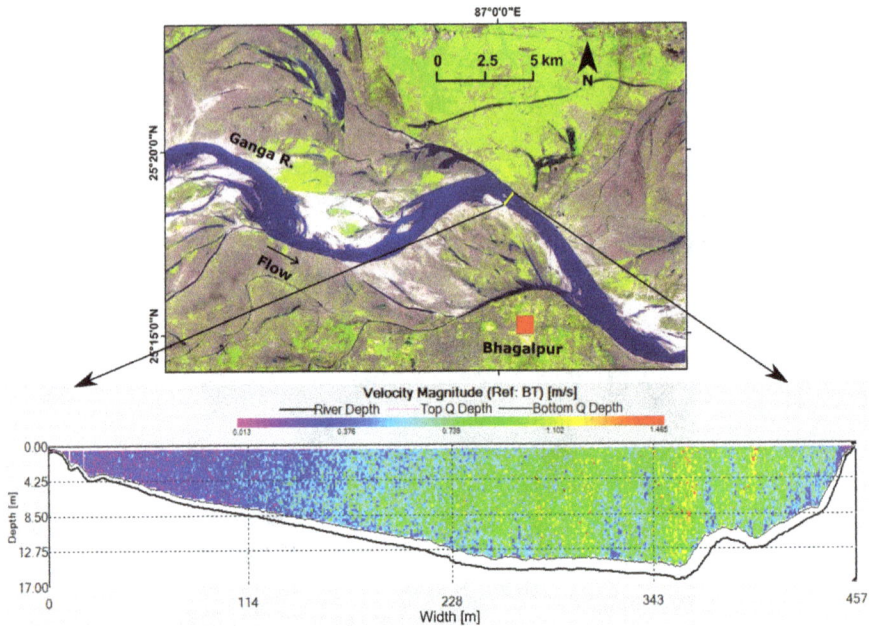

FIGURE 9.6   A cross-sectional profile of the Ganga River near Bhagalpur, in the lower Ganga. The deep and wide cross sections are the most preferred habitat of the Ganga dolphin (image source: Sentinel 2).

TABLE 9.2
Depth is one of the most important limiting factors of habitat preference and distribution of the Ganga River cetacean. The table gives the record depth and river morphology of sighted dolphins (modified after: Sonkar and Gaurav 2020)

River	Depth (m)	Channel Pattern of the Recorded Habitat	Reference
Brahmaputra	4–12	Confluences with high fish assemblage, deep meanders, and tributaries with eddy-counter currents. Depth preference between 4 to 6 m.	Biswas and Boruah 2000; Wakid 2009
Ganga	1–14.2	Deep meanders and confluences with eddy countercurrents Dolphins of upstream reaches of the Ganga River are more adapt to a shallow depth than downstream populations.	Bashir et al. 2010; Behra, Singh, and Sagar 2014; Sinha 1997; Sinha and Kannan 2014

River	Depth (m)	Channel Pattern of the Recorded Habitat	Reference
Ganga between the Bijnor to Narora barrage	1–14 m	Deep pools, confluences, and side channels The entire reach has shallow depth with intermittent deep pools. Reach morphology: Meandering and wandering channels	Bashir et al. 2010; Gaur, Akolkar, and Arora 2009
Ganga at Farakka	No data	Deep pools with shallow nearby banks, confluences Reach morphology: Meandering and wandering channels upstream of the barrage. Single channel with bifurcation in the downstream reach of the barrage	Banarjee 1999; Gupta, Atkinson, and Carling 2013; Thakur, Laha, and Aggarwal 2012
Karnali	1.7–5.2	Single narrow sinuous channel with small tributaries	Paudel et al. 2020
Gandak	2.2–5	Meandering channels with deep pools	Choudhary et al. 2012
Kosi	5–8	Confluence, deep meanders, and eddy countercurrents Absence of population in deep confluence due to lack of prey availability	Sinha and Sharma 2003
Hooghly between Krishnanagar to Kolkata	6	Good concentration of dolphins at the confluence and behind sand bars Reach morphology: Wide straight channels with occasional deep meanders	Sharma 2010
Yamuna	No data	Confluence and deep pools Found downstream of the confluence with Chambal and Betwa River Reach morphology: Single channel with meanders	Behra, Singh, and Sagar 2014
Sundarban mangroves	4–12	Low salinity High turbidity, moderate depth, and confluences Upstream distribution limit constrained by interspecies competition	Smith et al. 2009

9.3.3.2 Hydrological and Geomorphic Impact

Physical barriers can sometimes have a contrasting impact on population distribution. Sonkar and Gaurav (2020) compared the two distinct habitats of the freshwater cetacean one in the middle Ganga between Bijnor and the Narora barrage, and the other in the lower Ganga at the Farakka barrage. From Bijnor to the Narora barrage, there is a reported rise in the dolphin population from 1980 to 2010. The physical isolation between the two barrages and the availability of minimum flow during the dry

season has aided the growth in the dolphin population in this reach. In contrast, in the lower Ganga at the Farakka barrage, the dolphin population has declined since the construction of the barrage in 1975 (Sinha 2000; Smith et al. 1998). The hydrological and geomorphic changes caused by the barrage operation and regulated flow have played a significant role in altering the cetacean habitat. In the immediate upstream of the Farakka barrage, excess siltation and bar formation have promoted macrophyte growth. Siltation and macrophyte growth are associated with poor dolphin habitat as observed in the Sone River (Sinha and Sharma 2003). Siltation has raised the channel bed, bank erosion, and channel shifting (Pal and Pani 2019; Sinha and Ghosh 2012; Thakur, Laha, and Aggarwal 2012). The elevated channel bed leads to an increase in the frequency of overbank flooding, and events leading to inundation of channel floodplains. Floodplain inundation induces retrogressive succession in macrophytes vegetation changing their community dynamics (Edwards 1968; Corenblit et al. 2007). Downstream of the Farakka barrage, the channel has incised and the width has reduced significantly from about 3.4 km in 1965 (pre-barrage period) to 735 m in 2018 (Sonkar and Gaurav 2020). During the pre-barrage period, the downstream reach at Farakka had a multichannel flow (Figure 9.7). The reduction in width has disconnected the Ganga River from its floodplain, which has significantly affected the habitat quality (Sonkar and Gaurav 2020).

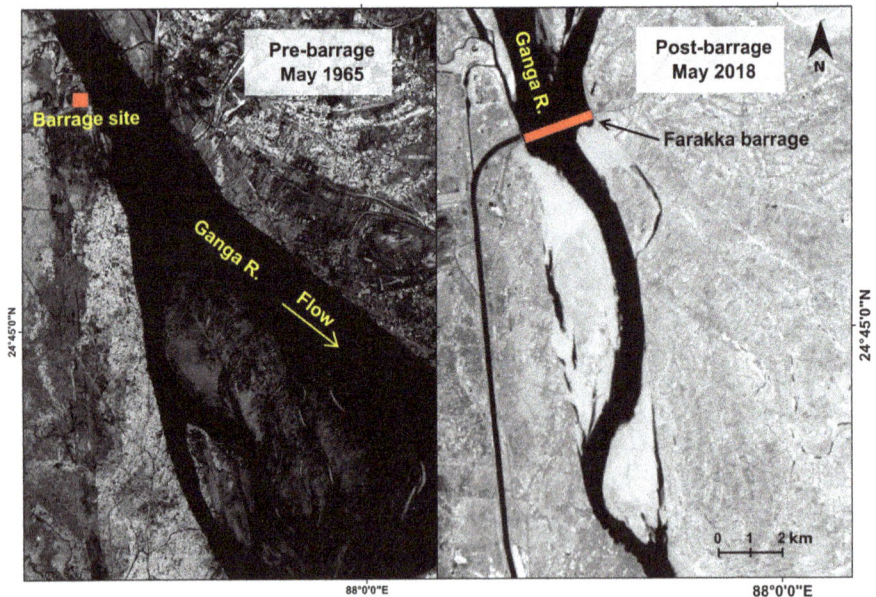

FIGURE 9.7 The Ganga River at Farakka barrage. Image on the left shows the Ganga River in the pre-barrage period of 1965 and image on the right shows the post-barrage condition (image source: CORONA and Landsat OLI).

9.4   CONCLUSION

Riverine species in the Himalayan foreland are under threat due to the habitat fragmentation and degradation brought on by human interventions. Flow regulation and poor sediment delivery near the structural barriers reduce the lateral (channel to flood plain) and along stream connectivity. The lateral connectivity is important for the nutrient exchange between the instream and its floodplain. On the other hand, excess siltation upstream directly contributes to channel bed aggradation and high turbidity that eventually degrades the habitat suitability. The ecological integrity of a riverine habitat is maintained by the hydro-geomorphic process and easily acclimatizes to short temporal perturbation in the system but is highly vulnerable to large changes. The indicator species show their vulnerability to high magnitude hydrogeomorphic changes. This emphasizes the need for an integrated approach to study the hydro-geomorphic changes in rivers due to anthropogenic interventions at different spatial scales, such as micro (cross section or biotope), meso (river reach or segment), and macro (catchment, sub-catchment, and network-centric).

Aquatic fauna occupies a niche based on the flow characteristics and available habitat features at a micro-scale, but alteration on meso- or macro-scale can influence changes on lower levels. Impact analysis for conservation and restoration purposes must employ a more comprehensive approach that includes the impact assessment of anthropogenic interventions at all three distinct scales. A much more radical mitigation measure of large dams and barrage removal may become the need of the hour in the future given the present conservation practices. Dam removal is an emerging practice in western nations to restore ecological integrity and ecosystem services. Systematic removal of anthropogenic obstruction and regulation points in rivers can be a way forward, which, of course, depends on public interest. In summary, to properly assess and maintain habitat, be it of an indicator species or the ecosystem as a whole, a multi-spatial scale study must be carried out with the integration of hydrological and sediment connectivity aspects. The way forward cannot be just through hydrological assessment but a holistic approach of hydrological, geomorphic, and ecological impact assessment of the human intervention.

REFERENCES

Alam, Shayer MahmoodIbney, and Noor Jahan Sarker. 2012. "Status and Distribution of the Gangetic Dolphin, Platanista Gangectica Gangetica (Roxburgh, 1801) in River Buriganga during 2003–2004 and Its Conservation." *Bangladesh Journal of Zoology* 40 (1): 21–31.
Arunachalam, M. 2000. "Assemblage Structure of Stream Fishes in the Western Ghats (India)." *Hydrobiologia* 430: 1–31. https://doi.org/https://doi.org/10.1023/A:1004080829388.
Bain, Mark B., and John T. Finn. 2013. "Streamflow Regulation and Fish Community Structure." *Ecological Society of America* 69 (2): 382–392.
Banarjee, M. 1999. "A Report on the Impact of Farakka Barrage on the Human Fabric." New Delhi.
Bashir, T., A. Khan, P. Gautam, and S. K. Behera. 2010. "Abundance and Prey Availability Assessment of Ganges River Dolphin (Platanista Gangetica Gangetica) in a Stretch of

Upper Ganges River, India." *Aquatic Mammals* 36 (1): 19–26. https://doi.org/10.1578/AM.36.1.2010.19.

Behera, S. K., H. Singh, and V. Sagar. 2014. "Indicator Species (Gharial and Dolphin) of Riverine Ecosystem: An Exploratory of River Ganga." In *Our National River Ganga: Lifeline of Millions*, edited by R Sanghi, 103–1213. Switzerland: Springer International Publishing. https://doi.org/10.1007/978-3-319-00530-0.

Best, J. 2019. "Anthropogenic Stresses on the World's Big Rivers." *Nature Geoscience* 12 (1): 7–21. https://doi.org/10.1038/s41561-018-0262-x.

Biswas, S. P., and S. Boruah. 2000. "Ecology of the River Dolphin (Platanista Gangetica) in the Upper Brahmaputra." *Hydrobiologia* 430: 97–111.

Braulik, G. T., and B. D. Smith. 2017. "Platanista Gangetica." *South Asian River Dolphin*, The IUCN Red List of Threatened Species 8235. http://doi.org/10.2305/IUCN.UK.2017-3.RLTS.T41758A50383612.en.

Choubey, Jitendra. 2012. "Bridge Over Kosi: Connecting People or Brainwashing Them?" In *Agony of Floods. Flood Induced Water Conflicts in India*, edited by Eklavya Prasad, K. J. Joy, Suhas Paranjape, and Shruti Vispute, 23–32. Pune: Forum for Policy Dialogue on Water Conflicts in India.

Choudhary, S., S. Dey, S. Dey, V. Sagar, T. Nair, and N. Kelkar. 2012. "River Dolphin Distribution in Regulated River Systems: Implications for Dry-Season Flow Regimes in the Gangetic Basin." *Aquatic Conservation: Marine and Freshwater Ecosystems* 22 (1): 11–25. https://doi.org/10.1002/aqc.1240.

Choudhury, Nazrana Begam, Muhammed Khairujjaman Mazumder, Himabrata Chakravarty, Amir Sohai Choudhury, Freeman Boro, and Imrana Begam Choudhury. 2019. "The Endangered Ganges River Dolphin Heads towards Local Extinction in the Barak River System of Assam, India: A Plea for Conservation." *Mammalian Biology* 95 (March): 102–111. https://doi.org/10.1016/j.mambio.2019.03.007.

Corenblit, Dov, Eric Tabacchi, Johannes Steiger, and Angela M. Gurnell. 2007. "Reciprocal Interactions and Adjustments between Fluvial Landforms and Vegetation Dynamics in River Corridors: A Review of Complementary Approaches." *Earth-Science Reviews* 45 (1–2): 56–86. https://doi.org/10.1016/j.earscirev.2007.05.004.

Das, A., A. K. Das, and S. K. Dutta. 2013. "An Assessment of Assisted Recovery of Gavialis Gangeticus in the River System of Northeast India." In *Proceeding of the World Crocodile Conference, 22nd Workshop Meeting of the Special Survival Commission of the IUCN*, edited by IUCN, 29–35. Negombo, Sri Lanka: International Union for Conservation of Nature.

Devkota, Lochan, Alessandra Crosato, and Sanjay Giri. 2012. "Effect of the Barrage and Embankments on Flooding and Channel Avulsion Case Study Koshi River, Nepal." *A Journal of Rural Infrastructure Development, Society of Engineers' for Rural Development, Nepal (SERDeN)* 3 (3): 124–132.

Dubey, Vineet Kumar, Uttam Kumar Sarkar, Ajay Pandey, Rupali Sani, and Wazir Singh Lakra. 2012. "The Influence of Habitat on the Spatial Variation in Fish Assemblage Composition in an Unimpacted Tropical River of Ganga Basin, India." *Aquatic Ecology* 46: 165–174. https://doi.org/10.1007/s10452-012-9389-9.

Edwards, D. 1968. "Some Effects of Siltation upon Aquatic Macrophyte Vegetation in Rivers." *Hydrobiologia* 34 (1): 29–37.

Gaur, A., P. Akolkar, and M. P. Arora. 2009. "Water Quality Assessment of River Ganga for Conservation of Gangetic Dolphins (Platanista Gangetica) at Garhmukteshwar." *Environment Conservation Journal* 10 (3): 57–62.

Gregory, K. J. 2006. "The Human Role in Changing River Channels." *Geomorphology* 79 (3–4): 172–191. https://doi.org/10.1016/j.geomorph.2006.06.018.

Gregory, K. J., R. J. Davis, and P. W. Downs. 1992. "Identification of River Channel Change to Due to Urbanization." *Applied Geography*, 12: 299–318.

Gupta, N., P. M. Atkinson, and P. A. Carling. 2013. "Decadal Length Changes in the Fluvial Planform of the River Ganga: Bringing a Mega-River to Life with Landsat Archives." *Remote Sensing Letters* 4 (1): 1–9. https://doi.org/10.1080/2150704X.2012.682658.

Gupta, Nishikant, Varun Tiwari, Mark Everard, Melissa Savage, Syed Ainul Hussain, Michael A. Chadwick, Jeyaraj Antony Johnson, Asghar Nawab, and Vinod K. Belwal. 2020. "Assessing the Distribution Pattern of Otters in Four Rivers of the Indian Himalayan Biodiversity Hotspot." *Aquatic Conservation: Marine and Freshwater Ecosystems* 30 (3): 601–610. https://doi.org/10.1002/aqc.3284.

Jensen, F. H., A. Rocco, R. M. Mansur, B. D. Smith, V. M. Janik, and P. T. Madsen. 2013. "Clicking in Shallow Rivers: Short-Range Echolocation of Irrawaddy and Ganges River Dolphins in a Shallow, Acoustically Complex Habitat." *PLoS One* 8 (4): e59284. https://doi.org/10.1371/journal.pone.0059284.

Jowett, I G. 1997. "Instream Flow Methods: A Comparison of Approaches." *Regulated Rivers: Research & Management* 13 (2): 115–127.

Kachari, Akash, Budhin Gogoi, Rashmi Dutta, Aran Kamhun, Pritha Ghosh, Maitra Sudipta, Samir Bhattacharya, and Debangshu N. Das. 2014. "Habitat Preference of an Endangered Hill Stream Catfish Olyra Longicaudata (McClelland) from Arunachal Pradesh, India." *International Journal of Fisheries and Aquatic Studies* 1 (3): 86–93.

Kelkar, N. 2008. "Patterns of Habitat Use and Distribution of Ganges River Dolphins Platanista Gangetica in a Human-Dominated RiverScape in Bihar, India." Manipal University.

Kelkar, N., J. Krishnaswamy, S. Choudhary, and D. Sutaria. 2010. "Coexistence of Fisheries with River Dolphin Conservation." *Conservation Biology* 24 (4): 1130–1140. https://doi.org/10.1111/j.1523-1739.2010.01467.x.

Kondolf, G. M. 1997. "Hungry Water: Effects of Dams and Gravel Mining on River Channels." *Environmental Management* 21 (4): 533–551. https://doi.org/10.1007/s002679900048.

Kuiper, Jan J., Jan H. Janse, Sven Teurlincx, Jos T. A. Verhoeven, and Rob Alkemade. 2014. "The Impact of River Regulation on the Biodiversity Intactness of Floodplain Wetlands." *Wetlands Ecology and Management* 22 (6): 647–658. https://doi.org/10.1007/s11273-014-9360-8.

Kumar, Shikhar, Pallavi Jha, and Klaus Baier. 2012. "Pollution of Ganga River Due to Urbanization of Varanasi Adverse Conditions Faced by the Slum Population." *Environment and Urbanization ASIA* 3 (2): 343–352. https://doi.org/10.1177/0975425312473229.

Lal Mohan, R. S. 1996. "Mortality of Ganges River Dolphin (Platanista Gangetica) in Gillnets of Ganges and Brahmaputra in India." *Tigerpaper (FAO)* 22 (1): 11–13.

Latrubesse, Edgardo M., Eugenio Y. Arima, Thomas Dunne, Edward Park, Victor R. Baker, Fernando M. D'Horta, Charles Wight, et al. 2017. "Damming the Rivers of the Amazon Basin." *Nature* 546 (7658): 363–369. https://doi.org/10.1038/nature22333.

Ligon, Franklin K., William E. Dietrich, and William J. Trush. 1995. "Downstream Ecological Effects of Dams." *BioScience* 45 (3): 183–192. https://doi.org/10.2307/1312557.

Lytle, David A., and N. Le Roy Poff. 2004. "Adaptation to Natural Flow Regimes." *Trends in Ecology and Evolution* 19 (2): 94–100. https://doi.org/10.1016/j.tree.2003.10.002.

Mathur, R. P., and Vishal Kapoor. 2015. "Concept of Keystone Species and Assessment of Flows (Himalayan Segment- Ganga River)." In *Hydropower for Sustainable Development*, 387–397. Dehradun: International Conference on Hydropower for Sustainable Development.

Mazumder, Muhammed Khairujjaman. 2014. "Diversity, Habitat Preferences, and Conservation of the Primates of Southern Assam, India: The Story of a Primate

Paradise." *Journal of Asia-Pacific Biodiversity* 7 (4): 347–354. https://doi.org/10.1016/j. japb.2014.10.001.

Mazumder, Muhammed Khairujjaman, Freeman Boro, Badruzzaman Barbhuiya, and Utsab Singha. 2014. "A Study of the Winter Congregation Sites of the Gangetic River Dolphin in Southern Assam, India, with Reference to Conservation." *Global Ecology and Conservation* 2 (March): 359–366. https://doi.org/10.1016/j.gecco.2014.09.004.

Nale, J. P., A. K. Gosain, and R. Khosa. 2017. "Environmental Flow Assessment of River Ganga—Importance of Habitat Analysis as a Means to Understand Hydrodynamic Imperatives for a Sustainable Ganga Biodiversity." *Current Science* 112 (11): 2187–2188. https://doi.org/10.18520/cs/v112/i11/2187-2188.

Nautiyal, P. 2010. "Food Chains of Ganga River Ecosystems in the Himalayas." *Aquatic Ecosystem Health & Management* 13 (4): 362–373. https://doi.org/10.1080/14634988. 2010.528998.

O'Keeffe, J., N. Kaushal, V. Smakhtin, and L. Bharati. 2012. "Assessment of Environmental Flows for the Upper Ganga Basin." *WWF- India*. New Delhi. https://doi.org/10.1007/ 978-3-319-00530-0.

Overeem, I., A. J. Kettner, and J. P.M. Syvitski. 2013. "Impacts of Humans on River Fluxes and Morphology." *Treatise on Geomorphology* 9: 828–842. https://doi.org/10.1016/ B978-0-12-374739-6.00267-0.

Paine, R. T. 1995. "A Conversation on Refining the Concept of Keystone Species." *Conservation Biology* 9 (4): 962–964. https://doi.org/10.1046/j.1523-1739.1995.09040962.x.

Pal, Raghunath, and Padmini Pani. 2019. "Remote Sensing and GIS-Based Analysis of Evolving Planform Morphology of the Middle-Lower Part of the Ganga River, India." *Egyptian Journal of Remote Sensing and Space Science* 22 (1): 1–10. https://doi. org/10.1016/j.ejrs.2018.01.007.

Pal, Swades. 2017. "Impact of Tilpara Barrage on Backwater Reach of Kushkarni River: A Tributary of Mayurakshi River." *Environment, Development, and Sustainability* 19 (5): 2115–2142. https://doi.org/10.1007/s10668-016-9833-4.

Paudel, S., and J. L. Koprowski. 2020. "Factors Affecting the Persistence of Endangered Ganges River Dolphins (Platanista Gangetica Gangetica)." *Ecology and Evolution* 10 (6): 3138–3148. https://doi.org/10.1002/ece3.6102.

Paudel, Shambhu, John L. Koprowski, and Michael V. Cove. 2020. "Seasonal Flow Dynamics Exacerbate Overlap between Artisanal Fisheries and Imperiled Ganges River Dolphins." *Scientific Reports* 10 (1): 1–12. https://doi.org/10.1038/s41598-020-75997-4.

Paudel, Shambhu, John L. Koprowski, Usha Thakuri, Rajesh Sigdel, and Ram Chandra Gautam. 2020. "Ecological Responses to Flow Variation Inform River Dolphin Conservation." *Scientific Reports* 10 (1): 1–14. https://doi.org/10.1038/s41598-020-79532-3.

Petts, Geoffrey E. 2009. "Instream Flow Science for Sustainable River Management." *Journal of the American Water Resources Association* 45 (5): 1071–1086. https://doi. org/10.1111/j.1752-1688.2009.00360.x.

Phillips, Jonathan D. 1995. "Biogeomorphology and Landscape Evolution: The Problem of Scale." *Geomorphology* 13 (1–4): 337–347. https://doi.org/10.1016/0169-555X(95) 00023-X.

Poff, N. L., J. D. Allan, M. B. Bain, J. R. Karr, K. L. Prestegaard, B. D. Richter, R. E. Sparks, and J C. Stromberg. 1997. "The Natural Flow Regime; a Paradigm for River Conservation and Restoration." *BioScience* 47 (11): 769–784. https://doi.org/10.2307/1313099.

Press Trust of India. 2018. "940 Dams, Barrages Built on Ganga Restricting Its Flow: Environmentalists." *The Economic Times*. https://economictimes.indiatimes.com/ news/politics-and-nation/940-dams-barrages-built-on-ganga-restricting-its-flow-environmentalists/articleshow/66784449.cms.

Rao, R. J., S. Tagor, H. Singh, and N. Dasgupta. 2013. "Monitoring of Gharial (Gavialis Gangeticus) and Its Habitat in the National Chambal Sanctuary, India." In *Proceeding of the World Crocodile Conference, 22nd Workshop Meeting of the Special Survival Commission of the IUCN*, edited by IUCN, 66–73. Gland, Switzerland: International Union for Conservation of Nature.

Rudra, K. 2010. "Dynamics of the Ganga in West Bengal, India (1764–2007): Implications for Science-Policy Interaction." *Quaternary International* 227 (2): 161–169. https://doi.org/10.1016/j.quaint.2009.10.043.

Santy, Sneha, Pradeep Mujumdar, and Govindasamy Bala. 2020. "Potential Impacts of Climate and Land Use Change on the Water Quality of Ganga River around the Industrialized Kanpur Region." *Scientific Reports* 10 (1): 1–13. https://doi.org/10.1038/s41598-020-66171-x.

Sharma, G. 2010. "Current Status of Susu (Palntanista Gangetica, Roxburgh, 1801) in River Hooghly in West Bengal, India." *Records of the Zoological Survey of India* 110 (1): 61–69.

Silva, P. De, W. A. Khan, B. Kanchanasaka, I. Reza Lubis, M. M. Feeroz, and O. F. Al-Sheikhly. 2015. "Lutrogale Perspicillata. The IUCN Red List of Threatened Species." 8235. https://doi.org/10.2305/IUCN.UK.2015-2.RLTS.T12427A21934884.en.

Singh, Harendra, Dharmveer Singh, Sudhir Kumar Singh, and D. N. Shukla. 2017. "Assessment of River Water Quality and Ecological Diversity through Multivariate Statistical Techniques, and Earth Observation Dataset of Rivers Ghaghara and Gandak, India." *International Journal of River Basin Management* 15 (3): 347–360. https://doi.org/10.1080/15715124.2017.1300159.

Singh, M., R. Sinha, and S. K. Tandon. 2020. "Geomorphic Connectivity and Its Application for Understanding Landscape Complexities: A Focus on the Hydro-Geomorphic Systems of India." *Earth Surface Processes and Landforms*: 1–21. https://doi.org/10.1002/esp.4945.

Singh, S., K. Prakash, and U. K. Shukla. 2018. "Decadal Scale Geomorphic Changes and Tributary Confluences within the Ganga River Valley in Varanasi Region, Ganga Plain, India." *Quaternary International* 507 (February): 124–133. https://doi.org/10.1016/j.quaint.2018.05.022.

Sinha, R. K. 1997. "Status and Conservation of Ganges River Dolphin in Bhagirathi-Hooghly River Systems in India." *International Journal of Ecology and Environmental Sciences* 23: 343–355.

Sinha, R. K. 2000. "Status of the Ganges River Dolphin (Platanista Gangetica) in the Vicinity of Farakka Barrage, India." In *Biology and Conservation of Freshwater Cetaceans in Asia*, edited by R. R. Reeves, B. D. Smith, and T. Kasuya, 23rd ed., 42–48. Gland, Switzerland: IUCN.

Sinha, R. K. 2002. "An Alternative to Dolphin Oil as a Fish Attractant in the Ganges River System: Conservation of the Ganges River Dolphin." *Biological Conservation* 107: 253–257.

Sinha, R. K., K. Gaurav, S. Chandra, and S. K. Tandon. 2013. "Exploring the Channel Connectivity Structure of the August 2008 Avulsion Belt of the Kosi River, India: Application to Fl Ood Risk Assessment." *Geology* 41 (10): 1099–1102. https://doi.org/10.1130/G34539.1.

Sinha, R. K., and Santosh Ghosh. 2012. "Understanding Dynamics of Large Rivers Aided by Satellite Remote Sensing: A Case Study from Lower Ganga Plains, India." *Geocarto International* 27 (3): 207–219. https://doi.org/10.1080/10106049.2011.620180.

Sinha, R K., and K. Kannan. 2014. "Ganges River Dolphin: An Overview of Biology, Ecology, and Conservation Status in India." *AMBIO: A Journal of the Human Environment* 43: 1029–1046. https://doi.org/10.1007/s13280-014-0534-7.

Sinha, R. K., H. Mohanta, V. Jain, and S. K. Tandon. 2017. "Geomorphic Diversity as a River Management Tool and Its Application to the Ganga River, India." *River Research and Applications* 33 (7): 1–27.

Sinha, R. K., and G. Sharma. 2003. "Current Status of the Ganges River Dolphin, Platanista Gangetica in the Rivers Kosi and Son, Bihar, India." *Journal of the Bombay Natural History Society* 100 (1): 27–37.

Sinha, R. K., K. Sripriyanka, Vikrant Jain, and Malay Mukul. 2014. "Avulsion Threshold and Planform Dynamics of the Kosi River in North Bihar (India) and Nepal: A GIS Framework." *Geomorphology* 216: 157–170. https://doi.org/10.1016/j.geomorph.2014.03.035.

Smakhtin, V., and M. Anputhas. 2006. "An Assessment of Environmental Flow Requirments of Indian River Basins." Colombo, Sri Lanka.

Smith, B. D., A. K. M. Aminul Haque, M. S. Hossain, and A. Khan. 1998. "River Dolphins in Bangladesh: Conservation and the Effects of Water Development." *Environmental Management* 22 (3): 323–335. https://doi.org/10.1007/s002679900108.

Smith, B. D., G. Braulik, S. Strindberg, R. Mansur, and B. Ahmed. 2009. "Habitat Selection of Freshwater-Dependent Cetaceans and the Potential Effects of Declining Freshwater Flows and Sea-Level Rise in Waterways of the Sundarbans Mangrove Forest, Bangladesh." *Aquatic Conservation: Marine and Freshwater Ecosystem* 225 (October 2008): 209–225. https://doi.org/10.1002/aqc.

Smith, B. D., and R. R. Reeves. 2000. "Report of the Workshop on the Effects of Water Development on River Cetaceans, 26–28 February 1997, Rajendrapur, Bangladesh." In *Biology and Conservation of Freshwater Cetaceans in Asia*, edited by R. R. Randall, B. D. Smith, and T. Kasuya, 15–21. Gland, Switzerland: IUCN.

Sonkar, G. K., and K. Gaurav. 2020. "Assessing the Impact of Large Barrages on Habitat of the Ganga River Dolphin." *River Research and Applications* 36 (9): 1916–1931. https://doi.org/10.1002/rra.3715.

Srivastava, V. K. 2007. "River Ecology in India: Present Status and Future Research Strategy for Management and Conservation." *Proceedings of the Indian National Science Academy* 73.

Tare, Vinod, Suresh Kumar Gurjar, Haridas Mohanta, Vishal Kapoor, Ankit Modi, R. P. Mathur, and Rajiv Sinha. 2017. "Eco-Geomorphological Approach for Environmental Flows Assessment in Monsoon-Driven Highland Rivers: A Case Study of Upper Ganga, India." *Journal of Hydrology: Regional Studies* 13 (November 2016): 110–121. https://doi.org/10.1016/j.ejrh.2017.07.005.

Thakur, P. K., C. Laha, and S. P. Aggarwal. 2012. "River Bank Erosion Hazard Study of River Ganga, Upstream of Farakka Barrage Using Remote Sensing." *Natural Hazards* 61 (1): 967–987. https://doi.org/10.1007/s11069-011-9944-z.

Tharme, R. E. 2003. "A Global Perspective on Environmental Flow Assessment: Emerging Trends in the Development and Application of Environmental Flow Methodologies for Rivers." *River Research and Applications* 441 (September 2002): 397–441. https://doi.org/10.1002/rra.736.

Wakid, A. 2009. "Status and Distribution of the Endangered Gangetic Dolphin (Platanista Gangetica Gangetica) in the Brahmaputra River within India in 2005." *Current Science* 97 (8): 1143–1151.

Ward, J. V. 1998. "Riverine Landscapes: Biodiversity Patterns, Disturbance Regimes, and Aquatic Conservation." *Biological Conservation* 83 (3): 269–278.

WII-GACMC. 2017. "Aquatic Fauna of the Ganga River. Status and Conservation." Dehradun.

Yadav, Sanjeev Kumar, Asghar Nawab, and Afifullah Khan. 2013. "Conserving the Critically Endangered Gharial Gavialis Gangeticus in Hastinapur Wildlife Sanctuary, Uttar Pradesh: Promoting Better Coexistence for Conservation." In *Proceeding of the World Crocodile Conference, 22nd Workshop Meeting of the Special Survival Commission of the IUCN*, edited by IUCN, 78–82. Gland, Switzerland: International Union for Conservation of Nature.

10 Modeling the Potential Impact of Land Use/ Land Cover Change on the Hydrology of Himalayan River Basin
A Case Study of Manipur River, India

Vicky Anand* and Bakimchandra Oinam
Department of Civil Engineering, National
Institute of Technology Manipur, India.

CONTENTS

10.1 Introduction 190
10.2 LULC of the Himalayas 191
10.3 Hydrology of the Himalayas..192
10.4 Materials and Methodology..193
    10.4.1 Study Area...193
    10.4.2 Data Sets Used in the Study 193
    10.4.3 LULC Simulation Technique...194
    10.4.4 SWAT Hydrological Model and Its Performance.............................195
10.5 Results...197
    10.5.1 LULC Change Analysis..197
    10.5.2 Sensitivity Analysis, Calibration, and Validation............................199
    10.5.3 Variability in Hydrological Parameters under Different LULC
        Scenarios.. 200
10.6 Discussion.. 200
10.7 Conclusion 201
Acknowledgment 201
References... 202

10.1   INTRODUCTION

The availability of water resources is an essential component for the survival of living organisms. Scarcity and access to freshwater are serious issues in various parts of the world (Nasiri et al. 2020). Thus, the sustainable use and availability of water resources turn out to be the nucleus of national and local strategies for such regions (Nasiri et al. 2020). Hydrological models play a major role in modeling the water resources of a river and the lake (Dutta and Sarma 2020). Hence, the use of hydrological models becomes an important tool for the proper planning of water resources. Hydrological models have a wide range of applications in water resources planning and management such as climate impact assessments (Das et al. 2021), flood forecasting (Unduche et al. 2018), water quality assessment (Shinde et al. 2017), and nutrient flow analysis (Somaye et al. 2019).

Several factors such as anthropogenic activities, change in climate, increasing population, and change in hydrological processes have created a serious threat to the river system and have made it more complex (Wang et al. 2013). Complex factors like climate, topographic condition, land use land cover (LULC) of the region, and soil properties often affect the hydrological processes (Jonoski et al. 2019). In particular, the acceleration of evaporation is due to an increase in temperature, which leads to a lowering of water availability, whereas an increase in water availability can be due to increasing precipitation without a prominent change in evapotranspiration (IPCC 2007). In recent times all across India, changes in the hydrological processes have been predominantly observed (Anand et al. 2020). Hence, it becomes essential to predict the hydrological processes and get a clear understanding of anthropogenic and natural disturbances in the hydrological regime of rivers (Wang et al. 2019).

In the past few years, several hydrological models have been developed to assist the complexity associated with the hydrological processes (Anand et al. 2020; Arnold and Allen 1996). A hydrological model simplifies the hydrological systems by forecasting hydrological phenomena and changes to compute the hydrological effects which aid in formulating various policies on the management of water resources (Mirza 2003). Mathematical relationships are established based on the mutual relationship between surface and climate parameters (Gosain et al. 2009). A semi-distributed model is comprehensive and versatile among the various hydrological models (Gosain et al. 2009). The models like Soil and Water Assessment Tool (SWAT; Arnold et al. 1993), Water Erosion Prediction Project (WEPP; Laflen et al. 1991), MIKE-SHE (Abbott et al. 1986), TOPMODEL (Kirkby 1997), and Agricultural Non-Point Source Pollution (AGNPS; Young et al. 1989) have been extensively used in the past. The SWAT model has been extensively used by hydrologists to assess the effect of change in climate on water resources (Bouraoui et al. 2005). This model is widely used to assess water quality and the impact of change in LULC and estimate soil loss to predict surface runoff (Gassman et al. 2007). In the past years, several studies have been done across India using the SWAT model. Pervez and Henebry (2015) analyzed the impact of change in LULC and climate on the freshwater availability in the Brahmaputra River basin, Das et al. (2021) assessed the impact of change in climate in the Gomti River basin under different representative concentration pathways scenarios, Sana et al. (2018) examined the flows and sediment dynamics in the Ganga River under present and future climate scenarios.

Manipur River basin lies in the northeastern part of India nestled in the Himalayan ranges in the state of Manipur. In the past few decades, various natural and anthropogenic

factors have led to significant changes in the climatic condition of the region (Anand and Oinam 2019) that has led to change in the hydrology of the region that has further resulted in frequent landslides, floods, and other natural hazards (Singh et al. 2011; Singh et al. 2011). Future changes in LULC in the basin will bring new challenges to sustainable use of the resources and water resources management with the development of the economy. Hydrological models like SWAT can aid in quantifying the challenges due to changes in LULC in the future. This study discusses the impact of change in LULC on the hydrology of the Manipur River basin in the future.

10.2 LULC OF THE HIMALAYAS

The Himalayan ranges are home to several indigenous plant and animal species, and it is also one of the hot spots of biodiversity. The Himalayan mountain ranges are dominated by diverse LULC. Several studies have been done in the various stretches of the Himalayas to investigate the LULC and their impacts as a baseline requirement for planning and sustainable management of natural resources. Table 10.1 shows a review of research studies carried out in the Himalayas in past decades.

TABLE 10.1
Review of Research Studies Carried Out in the Himalayas

Study Area	Satellite/Sensor	LULC Classes	Source
North Kashmir Himalayas	Landsat-TM, Landsat-8 OLI, IRS-P6 LISS III	Built-up, agriculture, plantation, dense forests, sparse forests, scrub lands, pasture lands, water bodies, aquatic vegetation, aquatic vegetation, and snow and glaciers	Fayaz et al. (2020)
Solan Forest, Himachal Pradesh	IRS-1D (LISS-III)	Chir pine (*Pinus roxburghii*) forests, ban oak (*Quercus leucotricophora*) forests, khair (*Acacia catechu*) forests, bamboo (*Dendrocalamus strictus*) forests, broad-leaved forests, cultivation, and culturable blank	Shah and Sharma (2015)
Garhwal Himalayan Ranges	Landsat-2 MSS, Landsat-5 TM, Landsat-8 OLI	Dense forest, open forest, pastureland, agriculture land, built-up area, scrub land, barren land, water bodies, and snow and glacier	Batar et al. (2017)
Central Himalayan Region	Landsat scenes	Built-up, cropland, fallow land, plantation, deciduous forest (dense/closed), deciduous forest (open), pine forest (dense/closed), oak forest (dense/closed), mixed forest (dense/closed), pine forest (open), oak forest (open), mixed forest (open), forest plantation, scrub forest, tree clad, alpine/sub-alpine grassland, temperate/subtropical grassland, wastelands, water bodies, permanent snow and ice	Chakraborty et al.(2016)

(*Continued*)

TABLE 10.1
(Continued)

Study Area	Satellite/Sensor	LULC Classes	Source
Sikkim Himalayas	Landsat-5 TM, Sentinel-2A MSI	Agricultural land, barren land, built-up land, dense forest, open forest, and waterbodies	Mishra et al. (2020)
Mehao Wildlife Sanctuary in Arunachal Pradesh (Eastern Himalayas), India	IRS-P6 LISS III	Dense forest, open forest, scrubland, cultivated areas, riverbed, water	Areendran et al.(2018)
Eastern Himalayan landscape of Nagaland, North East India	Landsat 4–5 TM, Landsat-8 OLI	Agricultural land, built-up land, forest land, water body	Ritse et al. (2020)
Western parts of Mizoram, Northeast India (Eastern Himalayas)	L-band of ALOS PALSAR-1	Built-up, deciduous forest, evergreen forest, scrubland, bare land and water body	Parida and Mandal (2020)

10.3 HYDROLOGY OF THE HIMALAYAS

The two circulation systems, namely, western disturbances and the Asian summer monsoon, in the Himalayan region make its hydrology more complex (Bookhagen and Burbank 2010). The strength of western disturbances decreases from west to east along with the Himalayan ranges, whereas the strength of the Asian summer monsoon decreases as it moves toward the west from the east (Qazi et al. 2020). The central part of the Himalayas receives precipitation from both the western disturbance from the west and the Asian summer monsoon from the east (Azam et al. 2016). In the eastern parts of the Himalayas (Brahmaputra and Ganges), due to high elevations and high amounts of precipitation during the monsoon, glaciers experience a high amount of snow accumulation (Qazi et al. 2020). The Himalayas serves as a barricade to the southwest monsoon winds from escaping it to Central Asia, causing maximum precipitation on the southern side, with a decrease in rainfall intensity as it moves from east to west (Shrestha et al. 1999). The glaciers in the Himalayas are subjected to varying climates depending on regional orography and geographical position. Behaviors and the types of glaciers vary over short distances due to variations in the precipitation regimes along the Himalayas (Qazi et al. 2020; Maussion et al. 2014). The intensity, magnitude, and timing of the precipitation during the monsoon vary from east to west. The northeastern states of India, namely, Manipur, Meghalaya, Sikkim, Assam, Arunachal Pradesh, Mizoram, and Nagaland receive the maximum amount of precipitation due to southwest monsoon winds as compared to the northwestern parts of India (Qazi et al. 2020). Snow plays a major in the hydrology of the Himalayas. Snow-cover dynamics in the high Himalayas influence the water availability and timing in downstream basins (Bookhagen and Burbank 2010).

10.4   MATERIALS AND METHODOLOGY

10.4.1   STUDY AREA

The Manipur River basin lies in the northeastern state of Manipur in the Himalayan ranges. The basin is located between 24° N to 25°25′ N latitude and 93°36′ E to 94°27′ E longitude (Figure 10.1). Manipur River basin is hydrologically divided into nine sub-basins: Western, Khuga, Sekmai, Heirok, Khuga, Nambul, Kongba, Iril, and Imphal. The basin comprises varying topographical slopes and altitudes. Soils found in the catchment are mostly clayey and silty by texture (NBSS and LUP 2001). The study area is characterized as tropical to subtropical type of climate. The basin receives an average rainfall of 1350 mm, with an annual temperature ranging between 12°C to 31°C.

10.4.2   DATA SETS USED IN THE STUDY

Comprehensive SWAT hydrological model was used in this study. The SWAT hydrological model requires LULC data, the digital elevation model (DEM), meteorological data, and soil data as input data for model setup. Based on the input parameter the hydrological response units (HRUs) were generated. To calibrate and validate the hydrological model, the observed discharge dataset was used. The observed

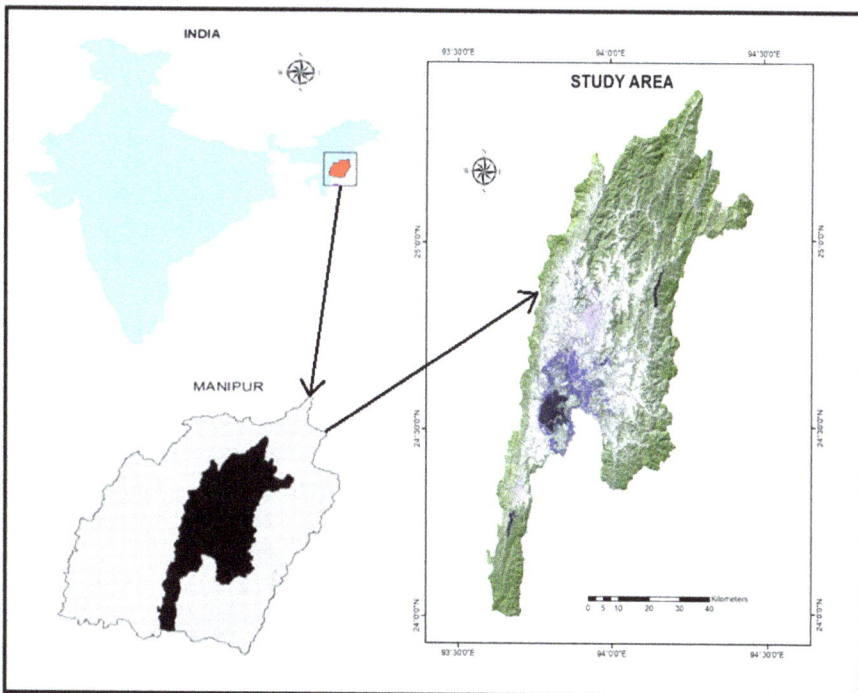

FIGURE 10.1   The Manipur River basin.

TABLE 10.2

Input Data and Source

Data Sets Used	Resolution	Year	Sources
DEM	12.5 m	2010	ALOS PALSAR, Alaska Satellite Facility Link: https://asf.alaska.edu
Land use land cover*	30 m	2017	Landsat 8, Earth resource observation and science https://earthexplorer.usgs.gov
Meteorological data (includes rainfall, temperature, relative humidity, solar radiation, wind speed)	Daily	1999–2017	Directorate of Environment Manipur
Soil data*	1 km	–	National Bureau of Soil Survey and Land Use Planning (NBSS & LUP)
River discharge	–	2008–2017	Source: Anand et al, 2020

* Resampled to 12.5 m.

streamflow dataset computed by Anand et al. (2020) using the stage-discharge curve was used in this study. Using supervised classification on Landsat 8 Operational Land Imager (OLI) Level 1 satellite imagery, the LULC map was generated for 2017. The data sets obtained from the various organizations have been tabulated in Table 10.2.

10.4.3 LULC SIMULATION TECHNIQUE

There are seven basic forms of modeling techniques, like hybrid models, process-based models, data-mining models, machine learning models, and statistical models, that are being used for LULC modeling. Cellular automation (CA) and artificial neural networks (ANNs) are the most commonly used modeling techniques. CA is a dynamic model that was developed by von Neumann and Burks (1966) to simulate complex patterns. CA can capture global patterns and local behaviors (Wolfram 1984). The basic components of CA include (1) grid-gap, which can be represented as an irregular or a regular cell;(2) each cell can change based on cells in its neighboring areas;(3) the classification of data on transition rules;(4) the extent of influence of neighbor cells from the central cell; and (5) time step. The best sets of transition rules are defined to model the future LULC (Landis and Zhang 1998). Generally, there are two ways to model and calibrate CA models both being dependent on time and space (Pontius et al. 2004; Figure 10.2). The first is through the hit-and-try approaches, in which a comparison is made between the simulations using different sets of parameters (Clarke et al. 1997), and the other is by using several statistical methods and machine learning techniques (Wu 2002). The structure of the CA model is shown in Figure 10.3.

Parallel to a CA, an ANN was developed and is considered to be the fundamental principle of artificial intelligence (von Neumann 1951; McCarty 1959). The objective

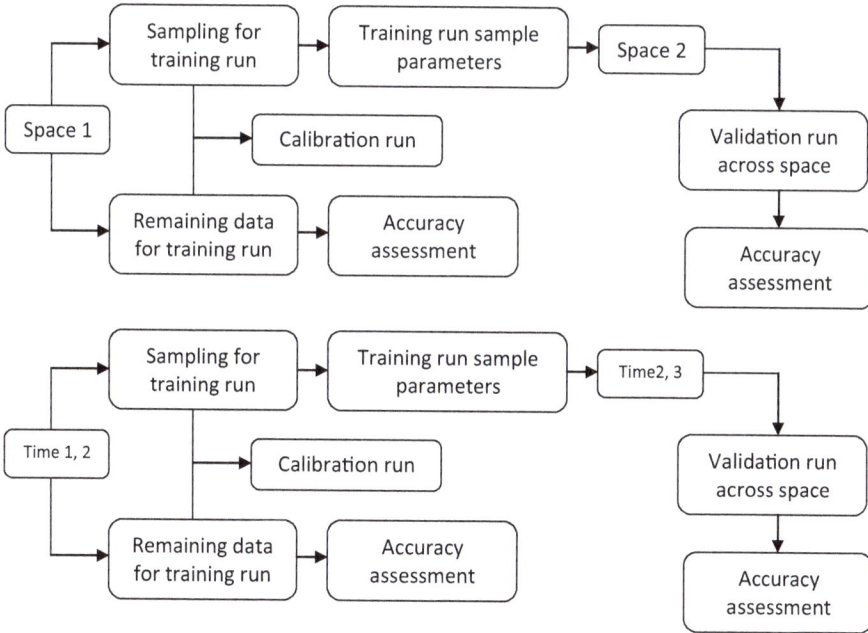

FIGURE 10.2   Calibration and validation across space and time.

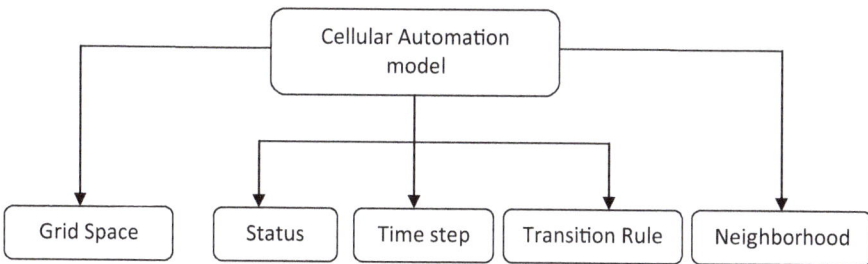

FIGURE 10.3   Structure of a CA model.

of artificial intelligence theory was to create a machine that is capable of mathematical logic that is similar to humans. This learning takes place through complex logical and probabilistic mathematical functions. The conceptual architecture of CA–ANN is shown in Figure 10.4. In this study, the CA–ANN simulation technique has been used for the future simulation of the LULC.

10.4.4   SWAT Hydrological Model and Its Performance

SWAT is a physical-based semi-distributed model with explicit parameterization (Neitsch et al. 2011). The SWAT model is a very flexible hydrological model and can be applied to a wide range of different environmental conditions (Arnold

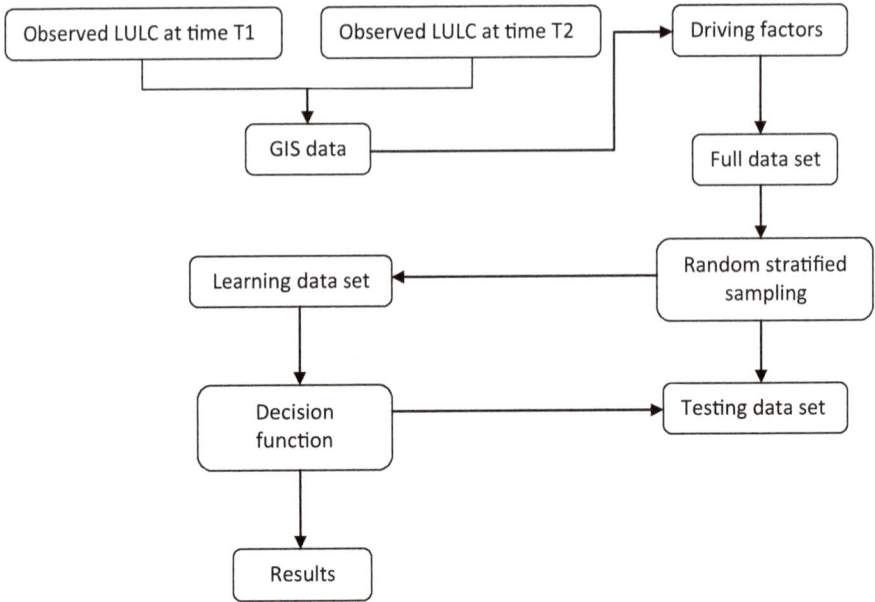

FIGURE 10.4   Conceptual architecture of CA–ANN.

and Fohrer 2005). The SWAT model, a freeware, was developed by the USDA-Agricultural Research Service to assist with the assessment of watersheds ranging in size from small to large watersheds (Neitsch et al. 2011). One of the major advantages of SWAT is the integration of the basin-scale model with GIS, providing much-improved modeling linkages within a management basin (Arnold and Fohrer 2005). In SWAT, the basin is divided into several smallest units called HRUs based on soil type and LULC. Water balance in SWAT is based on Equation 10.1 (Arnold and Fohrer 2005):

$$SW_t = SW_0 + \sum_{i=0}^{n} R_{day} + Q_{surf} + E_a + W_{seep} + Q_{gw}, \qquad (10.1)$$

where SW_t and SW_o are final and initial water contents on day t and i, respectively, in mm. R_{day}, Q_{surf}, E_a, W_{seep}, and Q_{gw} is the amount of precipitation, surface runoff, evapotranspiration, percolation entering the vadose zone, and return flow, respectively, on day i (mm). In the present study, the Soil Conservation Service Curve Number (SCS-CN) method (USDA 1972) was used for computing surface runoff.

To accept the dependability of any model, the model performance evaluation is necessary. If the statistical indices of the model performance fall within the permissible range then the model outputs are acceptable (Moriasi et al. 2007). There are several statistical model performance indices to evaluate the model reliability such as Nash-Sutcliffe efficiency (NSE; Nash and Sutcliffe 1970), coefficient of determinacy (R^2), and Kling-Gupta efficiency (KGE; Gupta et al. 2009).

1. NSE, computed as Equation 10.2.

$$NS = 1 - \frac{\sum_j (G_m - G_k)_j^2}{\sum_j (G_{m,j} - \overline{G_m})^2},$$ (10.2)

Where G is a variable, m is observed data, k is simulated data, and j is the jth simulated value (Nash and Sutcliffe 1970).

2. Coefficient of determination (R^2), computed as shown in Equation 10.3

$$R^2 = \frac{\left[\sum_j \left(G_{n,j} - \overline{G_m}\right)\left(G_{k,j} - \overline{G_k}\right)\right]^2}{\sum_j \left(G_{n,j} - \overline{G_n}\right)^2 \sum_j \left(G_{k,j} - \overline{G_k}\right)^2},$$ (10.3)

Where G is a variable, n is observed data, k is simulated data, and j is the jth simulated value.

3. KGE, computed as shown in Equation 10.4

$$KGE = 1 - \sqrt{\left((r-1)^2 + (\alpha-1)^2 + (\beta-1)^2\right)},$$ (10.4)

Where r is the coefficient of the linear regression between simulated and observed variable, β is the ratio of means of the simulated and the observed data, and α is the ratio of the standard deviation of the simulated and the observed data (Gupta et al. 2009).

10.5 RESULTS

10.5.1 LULC CHANGE ANALYSIS

The overriding objective of this study is to assess the impact of change in LULC on the hydrology of the Manipur River basin. Based on the past LULC for 2007, 2014, and 2017, the future LULC for 2030 was generated using the Land Change Modeller (LCM) embedded in TerrSet. The LULC map for 2017 and the future projected LULC map for 2030 are shown in Figure 10.5. For three different years in the past, LULC was mapped using maximum likelihood classifiers, and a change detection was performed. Future LULC for 2030 was projected by considering various government norms, driving factors, and environmental policies. Table 10.3 shows the percentage change in area under various LULC classes. Minor changes can be observed in forest and agriculture class but there was a major change in area under wetlands, herbaceous wetlands, area being builtup, and water bodies. A decrease of 1.46% of area under forest cover can be observed mainly concentrated in the foothills due to the conversion of the area from forests to agriculture. The drastic change in the area with an increment of 10.66% was observed under built-up mainly concentrated

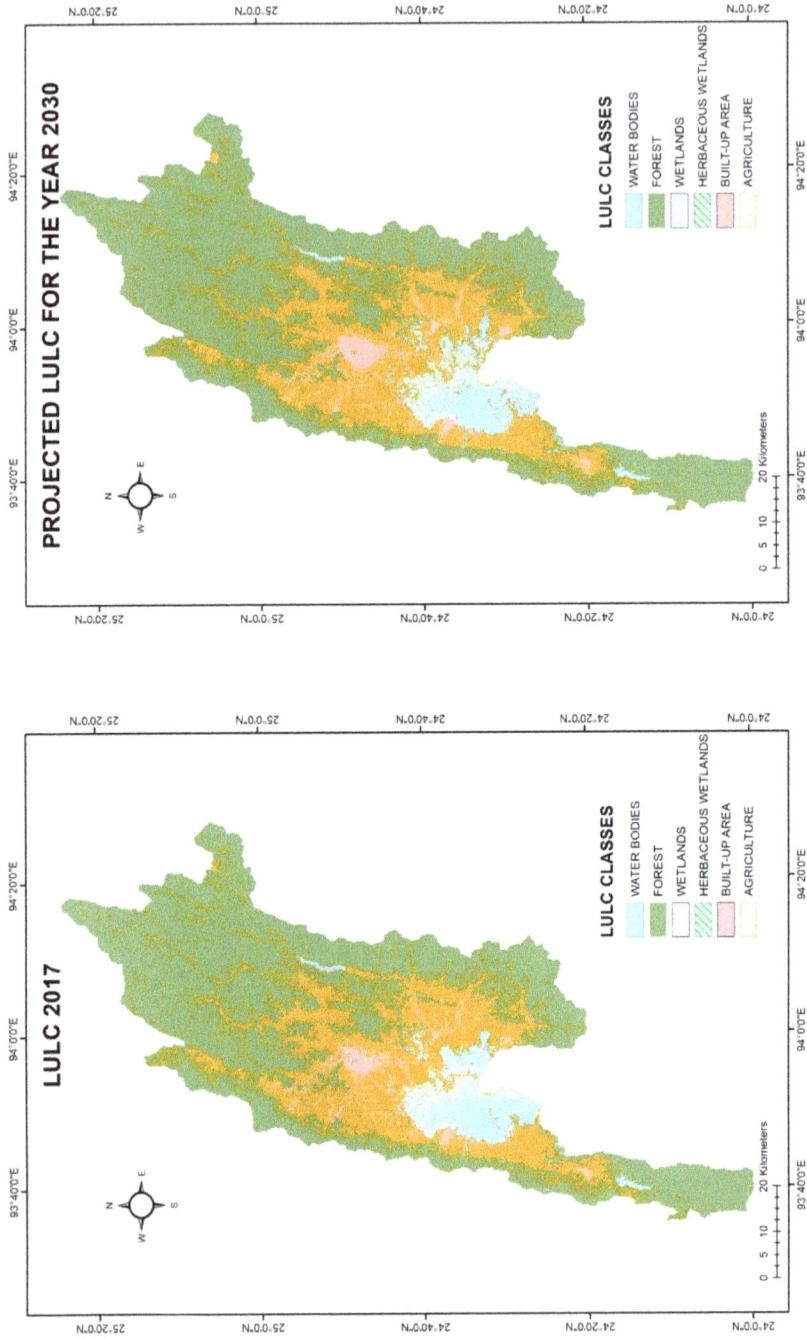

FIGURE 10.5   LULC map for 2017 and 2030.

TABLE 10.3
LULC Change Assessment between 2017 and 2030

LULC Classes	Area (km²) 2017	Area (km²) 2030	% Change
Water bodies	93.95	94.16	0.22
Agriculture	1556.45	1563.6	0.45
Wetlands	133.25	132.68	−0.43
Herbeceous wetlands	185.04	186.94	1.02
Built-up area	235.25	293.82	10.66
Forest	2786.19	2746.19	−1.46
Total	5017.39	5017.39	

TABLE 10.4
List of Sensitive Parameters Used for the Calibration and Validation of Model

S.No.	Parameter Name	Definition	Sensitivity
1	CN2	Initial SCS runoff curve number for moisture condition II	High
2	Alpha_Bf	Baseflow alpha factor	High
3	Gw_Delay	Groundwater delay	Low
4	Sol_Awc	Soil water capacity of the first soil layer	Medium
5	HRU_Slope	Average slope steepness	High
6	ESCO	Soil evaporation compensation factor	Medium
7	SolOrgn	Initial NO_3 concentration in soil layer	Negligible
8	Biomix	Biological mixing efficiency	Negligible

in the Imphal city region. Detailed analysis of future projected LULC for 2030 can be found in the literature (Anand and Oinam 2020).

10.5.2 SENSITIVITY ANALYSIS, CALIBRATION, AND VALIDATION

The SWAT model was setup using various input parameters, and a total of 31 sub-catchments were delineated. Before using the model outputs further for analysis, the sensitivity of parameters, model calibration, and validation are essential. Sensitivity analysis was carried out on different sets of parameters. Sensitive parameters used for the calibration and validation of the SWAT model are shown in Table 10.4. The model was calibrated between 2008 and 2014 and validated between 2015 and 2017 on a monthly time step (Figure 10.6). The model returned $R^2 = 0.78$, NSE = 0.71, KGE = 0.73 and $R^2 = 0.75$, NSE = 0.67, KGE =0.71 for calibration and validation period, respectively. The model results of water balance components, such as actual evapotranspiration (AET), runoff, and water yield, during 1999–2017 were used as a baseline scenario for the comparison with the future simulated result based on future LULC scenarios.

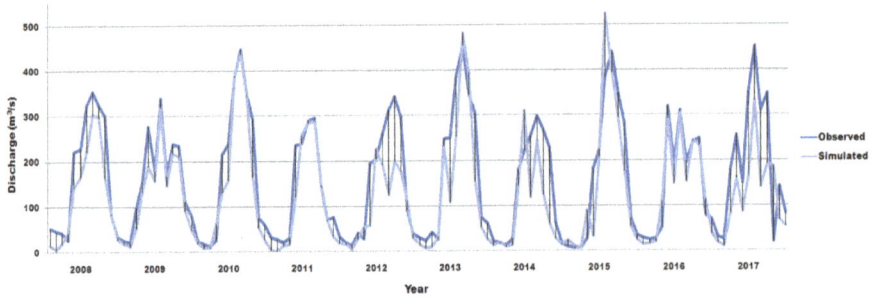

FIGURE 10.6 Calibration and validation result of SWAT model of Manipur River basin at Ithai (2008–2017).

TABLE 10.5

Variation in Hydrological Parameters under the Future Projected LULC Scenario

Hydrological Parameters	2017 LULC Scenario	Projected 2030 LULC Scenario
Runoff (in m³/s)	150.82	153.86
Water Yield (in mm)	701.06	690.71
Actual Evapotranspiration (in mm)	583.80	595.90
Potential Evapotranspiration (in mm)	1224.90	1229.20

10.5.3 VARIABILITY IN HYDROLOGICAL PARAMETERS UNDER DIFFERENT LULC SCENARIOS

In this research, the impact of change in LULC on the hydrology of the Manipur River basin has been projected. The model was setup and calibrated for 1999–2017 using the current LULC data, that is, 2017. Based on the current LULC scenario of 2017, various hydrological parameters, such as runoff, AET, potential evapotranspiration, and water yield, have been analyzed and evaluated. A future hydrological scenario was predicted based on the projected LULC scenario for 2030. Later, the predicted hydrological parameters for the year 2030 were compared with the baseline scenario of 2017 and it was observed that there is an increase of 2.02% and 2.07% in runoff and AET, respectively, and a decrease in water yield by 1.48%. The hydrological parameters under both LULC scenarios are tabulated in Table 10.5.

10.6 DISCUSSION

The impact of LULC change is of serious concern in the eastern Himalayan ranges in the northeastern part of India. From the future LULC predicted by the model for 2030, a decrease in the area under wetlands and forests by 0.43% and 1.46%, respectively, and an increase in the built-up area by 10.66% was observed as compared to 2017. The decrease in the wetlands is mainly concentrated in the region near Loktak

Lake. The decrease in the wetlands poses a serious threat to the endangered aquatic and terrestrial species found in Loktak Lake and peripheral zone of Loktak Lake. For the restoration and sustainability of wetlands around Loktak Lake, the change in LULC is of major concern. The fragile ecosystem is severely affected by the change in LULC conditions surrounding the lake. The increasing population has put immense pressure on the natural resources in the catchment for the livelihood of the people. Due to the increase in population has led to an increase in the built-up area mainly concentrated in the region surrounding Imphal city, Thoubal, Bishnupur, and Churachandpur and a decrease in the forest cover mainly in the foothills in the basin.

Changes in the LULC scenario will affect the water balance in the basin in the coming future. The overall hydrological cycle of the basin is contributed by various hydrological parameters like runoff, evapotranspiration, and water yield. The LULC of the watershed highly influences the water balance of the basin. As the built-up area is likely to increase in the near future, this will lead to an increase in the area under the paved surface. An increase in the paved surface will further lead to a decrease in the groundwater infiltration or the water yield and an increase in the surface runoff and evapotranspiration as predicted in the model result. As per the model results, the other reason behind the decrease in water yield and increase in surface runoff can be decreased in forest cover due to conversion of forest to the agricultural area mainly shifting agriculture commonly practiced in the hilly regions in India. A slight change in area under forest in the entire catchment may lead to a shift in water yield and runoff from the uplands. An increase in the runoff increases the chances of flooding in the low-lying area near Loktak Lake and landslides in the hilly regions.

10.7   CONCLUSION

Changing LULC patterns in the past few decades and its projected future trends due to various natural and anthropogenic factors is a serious concern in the Himalayan ranges. Deforestation on the steep slopes of the Himalayas for the purpose of agriculture and livelihood and rapid urbanizations due to an increase in population can have serious impacts on the Manipur River basin. Hydrological models like the SWAT can be used as a versatile tool to model and analyze the impact of change in LULC dynamics on the hydrology of a river basin. As this study shows the good performance of the SWAT model with $R^2 = 0.78$, NSE = 0.71, and KGE = 0.73 in the calibration and $R^2 = 0.75$, NSE =0.67, KGE =0.71 in the validation period, strongly suggests the applicability of SWAT even in the basins which have extremely diverse LULC. The outcomes of this study will aid the decision and policymakers in building policies for environment protection and conservation. The future scope can be done by integrating the combined effects of change in both LULC and climate based on the Global Climate Model (GCM) and Regional Climate Model (RCM).

ACKNOWLEDGMENT

The authors gratefully acknowledge the valuable databases from the Alaska Satellite Facility, NBSS &LUP, the Earth resource observation and science, Directorate of Environment, Govt. of Manipur.

REFERENCES

Abbott, M.B., Bathurst, J.C., Cunge, J.A., O'Connell, P.E., Rasmussen, J. 1986. "An introduction to the European Hydrological System—Système Hydrologique Européen, SHE, 2: Structure of a physically-based, distributed modelling system." *Journal of Hydrology*, 87: 61–77.

Anand, V., Oinam, B., Parida, B.R. 2020. "Uncertainty in hydrological analysis using multi-GCM predictions and multi-parameters under RCP 2.6 and 8.5 scenarios in Manipur River basin, India."*Journal of Earth System Science*, 129:223. https://doi.org/10.1007/s12040-020-01492-z

Anand, V., Oinam, B. 2020. "Future land use land cover prediction with special emphasis on urbanization and wetlands." *Remote Sensing Letters*, 11(3): 225–234.

Anand, V., Oinam, B. 2019. "Future climate change impact on hydrological regime of river basin using SWAT model."*Global Journal of Environmental Science and Management*, 5(4):471–484.

Areendran, G., Puri, K., Raj, K., Mazumdar, S., Joshi, R. 2018. "A geospatial study to assess the land use land cover of Mehao Wildlife Sanctuary in Arunachal Pradesh, India."*Asian Journal of Geographical Research*, 1(2): 1–8.

Arnold, J.G., Allen, P.M. 1996. "Estimating hydrologic budgets for three Illinois watersheds." *Journal of Hydrology*, 176(1–4): 57–77.

Arnold, J.G., Fohrer, N. 2005. "SWAT-2000: Current capabilities and research opportunities in applied water-shed modeling." *Hydrological Processes*, 19(3): 563–572.

Arnold, J.G., Allen, P.M., Bernhardt, G. 1993. "A comprehensive surface groundwater flow model." *Journal of Hydrology*, 142:47–69.

Azam, M.F., Ramanathan, A.L., Wagnon, P., Vincent, C., Linda, A., Berthier, E., Sharma, P., Mandal, A., Angchuk, T., Singh, V.B., Pottakkal, J.G. 2016. "Meteorological conditions, seasonal and annual mass balances of Chhota Shigri glacier, western Himalaya, India." *Annals of Glaciology*, 57(71):328–338.

Batar, A.K., Watanabe, T., Kumar, A. 2017. "Assessment of Land-Use/Land-Cover Change and forest fragmentation in the Garhwal Himalayan Region of India."*Environments*, 4: 34.

Bookhagen, B., Burbank, D.W. 2010. "Toward a complete Himalayan hydrological budget: Spatiotemporal distribution of snowmelt and rainfall and their impact on river discharge." *Journal of Geophysical Research Earth*, 115(F03019).

Bouraoui, F., Benabdallah, S., Jrad, A., Bidoglio, G. 2005. "Application of the SWAT model on the Medjerda river basin (Tunisia)." *Physics and Chemistry of Earth Parts A/B/C*, 30(8–10): 497–507.

Chakraborty, A., Sachdeva, K., Joshi, P.K. 2016. "Mapping long-term land use and land cover change in the central Himalayan region using a tree-based ensemble classification approach." *Applied Geography*, 74: 136–150.

Clarke, K.C., Hoppen, S., Gaydos, L. 1997. "A self-modifying cellular automaton model of historical urbanization in the San Francisco Bay area." *Environment and Planning B: Planning and Design*, 24: 247–261.

Das, B., Jain, S., Thakur, P., Singh, S. 2021. "Assessment of climate change impact on the Gomti River basin in India under different RCP scenarios." *Arabian Journal of Geosciences*, 14: 1–16.

Dutta, P., Sarma, A. 2020. "Hydrological modeling as a tool for water resources management of the data-scarce Brahmaputra basin." *Journal of Water and Climate Change*, 12. http://doi.org/10.2166/wcc.2020.186.

Fayaz, A., Shafiq, M., Singh, H., et al.2020."Assessment of spatiotemporal changes in land use/land cover of North Kashmir Himalayas from 1992 to 2018." *Modelling Earth System Environment*, 6:1189–1200.

Gassman, P.W., Reyes, M.R., Green, C.H., Arnold, J.G. 2007. "The soil and Water Assessment Tool: Historical development, applications, and future research directions." *Transactions of the American Society of Agricultural and Biological Engineers*, 50(4): 1211–1250.

Gosain, A.K., Mani, A., Dwivedi, C. 2009. *Hydrological Modeling Literature Review: Report No. 1*. Indo-Norwegian Institutional Cooperation Program, New Delhi, India.

Gupta, H.V., Kling, H., Yilmaz, K.K., Martinez, G.F. 2009. "Decomposition of the mean squared error and NSE performance criteria: Implications for improving hydrological modeling." *Journal of Hydrology*, 377: 80–91

IPCC. 2007. "Climate change 2007: Impacts, adaptation and vulnerability. Contribution of working group II to the fourth assessment report of the intergovernmental panel on climate change." Cambridge University Press, Cambridge, 976pp.

Jonoski, A., Popescu, I., Zhe, S., Mu, Y., He, Y. 2019. "Analysis of flood storage area operations in Huai River using 1D and 2D river simulation models coupled with global optimization algorithms." *Geosciences*, 9: 509.

Kirkby, M.J. 1997. "Topmodel: A personal view." *Hydrological Processes*, 11(9): 1087–1098.

Laflen, J.M., Elliot, W.J., Simanton, J.R., Holzhey, S., Kohl, K.D. 1991. "WEPP soil erodibility experiments for rangeland and cropland soils." *Journal of Soil Water Conservation*, 46(1): 39–44.

Landis, J., Zhang, M. 1998. "The second generation of the California urban futures model. Part 1: Model logic and theory." *Environment and Planning B: Planning and Design*, 30: 657–666.

McCarty, J. 1959. "Programs with common sense. In Mechanisation of thought processes." *Proceedings of the Symposium of the National Physics Laboratory*. London, pp. 77e84.

Maussion, F., Scherer, D., Molg, T., Collier, E., Curio, J., Finkelnburg, R. 2014. "Precipitation seasonality and variability over the Tibetan plateau as resolved by the high Asia reanalysis." *Journal of Climatology*, 27(5):1910–1927.

Mirza, M. M. Q. 2003. "Climate change and extreme weather events: Can developing countries adapt?" *Climate Policy*, 3(3): 233–248.

Mishra, P.K., Rai, A., Rai, S.C. 2020. "Land use and land cover change detection using geospatial techniques in the Sikkim Himalaya, India." *The Egyptian Journal of Remote Sensing and Space Science*, 23(2): 133–143.

Moriasi, D.N., Arnold, J.G., Van Liew, M.W., Bingner, R.L., Harmel, R.D., Veith, T.L. 2007. "Model evaluation guidelines for systematic quantification of accuracy in watershed simulations." *American Society of Agricultural and Biological Engineers*, 50(3):885–900.

Nash, J.E., Sutcliffe, J.V. 1970. "River flow forecasting through conceptual models. A discussion of principles." *Journal of Hydrology*, 10(3): 282–290.

Nasiri, S., Ansari, H., Ziaei, A. 2020. "Simulation of water balance equation components using SWAT model in Samalqan Watershed (Iran)." *Arabian Journal of Geosciences*, 13. http://doi.org/10.1007/s12517-020-05366-y.

NBSS and LUP (National Bureau of Soil Survey and Land Use Planning). 2001. "Land capability classes of catchment area of Loktak Lake, Manipur." NBSS and LUP, Jorhat, Assam, India.

Neitsch, S.L., Arnold, J.G., Kiniry, J.R., Williams, J.R. 2011. *Soil and Water Assessment Tool Theoretical Documentation Version 2009*. Texas Water Resources Institute, Texas, USA.

Parida, B.R., Mandal, S.P. 2020. "Polarimetric decomposition methods for LULC mapping using ALOS L-band PolSAR data in Western parts of Mizoram, Northeast India."*SN Applied Sciences*, 2: 1049.

Pervez, S., Henebry, G.M. 2015. "Assessing the impacts of climate and land use and land cover change on the freshwater availability in the Brahmaputra River basin." *Journal of Hydrology: Regional Studies*, 3: 285–311.

Pontius, R.G., Huffaker, D., Denman, K. 2004. "Useful techniques of validation for spatially explicit land-change models." *Ecological Modelling*, 179(4): 445–461.

Qazi, N.Q., Jain, S.K., Thayyen, R.J., Patil, P.R., Singh, M.K. 2020. "Hydrology of the Himalayas." In: Dimri A., Bookhagen B., Stoffel M., Yasunari T. (eds.), *Himalayan Weather and Climate and Their Impact on the Environment.* Springer, Cham. https://doi.org/10.1007/978-3-030-29684-1_21

Ritse, V., Basumatary, H., Kulnu, A.S., et al.2020. "Monitoring land use land cover changes in the Eastern Himalayan landscape of Nagaland, Northeast India."*Environmental Monitoring Assessment*, 192:711.

Sana, K., Sinha, R., Whitehead, P., Sarkar, a S., Jin, L., Futter, M.N. 2018. "Flows and sediment dynamics in the Ganga River under present and future climate scenarios." *Hydrological Sciences Journal*, 63(5): 763–782.

Shah, S., Sharma, D. 2015. "Land use change detection in Solan Forest Division, Himachal Pradesh, India."*Forest Ecosystem*, 2: 26.

Shinde, V., Tiwari, K., Nandgude, S., Singh, M. 2017. "Water quality assessment and application of SWAT model for hydrologic simulations in a mined watershed." *Climate Change and Environmental Sustainability*, 5: 111.

Shrestha, A.B., Wake, C.P., Mayewski, P.A., Dibb, J.E. 1999. "Maximum temperature trends in the Himalaya and its vicinity: An analysis based on temperature records from Nepal for the period 1971–94."*Journal of Climatology*, 12(9): 2775–2786.

Singh, C., Behera, K., Rocky, W. 2011. "Landslide susceptibility along NH-39 between Karong and Mao, Senapati District, Manipur." *Journal of the Geological Society of India*, 78. http://doi.org/10.1007/s12594-011-0120-6

Singh, C.R., Thompson, J.R., Kingston, D.G., French, J.R. 2011. "Modelling water-level options for ecosystem services and assessment of climate change, Loktak Lake, northeast India." *Hydrological Sciences Journal* 56(8): 1518–1542.

Somaye, I., Delavar, M., Niksokhan, M.H. 2019. "Identification of nutrients critical source areas with SWAT model under limited data condition."*Water Resource*, 46: 128–137.

Unduche, F., Tolossa, H., Senbeta, D., Eric, Z. 2018. "Evaluation of four hydrological models for operational flood forecasting in a Canadian Prairie watershed." *Hydrological Sciences Journal*, 63(8): 1133–1149.

USDA Soil Conservation Service. 1972. "Section 4. Hydrology. In National engineering handbook, US." Department of Agriculture-Soil Conservation Service, Washington, DC.

von Neumann, J., Burks, A.W. 1966. *Theory of Self Reproducing Automata.* University of Illinois Press, Urbana, IL.

von Neumann, J. 1951. "The general and logical theory of automata." In Jeffress L.A. (ed.), *Cerebral Mechanisms in Behaviors The Hixson Symposium, 1948* (pp. 1e41). Willey, Pasadena, CA, New York.

Wang, M., Du, L., Ke, Y., Huang, M., Zhang, J., Zhao, Y., Gong, H. 2019. "Impact of climate variabilities and human activities on surface water extents in reservoirs of Yongding River Basin, China, from 1985 to 2016 based on Landsat observations and time series analysis." *Remote Sensing*, 11: 560.

Wang, Y., Ding, Y.J., Ye, B.S., Liu, F.J., Wang, J., Wang, J. 2013. "Contributions of climate and human activities to changes in runoff of the Yellow and Yangtze rivers from 1950 to 2008." *Science China Earth Sciences*, 56(8): 1398–1412.

Wolfram, S. 1984. "Computer Software in Science and Mathematics." *Scientific American*, September, pp. 188–203.

Wu, F. 2002. "Calibration of stochastic cellular automata: The application to rural-urban land conversions." *International Journal of Geographical Information Science*, 16(8): 795–818.

Young, R.A., Onstad, C.A., Bosch, D.D., Anderson, W.P. 1989. "AGNPS: A nonpoint source pollution model for evaluating agricultural watershed."*Journal of Soil and Water Conservation*, 44: 168–173.

11 The Application of Remote Sensing for Water Resources Management in Data-Scarce Watersheds in the Hindu Kush Himalaya Region
A Case of Kabul River Basin

Fazlullah Akhtar*, Abdul Haseeb Azizi,
Usman Shah, Christian Borgemeister,
Bernhard Tischbein, and Usman Khalid Awan

CONTENTS

11.1 Introduction 206
11.2 Materials and Methods 207
    11.2.1 Study Area 207
    11.2.2 Analysis of the Spatiotemporal Variation of
        the SCA in the KRB 209
    11.2.3 Analysis of the Spatiotemporal Variation of Actual
        Evapotranspiration across the KRB 210
        11.2.3.1 The Key Data Set Used for the Actual
               Evapotranspiration Estimation 210
    11.2.4 Trend Analysis Using the Mann–Kendall Test 212
    11.2.5 Sen's Slope Estimator 212
11.3 Results 212
    11.3.1 Analysis of the Spatiotemporal Variation in
        the SCA across the KRB 212

DOI: 10.1201/9781003265160-13

205

    11.3.1.1 Trend Analysis of the SCA over the KRB 213
  11.3.2 Analysis of the Spatiotemporal Variation in the Actual
    Evapotranspiration across the KRB 215
    11.3.2.1 Trend Analysis of the Actual Evapotranspiration over
     the Sub-Basins of the KRB...216
11.4 Discussion and Conclusion 217
References 219

11.1 INTRODUCTION

The global population has been estimated to be about 7.8 billion in 2021 (UNFPA 2021), which is projected to reach 9.7 billion by the end of 2050. Population growth and increasing income levels in emerging economies, which is associated with dietary changes such as increased food intake, is driving up the global food demand. According to Elferink and Schierhorn (2016), by 2050, food demand is anticipated to rise by 59–98%, while by the end of 2100, the global population is projected to hit 11.2 billion (Roser 2013). This will have far-reaching consequences for agricultural markets. Farmers worldwide will need to increase crop yields, either by expanding the land dedicated to agriculture or by boosting productivity of the existing farming area through fertilizer and irrigation, as well as embracing new technology like precision farming. This already challenging situation will be further complicated by several additional processes, such as climate change, urbanization, and a lack of investment; consequently, this leads to the concern that it might become impossible to produce enough food (Elferink and Schierhorn 2016).

  Snowmelt runoff and overall streamflow rise linearly with increases in precipitation (Singh and Kumar 1997). However, runoff from glacier melt is negatively correlated to precipitation variations. Snowmelt runoff is thought to be more responsive to changes in precipitation than glacier melt runoff. A long-term rise in air temperature or a reduction in precipitation will limit the size of glaciers due to increased melt-runoff. Singh and Kumar (1997) further highlight that combination of higher temperature and lower precipitation will have a compound effect on the glacier's size. However, in greater precipitation conditions, the glaciers in tendency will grow in size while increasing the likelihood of their progress.

  Wake (2021) report on "Water Wars" shows that due to expected climate change, crop yields, and food supply will be affected; snow and ice, which melt to feed vast river basins, are also at risk as a result of global warming. More than one-sixth of the world's population relies presently on melt-fed river catchments. However, this potential is dwindling as climate change exacerbates weather variability, resulting in higher winter flows due to earlier melt. The temperature has a significant impact on biological activities and the growth of organisms because some species, particularly aquatic plants, thrive at higher temperatures (Jiang et al. 2008; Fondriest Environmental, Inc. 2014). Therefore, increased temperatures and the subsequent enhanced biological activities may result in waterborne pathogens, which deteriorate the drinking water quality (Fu et al. 2021). These changes might occur in line with an increase in population and eventual increasing demand for a finite resource. The

water scarcity and its consequent effects are projected to worsen amid the projected increase in population and increased water demand (Akhtar and Shah 2020). To efficiently use the available water resources for the consuming sectors, there is a need to analyze water resources in terms of quality and quantity and examine their spatial and temporal distribution.

For sustainable water resource management, it is essential to estimate different biophysical processes which are based on detailed information that is often very complex, challenging, and costly due to a high degree of heterogeneity regarding the (1) features of the basins relevant to processes on runoff generation providing water supply and referring to the (2) management practices determining spatio-temporal water demand (Beach et al. 2008). The use of remote-sensing technology is a cost-effective way for flood hazard assessment that can be helpful in regions where physical data are missing (Kim et al. 2019). Similarly, remote-sensing data and techniques can be used to predict yield (You et al. 2017), soil salinity (Gorji et al. 2017), and snow water equivalent (Collados-Lara et al. 2020). According to Akhtar et al. (2018), the use of remote-sensing-based algorithms to estimate actual evapotranspiration (ET_a) in data-scarce regions can be the best alternative for initiatives aimed at water resources management and investment plans. Akhtar et al. (2018) used remote sensing products, and SEBS to estimate the ET_a, which was later used to evaluate the performance of the irrigation systems in the data-scarce Kabul River Basin (KRB).

Water resources management and associated plans and strategies need extensive spatiotemporal data, which is frequently lacking in the case of Afghanistan owing to widespread insecurity and a lack of attention by governing authorities to the construction and installation of the representative hydro-meteorological monitoring system. In such circumstances, the data scarcity issue can be resolved by utilizing advanced technologies. This study highlights how remote sensing techniques and products were used to estimate the long-term spatiotemporal variation, from 2003–2018, in the snow-covered area (SCA) across the KRB, the key feeding source of the rivers being exploited for several purposes. Furthermore, crop water use (i.e., ET_a) has been estimated at different spatiotemporal scales across the KRB. It is noteworthy that the ET_a estimates used in this chapter have been derived from research conducted by Akhtar (2017), which covers only a period of 2003–2013, while the SCA data have been presented newly in this data-scarce watershed covering a period of 2003–2018. The variation in the SCA estimates in the KRB was further related to those of the Upper Indus Basin (UIB), which is geographically connected to the KRB and, in some cases, experiences similar climatic conditions.

11.2   MATERIALS AND METHODS

11.2.1   Study Area

The KRB is a data-scarce transboundary river basin, home to roughly one-third of the country's population (Figure 11.1). The basin is highly utilized in terms of water withdrawals from both surface and groundwater. Since 2001, groundwater

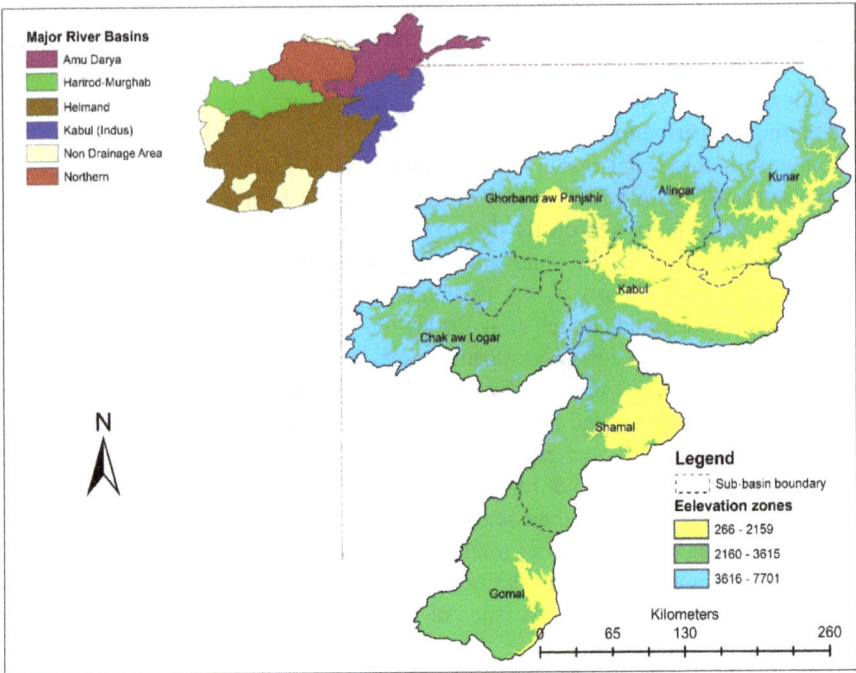

FIGURE 11.1 Location of the KRB in Afghanistan; the dotted lines show the boundaries of the seven sub-basins that were used for the actual evapotranspiration estimation and the three distinct elevation zones (colored) were used for the SCA estimation.

extraction has risen manifold, and it is being used not just for municipal consumption but also for industrial purposes. Conversely, rapid urbanization, land surface pavement, and a lack of infrastructure for artificial aquifer recharge have reduced groundwater recharge. The irrigation system's performance has been rated poor throughout the KRB (Akhtar et al. 2018). About 7% of the basin's land area is irrigated, 0.7% is rainfed, 1.5% is water bodies and marshland, 59.7% rangeland, and 17.5% forest and shrubs, while the rest is barren and has other land cover types (FAO 2016). The geographic heterogeneity between upstream and downstream locations is highly evident in many ways. Upstream receives snowfall while downstream receives precipitation predominantly in the form of rainfall. The mean annual precipitation received downstream of the KRB with a fringe effect of Monsoon is approximately 327 mm, while upstream receives about 418 mm precipitation mainly in the form of snow. Similarly, upstream is a mono-cropped region, whereas downstream is frequently double cropped (Akhtar et al. 2018). The KRB experiences four distinct seasons: winter, spring, summer, and autumn, which range from January–March, April–June, July–September, and October–December, respectively. The majority of soils of the KRB are rocky alkaline with undulating terrain.

11.2.2 ANALYSIS OF THE SPATIOTEMPORAL VARIATION OF THE SCA IN THE KRB

Precipitation falls mainly in the form of snow in the KRB; accumulation and snowmelt are thus the primary sources of streamflow and groundwater recharge. The principal use of water from snowmelt is mainly intended for agricultural, industrial, and municipal purposes, either directly or indirectly. For being the key water source, it is vital to estimate the spatiotemporal changes in the SCA across the different spatial units of the KRB.

In this study, variation in the SCA was estimated at various elevation zones across the KRB; the topographic details of these zones were derived from a 30m Shuttle Radar Topography Mission (SRTM) digital elevation model (DEM) data that were retrieved from the Earth Explorer domain (http://earthexplorer.usgs.gov/). Three distinct elevation zones were established for a comprehensive spatiotemporal investigation of SCA in the KRB. Based on the geographic distribution of SCA in the KRB, the altitudinal zonal boundaries were defined as low-, mid-, and high-elevation zones. The different elevation zones concerning their elevation range and spatial coverage (%) are given below in detail; these zones are, namely, low-elevation zone (LEZ), mid-elevation zone (MEZ), and high-elevation zone (HEZ) (Table 11.1).

Remote-sensing products based on MODIS have been commonly used in remote areas for hydrological forecasts and scenario analyses (Azizi and Asaoka 2019, 2020a, b). There are different MODIS products available for snow cover with spatial resolutions of 500 m and 5 km and temporal resolutions of daily, 8-day, and monthly (Hall et al. 2006; Hall and Riggs 2021). These snow-cover products from Aqua and Terra satellites are available from December 1999 and May 2002, respectively.

For this study, we utilized a combined and improved snow-cover product (i.e., MOYDGL06) with a spatial resolution of 500 m to assess the spatiotemporal variation of SCA from 2003–2018 over the KRB (Sher and Thapa 2020). The 8-day combined and improved maximum snow-cover product of the MODIS from both Terra (MOD10A2.006) and Aqua (MYD10A2.006) sensors are available only from 2002–2018 (Sher and Thapa 2020). Both the products have been combined to reduce the uncertainties in the snow-cover estimation. The seasonal, spatial, and temporal filters have also been applied to the MODIS snow products to reduce the underestimation caused by the cloud cover. Combining these products helped reduce the overestimation in that the corresponding pixel in both data sets is considered snow unless it is identified as no snow (Sher and Thapa 2020).

TABLE 11.1
Characteristics of the Elevation Zones Derived from the SRTM-DEM of the KRB

Zone	Elevation Range (meter a.m.s.l.)	Mean Elevation (meter a.m.s.l.)	Zonal Area (km²)	Zonal Area (%)
LEZ	380–2159	1460	20,436	30
MEZ	2160–3615	2860	29,670	43
HEZ	3616–7701	4371	18,396	27
Total			68,502	100

Before the trend analysis, the complete data set of the KRB and zone-wise SCA was adjusted for seasonality; it is a time series of data that can be described as a pattern that systematically repeats over a specific period, in this example, every year. The presence of seasonal variability in time-series data makes it difficult to evaluate trends in data over a particular time to determine whether changes reflect a significant rise or reduction in the size of the data or are due to regularly occurring seasonal fluctuation. For this purpose, the means of the same months throughout the time series were calculated and then subtracted from the relevant monthly SCA (%), through this way, seasonality was removed from the SCA time series. The seasonal adjustment (de-seasonality) was performed to the full SCA data set to analyze seasonal change and remove its influence on time-series data. Furthermore, trend analysis for the basin-wide and zone-wise SCA was done at a 5% significance level using Kendall's tau (s) and Sen's slope (S).

11.2.3 ANALYSIS OF THE SPATIOTEMPORAL VARIATION OF ACTUAL EVAPOTRANSPIRATION ACROSS THE KRB

The Surface Energy Balance System (SEBS; Su 2002) was used to estimate the ET_a by combining data from the Global Land Data Assimilation System (GLDAS) (Fang et al. 2008) and MODIS. The SEBS is a single-source model that estimates atmospheric turbulent fluxes and a surface evaporative fraction using remote-sensing data. It has already been successfully used in a wide range of ecosystems, and it is a very trustworthy evapotranspiration model based on remote-sensing data (Liaqat et al. 2015; Akhtar 2017; Akhtar et al. 2018). Liaqat and Choi (2017) compared the SEBS-modeled ET_a with flux tower measurements; their comparison yielded the coefficient of determination (i.e., R^2) of 0.72 and 0.51, respectively at cropland and forestland. The SEBS calculate the energy necessary for water to change phases from liquid to gas and estimates daily ET_a using remotely sensed and meteorological data, as shown:

$$\lambda E = R_n + G_0 + H,$$

where λE is the turbulent latent heat flux (Wm^{-2}), λ is the latent heat of vaporization (Jkg^{-1}), and E is evapotranspiration, R_n is net radiation (Wm^{-2}), G_0 is the soil heat flux (Wm^{-2}), and H is the sensible heat flux (Wm^{-2}).

Trends in the estimated ET_a time-series data for the seven sub-basins (sub-basin-wise) were analyzed at monthly timescales using the Mann–Kendall test and Sen's slope estimator.

11.2.3.1 The Key Data Set Used for the Actual Evapotranspiration Estimation

11.2.3.1.1 GLDAS Data

The GLDAS has generated multiple surface land models such as Mosaic, Noah, Community Land Model (CLM), and the Variable Infiltration Capacity (VIC). Land surface variables (e.g., soil moisture and surface temperature) and fluxes, such as evaporation and sensible heat flux parameters are also included in GLDAS products.

The GLDAS product used in this study has a 3-h temporal resolution and a 25-km spatial resolution (Rodell et al. 2004). In areas characterized by data scarcity or a lack of climatic data monitored at the ground, GLDAS meteorological values can give reliable alternative observation data (Kiptala et al. 2013; Armanios and Fisher 2014; Akhtar 2017). The meteorological variables required as the input data for running the SEBS algorithms were collected from the Goddard Earth Sciences Data and Information Services Center (GES DISC, http://disc.sci.gsfc.nasa.gov/hydrology). The input variables include wind speed (m/s), short-wave radiation (W/m^2), long-wave radiation (W/m^2), air temperature (K), air pressure (Pa), and specific humidity (Kg/Kg) as shown in Table 11.2.

11.2.3.1.2   MODIS Data

Land surface temperature, vegetation cover, and land surface albedo play key roles in influencing heat and water exchanges between the land surface and the surrounding atmosphere and the allocation of available energy between soil and vegetation (Jia et al. 2009). The MODIS data were extracted from the Land Processes Distributed Active Archive Center (LP DAAC) of the U.S. Geological Survey (USGS; (https://lpdaac.usgs.gov/products/modis_products_table) for 11 years (2003–2013) to estimate the ET$_a$ by using the SEBS algorithm. The land surface variables used in the SEBS algorithm are presented in table 11.3. The extracted MODIS data sets were

TABLE 11.2

Characteristic of the Climate Parameters Retrieved from GLDAS–NOHA Model

No.	Variable Name	Unit	Temporal Resolution	Temporal Coverage	Spatial Resolution
1	Air Temperature	(K)	3 hours	2003–2013	25 km
2	Wind Speed	(m/s)	3 hours	2003–2013	25 km
3	Specific Humidity	(Kg/Kg)	3 hours	2003–2013	25 km
4	Long-Wave Radiation	(W/m^2)	3 hours	2003–2013	25 km
5	Short-Wave Radiation	(W/m^2)	3 hours	2003–2013	25 km
6	Air Pressure	(Pa)	3 hours	2003–2013	25 km

TABLE 11.3

Land Surface Products' Details Used in the Surface Energy Balance System for Estimating ET$_a$ across the KRB

No.	Variable	Spatial Resolution	Temporal Resolution
1	Emissivity/LST (MOD11A1)	1 km	Instantaneous
2	NDVI (MOD13A2)	1 km	16 days
3	LAI (MCD15A2)	1 km	8 days
4	Albedo (MCD43B3)	1 km	8 days
5	Land Cover (MCD12Q1)	0.5 km	Annual

resampled using the nearest neighbor interpolation approach, facilitated by the MODIS reprojection tool (https://lpdaac.usgs.gov/tools/modis_reprojection_tool). The evaporative behavior of various land-cover types was studied by applying the land-cover map in the KRB.

11.2.4 TREND ANALYSIS USING THE MANN–KENDALL TEST

To identify trends in time-series data, a widely applied nonparametric Mann–Kendall (MK) trend test was used (Mann 1945); the values of the trend test in the snow cover and ET_a time series were then defined by using Kendall's tau (τ) coefficient. The statistical significance limit or p-value used in this analysis was 5% (P = 0.05). The monotonic trend upward or downward indicates that a variable regularly increases or decreases over time, yet the trend may or may not be linear. The positive MK would mean an increase with time, while a negative value indicates a decrease over time. Furthermore, there would be no trend if the calculated probability is lower than the level of significance. The data are normally distributed for a time series equal to or more than 10 years (n ≥ 10).

11.2.5 SEN'S SLOPE ESTIMATOR

A nonparametric approach was developed by Theil and Sen (Sen 1968; Gilbert 1987), the Theil–Sen estimator (also known as Sen's slope). The magnitude of the trend slope in the data time series of snow cover and ET_a was calculated using Sen's slope (S) estimator. Sen's slope approach determines how changes happen in data between two successive intervals. The median of all the changes is used to adjust the value of change over the entire period.

11.3 RESULTS

11.3.1 ANALYSIS OF THE SPATIOTEMPORAL VARIATION IN THE SCA ACROSS THE KRB

The temporal variability of the SCA during 2003–2018 has been high; the mean annual SCA during the study period was 30 ± 27%. Similarly, the mean annual minimum and maximum SCA during the period mentioned earlier were 23 km^2 and 36 km^2. The maximum SCA (i.e., 36%) was observed in 2012 while the minimum SCA (i.e., 23%) was observed in 2016 (Figure 11.2). Since the snowfall usually begins in late September/October, the peak snow cover is normally achieved in the month of January to late February. Therefore, the inter-annual variability shows that the mean monthly SCA has been the highest in February, which was 81%. Contrary to that, the mean minimum SCA was observed in August and early September, which was 2%. Snowmelt usually begins in late February at the onset of the spring season and continues until late August or early September when the snow cover is reduced to 2% (Figure 11.2).

Elevation plays a vital role in snow accumulation (Azizi and Akhtar 2021); therefore, we have considered evaluating the SCA coverage for three distinct elevation

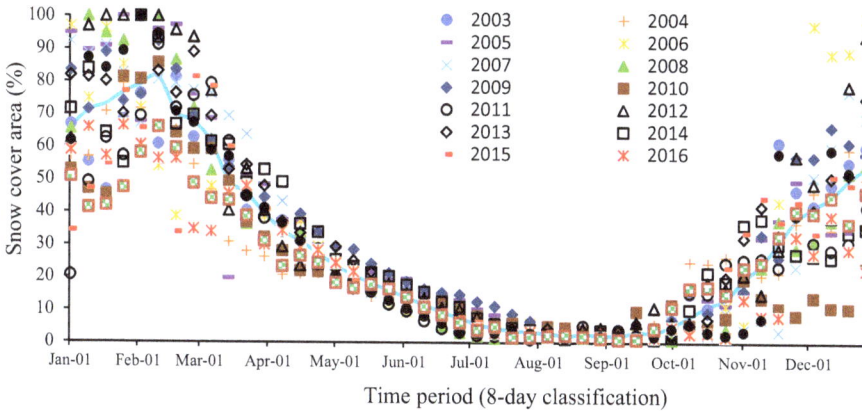

FIGURE 11.2   Annual snow cover estimated over 16 years (i.e., 2003–2018) in the KRB; the SCA here depicts the share of the area (%).

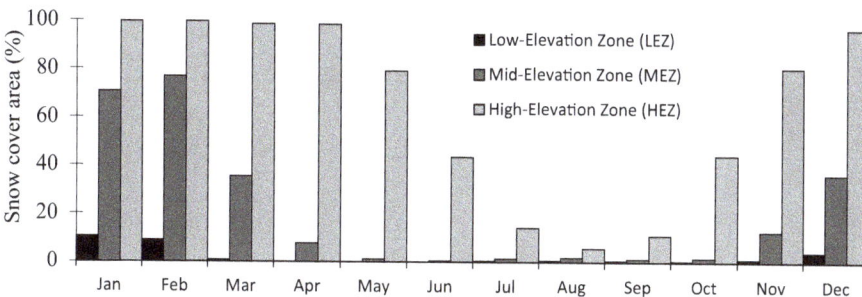

FIGURE 11.3   Intra-annual pattern of the mean monthly SCA of the three elevation zones in the KRB.

zones. Figure 11.3 shows the variation in the spatial extent of SCA for the three elevation zones (LEZ, MEZ, HEZ) of the KRB. The mean minimum SCA was observed in August in the LEZ, MEZ, and HEZ with a value of 1%, 2%, and 6%, respectively. The mean maximum SCA was found in the month of January (i.e., 10%) for LEZ, while it was 77% and 100% (in February) for MEZ and HEZ, respectively. Generally, more than 90% of HEZ remains under snow cover during the winter and spring months (December–April).

11.3.1.1   Trend Analysis of the SCA over the KRB

The trend analysis of monthly SCA during the study period for the basin-wide approach shows a slightly decreasing trend (Figure 11.4). A non-significant decreasing trend with MK value of $\tau = -0.045$ ($p > 0.05$) and Sen's slope (S) = -0.004 (%/year) was the result for the basin-wide monthly SCA. The trend analysis of three elevation zones (LEZ, MEZ, HEZ) of the KRB (Figure 11.4) indicated a slightly decreasing trend except for the LEZ. The Kendall's tau coefficient (τ) values estimated for the

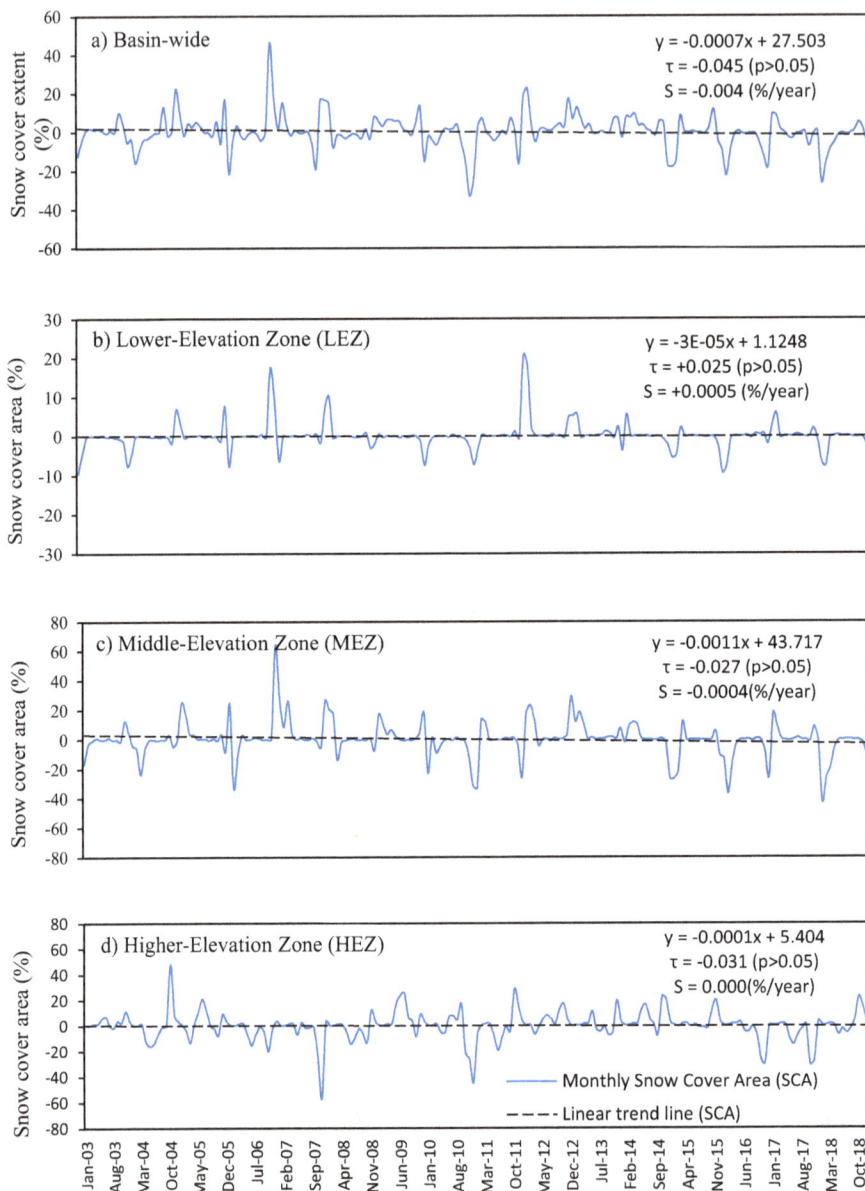

FIGURE 11.4 Time series of mean monthly SCA (seasonally adjusted) during 2003–2018 (a) across the KRB (b) LEZ, (c) MEZ, and (d) HEZ. The linear trend line equation, MK's trend test, and Sen's slope estimator are used to examine the trend.

LEZ, MEZ and HEZ were +0.025, −0.027, and −0.031, respectively. Sen's slope (S) values (%/year) of +0.0005, −0.0004, and 0.000 were obtained for the LEZ, MEZ, and HEZ, respectively. The MK values were not significant at a 5% significance level in all elevation zones.

11.3.2 ANALYSIS OF THE SPATIOTEMPORAL VARIATION IN THE ACTUAL EVAPOTRANSPIRATION ACROSS THE KRB

The agriculture and livestock sector in Afghanistan consumes around 98% of the surface water supplies (FAO 2012); actual water consumption by crops and plants (i.e., ET_a) therefore ranks the most crucial factor to be estimated for designing any water management relevant initiative. In this study, ET_a was estimated with SEBS across the KRB and its sub-basins (Alingar, Chak aw Logar, Ghorband aw Panjshir, Gomal, Kabul, Kunar, Shamal) have been derived from Akhtar (2017) and Akhtar et al. (2018). The estimates of ET_a through SEBS have been compared by Akhtar (2017) with the ET_a through advection aridity method (Brutsaert and Stricker 1979) at two sites, namely, Nawabad and Sultanpur; the comparison shows a coefficient of determination of 0.81 and 0.77, respectively. The mean annual maximum and minimum ET_a across the KRB was 518 mm and 418 mm, respectively. While analyzing the ET_a across individual sub-basin, we found out that the Shamal sub-basin experienced the highest mean annual ET_a, which was 551 mm ± 45 mm (Figure 11.5). Similarly, the Kunar, Kabul, Alingar, Gomal, and Ghorband aw Panjshir sub-basins experienced a mean annual ET_a of 521 ± 37 mm, 503 ± 48 mm, 491 ± 21 mm, 465 ± 31 mm, and 447 ± 23 mm, respectively; the lowest mean annual ET_a (i.e., 421 ± 32 mm) was experienced by Chak aw Logar sub-basin.

Analysis of the mean monthly ET_a during 2003–2013 across the KRB, as shown in Figure 11.6, reveals July experienced an ET_a of 68 ± 6 mm followed by May and June. with 67 ± 5 and 66 ± 4 mm, respectively. Overall, all the sub-basins experienced the highest ET_a during May–August. The least ET_a was experienced in the month of December (i.e., 7 ± 1 mm) and January (i.e., 9 ± 3 mm), whereby most parts of the KRB experience freezing temperatures. The ET_a has been typically higher in the Shamal and Kunar sub-basins than the rest of the sub-basins of the KRB. The seasonal evaluation of the ET_a reveals that the spring season (April–June) experienced the highest ET_a, which accounts for 37–41% of the annual ET_a across all the sub-basins of the KRB.

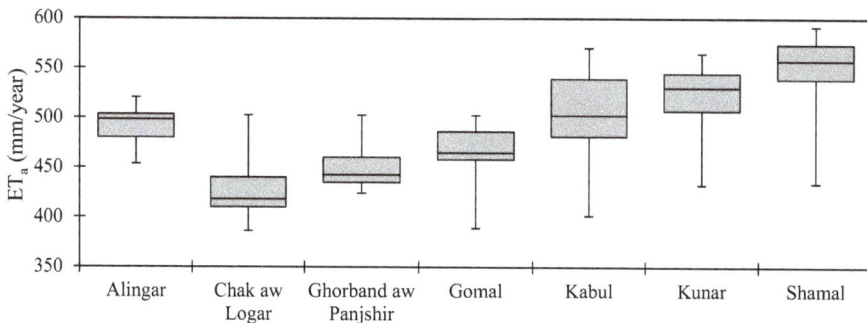

FIGURE 11.5 The boxplot indicates variation in the mean annual ET_a in the sub-basins of the KRB from 2003–2013.

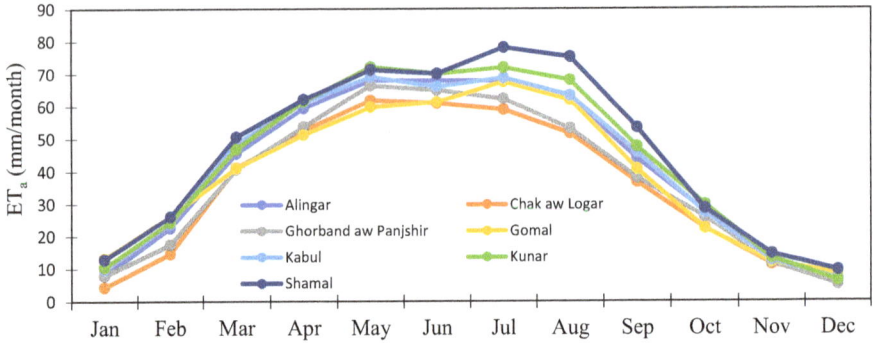

FIGURE 11.6 Analysis of mean monthly ET_a across the sub-basins of the KRB.

TABLE 11.4
Trend analysis of monthly ET_a for the sub-basins of the KRB over 2003–2013. The bold values indicate a significant trend at 5% significance level

Sub-Basin	Test	Jan.	Feb.	Mar.	Apr.	May	Jun.	Jul.	Aug	Sep.	Oct	Nov	Dec
Alingar	τ	−0.05	0.02	0.05	−0.02	0.38	0.38	0.05	0.38	0.38	0.31	−0.13	−0.20
	S	−0.04	0.14	0.23	−0.03	0.74	0.47	0.19	0.69	0.60	0.55	−0.05	−0.19
Chak aw	τ	0.42	−0.31	0.13	0.42	0.45	**0.53**	0.09	0.13	**0.53**	0.42	0.20	−0.02
Logar	S	0.65	−1.67	0.28	0.52	1.17	0.59	0.23	0.10	0.38	0.32	0.08	−0.02
Ghorband	τ	0.02	−0.05	−0.05	0.05	0.45	0.31	0.13	−0.31	0.09	−0.09	−0.45	−0.16
aw Panjshir	S	0.05	−0.09	−0.06	0.11	1.33	0.43	0.32	−0.27	0.08	−0.08	−0.50	−0.06
Gomal	τ	0.38	0.24	0.24	0.45	**0.53**	**0.56**	−0.13	0.31	0.31	**0.67**	**0.60**	0.38
	S	0.20	0.52	0.24	0.64	0.87	1.01	−0.37	0.57	0.48	0.66	0.47	0.23
Kabul	τ	−0.02	0.16	0.09	0.42	**0.56**	**0.75**	**0.67**	**0.64**	**0.49**	**0.67**	0.31	−0.05
	S	−0.04	0.66	0.32	1.64	1.33	1.53	1.21	1.91	1.26	1.13	0.29	−0.06
Kunar	τ	0.16	0.13	**0.56**	**0.60**	**0.71**	**0.56**	0.42	0.38	0.35	0.38	0.05	−0.09
	S	0.18	0.21	0.63	1.26	0.98	1.23	0.93	1.24	0.49	0.68	0.11	−0.04
Shamal	τ	−0.05	−0.27	0.09	0.13	**0.60**	**0.56**	0.16	0.24	0.31	**0.71**	**0.67**	−0.02
	S	−0.08	−0.43	0.27	0.36	1.31	1.94	0.15	0.75	0.52	0.92	0.58	−0.03

Note: Trend was analyzed using the MK trend test (τ) and Sen's slope estimator (S).

11.3.2.1 Trend Analysis of the Actual Evapotranspiration over the Sub-Basins of the KRB

The nonparametric statistical methods (Kendall's tau and Sen's Slope) for monthly ET_a during 2003–2013 show that Alingar and Ghorband aw Panjshir sub-basins featured a mix of increasing and decreasing trends in monthly ET_a during the study period. However, none of them showed significant trends. At Chak aw Logar sub-basin a significant increasing trend with MK value of τ = +0.53 (p < 0.05) and Sen's Slope (S) = +0.59 (mm/month) and τ = +0.53 (p < 0.05) and S = +0.38 (mm/month) resulted for monthly ET_a. A significant increasing trend was found at the Gomal sub-basin with MK values ranging from τ = +0.53 to +0.67 (p < 0.05) and S = +0.47 to 1.01 (mm/month) for the months of May, June, October, and November (Table 11.4).

TABLE 11.5

Trend analysis of mean annual evapotranspiration for the sub-basins over 2003–2013 in the KRB. The trend was analyzed using MK's trend test (τ) and Sen's slope estimator (S)

S. No	Sub-basin	Kendall's tau	Sen's slope	Trend
1	Alingar	0.418	4.593	Absent
2	Chak aw Logar	0.345	3.796	Absent
3	Ghorband aw Panjshir	0.091	0.715	Absent
4	Gomal	0.455	4.549	Absent
5	Kabul	**0.600**	11.00	Present
6	Kunar	**0.709**	7.395	Present
7	Shamal	**0.636**	7.634	Present

Similarly, at the Shamal sub-basin, a significant increasing trend resulted for May, June, October, and November. with MK values ranging from τ = +0.56 to +0.71 (p < 0.05) and S = +0.58 to 1.94 (mm/month). A significant increasing trend at Kunar sub-basin was observed from March–June with MK values ranging from τ = +0.56 to +0.71 (p<0.05) and S = +0.63 to 1.23 (mm/month). Unlike other regions, the Kabul sub-basin has significantly increasing trends from May–October in more cases. In December, all sub-basins had a nonsignificant decreasing trend except the Gomal sub-basin, which showed an increasing trend.

Further to the trend analysis in monthly time steps, we also evaluated annual trends in the ET_a from 2003–2013 (Table 11.5). The trend in annual ET_a as detected by MK was significant only for Kabul, Kunar, and Shamal sub-basins, the MK values are 0.600, 0.709, and 0.636, respectively, representing a positive trend in evapotranspiration being statistically significant at a 5% confidence level. Trends in annual evapotranspiration for all sub-basins resulted in a positive trend.

11.4 DISCUSSION AND CONCLUSION

To better understand the hydrological system and the function of snowmelt runoff, as well as its impact on irrigated agriculture in the KRB, it is essential to investigate the statistical characteristics of snow cover and its seasonal and inter-annual variation (Hasson et al. 2014). We used an improved MODIS snow-cover product (i.e., MOYDGL06) to estimate spatiotemporal variation in the SCA in the KRB. The SCA shows high variability during the accumulation season and moderate snow cover variability during the ablation season, consistent with research done in other studies on the UIB (Hasson et al. 2014; Bilal et al. 2019). The comparison to the UIB is based on the physiographic and climatic homogeneity of the KRB and UIB. The SCA analyzed in three different elevation zones indicate a slightly negative trend except for LEZ, where the annual SCA increases by 0.0005% per year. This might be attributed to wind velocity and snowdrift redistributing snowfall at higher elevations (Azizi and Asaoka 2020a). The trend for all the three elevation zones (LEZ, MEZ, HEZ) was not statistically significant.

According to the current study, the mean annual SCA in the KRB dropped by 5% in 2018 compared to 2003; however, there has been an increase in the SCA too; but in general, the mean annual SCA across the KRB during the study period was 29.3 ± 26.8%. The decrease in SCA can be attributed to decreasing precipitation trends, especially in winter (December–March; Azizi and Asaoka 2020a). This finding is in line with the previous research conducted in the UIB from 2001–2012 (Hasson et al. 2014). However, Tahir et al. (2015), Tahir et al. (2016), and Ahmad et al. (2018) observed an increasing snow cover within the sub-basins of the UIB (Astore, Hunza, Gilgit, Chitral), which is geographically connected and experience similar climatic conditions as that of KRB. This contrast is mainly because Astore, Hunza, Gilgit, and Chitral are small tributaries with different hydro-meteorological characteristics that do not indicate the whole UIB. The total area of UIB is 165,000 km², whereas 7%, 5%, and 33% of Gilgit, Asotre, and Hunza, respectively are located above 5000 m ASL, and around 8%, 6%, and 25% of these areas are, respectively, covered with snow and glaciers.

Based on the elevation zone, there is a declining trend in snow cover, notably in the MEZ and HEZ portions of the KRB, which could be attributed to increasing temperature and decreasing precipitation (Azmat et al. 2017). These findings are consistent with prior research covering a period of 2000–2009 in the UIB sub-basins, which goes through half of the period as considered under the current study (Azmat et al. 2017). Western disturbances are extratropical synoptic-scale weather systems that provide heavy precipitation to the Hindukush Himalaya (HKH) throughout the winter (Dimri and Chevuturi 2014). Due to the heavy snowfall throughout the winter season, permanent snowfields have been formed in the higher regions of the HKH range (Azizi and Akhtar 2021). The increased melting of snow and ice packs and snow accumulation because less precipitation in the form of snowfall may cause snow and glacier-covered areas to reduce as a result of rising temperatures (Azizi and Asaoka 2020a). Climate change is also expected to alter upstream snow-cover extent and ice in the HKH region, especially in the Brahmaputra, Ganges, and Indus River basins, threatening the water supply and food security for many people (Azizi and Asaoka 2020a).

Since in this study we focused on SCA rather than the height or volume of snow, a higher SCA does not always imply a higher amount of water retained as snow and vice versa; however, other parameters need to be explored concerning the SCA. In terms of management, variability in precipitation and SCA (provided the snow water equivalent and snow depth, etc., stays unaltered) may necessitate a basin-wide water management plan, as a long-term decreasing snow trend would result in a relative shortage of water availability in the basin. The anticipated decrease may affect opportunities to reach a demand that has not been prioritized in regional management initiatives and programs. However, future work must incorporate data on snow depth and cover measured in situ, which are currently unavailable for the research region. This would enhance current estimates related to SCA performed as a first step by increasing the snow volume assessment.

Besides SCA being analyzed in detail, ET_a analysis was also carried out at the sub-basin scale in the KRB. For the strategically important transboundary KRB,

the spatial information received is highly precious. The ET_a data during the period 2003–2013 in these spatial units are provided for monthly, seasonal, and yearly time steps essential for water management. Such detailed information will assist policy-makers in the region to manage their water resources strategically and operationally and monitor water allocation during strategic periods (Akhtar 2017). Besides, remote-sensing-based ET_a estimation could be helpful in the performance evaluation of the irrigation systems in the data-scarce basins, for example, the KRB (Akhtar et al. 2018). The higher ET_a estimated for Shamal, Kunar, and Kabul sub-basins is because these watersheds host relatively large irrigated areas and experience high temperatures compared to the rest of the sub-basins (Akhtar 2017).

Trend analysis results indicated a significantly increasing trend in ET_a at the monthly timescale for all the sub-basins except the Alingar and Ghorband aw Panjshir sub-basins. The monthly ET_a trend analysis showed that the highest number of sub-basins with a significant increasing trend was observed in May and June, while the lowest numbers of the sub-basins with significant increasing trends were found in March, April, July, and August. The results of the ET_a series on an annual scale indicated a significant increasing trend for Kabul, Kunar, and Shamal sub-basins; such an increasing trend may be justified with a partial increase in irrigated areas and cropping intensity due to rehabilitation of the irrigation system and on-farm water management interventions. However, additional field-based analyses and surveys may (dis)approve this increasing trend in the sub-basins mentioned earlier because field-based experiments and flux towers may bring accuracy in the ET_a estimation, which is usually not possible due to high costs and local deficient technical capacity.

The detailed spatial and temporal estimates of SCA and ET_a can effectively assist region-specific water management and planning initiatives as they consider significant factors of the water supply-side (SCA), as well as the demand (ET_a), and therefore may be used as a measure for assessing water allocation and support irrigation performance assessment in data-scarce regions.

REFERENCES

Ahmad, Shakeel, Muhammad Israr, Shiyin Liu, Huma Hayat, Jawaria Gul, Sara Wajid, Muhammad Ashraf, Siddique Ullah Baig, and Adnan Ahmad Tahir. 2018. "Spatio-temporal trends in snow extent and their linkage to hydro-climatological and topographical factors in the Chitral River Basin (Hindukush, Pakistan)." *Geocarto International* 35 (7): 711–734.

Akhtar, F. 2017. "Water availability and demand analysis in the Kabul River Basin." Ph.D., Zentrum für Entwicklungsforschung, University of Bonn. https://hdl.handle.net/20.500.11811/7031.

Akhtar, Fazlullah, Usman Khalid Awan, Bernhard Tischbein, and Umar Waqas Liaqat. 2018. "Assessment of irrigation performance in large river basins under data scarce environment—A case of Kabul river basin, Afghanistan." *Remote Sensing* 10 (6): 972.

Akhtar, Fazlullah, and Usman Shah. 2020. "Emerging water scarcity issues and challenges in Afghanistan." In *Water Issues in Himalayan South Asia*, 1–28. Springer.

Armanios, Daniel Erian, and Joshua B. Fisher. 2014. "Measuring water availability with limited ground data: Assessing the feasibility of an entirely remote-sensing-based

hydrologic budget of the Rufiji Basin, Tanzania, using TRMM, GRACE, MODIS, SRB, and AIRS." *Hydrological Processes* 28 (3): 853–867.

Azizi, Abdul Haseeb, and Fazlullah Akhtar. 2021. "Analysis of spatiotemporal variation in the snow cover in Western Hindukush-Himalaya region." *Geocarto International*: 1–17.

Azizi, Abdul Haseeb, and Asaoka Yoshihiro. 2020a. "Assessment of the impact of climate change on snow distribution and river flows in a snow-dominated mountainous watershed in the western Hindukush–Himalaya, Afghanistan." *Hydrology* 7 (4): 74.

Azizi, Abdul Haseeb, and Asaoka Yoshihiro. 2020b. "Incorporating snow model and snow-melt runoff model for streamflow simulation in a snow-dominated mountainous basin in the western Hindukush-Himalaya region." *Hydrological Research Letters* 14 (1): 34–40.

Azizi, Abdul Haseeb, and Yoshihiro Asaoka. 2019. "Estimating spatial and temporal snow distribution using numerical model and satellite remote sensing in the western Hindukush-Himalaya region." *Journal of Japan Society of Civil Engineers, Ser. G (Environmental Research)* 75 (5): I_125–I_134.

Azmat, Muhammad, Umar Waqas Liaqat, Muhammad Uzair Qamar, and Usman Khalid Awan. 2017. "Impacts of changing climate and snow cover on the flow regime of Jhelum River, Western Himalayas." *Regional Environmental Change* 17 (3): 813–825.

Beach, Robert H., Benjamin J. DeAngelo, Steven Rose, Changsheng Li, William Salas, and Stephen J. DelGrosso. 2008. "Mitigation potential and costs for global agricultural greenhouse gas emissions 1." *Agricultural Economics* 38 (2): 109–115.

Bilal, Hazrat, Siwar Chamhuri, Mazlin Bin Mokhtar, and Kasturi Devi Kanniah. 2019. "Recent snow cover variation in the Upper Indus Basin of Gilgit Baltistan, Hindukush Karakoram Himalaya." *Journal of Mountain Science* 16 (2): 296–308.

Brutsaert, Wilfried, and Han Stricker. 1979. "An advection-aridity approach to estimate actual regional evapotranspiration." *Water Resources Research* 15 (2): 443–450.

Collados-Lara, Antonio-Juan, David Pulido-Velazquez, Eulogio Pardo-Igúzquiza, and Esteban Alonso-González. 2020. "Estimation of the spatiotemporal dynamic of snow water equivalent at mountain range scale under data scarcity." *Science of the Total Environment* 741: 140485.

Dimri, A.P., and Amulya Chevuturi. 2014. "Model sensitivity analysis study for western disturbances over the Himalayas." *Meteorology and Atmospheric Physics* 123 (3): 155–180.

Elferink, Maarten, and Florian Schierhorn. 2016. "Global demand for food is rising. Can we meet it." *Harvard Business Review* 7 (4): 2016.

Fang, Hongliang, Patricia Hrubiak, Hiroko Kato, Matthew Rodell, William L. Teng, and Bruce E. Vollmer. 2008. "Global land data assimilation system (GLDAS) products from NASA hydrology data and information services center (HDISC)." In *American Society for Photogrammetry and Remote Sensing (ASPRS)*.

FAO. 2012. "Country profile –Afghanistan." *Food and Agriculture Organization of the United Nations (FAO)*, Rome, Italy. Accessed 04 Aug 2021.

FAO. 2016. "Afghanistan land cover atlas." Accessed 03 Jul. 2021. www.fao.org/3/i5043e/i5043e.pdf.

Fondriest Environmental Inc. 2014. "Water temperature." *Fundamentals of Environmental Measurements*. www.fondriest.com/environmental-measurements/parameters/water quality/water-temperature/. 7th Feb. Accessed 04 Aug. 2021.

Fu, Yulong, Hongxi Peng, Jingqing Liu, Thanh H. Nguyen, Muhammad Zaffar Hashmi, and Chaofeng Shen. 2021. "Occurrence and quantification of culturable and viable but non-culturable (VBNC) pathogens in biofilm on different pipes from a metropolitan drinking water distribution system." *Science of the Total Environment* 764: 142851.

Gilbert, Richard O. 1987. *Statistical Methods for Environmental Pollution Monitoring*. John Wiley & Sons.

Gorji, Taha, Elif Sertel, and Aysegul Tanik. 2017. "Monitoring soil salinity via remote sensing technology under data scarce conditions: A case study from Turkey." *Ecological Indicators* 74: 384–391.

Hall, D. K., G. A. Riggs, and V. V. Salomonson. 2006. "MODIS/Terra snow cover 5-min L2 swath 500m." *NASA National Snow and Ice Data Center Distributed Active Archive Center*. https://doi.org/10.5067/ACYTYZB9BEOS.

Hall, D. K., G. A. Riggs, and V. V. Salomonson. 2021. "MODIS/Terra snow cover 8-day L3 global 500m SIN grid." *NASA National Snow and Ice Data Center Distributed Active Archive Center*. https://doi.org/10.5067/MODIS/MOD10A2.061.

Hasson, Shabeh, Valerio Lucarini, Mobushir R. Khan, Marcello Petitta, Tobias Bolch, and Giovanna Gioli. 2014. "Early 21st century snow cover state over the western river basins of the Indus River system." *Hydrology and Earth System Sciences* 18 (10): 4077–4100.

Jia, L., G. Xi, S. Liu, C. Huang, Y. Yan, and G. Liu. 2009. "Regional estimation of daily to annual regional evapotranspiration with MODIS data in the Yellow River Delta wetland." *Hydrology and Earth System Sciences* 13 (10): 1775–1787.

Jiang, Xia, Xiangcan Jin, Yang Yao, Lihe Li, and Fengchang Wu. 2008. "Effects of biological activity, light, temperature and oxygen on phosphorus release processes at the sediment and water interface of Taihu Lake, China." *Water Research* 42 (8–9): 2251–2259.

Kim, Dong-Eon, Philippe Gourbesville, and Shie-Yui Liong. 2019. "Overcoming data scarcity in flood hazard assessment using remote sensing and artificial neural network." *Smart Water* 4 (1): 1–15.

Kiptala, J. K., Y. Mohamed, Marloes L. Mul, and P. Van der Zaag. 2013. "Mapping evapotranspiration trends using MODIS and SEBAL model in a data scarce and heterogeneous landscape in Eastern Africa." *Water Resources Research* 49 (12): 8495–8510.

Liaqat, Umar Waqas, and Minha Choi. 2017. "Accuracy comparison of remotely sensed evapotranspiration products and their associated water stress footprints under different land cover types in Korean peninsula." *Journal of Cleaner Production* 155: 93–104.

Liaqat, Umar Waqas, Minha Choi, and Usman Khalid Awan. 2015. "Spatio-temporal distribution of actual evapotranspiration in the Indus Basin Irrigation System." *Hydrological Processes* 29 (11): 2613–2627.

Mann, Henry B. 1945. "Nonparametric tests against trend." *Econometrica: Journal of the Econometric Society*: 245–259.

Muhammad, Sher, and Amrit Thapa. 2020. "An improved Terra–Aqua MODIS snow cover and Randolph Glacier Inventory 6.0 combined product (MOYDGL06*) for high-mountain Asia between 2002 and 2018." *Earth System Science Data* 12 (1): 345–356.

Rodell, Matthew, P. R. Houser, U. E. A. Jambor, J. Gottschalck, K. Mitchell, C.-J. Meng, K. Arsenault, B. Cosgrove, J. Radakovich, and M. Bosilovich. 2004. "The global land data assimilation system." *Bulletin of the American Meteorological Society* 85 (3): 381–394.

Roser, Max. 2013. "Future population growth." *Our World in Data*. Accessed 04 Aug. 2021. https://ourworldindata.org/future-population-growth.

Sen, Pranab Kumar. 1968. "Estimates of the regression coefficient based on Kendall's tau." *Journal of the American statistical association* 63 (324): 1379–1389.

Singh, Pratap, and Naresh Kumar. 1997. "Impact assessment of climate change on the hydrological response of a snow and glacier melt runoff dominated Himalayan River." *Journal of Hydrology* 193 (1–4): 316–350.

Su, Zhongbo. 2002. "The Surface Energy Balance System (SEBS) for estimation of turbulent heat fluxes." *Hydrology and Earth System Sciences* 6 (1): 85–100.

Tahir, Adnan Ahmad, Jan Franklin Adamowski, Pierre Chevallier, Ayaz Ul Haq, and Silvia Terzago. 2016. "Comparative assessment of spatiotemporal snow cover changes and hydrological behavior of the Gilgit, Astore and Hunza River basins (Hindukush–Karakoram–Himalaya region, Pakistan)." *Meteorology and Atmospheric Physics* 128 (6): 793–811.

Tahir, Adnan Ahmad, Pierre Chevallier, Yves Arnaud, Muhammad Ashraf, and Muhammad Tousif Bhatti. 2015. "Snow cover trend and hydrological characteristics of the Astore River basin (Western Himalayas) and its comparison to the Hunza basin (Karakoram region)." *Science of the Total Environment* 505: 748–761.

UNFPA. 2021. "World Population Dashboard." www.unfpa.org/data/world-population-dashboard. Accessed 04 Aug. 2021.

Wake, Bronwyn. 2021. "Water wars." *Nature Climate Change* 11 (2): 84–84.

You, Jiaxuan, Xiaocheng Li, Melvin Low, David Lobell, and Stefano Ermon. 2017. "Deep Gaussian process for crop yield prediction based on remote sensing data." *Thirty-First Association for the Advancement of Artificial Intelligence (AAAI) Conference on Artificial Intelligence.*

Section III

Geospatial Approaches for Monitoring Urban Ecosystem

12 Urban Sprawl and Future Growth Projection vis-à-vis Groundwater Resource Availability in Hill Township in Shillong (Meghalaya), India

Arvind Chandra Pandey[1,], Sudipta Hansda[1], and Navneet Kumar[2]*

1 Department of Geoinformatics, School of Natural Resource Management Central University of Jharkhand, Ranchi-835222, India

2 Department of Ecology and Natural Resources Management, Center for Development Research (ZEF), University of Bonn, Germany

CONTENTS

12.1 Introduction 226
12.2 Location of Study Area.. 227
12.3 Data Sources and Methodology.. 228
    12.3.1 Methodology 229
12.4 Results..231
    12.4.1 LULC Analysis and Urban Growth..231
    12.4.2 Urban Growth Simulation by CA for Prediction 234
    12.4.3 Spatial Pattern of Rainfall 234
    12.4.4 Various Thematic Maps (Soil, Drainage Density, Lineament
        Density).. 236
    12.4.5 Various Thematic Maps (Slope, Geomorphology, Lithology)...........237
    12.4.6 Integration of Thematic Layers and Modeling through GIS 241
    12.4.7 GWP Zone 241
    12.4.8 Urban Sprawl versus Groundwater Potential................................. 244

DOI: 10.1201/9781003265160-15

12.5 Conclusion 246
Acknowledgment 247
Conflicts of Interest.. 247
References.. 247

12.1 INTRODUCTION

Groundwater is one of the world's most renewable and widely distributed sources of water and is an essential source of water supply worldwide (Mahalingam et al., 2014). In India, more than 60% of irrigated agriculture and 85% of drinking water supplies rely on groundwater (The World Bank, 2012). Urban growth is increasing rapidly in densely populated areas and putting undue pressure on natural resources. There are natural conflicts, especially in the fast-growing big cities across the country. Urbanization has a significant impact on both the quantity and quality of groundwater resources (Kumar et al., 2017). As urban migration and groundwater resources are strongly connected (Kalhor and Emaminejad, 2019), the water demand is constantly increasing. Without efforts to provide adequate water, the water storage systems introduced over the years are not sufficient to meet the rapidly growing demand for water (Naik et al., 2008). The groundwater level is deteriorating due to moderate rainfall, which ultimately results in a decrease in groundwater recharge. The impact of urban sprawl on groundwater in a particular urban area depends on its location and the economic situation of the city (Patra et al., 2018). The steady increase in the built-up environment and its impact on potential groundwater needs to be considered to minimize its negative impacts (Khan et al., 2019). The use of Geographic Information System (GIS) technology has made it much easier to assess groundwater (Khan et al., 2017). Mapping of potential groundwater resources is important for proper resource utilization considering an increase in future requirements (Suganthi et al., 2013). The groundwater level is deteriorating due to moderate rainfall, which ultimately decreases groundwater recharge.

Urbanization is both a blessing and a curse for the environment, but it is also an important development indicator. Development cannot be halted, but it necessitates maintaining a balance between fast-rising urbanization and pollution rates, particularly in terms of groundwater quality. In Hyderabad city (India), the effects of growing urbanization on groundwater resources are twofold: effects on natural aquifer recharge due to concrete sealing and pollution of groundwater due to drainage leakage and industrial waste and effluents (Wakode et al., 2014). Rapid urbanization, combined with a lack of good planning and significant rural–urban mobility, constitutes the primary driver of those changes linked to a loss of ecosystem services and negatively influences the city's human well-being (Wang et al., 2020). Built-up growth of 473% in Ranchi township of Jharkhand from 1927–2005 occurred mostly at the expense of agricultural land and reflected a negative environmental impact and harmful effects of built-up expansion (Kumar et al., 2011). In western parts of Mizoram, Northeast India, the land use land cover (LULC) was assessed, and it was reported that forest clearance due to Jhum cultivation (i.e., shifting cultivation) was prominent (Parida and Mandal, 2020). Climate change can cause adverse impacts on hydrological processes in a river basin of Northeast India, by altering runoff due to modifications in LULC, catchment hydrology, and water balance (Anand et al., 2020) which would affect groundwater.

For predicting future urban growth in an Indian metropolis, some researchers employed a cellular automata (CA) model (Maithani, 2010) and multicriteria evaluation (MCE) approach to uncover future urban expansion potential. The CA model was used to distribute land for future urban development based on MCE's urban suitability image, a site's neighborhood information, and the amount of land projected by the Markov chain process. The Kappa coefficient was used to assess the model's output, and the calibrated model was used to simulate future urban expansion (Maithani, 2010). Based on Shannon's entropy and multi-temporal satellite data analysis, previous studies indicated unplanned and haphazard growth, resulting in urban densification in many Himalayan cities during 1972–2015 (Diksha and Kumar, 2017). Many urbanization research uses GIS as a tool for understanding the effects of urban sprawl on the environment (Dhaoui, 2014). GIS outlines urban growth patterns by measuring the distances of new urban spaces from urban centers and streets, making GIS a decision support system to assist city planning. Satellite-based earth observation provided ample scope for effectively carrying out water resource assessment. The integrated GIS and remote-sensing techniques have proved to be a productive tool in groundwater studies (Ramamoorthy and Rammohan, 2015). Several researchers have used GIS and satellite data for water resource management, groundwater assessment, extreme rainfall events, and modeling (Arulbalaji et al., 2019; Parida et al., 2017). To introduce the geographical variability of hydro-geomorphological properties, a recharge estimation methodology has been applied in a GIS, which revealed that urbanization has resulted in a widespread drop in the water table and groundwater quality in Ajmer city (Jat et al., 2008). The greatest advantage of using geospatial hydrological methods is their potential to create data in time and space domains, critical to effective analysis, forecasting, and validation (Putranto and Aryanto, 2018). Delineation of groundwater potential (GWP) zones based on satellite data entails mapping geomorphology, lithology, slope, lineaments, soil, water, land use/land cover, relief, and rainfall (Shekhar and Pandey, 2015). Urbanization in Meghalaya is not only associated with industrialization but also due to the growth and expansion of establishment. The current study is an attempt to study the impact of urban sprawl and future built-up growth on the potential groundwater availability in Shillong, a fast-growing city in the Himalayas, northeastern India in the state of Meghalaya.

12.2   LOCATION OF STUDY AREA

Shillong is the study area, which is the capital city of Meghalaya (created in 1972 from the State of Assam with two hill districts of United Khasi and Jaintia Hills, and Garo Hills) and located in the Khasi Hills on the Meghalaya Plateau (Figure 12.1). Township elevation range lies within 1400–1900 m above mean sea level (amsl). The state has seven provinces, and Shillong is the headquarters of East-Khasi Hills district. Shillong, on the Continuum of Urbanisation, was a small village until it was made the capital of the Khasi and Jaintia Hills district in 1866. The area of Shillong was the habitat of the Khasis, who had been living in these hills from ancient times. The city started with a few scattered huts in the adjoining villages of Laban, Laitumkhrah, Nongkseh, and Lawsohtun. Shillong's growth can be traced to the establishment of the cantonment by the British in 1867. Since then, Shillong grew and its history of urbanization began, and later, it was made the capital of Assam in

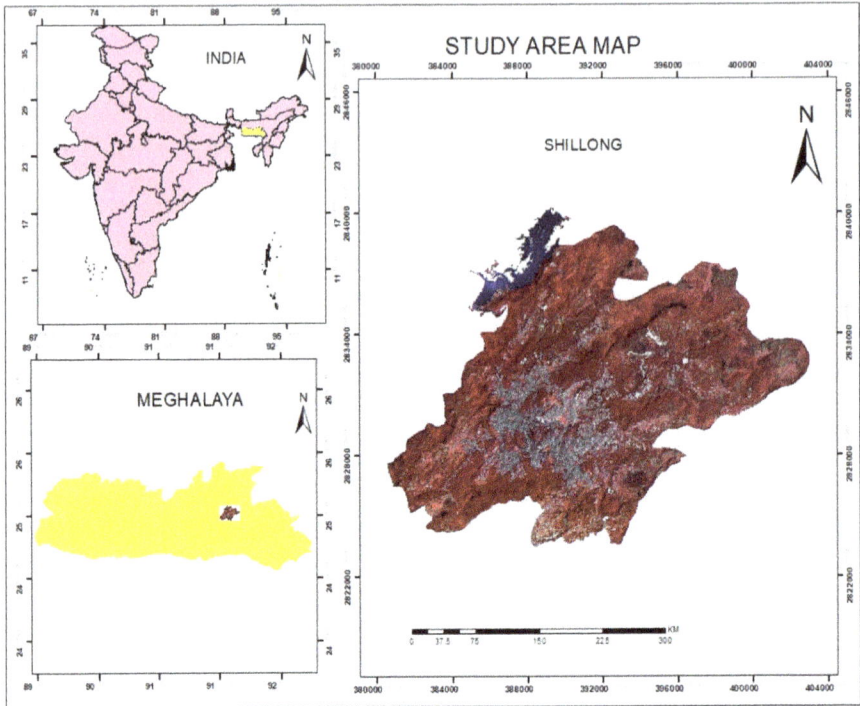

FIGURE 12.1   Location map of Shillong.

1874 and remained so until January 1972, after the establishment of the Meghalaya. The Census (1911) shows 13,639 persons, which gradually increased to 3,54,759 as per the census of 2011. There are two natural drainage channels in the Shillong city, such as the Umkhrah and Umshyrpi, which confluence at the Bijon-Bishop Falls and flow into the Umiam River system. The climate of Meghalaya is generally mild. In August, the temperature in Shillong (Khasi Hills) is about 21–23°C and then drops to 8–10°C in January. One of the wettest regions in the world called Cherrapunji is located in Meghalaya, which receives an average precipitation of about 11,430 mm in a year during the monsoon season (May–September). The annual rainfall in Shillong is about 2,290 mm and during the winter months (December–February), the weather is relatively dry.

12.3   DATA SOURCES AND METHODOLOGY

The most important component of a GIS is the data it uses that can be derived from numerous sources and used for spatial analysis. The present study employed various satellite data and published maps for creating a thematic database for spatial analysis pertaining to urban sprawl and GWP assessment (Table 12.1). The data sources used are as presented in Table 12.1.

TABLE 12.1
Data Used for the Present Study

Data Used	Sensor Characteristics	Acquisition Date	Purpose
Landsat-7	Spatial Resolution: 30 m Spectral Bands: 4, 3, 2 bands used Band 4 Near-Infrared (0.77–0.90 μm) 30 m Band 3 Visible (0.63–0.69 μm) 30 m Band 2 Visible (0.52–0.60 μm) 30 m	28.12.2000	Land use land cover map of 2000
Landsat-8	Spatial Resolution: 30 m Spectral Bands: 5, 4, 3 bands used Band 5 Near-Infrared (0.85–0.88 μm) 30 m Band 4 Red (0.64–0.67 μm) 30 m Band 3 Visible (0.53–0.59 μm) 30 m	22.12.2020	Land use land cover map of 2020
Carto DEM (CARTOSAT 1)	Spatial Resolution: 30 m	29.04.2015	Slope, Drainage Density and Lineament Density
PERSIANN rainfall data	Spatial Resolution: 0.25° × 0.25°	2000	Rainfall Map
PERSIANN-CCS rainfall data	Spatial Resolution: 0.04° × 0.04°	2020	Rainfall Map

Landsat-7 and -8 provide satellite data with a spatial resolution of 30 m, and these data were acquired from the U.S. Geological Survey (USGS) for mapping the LULC map for 2000 and 2020, respectively. CartoDEM is an Indian National DEM generated from CARTOSAT-1, which has been downloaded from the Bhuvan platform with a spatial resolution of 30 m. The current PERSIANN (Precipitation Estimation from Remotely Sensed Information using Artificial Neural Networks) system uses neural network separation processes/measurement methods to measure rainfall values at 0.25° × 0.25°. The PERSIANN-Cloud Classification System (PERSIANN-CCS) is a real-time global high-resolution (0.04° × 0.04°) satellite precipitation product (Nguyen et al., 2019) that was downloaded for 2020. Lithology and geomorphology maps of the study area are obtained from the Geological Survey of India. Furthermore, the soil map is obtained from National Bureau of Soil Survey & Land Use Planning and the distribution of geological units was obtained from published maps from the Geological Survey of India.

12.3.1 Methodology

In this study, satellite data acquired in 2000 and 2020 have been used to map the urban growth of Shillong capital city to measure the patterns of urban sprawl in conjugation with population changes (Figure 12.2). Landsat-7 and Landsat-8 data for 2000 and 2020, respectively, were used for mapping land use/land cover (LULC) changes using the supervised classification technique. The six LULC classes, such

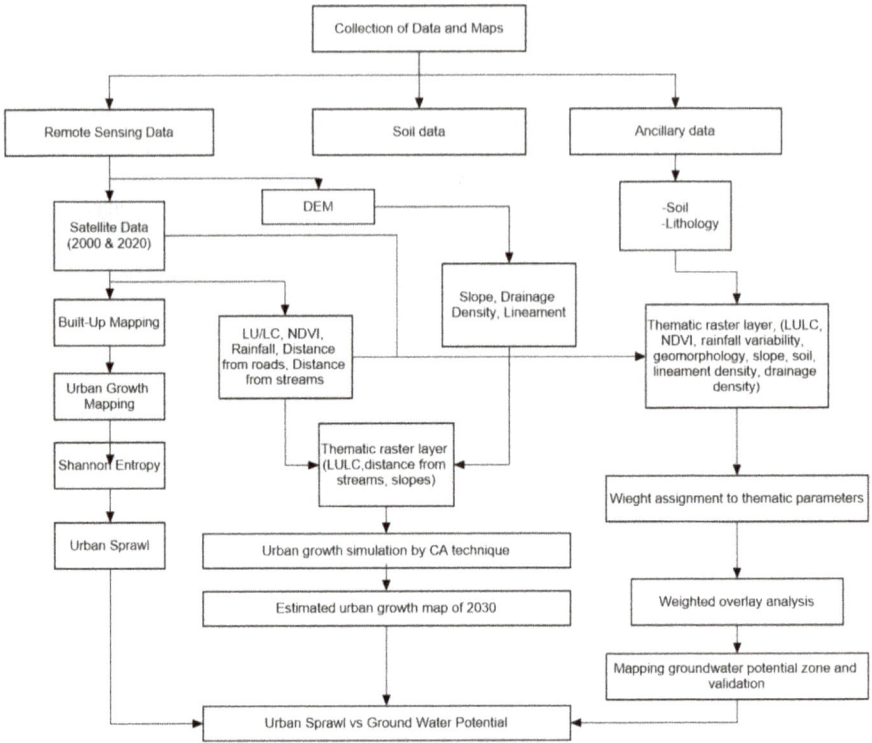

FIGURE 12.2   Methodology flowchart adapted in the study.

as agricultural land, built-up, forest, degraded forest, water bodies, and others, were mapped. To authenticate the urban landscape, Google Earth images were used followed by computation of overall accuracy and kappa value. Built-up land was mapped to specify urban growth, and further, Shannon's entropy was calculated for assessing the city expansion between 2000 to 2020. The study area comprising Shillong city was divided into four parts, namely, northwest (zone 1), southwest (zone 2), southeast (zone 3), and northeast (zone 4), for computing Shannon's entropy.

Shannon's entropy (**Sn**) was calculated using the following formula:

$$Sn = -\Sigma\ Vi\ log_e\ (Vi),$$

where **Vi** is the value of the variable in the ith zone, for example, the ratio of the urban area (as a percentage) for each area, calculated as a percentage of the local urban area / total percentage of the urban area of all segments. **n** = total number of areas, that is, 4. The number of entropy distances from 0 to log_e (n).

This study initially focused on understanding the urban system's dynamics and characteristics of growth, with a more complex perspective, such as a background and the need to model the urban system. Some modern understanding of the existing

modeling techniques is aimed at gaining a deeper understanding of modeling urban systems. Accuracy assessment methods are designed to assess the performance of the model, which helps us to estimate LULC change by 2030.

For groundwater resource assessment, this study covered a wide range of issues, such as basic map editing, digitization, and image processing using software and interpreting the results. GIS and remote-sensing data have been used to prepare various factor maps affecting GWPs, such as lithology, lineament, geomorphology, soil, LULC, slope, and drainage network. DEM data have been downloaded from Bhuvan (https://bhuvan-app3.nrsc.gov.in/), and this has been used as a basis for building slope, aspect, relief, and drainage network maps. Drainage density evaluates the features of runoff and groundwater discharge in the area. The natural break method was employed to calculate the drainage density (i.e., length of total drainage channels per unit area) values (Jenks, 1967). Drainage density (D_D) is computed using the following formula:

$$D_D = L_S/A_B,$$

where L_S = length of total stream of all orders and A_B = area of the basin.

All vector and raster maps were prepared using ArcGIS software, and those thematic maps were incorporated in weighted overlay analysis in the GIS platform.

12.4 RESULTS

12.4.1 LULC Analysis and Urban Growth

LULC plays a vital role in deciphering the regional land-use transformation. Land-use practices are largely affected due to forest degradation affecting biodiversity as well as climate change–induced disasters. The present study has analyzed the change in the LULC pattern based on satellite images of 2000 and 2020. The result of the six categorized LULC mapped with an overall accuracy of 88.36% and a kappa coefficient of 0.82 for 2000 is shown in Figure 12.3a. The LULC map (2020) is shown in Figure 12.3b prepared with an overall accuracy of 95.05% and a kappa coefficient of 0.93. Figure 12.3 shows the amount of "from–to" change that has taken place over 20 years.

Urbanization in Shillong has maintained a steady growth, which indicated an increasing population and a reduced water quality and quantity of the region. The temporal satellite data was used to monitor and analyze the urban growth pattern in the city of Shillong for 2000 and 2020. Our analysis revealed that the urban area has increased from 43.08 km² to 60.08 km² between 2000 and 2020, with a 16.95-km² increase in urban area and a 39.34% urban growth rate. The urban growth percentage is calculated as (Urban Growth/Year 2000 Urban Area) * 100 = 39.34%.

The city has been divided into four parts, namely, northwest (zone 1), southwest (zone 2), southeast (zone 3), and northeast (zone 4) and is shown in Figure 12.4. In the current study, the number of entropy distances was taken from 0 to 1.386. The value of 0 indicates that the allotment of built-up areas is very small, and the values around the property (n) indicate that the allotment of urban areas is very widespread. High entropy values indicate the presence of sprawl. The results based on Shannon's

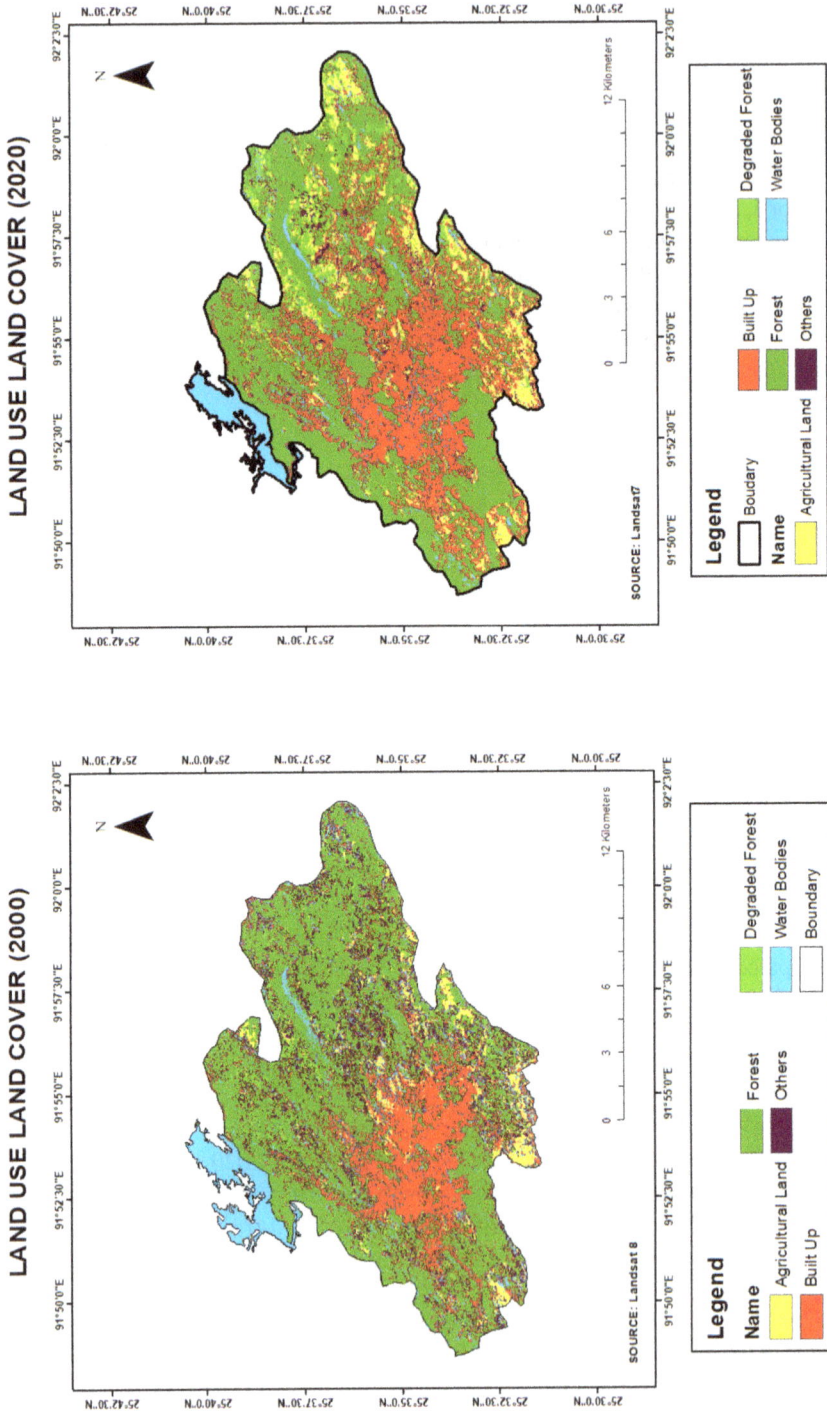

FIGURE 12.3   LULC map of Shillong for (a) 2000 and (b) 2020.

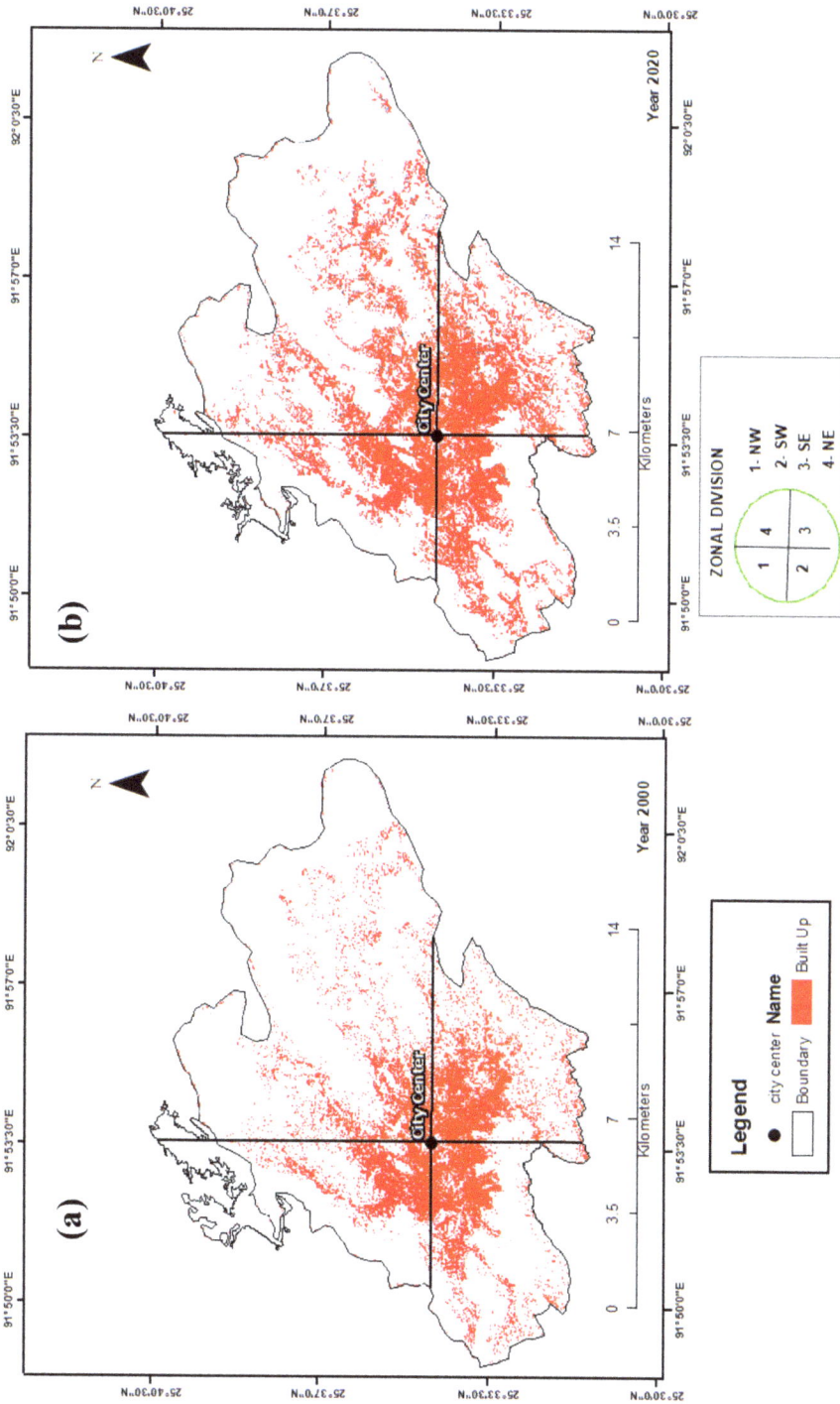

FIGURE 12.4 Map showing the zonal division of Shillong and urban sprawl for (a) 2000 and (b) 2020.

entropy show a high-dispersion (1.356) urban growth pattern during 2000 and a comparatively low-dispersion (1.329) urban growth pattern for 2020 in Shillong.

12.4.2   URBAN GROWTH SIMULATION BY CA FOR PREDICTION

The visualization of the futuristic changes was predicted through the application of CA considering the LULC dynamics during 2000 and 2020. The result indicated the probabilities of the increasing area that will be covered by the built-up area. The basis for transition is that the method computes the past transformation in pixels values with reference to 2020. Also, this method was affected by the spatial variables that are used such as slope, relief, road network, and streams. The study also exhibited land cover transition and it is predicted that built-up area, degraded forest, and agricultural land would increase while water bodies, forests, others feature classes would decrease by 2030. The predicted LULC map for 2030 was shown in Figure 12.5.

As per the present estimation, it is perceived that the built-up area would increase from 60.08 km^2 to 70.28 km^2 (an increase of 10.2 km^2) from 2020 to 2030, with a growth percentage of 16.97%.

12.4.3   SPATIAL PATTERN OF RAINFALL

Meghalaya receives plenty of rainfall during the monsoon season every year. However, the rainfall is having high temporal variability and due to the impact of climate changes, there are significant changes in the rainfall pattern. The observation of change in rainfall is presented in Figure 12.6 based on the PERSIANN and PERSIAN-CCS rainfall data for 2020 and 2000, respectively. The analysis indicated

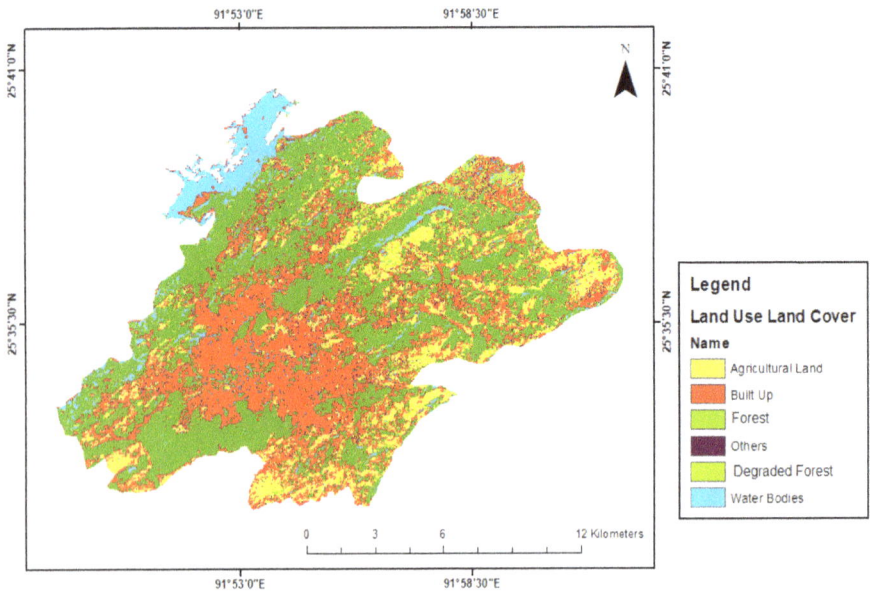

FIGURE 12.5   Predicted LULC map of Shillong for 2030.

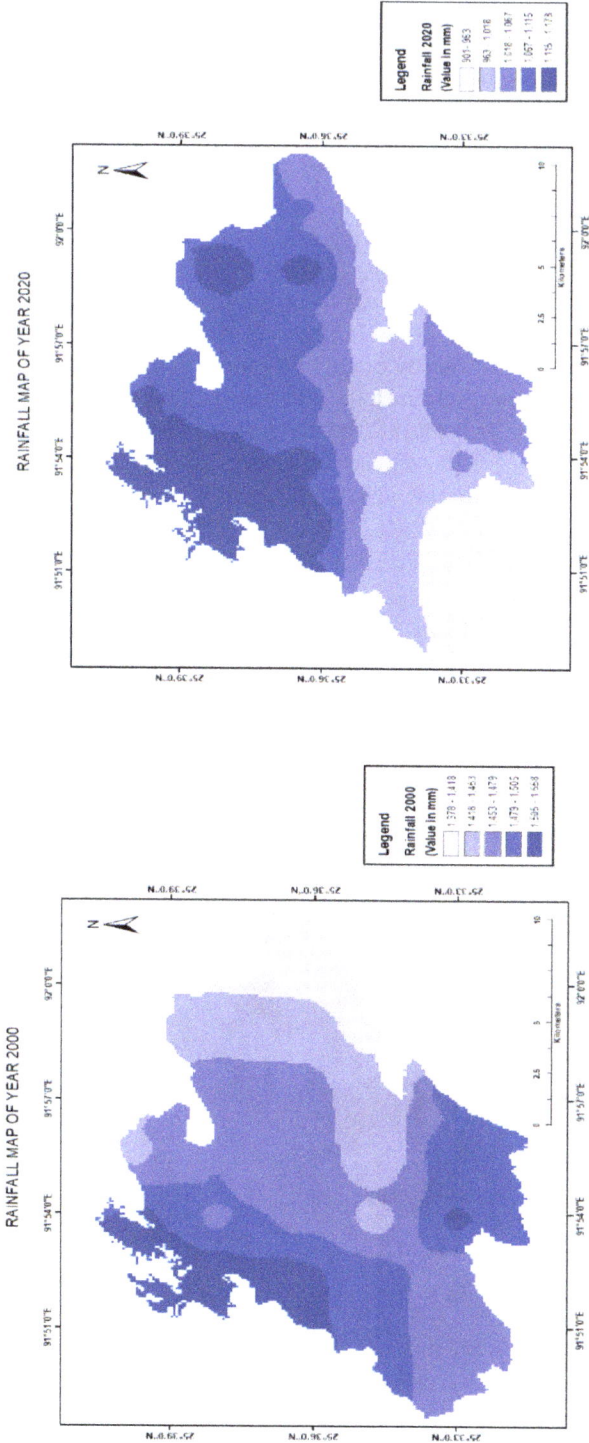

FIGURE 12.6 Rainfall maps of Shillong for 2000 and 2020.

that in 2000 and 2020, the maximum rainfall received was 1558 mm and 1178 mm, respectively, in Shillong. The minimum rainfall received was 1378 mm in 2000 and 901 mm in 2020. The rainfall shows a prominent decreasing spatial pattern as shown in Figure 12.6. As rainfall constitutes the prime factor of groundwater recharge in the area, rainfall information is vital from a GWP point of view in the Shillong region.

12.4.4 VARIOUS THEMATIC MAPS (SOIL, DRAINAGE DENSITY, LINEAMENT DENSITY)

As per the U.S. Natural Resource Conservation Service (NRCS), soils are categorized into four hydrologic groups, such as A, B, C, and D, based on infiltration characteristics of the soils. The NRCS Soil Survey Staff (1996) mention the hydrological group that has similar runoff potential. Soil structures contribute to the flow of runoff where a low penetration rate exists, as on bare soil. The different soil classes present in the area are shown in Figure 12.7, which showed the dominance of C-type soils with low penetration rates, when well watered, with high runoff potential, reflecting a poor groundwater recharge potential of soils in the region. Terrain exhibiting sparse drainage development shows prominent infiltration and sluggish runoff. Areas with moderate drainage density

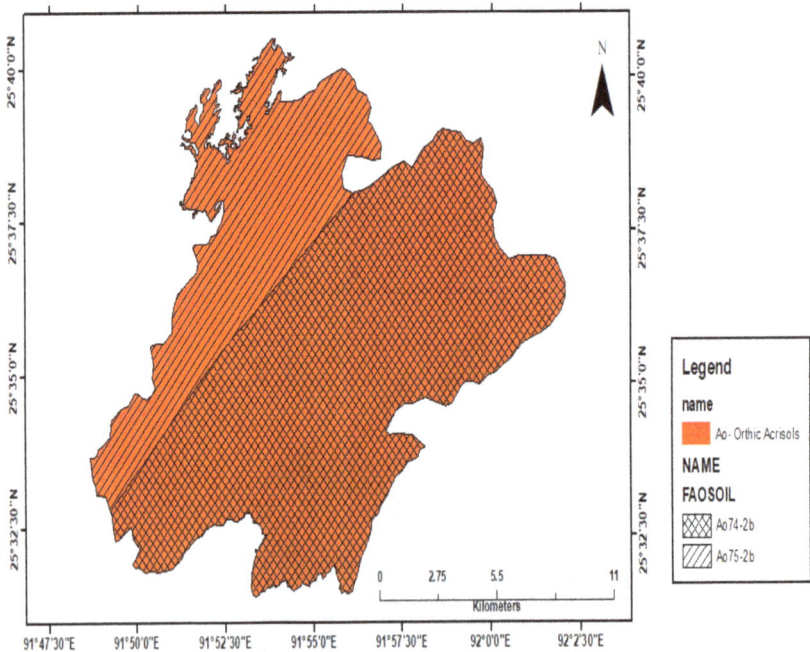

SEQN	FAO SOIL NAME	NLAYERS	HYDGRP	SOL_ZMX	TEXTURE	SOL_Z1	CLAY1	SILT1	SAND1	ROCK1	SOL_Z2	CLAY2	SILT2	SAND2	ROCK2
3647	Ao75-2b	2	C	1000	SANDY_CLAY_LOAM	300	23	25	52	0	1000	36	23	40	0
3646	Ao74-2b	2	C	1000	SANDY_CLAY_LOAM	300	22	27	51	0	1000	34	25	41	0

FIGURE 12.7   Soil map of Shillong.

MAP SHOWING WATERSHED ZONES , STREAM ORDERS AND DRAINAGE DENSITY

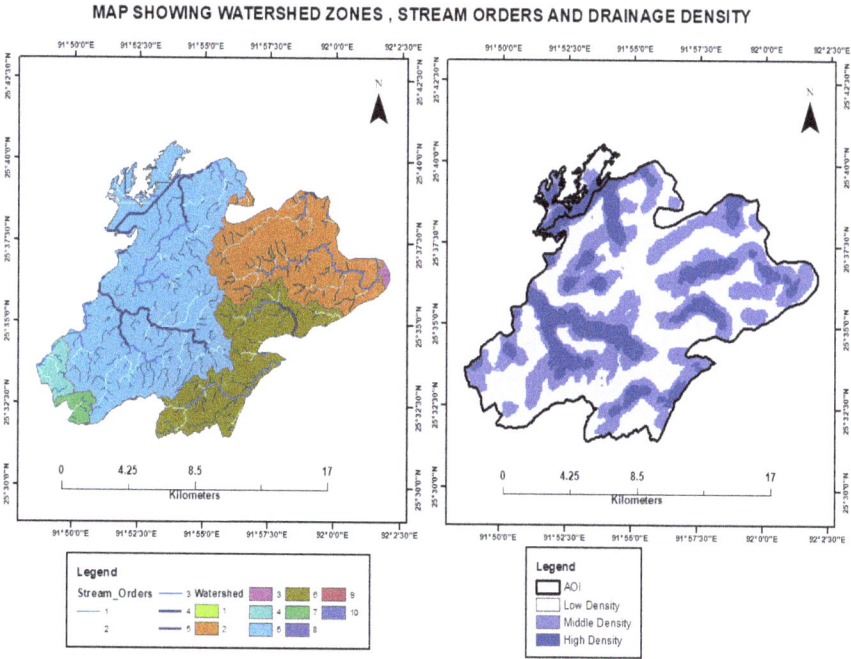

FIGURE 12.8 Watershed zones, stream orders, and drainage density of Shillong.

are considered to exhibit excellent groundwater recharge potential zones and areas with high drainage density indicate higher runoff and low infiltration as shown in Figure 12.8.

The lineaments are linear or curvilinear natural elements on the terrain largely developed by tectonic activity. Lineaments are often associated with a linear ridge topography, pattern, straight drainage course, and the like. Generally, groundwater flow and storage are controlled by lineament. The study revealed a close relationship between the lineaments and the groundwater flow and groundwater recharge potential. Developing a lineament map that is closely related to GWP and yields is important for groundwater targeting, development, and management. The existing lineaments in the study area were extracted from the conjunctive use of FCC of satellite image and CartoDEM (Figure 12.9). Most of the lineaments were delineated over northern, as well as eastern, parts, reflecting that higher lineament density induces more recharge to groundwater.

12.4.5 Various Thematic Maps (Slope, Geomorphology, Lithology)

Terrain slope information is one of the GIS layers that play an important role in natural resources planning. The present study area has been classified into six slope categories (Figure 12.10a) based on Carto-DEM using the terrain analysis module in ArcGIS. The majority of the landscape exhibit is very low to moderate slopes as shown in Figure 12.10a. Moderately sloping areas are observed near the built-up areas. It is presumed that high-slope areas render less infiltration of rainwater, resulting in less recharge of groundwater and vice versa.

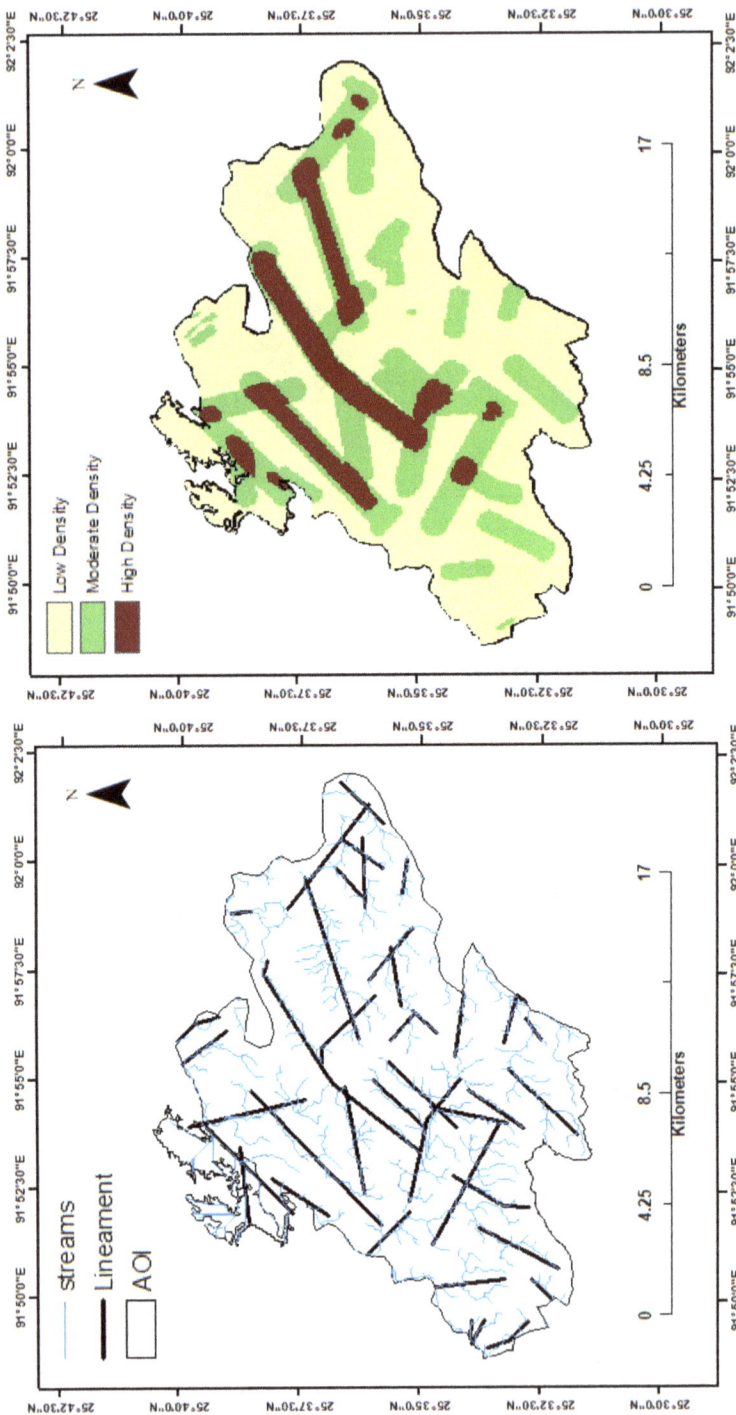

FIGURE 12.9 Lineament (left) and lineament density (right) of Shillong.

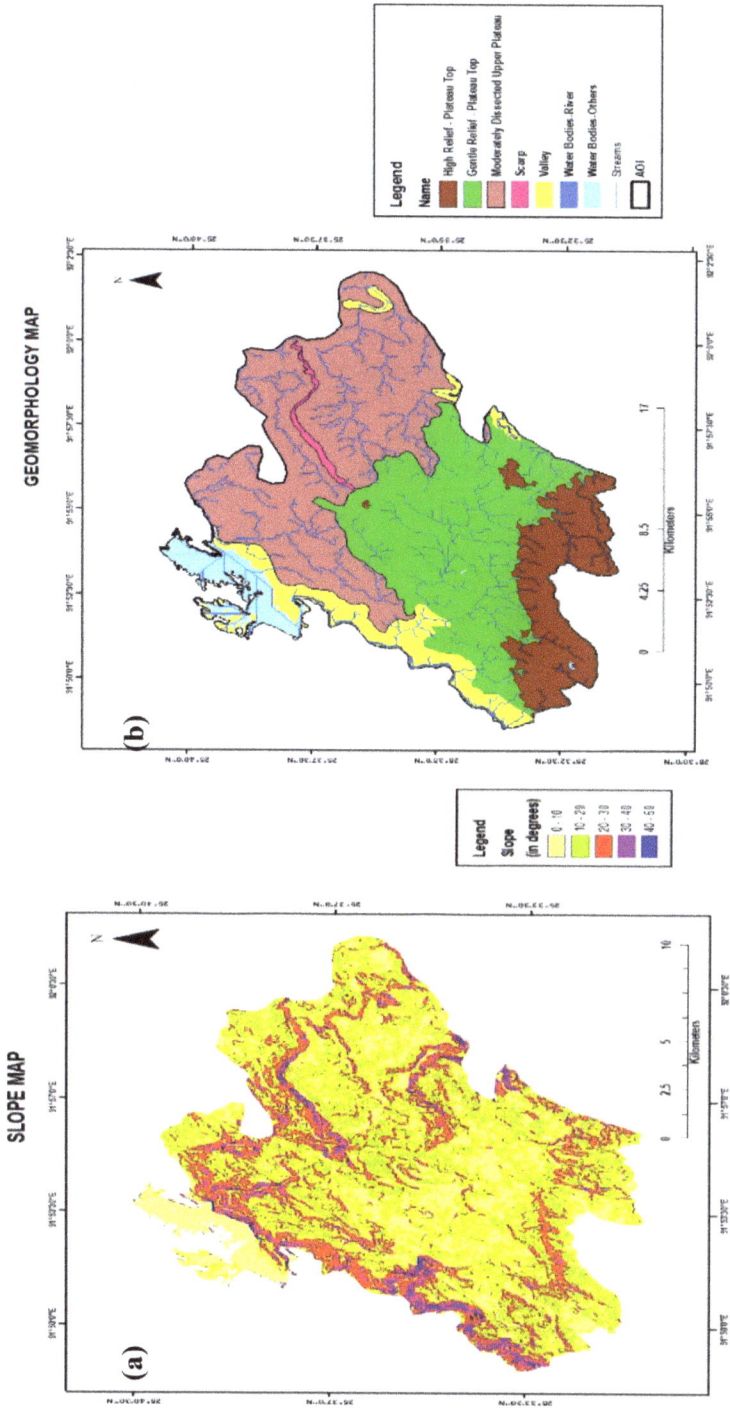

FIGURE 12.10   Slope and geomorphological maps of Shillong.

Terrain landform as a unit reflects the interplay between the geomorphological process and the hydrogeological process to control GWP. The use of geomorphological techniques and concepts in the analysis of groundwater systems plays an important role in classifying GWP. The hydro-geomorphological units of this area were mapped using satellite data in conjunction with other terrain parameters. The geomorphological units mapped are composed of plateau top, moderate dissected upper plateau, valley, and scrap as shown in Figure 12.10b. In valleys, unconsolidated material with the dominance of boulders, along with sand-silt, comprises the valley fills and at places exhibit alignment along lineament. Valley flats and surface water bodies permit adequate infiltration resulting in good potential. On the contrary plateau top and moderate dissected upper plateau landforms consist of nonfractured rock with low infiltration capacity, therefore, exhibit poor potential.

Lithology reflects the physical properties of rock in terms of its mineral composition affecting its compactness and weathering. In the study area, broadly three lithological units namely, conglomerate, epidiorite, and quartzite with a thin phyllite interband, are recognized as deduced from the Survey of India geological map (Figure 12.11). The majority area of the study area belongs to quartzite with a thin

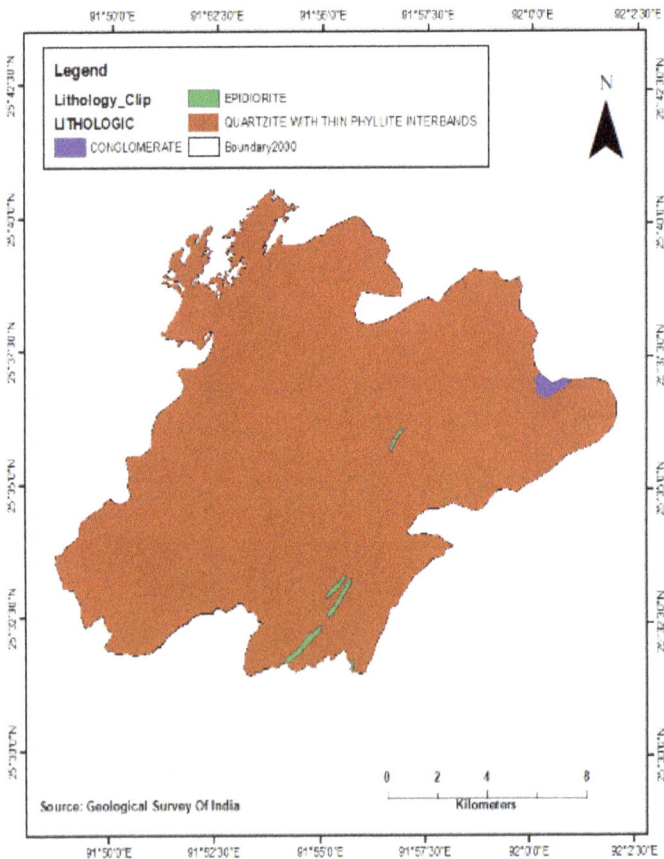

FIGURE 12.11   Lithology map of Shillong.

phyllite interband lithology reflecting fewer weathering zones. The highest infiltration rate takes place in conglomerate whereas the quartzite with thin phyllite interbands has a moderate infiltration rate and epidiorite has the lowest infiltration rate.

12.4.6 INTEGRATION OF THEMATIC LAYERS AND MODELING THROUGH GIS

About eight factors such as geomorphology, slope, lineament, drainage, lithology, soil, LULC, and precipitation were designated to define the potential of groundwater in the study area. Since the influence of these factors on GWP is not the same, the weight of each factor is divided according to the effect of the flow and storage of groundwater (Chaudhary and Kumar, 2018). The weight of a factor represents the ratio of its value to the potential for groundwater, where the maximum weight value is given to the most influencing factor.

Weighted Index Overlay Model: Depending on the groundwater potentiality, every class is appropriately placed in one of the suitable factor categories. The appropriate weight in the eighth scale is given to each category of a certain thematic layer according to their role in groundwater potentiality (Table 12.2). The rank of each thematic map is reduced by the weight of the article (Chaudhary and Lal, 2018). The weight given to the different classes of all the thematic layers and the rank of each feature is given in Table 12.3. The final GWP map (Figure 12.12) is prepared based on the previously mentioned AHP procedure as described in Shekhar and Pandey (2015). The integration and analysis of various thematic databases, such as geomorphology, lithology, soil, slope, lineament density, weather zone thickness, drainage density, and rainfall, proved useful in defining and deducing the GWP of areas.

12.4.7 GWP ZONE

A potential groundwater map was developed using a weighted overlay analysis by summarizing the weights of every thematic layer. Location points with a higher weight are considered to have an area with a high groundwater potential zone (Ibrahim-Bathis and Ahmed, 2016). The resultant GWP map was subdivided into five categories from excellent to very poor potential following equal interval classification (Figure 12.12). The various classes have been identified as excellent, very good, good, poor, and very poor (Table 12.4).

In hard rocky terrain, lineament density and weathered zone thickness were shown to be one of the most important factors while water resource development in good groundwater potential zones (Ibrahim-Bathis and Ahmed, 2016). The poorest areas are still distributed, especially in areas with dense drainage networks where the groundwater would not be extracted owing to higher hydraulic conductivity due to high terrain relief and slope and therefore wouldn't be adequate for irrigation and other domestic uses. This entails groundwater development prioritization at higher groundwater level zones.

TABLE 12.2
Normalized Weight Theme Table

	Geomorphology	Slope	Lineament	Drainage	Lithology	Soil	LULC	Rainfall	Weight
Geomorphology	8.00	7.00	6.00	5.00	4.00	3.00	2.00	1.00	0.37
Slope	4.00	3.50	3.00	2.50	2.00	1.50	1.00	0.50	0.18
Lineament	2.67	2.33	2.00	1.67	1.33	1.00	0.67	0.33	0.12
Drainage	2.00	1.75	1.50	1.25	1.00	0.75	0.50	0.25	0.09
Lithology	1.60	1.40	1.20	1.00	0.80	0.60	0.40	0.20	0.07
Soil	1.33	1.17	1.00	0.83	0.67	0.50	0.33	0.17	0.06
LULC	1.14	1.00	0.86	0.71	0.57	0.43	0.29	0.14	0.05
Rainfall	1.00	0.88	0.75	0.63	0.50	0.38	0.25	0.13	0.05
Total	21.74	19.03	16.31	13.59	10.87	8.15	5.44	2.72	1.00

TABLE 12.3

Assigned and Normalized Weight of the Different Classes of Each Theme

Factors	Weight	Rank	Overall
Geomorphology			
Water Body (River)	36.79	5	183.97
Water Body (Other)		5	183.97
Valley		4	147.17
Moderate Dissected Plateau		3	110.38
Gentle Plateau Top		2	73.59
High-Relief Plateau Top		1	36.79
Scarp		1	36.79
Slope (degree)			
<14	18.40	5	91.98
14–28		4	73.59
28–42		3	55.19
42–56		2	36.79
>56		1	18.40
Lineament			
High	12.26	5	61.32
Moderate		3	36.79
Low		1	12.26
Drainage			
High	9.20	1	9.20
Moderate		3	27.60
Low		5	45.99
Lithology			
Conglomerate	7.36	1	7.36
Epidiorite		3	22.08
Quartzite with Phyllite Interbands		5	36.79
Soil			
Orthic Acrisols	6.13	3	18.40
LULC			
Water Bodies	5.26	5	26.28
Agriculture		4	21.02
Forest		3	15.77
Degraded Forest		2	10.51
Built Up		1	5.26
Others		1	5.26
Rainfall (mm)			
901–963	4.60	1	4.60
963–1018		2	9.20
1018–1067		3	13.80
1067–1115		4	18.40
1115–1178		5	23.00

GROUNDWATER POTENTIAL ZONE MAP

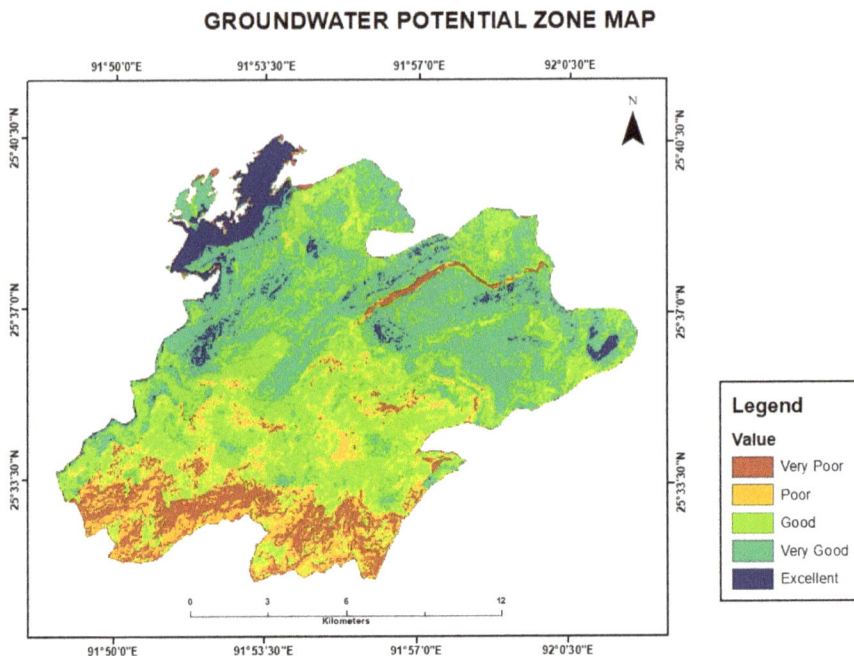

FIGURE 12.12   GWP zone map of Shillong

TABLE 12.4
GWP Zone Distribution

GWP Zones	Area in Sq. Km	Area in Percentage
Very Poor	18.26	9.37
Poor	26.94	13.82
Good	73.13	37.53
Very Good	66.37	34.06
Excellent	10.14	5.22
Total	194.84	100

12.4.8   URBAN SPRAWL VERSUS GROUNDWATER POTENTIAL

In Shillong city and its environs, the excellent groundwater potential zone can be observed in the northwestern parts in contrast to very poor GWP occupying the southernmost parts (Figure 12.13). The built-up area mapped for 2000 covers 22.6 km^2 of good and 10.46 km^2 of very good GWP zone, although some part of the built-up area is also lying over 7.39 km^2 of poor and 1.81 km^2 of very poor GWP zones (Table 12.5, Figure 12.13a). In contrast, the built-up area mapped for 2020

TABLE 12.5

Urban Area and Its Percentage Lying over the Various GWP Zones

GWP Zones	Urban Area in km² (2000)	Urban Area in % (2000)	Urban Area in km² (2020)	Urban Area in % (2020)	Urban Area in km² (2030)	Urban Area in % (2030)
Very Poor	1.81	4.2	4.2	6.99	6.72	9.56
Poor	7.39	17.15	11.4	18.97	13.62	19.4
Good	22.6	52.46	28.28	47.09	30.16	42.91
Very Good	10.46	24.28	15.3	25.46	17.38	24.72
Excellent	0.82	1.91	0.9	1.49	2.4	3.41
Total	43.08	100	60.08	100	70.28	100

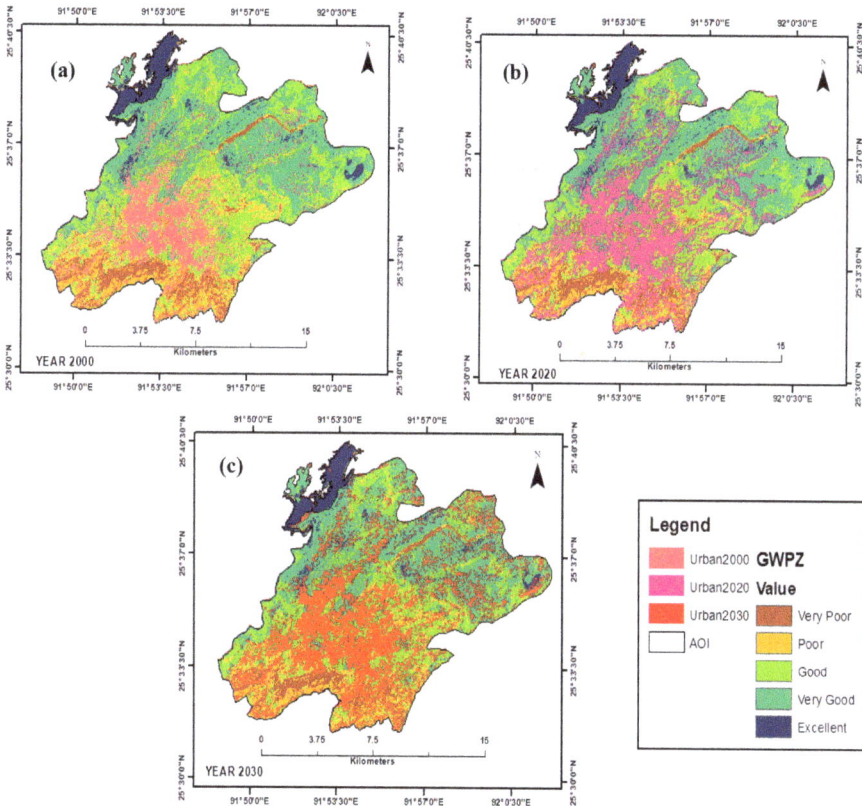

FIGURE 12.13 GWP zone and urban area for (a) 2000, (b) 2020, and (c) 2030 of Shillong.

covers 28.28 km² of good and 15.3 km² of very good GWP zones, while a major part of built-up area also covers 11.4 km² of poor and 4.2 km² of very poor GWP zones (Figure 12.13b). This clearly indicates the possible scarcity of groundwater under the built-up areas developed on poor and very poor GWP zones. By 2030, it was perceived that the projected built-up area might cover up to 6.72 km² of the very poor GWP zone (Figure 12.13c), which indicates that these areas would suffer in the future from low availability of groundwater resources if sustainable water resource planning through the development of surface water harvesting structure is not done.

12.5   CONCLUSION

This comprehensive study has utilized the satellite data between 2000 and 2020 to analyze the land-use pattern of Shillong city, and the following conclusions were drawn:

- Space-borne data employed for visualizing the urban growth of Shillong for the year 2000–2020 revealed that the total urban area increased from 41.07 km² to 56.18 km² during the said period with an urban growth of 39.36%.
- The study based on Shannon's entropy (1.356) shows that the most wide-spread urban growth pattern occurred in 2000 compared to 2020 (1.329) in Shillong. Over the past 20 years, the built-up development, agricultural extension, and deforestation have grown exponentially at the expense of other land-use areas.
- The study predicted urban growth up to 16.97% by 2030 from the existing built-up pattern in 2020. LULC changes need to be considered to ensure environmental sustainability as changes due to construction activities associated with urban sprawl and population growth would increase degraded forest, and reduce groundwater recharge, as well as the extension of agricultural land while water bodies, forest, and other land-use types will decline by 2030.
- The GWP zone map identified presence of excellent, very good, good, poor and very poor zones covering 10.14km² (5.22%), 66.37km² (34.06%), 73.13 km² (37.53%), 26.94 km² (13.82%), and 18.26km² (9.37%) area respectively.
- In the GWP zones for three years, that is, 2000, 2020, and 2030, indicated that in 2000, the urban area covered 17.15% of the poor zone and 4.2% of very poor zone, which increased to 18.97% and 6.99%, respectively, in 2020 and would further increase to 19.4% and 9.76%, respectively, in similar zones, indicating groundwater scarcity in these regions.
- By 2030, many areas may suffer from low availability of groundwater resources as population pressure would increase and groundwater recharge would decrease on account of urban expansion. Hence, it is very necessary to adopt judicious use of water resources, as well surface water conservation planning, along with urban expansion for sustainable water resource management in the groundwater scarce region of Shillong and its environs.

ACKNOWLEDGMENT

Authors thanks to U.S. Geological Survey (USGS) and Center for Hydrometeorology and Remote Sensing (CHRS) for providing Landsat satellite data and rainfall.

CONFLICTS OF INTEREST

The authors declare no conflicts of interest.

REFERENCES

Anand, Vicky, B. Oinam, and B.R. Parida. 2020. "Uncertainty in hydrological analysis using multi-GCM predictions and multi-parameters under RCP 2.6 and 8.5 scenarios in Manipur River Basin, India." *Journal of Earth System Science* 129 (1): 223. https://doi.org/10.1007/s12040-020-01492-z

Arulbalaji, P., D. Padmalal, and K. Sreelash. 2019. "GIS and AHP techniques based delineation of groundwater potential zones: A case study from Southern Western Ghats, India." *Scientific Reports* 9 (1): 2082. https://doi.org/10.1038/s41598-019-38567-x

Chaudhary, B.S., and S. Kumar. 2018. "Identification of groundwater potential zones using remote sensing and GIS of KJ Watershed, India." *Journal of the Geological Society of India* 91: 717–721.

Dhaoui, I. 2014. "Urban sprawl: The GIS and remote sensing data assessments." MPRA Paper No. 87650. International Conference on natural Hzards and Geomatics, Hammamet, May, 17–20th 2013. Available Online: https://mpra.ub.uni-muenchen.de/87650/

Diksha, and A. Kumar. 2017. "Analysing urban sprawl and land consumption patterns in major capital cities in the Himalayan region using geoinformatics." *Applied Geography* 89: 112–123.

Ibrahim-Bathis, K., and S.A. Ahmed. 2016. "Geospatial technology for delineating groundwater potential zones in Doddahalla Watershed of Chitradurga District, India." *The Egyptian Journal of Remote Sensing and Space Science* 19 (2): 223–234. https://doi.org/10.1016/j.ejrs.2016.06.002

Jat, M. K., P. K. Garg, and D. Khare. 2008. "Modelling of urban growth using spatial analysis techniques: A case study of Ajmer City (India)." *International Journal of Remote Sensing* 29 (2): 543–567. https://doi.org/10.1080/01431160701280983

Jenks, G.F. 1967. "The data model concept in statistical mapping." *International Yearbook of Cartography* 7: 186–190.

Kalhor, K., and N. Emaminejad. 2019. "Sustainable development in cities: Studying the relationship between groundwater level and urbanization using remote sensing data." *Groundwater for Sustainable Development* 9 (2019): 100243.

Khan, A., H.H. Khan, and R. Umar. 2017. "Impact of land-use on groundwater quality: GIS-based study from an alluvial aquifer in the western Ganges basin." *Applied Water Science* 7 (8): 4593–4603.

Khan, A., Samiullah Atta-ur-Rahman, and Md. Ali. 2019. "Impact of built environment on groundwater depletion in Peshawar, Pakistan." *Journal of Himalayan Earth Sciences* 52: 86–105.

Kumar, A., A.C. Pandey, N. Hoda, and A.T. Jeyaseelan. 2011. "Evaluating the long-term urban expansion of Ranchi urban agglomeration, India using geospatial technology." *Journal of the Indian Society of Remote Sensing* 39 (2): 213–224.

Kumar, N., B. Tischbein, J. Kusche, M.K. Beg, and J.J. Bogardi. 2017. "Impact of land-use change on the water resources of the Upper Kharun Catchment, Chhattisgarh, India." *Regional Environmental Change* 17: 2373–2385. https://doi.org/10.1007/s10113-017-1165-x

Mahalingam, B., M.D. Bhauso Ramu, and P. Jayashree. 2014. "Assessment of groundwater quality using GIS techniques: A case study of Mysore City." *International Journal of Engineering and Innovative Technology (IJEIT)* 3(8): 117–122.

Maithani, S. 2010. "Cellular automata based model of urban spatial growth." *Journal of the Indian Society of Remote Sensing* 38(4): 604–610.

Naik, P.K., A.J.A. Tambe, B.N. Dehury, and A.N. Tiwari. 2008. "Impact of urbanization on the groundwater regime in a fast growing city in central India." *Environmental monitoring and assessment* 146(1): 339–373.

Nguyen, Phu, Eric J. Shearer, Hoang Tran, Mohammed Ombadi, Negin Hayatbini, Thanh Palacios, Phat Huynh, et al. 2019. "The CHRS data portal, an easily accessible public repository for persiann global satellite precipitation data." *Scientific Data* 6 (1): 180296. https://doi.org/10.1038/sdata.2018.296

Patra, S, S. Sahoo, P. Mishra, and S.C. Mahapatra. 2018. "Impacts of urbanization on land use /cover changes and its probable implications on local climate and groundwater level." *Journal of Urban Management* 7 (2): 70–84. https://doi.org/10.1016/j.jum.2018.04.006

Parida, B.R., S. Behera, O. Bakimchandra, A.C. Pandey, and N. Singh. 2017. "Evaluation of satellite-derived rainfall estimates for an extreme rainfall event over Uttarakhand, Western Himalayas." *Hydrology* 4 (2): 22. https://doi.org/10.3390/hydrology4020022

Parida, B.R., and S.P. Mandal. 2020. "Polarimetric decomposition methods for LULC mapping using ALOS L-band PolSAR data in Western Parts of Mizoram, Northeast India." *SN Applied Sciences* 2 (6): 1049. https://doi.org/10.1007/s42452-020-2866-1

Putranto, T.T., and D.E. Aryanto. 2018. "Spatial analysis to determine groundwater recharge area in purworejo regency, central Java province/Indonesia." *E3S Web of Conferences*, 73: 03025. EDP Sciences.

Ramamoorthy, P., and V. Rammohan. 2015. "Assessment of groundwater potential zone using remote sensing and GIS in Varahanadhi watershed, Tamilnadu, India." *International Journal for Research in Applied Science and Engineering Technology* 3(5): 695–702.

Shekhar, S., and A.C. Pandey. 2015. "Delineation of groundwater potential zone in hard rock terrain of India using remote sensing, geographical information system (GIS) and analytic hierarchy process (AHP) techniques." *Geocarto International* 30(4): 402–421.

Suganthi, S., L. Elango, and S.K. Subramanian. 2013. "Groundwater potential zonation by Remote Sensing and GIS techniques and its relation to the Groundwater level in the Coastal part of the Arani and Koratalai River Basin, Southern India." *Earth Sciences Research Journal* 17(2): 87–95.

Wakode, H.B., K. Baier, R. Jha, S. Ahmed, and R. Azzam. 2014. "Assessment of impact of urbanization on groundwater resources using GIS techniques: Case study of Hyderabad, India." *International Journal of Environmental Research* 8 (4). https://doi.org/10.22059/ijer.2014.808

Wang, S.W, B.M. Gebru, M. Lamchin, R.B. Kayastha, and W.K. Lee. 2020. "Land use and land cover change detection and prediction in the Kathmandu district of Nepal using remote sensing and GIS." *Sustainability* 12 (9): 3925. https://doi.org/10.3390/su12093925

The World Bank. 2012. *India groundwater: A valuable but diminishing resource.* www.worldbank.org/

13 Land Surface Temperature Responses to Urban Landscape Dynamics

Nimish Gupta* and Bharath Haridas Aithal
Ranbir and Chitra Gupta School of Infrastructure
Design and Management, Indian Institute of
Technology Kharagpur, West Bengal – 721302, India

CONTENTS

13.1 Climate—The Ever-Changing Phenomenon 249
13.2 Urbanization—A Boon or a Curse?... 250
13.3 LST 251
13.4 Effect on the Residents 254
13.5 State-of-the-Art Technology to Monitor and Quantify the Changes............255
13.6 Urban LST 257
    13.6.1 Data Collection and Preprocessing..257
    13.6.2 Creation of Temporal Land-Use Maps 257
    13.6.3 Quantification of LST 258
    13.6.4 Relating the Landscape Alteration With LST 261
    13.6.5 Understanding the Changes—Spatial Indices.......................................261
13.7 Applying the Methodological Framework...262
    13.7.1 Analysis of Landscape—Land Use 263
    13.7.2 LST 264
    13.7.3 Temperature Profile Graph 265
    13.7.4 Spatial Indices... 265
13.8 Better Planning and Proactiveness Leads to Comfortable Urban
Environment 267
    13.8.1 Forecasting LST 267
    13.8.2 Mitigation Strategies.. 269
13.9 Conclusion 270
Acknowledgment 270
References.. 270

13.1 CLIMATE—THE EVER-CHANGING PHENOMENON

Climate and weather, like any other physical phenomena, are highly dynamic and have been changing since the dawn of time. Historical records, scientific discoveries,

and paleontological analysis suggest that climate variations can be traced back to pre-Cambrian times, the Paleozoic era, the Mesozoic era, and the Cenozoic era. Planet Earth has experienced significant climatic variations during the Quaternary ice age, pluvial periods and postglacial periods (Lal, 2017). However, with an increase in human interventions, it has become one of the most significant and challenging issues that have grabbed the interest of the scientific community and researchers worldwide. Although humans have brought numerous innovative and positive changes in many sectors, thanks to their wisdom and knowledge, they have also created a mindset of considering themselves superior to all the other living beings and have continued to exploit nature for obtaining immediate short-term benefits. This mindset and actions are wreaking havoc on the planet in a variety of ways, and it is no longer simply a minor environmental concern at the neighborhood level but has eventually evolved into one of humanity's greatest developmental challenges at micro-, meso- and macro-levels (Karoly et al., 2003; Stott et al., 2006). The fourth, fifth and sixth assessment reports of the Intergovernmental Panel for Climate Change (IPCC) acknowledge that anthropogenic sources have resulted in a severe and nonnatural increase in atmospheric concentrations of greenhouse gases with high global warming potentials (CO_2, CH_4, SF6, and N_2O). Moreover, the tropospheric ozone concentration has increased, while the stratospheric ozone concentration has declined. All this has triggered a rise in global mean temperatures, initiated an increase in the number of hot days and nights, and altered the patterns of global continental precipitation (Pachauri, 2008).

India is one of the countries that are highly susceptible to the harsh effects of climate alterations owing to its vast population depending on the agrarian economy, coastal areas, the Himalayan regions, and islands. Every year, alterations in climate and increased extreme events, including heat strokes, drowning, water scarcity, submerging under landslides owing to rising temperature, change in rainfall patterns, landslides, floods, droughts, and crop failures, lead to the loss of lives (victims and farmer suicides). In the past few decades, India had faced enormous environmental challenges due to climatic alterations that have severely affected the natural ecosystem and resulted in food and water security issues. These indicators of climate change and related concerns are consequences of augmented demographic pressure, causing urban areas to expand and outgrow beyond their administrative boundaries. Inadvertent urbanization is affecting the natural climate at local, regional, and global levels, and leading to a faster rate of natural resource depletion.

13.2   URBANIZATION—A BOON OR A CURSE?

Urbanization is defined as any physical growth in the impervious surface induced by an increase in the population of an urban region due to increased migration or merging of sub-urban areas into the main city (Ramachandra et al., 2012a). It can be described as a highly dynamic and irreparable process that is defined by an urban area's lateral and vertical expansion. The process of urbanization is crucial since it has numerous beneficial characteristics associated with it, including better socioeconomic status, higher gross domestic product per capita, and economic stability (Henderson, 2003; Chen et al., 2014). However, if the population exceeds the region's carrying capacity

due to a lack of adequate planning (available space and natural resource distribution), it can have several undesirable and harmful effects on the residents and the environment (Weerakoon, 2017; Chandan et al., 2017). Currently, 55% of the world's population lives in cities, and this ratio is predicted to rise to 68% by 2050, with Asia and Africa accounting for about 90% of the increase (UNDESA, 2018). Higher living standards, better educational facilities, enhanced occupational possibilities, ease of traveling, and better health care facilities are some of the major causative factors of increasing urbanization. All these factors contribute to an increase in the rate of migration/movement of people from rural areas to the city's core and periphery, causing issues such as basic housing, a lack of fresh and drinkable water, inadequate infrastructure, and higher pollution levels, resulting in a deteriorated quality of life (Ramachandra et al., 2012b).

The prehistoric towns of Mohenjo-Daro and Harappa marked the beginning of Indian urbanization. Cities like Nasik, Meerut, Ujjain, and Delhi grew up to become business centers. Kolkata grew to prominence as India's flagship city before independence, with Mumbai, Kanpur, Delhi, and Chennai facing rapid urbanization. The onset of liberalization and globalization in the 1990s, resulted in the rise of the private sector, and a gigantic surge in urban areas in Indian cities was noticed. Since then, infrastructure development in cities and towns across India has increased substantially to bridge the gap between demand and supply of essentials. Socioeconomic policies, political influence, cultural factors, and inhabitants' basic mindsets all significantly impact urbanization. A considerable increase in the intensity and size of urban areas leads to fragmented and isolated settlements, contributing to unplanned and uncontrolled urbanization, also known as urban sprawl (Bharath et al., 2017). Although urban sprawl can relieve pressure on the city's core, it also has several negative consequences, including increased resource requirements, improper waste management, increased energy demands, higher pollutant discharge, and an increase in GHGs (greenhouse gases) and PMs (particulate matter—suspended, PM10, PM2.5, and PM1), all of which contribute to global warming. Landcover and land use are two quantitative indicators that may be used to characterize urbanization in terms of landscape changes. The conversion of vegetated spaces, agricultural fields, and open spaces into urban pockets are examples of landscape changes (Kandlikar and Sagar, 1999). Evaporative rates, surface albedo, heat storage capacity, moisture content, wind turbulence, and solar radiation are all affected by these variations (Pal and Ziaul, 2017). In terms of thermal comfort, air temperatures, and land surface temperatures, all these factors are closely related to the city's living conditions. One of the most critical parameters to understand urban climate change is land surface temperature (LST).

13.3   LST

The radiative skin temperature of the earth's surface as measured by a remote sensor is known as LST (Copernicus, 2018). In simpler terms, the warmth felt on touching a feature is the surface temperature of that particular feature. LST is the primary determinant of thermal behavior, as it drives the effective scorching temperature of

the Earth's surface (AATSR and SLSTR, 2018). It is an imperative index that depicts energy balance and climatic change and acts as a critical parameter for land surface processes and atmospheric cycles at local, regional, and global scales while playing a key role in quantitatively indicating global warming (Li et al., 2013). LST estimation largely depends upon albedo, surface characteristics, moisture content at the ground level, health, and density of vegetation season of year, and the time of day (Copernicus, 2018).

The evapotranspiration rates, as well as the latent and sensible heat patterns, are affected by changes in the landscape (Mojolaoluwa et al., 2018). If the same developmental trend in terms of urban population, construction activity, automobile counts, and electricity consumption persists, more significant climate fluctuations are almost likely, with a vast population experiencing thermal discomfort. This growing trend will escalate the concentration of pollutants at ground level and atmosphere in the form of PMs and GHGs, leading to an increase in the global and regional surface temperatures (Ministry of Statistic and Programme Implementation, 2015). According to scientific understanding, the water cycle, energy exchanges, biogeochemical cycle, crop patterns, wind turbulence and intensity, ecology, and rainfall pattern have a direct correlation with the LST of a given location. Climatology, urban climate change, hydrological analysis and modeling, agrometeorology, disaster and risk management, and vegetation health monitoring all use LST as a significant input or are closely connected to it. Some of the vital factors that affect the retrieval of LST include atmospheric composition and parameters, sensor characteristics, and surface conditions.

A rise in LST triggers heat waves, which are extremely harmful to all lifeforms. Due to a lack of open and vegetated places to mitigate its severity, urban regions are more sensitive to this, as it affects a large group of inhabitants. With the increased density of concrete structures and a reduction in the pervious surfaces in a large urban area, the surface temperature of the core city becomes higher when compared to the nearby rural areas, and this entire phenomenon is termed as the urban heat island (UHI) effect (Figure 13.1; Liang and Shi, 2009). This happens because synthetic materials utilized in urban buildings have a significantly higher heat retention capacity than natural ones. Furthermore, urban regions are densely packed with structures with high surface temperatures, whereas rural or sub-urban regions have fragmented developments with abundant vegetated or open spaces that allow heat to be dissipated, thus lowering surface temperatures. The study of UHI has become a subject of much interest in the scientific community, and it must be addressed with methods and legislative interventions that can assist inhabitants to achieve maximum thermal comfort. The upsurge in surface temperature driven by the UHI effect has a significant impact on urban ecology and urban climate and puts pressure on flora and fauna, as not all organisms can acclimatize to the fluctuating temperatures. It also influences the soil moisture and characteristics and the sensible and latent heat fluxes (Yang et al., 2016). The vicious cycle of temperature rise and UHI impact is depicted in Figure 13.2. The effects of UHI and rising LST are not limited to the atmosphere; they also have severe implications for inhabitants' health and well-being.

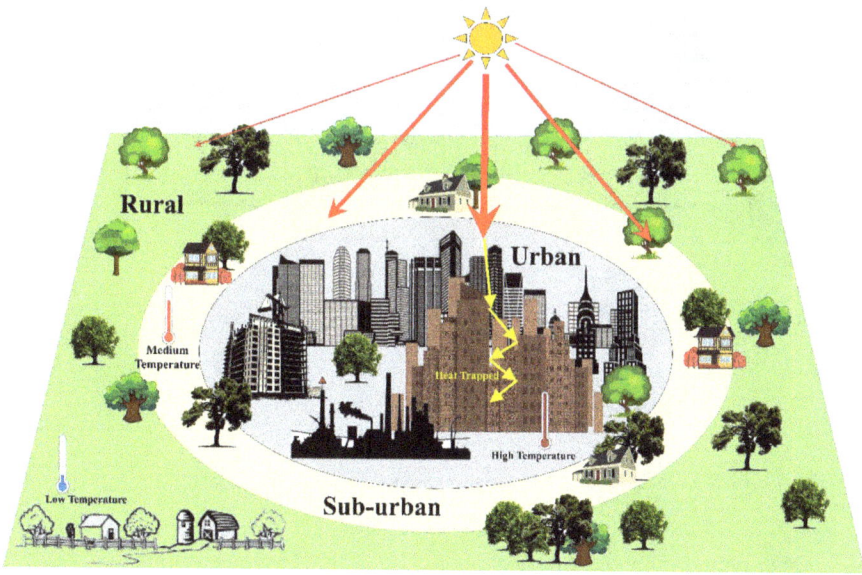

FIGURE 13.1   Temperature distribution among urban, suburban, and rural areas signifying UHI.

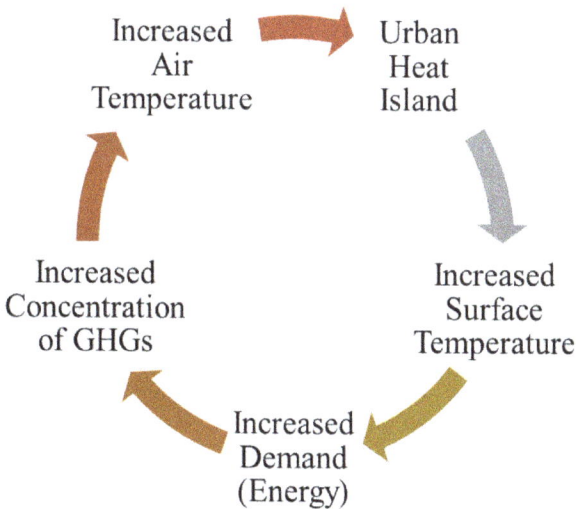

FIGURE 13.2   Temperature and UHI cycle.

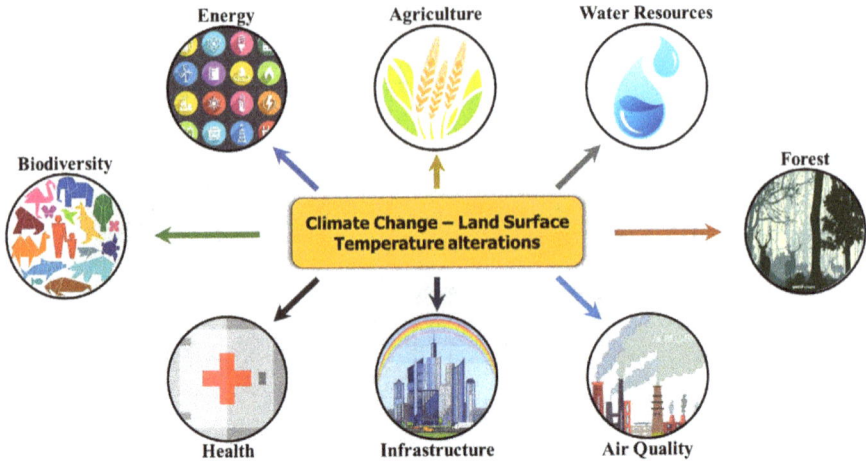

FIGURE 13.3   Sectors affected because of a rise in surface temperatures

13.4   EFFECT ON THE RESIDENTS

Climate change in the longer term can be potentially troublesome when looked upon from the societal aspect. An increase in daily temperatures, for example, might result in greater costs associated with the cooling devices for thermal comfort and even higher costs linked with an increase in the number of heat-related ailments. The increase in LST, as well as growing concerns of heatwaves and UHI, has an adverse influence on the planet's health, as well as human health. Heat waves affect city dwellers differently depending on their age, gender, location, social, and economic position (National Geographic, 2020). According to a wide range of scientific studies, the number of individuals who will be exposed to extreme weather disasters due to changing surface temperatures would increase dramatically in the future. As illustrated in Figure 13.3, elevated LST can impair a resident's quality of life in a variety of ways (Nuruzzaman, 2015; Lal, 2017; EPA, 2019) and affects virtually all the sectors on which our daily lives depend.

Some of the direct and indirect impacts associated with rising land surface temperature include the following:

- *Power Outage:* Increased energy demand often results in load shading or power cuts, causing supply disruptions even in essential sectors (hospitals). In extreme heat events, the elevated demands can lead to overloading, resulting in brownouts or blackouts.
- *Deterioration of air quality:* As the temperature rises, so does the need for energy for cooling, resulting in more GHG emissions from cooling equipment and more fuel burned, resulting in increased air pollution and worsening air quality.
- *Poor health:* The surge in surface temperatures have some serious insinuations on the health of the dwellers. It can result in an increased number

of hotter days with high humidity, resulting in an incremental spread of infectious diseases; an extended season with allergens that can infect many residents with pulmonary disease and respiratory problems; an increase in the number of people suffering a lethal/nonlethal stroke.

- *Thermal discomfort:* Increased surface temperatures combined with higher humidity levels lead to sultry weather, which is not suitable for any mental or physical work and affects the physiological health of the residents.
- *Food and water crisis:* Higher surface temperatures are primarily associated with crop failures that can lead to food shortages, resulting in under-nutrition, hunger, and an upsurge in the price of food and products. The growth in LST is also linked to sea-level rise, which poses a danger to people of low-lying regions' access to freshwater.
- *Rise in extreme events and disasters:* A rise in the surface temperatures of a region is closely related to an increment in the number of extreme events and disasters such as droughts, floods, heat waves, cyclones, and many more.
- *Increased mortality:* The number of people who have lost lives concerning heat-related ailments has increased dramatically. According to the World Health Organization, the number of individuals who have been exposed to heatwaves has grown by 0.13 billion in just 16 years (2000–2016). Over the last two decades, high temperatures in the form of heat waves have claimed more than 12% of disaster deaths globally. The 2003 heat wave in Europe killed more than 72,000 people, the 2010 summer heat wave in Russia killed more than 55,000 people, and the recent incident in Canada, where a heat dome formed, which is a cap-like structure that restricts the movement of hot air due to the development of high pressure, conceived as many as 500 lives (BBC News, 2021). Every year, more than 83,000 individuals in India die because of abnormally high surface temperatures and heat-related problems. According to a report published in Reuters (2018), if no measures are taken to address the present trend of rising temperatures, the death toll from heat strokes is predicted to soar by more than 2000% for a few locations throughout the world by 2080.

13.5 STATE-OF-THE-ART TECHNOLOGY TO MONITOR AND QUANTIFY THE CHANGES

Traditionally, temperature-sensing equipment (thermometers) at meteorological stations were used to monitor changes in a region's warmth. However, these measurements are unable to distinguish between temperature variations in different parts of a city. The point data collected must be extrapolated over the whole region, resulting in temperature errors and the inability to record any sudden changes. With the development and advancement of passive thermal remote sensing (RS) and the Geographic Information System (GIS) in the recent past, it is now possible to resolve this anomaly, as RS devices are capable of collecting the surface temperature spatially and

temporally across a more extensive area. Because of its simplicity and technology, the RS GIS has found its way into almost all disciplines and sectors and serves as a key player in urban climate research. With the relevant remotely sensed or ground-observed data, the GIS may be an excellent platform for monitoring, visualizing, analyzing, and predicting urban climate changes with altering landscape. It is one of the most efficient, quick, and cost-effective methods to monitor large areas. Earth observation-based monitoring has become easier for urban planners and government agencies as RS technology has advanced and high-resolution data has been made available (Chen et al., 2000; Ji et al., 2001). Apart from urban climate, different approaches and algorithms on the RS GIS platform may be used to produce and evaluate landscape modifications and characteristics like land use, height, topography, and many more.

The first meteorological satellite was launched in 1960, and since then, the discipline of meteorology and climatology using RS GIS has grown in popularity (Tomlinson et al., 2011). Compared to meteorological stations on a worldwide scale, the methods developed by researchers provide better and more precise information, thereby enhancing the quality of information (Mendelsohn et al., 2007). Remote sensing enabled LST measurement across the earth spatially rather than using ground-based point measurements, thanks to the development of sensors in the thermal infrared region. Efforts have been carried out in this field and several algorithms have been developed for estimating LST using polar and geostationary satellites. Sensors capable of acquiring data in the electromagnetic radiation (EMR) range from 8–15 μm are required for LST estimation. Table 13.1 provides satellites/sensors that incorporate this spectrum of RS, data availability timelines, and resolution details.

TABLE 13.1

Resolution and Data Availability of Satellites Incorporating Thermal Sensors

S No.	Satellite	Sensor	Data Availability	Spatial Resolution	Temporal Resolution	Spectral Resolution (thermal)
1.	GOES	GOES	1974–2018	4 km	Geostationary	2
2.	NOAA	AVHRR	1979–2018	1.1 km	Twice daily	2
3.	Landsat 4, 5	TM	1983–2013	120 m*	16 days	1
4.	Landsat 7	ETM+	1999–2013	60 m*	16 days	1
5.	ASTER	Terra	1999–present	90 m	Twice daily	5
6.	MODIS	Terra	2000–present	1 km	Twice daily	2
7.	MODIS	Aqua	2002–present	1 km	Twice daily	2
8.	AATSR	Envisat	2004–present	1 km	35 days	2
9.	SEVIRI	Meteosat – 8	2005–present	3 km	Geostationary	2
10.	MetOP	AVHRR	2007–present	1.1 km	29 days	2
11.	Landsat 8	OLI/TIRS	2013–present	100 m*	16 days	2

13.6   URBAN LST

Detailed LSTs of a region can be understood by overlaying it on the land-use map over various periods and correlating their impacts. These spatial maps can help us understand how landscape modification from one feature to another can affect the surface temperature.

13.6.1   DATA COLLECTION AND PREPROCESSING

Remotely sensed images captured by Landsat series (5—Thematic Mapper; 7— Enhanced Thematic Mapper+; 8—Operational Land Imager/Thermal Infrared Sensor) serve as the primary or major data source. The Landsat series offers data with a spatial resolution of 30 m, currently available for thermal RS. The data can be downloaded from various public domains such as U.S. Geological Survey (USGS; Earth Explorer and GLOVIS) at no cost. Secondary data sources include Bhuvan, Google Earth, Indian Meteorological Department (IMD), Survey of India maps, and field data collection. The collection involves procurement of ground control points (GCPs) using the Global Positioning System (GPS).

Raw image acquired by the satellite has certain inaccuracies induced due to the motion of satellite and atmospheric components such as trace gases, particles, and water vapor. Before using the data for analysis, it is first processed. The satellite data is also geo-corrected using GCPs collected via field visits and Google Earth. Furthermore, data related to the region of interest are clipped. A buffer region around the area of interest should be examined if the study entails comprehending sprawl and UHI.

13.6.2   CREATION OF TEMPORAL LAND-USE MAPS

Unplanned urbanization, the unrestrained movement of people from rural to urban areas, and the development of slums are some of the major problems. Comprehensive mapping and monitoring of spatial and temporal changes in urban areas, as well as reductions in green spaces and water bodies, can help design sustainable solutions. Land-use maps may be used to monitor, quantify, illustrate, and interpret these changes. Land-use classification is performed by integrating a set of spectral signatures to assign classes to each pixel in a stack of multiband data. The analysis offers crucial information and a greater understanding of land use. It plays an important role in the creation and adjustment of policies and programs that are beneficial for developmental planning and serve as a vital input for change detection investigations (Sanborn, 2021). Land-use analysis is a method of creating thematic maps in which each pixel is allocated a distinct category based on its spectral response derived from a series of bands collecting data in the electromagnetic (EM) spectrum's optical range (Gonzalez and Woods, 2007). The process of classification can be performed by either supervised classification algorithms or unsupervised classification techniques. This analysis focuses on the Gaussian Maximum Likelihood Classifier (GMLC), which is a widely used and proven supervised classification technique that evaluates the probability density function of each pixel under consideration for

belonging to a specific land-use class (Ramachandra et al., 2014) and involves the following steps:

a. *Creation of a false color composite (FCC):* When FCC is considered urban, water bodies and vegetation may be effectively distinguished based on visualization, assisting in the identification of heterogeneous patches. Standard FCC was considered for the analysis.

b. *Developing the training polygons:* The heterogeneous and representative training polygons for each land-use class from FCC should be digitized. The number and types of land-use classifications are completely determined by the needs of the users and the application. The polygons are digitized by overlaying an FCC on Google Earth for cross-validation and selecting the appropriate land-use class to be represented.

c. *Training and classification:* The polygons should next be converted into signatures by converting them to raster and assigning statistical values (mean and covariance) for each class generated by stacking multispectral bands. The classifier should be fed these signatures for training purposes. Following training, the GMLC algorithm is applied to the entire region of interest, yielding a four-class thematic map. A mathematical representation of GMLC is explained in Equation 13.1:

$$X \in C_j \text{ if } p\left(C_j / X\right) = \max[\, p\left(C_1 / X\right), p\left(C_2 / X\right), \ldots, p\left(C_m / X\right)] \qquad (13.1)$$

Here,

$p(C_j/X)$ *represents the conditional probability of pixel* X *being a member of class* C_j.

d. *Accuracy assessment:* The standard procedure includes the construction of an error/confusion matrix that includes all the classes. Using this matrix descriptive and quantitative measurement, such as overall accuracy and kappa coefficient, is estimated. On the other hand, the kappa coefficient takes into account the error, that is, the pixels that were omitted and committed, as well as the correctly categorized pixel, to give a normalized number ranging from 0 to 1. Kappa can be defined as a statistical measure of the difference between actual (reference data and automated classifier) and chance agreement (reference data and random classifier; Lillesand et al., 2003). The classification map uses training polygons obtained from an FCC, and the reference map is generated by using Google Earth merged with ground truth points collected from field visits.

13.6.3 QUANTIFICATION OF LST

Persistent changes in land use are disrupting the natural flow of energy in the environment, and an imbalance in diverse biogeochemical processes in the atmosphere has resulted in severe environmental changes. The LST is an important factor in determining a city's climate and health. Calculating the LST aids in detecting

issues related to changing land use as a result of anthropogenic activities. Some of the most common, efficient, and extensively used algorithms to quantify LST are the single-channel algorithm (SCA), the split-window algorithm (SWA), the radiative transfer equation (RTE), and temperature emissivity separation (TES). Each algorithm has its own set of advantages and disadvantages. We focus on LST estimation using the RTE. This algorithm requires only one thermal band data with emissivity values and considers the influence of atmospheric parameters such as transmittance, upwelling, and downwelling radiances. A simplified RTE is expressed in Equation 13.2.

$$B_i\left(T_i\right) = \tau_i\left[\varepsilon_i B_i\left(T_s\right) + \left(1 - \varepsilon_i\right)L_i^{\downarrow}\right] + L_i^{\uparrow} \tag{13.2}$$

Here,

$B_i(T_i) \rightarrow$ *top-of-atmosphere radiance received at the sensor for channel i having the at-satellite brightness temperature of* T_i,

$\tau_i \rightarrow$ *atmospheric transmittance measured at channel* i,

$\varepsilon_i \rightarrow$ *emissivity for channel* i,

$L_i^{\downarrow} \rightarrow$ *downwelling radiance measured at channel* i, *and*

$L_i^{\uparrow} \rightarrow$ *upwelling radiance measured at channel* i.

The top of atmospheric radiance may be calculated using the precalibrated parameters and the digital numbers (DNs) measured at the sensor is per Equation 13.3.

$$B_i\left(T_i\right) = \left(Gain * DN\right) + Offset \tag{13.3}$$

Here,

Gain $\rightarrow$ *band-specific multiplicative factor,*

Offset $\rightarrow$ *band-specific additive factor, and*

DN $\rightarrow$ *digital number.*

Planck's law defines ground radiance as a function of wavelength and surface temperature as illustrated in Equation 13.4:

$$B_i\left(T_s\right) = \frac{2hc^2}{\lambda_i^5 \times \left(e^{\frac{hc}{\lambda_i k T_s}} - 1\right)} \tag{13.4}$$

Here,

h $\rightarrow$ *Planck's constant = 6.626 × 10⁻³⁴ J-s,*

c $\rightarrow$ *speed of light = 2.98 × 10⁸ m/s,*

$\lambda_i \rightarrow$ *effective wavelength for channel* i,

k $\rightarrow$ *Boltzmann constant = 1.3806 × 10⁻²³ J/K, and*

$T_s \rightarrow$ *LST.*

$B_i(T_i)$ was estimated using Equation 13.3, and T_s was derived by Equation 13.5 that is formulated by rearranging Equations 13.2 and 13.4:

$$T_s = \frac{C_2}{\lambda_i \times \ln\left(\dfrac{C_1}{\lambda_i^5 \left(\dfrac{B_i(T_i) - L_i^{\uparrow} - \tau_i(1 - \varepsilon_i)L_i^{\downarrow}}{\tau_i \varepsilon_i}\right)} \right)} \tag{13.5}$$

Here,
$C_1 \rightarrow 1.19104 \times 10^8$ *W-μm⁴-m⁻²-s⁻¹ and*
$C_2 \rightarrow 14,387.7$ *μm-K.*

- *Land surface emissivity:* Land surface emissivity (LSE) is one of the critical parameters required to quantify LST. LSE is the basic character-istic of the material, which is equivalent to the amount of radiant energy emitted by a surface/feature when compared to the amount of energy radiated by the blackbody at a similar temperature and wavelength. It is greatly reliant on the material's composition, roughness, and view-ing angle (Li et al., 2013). Any small uncertainty in LSE can result in large-term inaccuracies in LST estimation. As per the Dash et al. (2010), 1% inaccuracy in LSE can induce as much as 0.3–0.7 K error in the LST depending on the heterogeneity in the LST of the region. Several techniques for estimating the LSE have been developed over the years: classification-based emissivity method, normalized emissivity method, Normalized Difference Vegetation Index (NDVI)-based emissivity method, NDVI threshold approach, and the TES method. This study focuses on estimation based on the NDVI threshold method because of its simplicity, reliability, and accuracy. To begin, the NDVI should be calculated using the red and near-infrared bands, and then pure classes of water, soil, and vegetation, as well as a mixed class of soil and veg-etation, should be segregated using thresholds. Emissivity is computed using Equation 13.6:

$$\varepsilon = \begin{cases} \varepsilon_w = \varepsilon_{w6/10/11}, \text{NDVI}_i \leq \text{NDVI}_{\text{water}} \\[2mm] \varepsilon_s = \begin{cases} \varepsilon_{s6/10} \\ \varepsilon_{s11} \end{cases}, \text{NDVI}_{\text{water}} \leq \text{NDVI}_i \leq \text{NDVI}_{\text{soil}} \\[3mm] \varepsilon_{SV} = \varepsilon_V P_V + \varepsilon_S(1 - P_V) + C, \text{NDVI}_{\text{soil}} \leq \text{NDVI}_i \leq \text{NDVI}_{\text{vegetation}} \\[3mm] \varepsilon_v = \begin{cases} \varepsilon_{v6/10} \\ \varepsilon_{v11} \end{cases}, \text{NDVI}_{\text{vegetation}} \leq \text{NDVI}_i \end{cases} \tag{13.6}$$

Here,

NDVI$_i$ is the NDVI of the pixel under consideration,

$\varepsilon_{j6/10}$ *is the emissivity for jth class associated with band 6 for Landsat 5/7 and band 10 for Landsat 8,*

ε_{j11} *is the emissivity for jth class associated with band 11 for Landsat 8, and j represents water, pure soil, and vegetation class.*

NOTE: *The emissivity of pure water, soil, and vegetation can be obtained via literature review. Usually, it is considered 0.991 for water, 0.9668 (band 6/10) and 0.9747 (band 11) for soil, and 0.9863 (band 6/10) and 0.9896 (band 11).*

- *Atmospheric parameters:* Atmospheric parameters including transmittance, upwelling radiance, and downwelling radiance can be estimated using the Atmospheric Correction Parameter Calculator provided by NASA (https://atmcorr.gsfc.nasa.gov/). The calculator uses MODTRAN codes to extract the required parameters from National Centers for Environmental Prediction (NCEP) database based on spectral response curves from the Landsat series and atmospheric profile. It uses input parameters such as year, month, day, image capture time, and geographic coordinates.

13.6.4 RELATING THE LANDSCAPE ALTERATION WITH LST

Class-wise analysis may be conducted after developing land-use and LST maps by overlaying them over each other, yielding comprehensive information about the influence of landscape modification on the region's LST. Creating a temperature profile graph across the area of interest is one of the finest ways to comprehend this.

13.6.5 UNDERSTANDING THE CHANGES—SPATIAL INDICES

Indices serve as the easiest mathematical tool to reduce large figures into smaller ones keeping the information intact. It can be considered the best way to analyze the changes. This communication discusses two spatial indices to comprehend the changes in urban LST.

- *Annual Rate of Change in LST (ARCLST):* This index provided information regarding the annual rise in surface temperature across the study area. Various scenarios can be considered wherein the growth of LST is presumed to be having linear, logistic, exponential, polynomial with varying orders/degrees, or any other form depending upon the urban growth and LST of the region of interest. A linear LST growth scenario is considered and is illustrated in Equation 13.7.

$$ARCLST_{Linear} = \frac{1}{time} \times \left(LST_{final\ year} - LST_{initial\ year} \right) \qquad (13.7)$$

- *Normalized Urban Heat Island Index (NUHII):* Cities around the world are now witnessing a rise in surface temperatures, causing increased phenomena of UHI. The NUHII is an index that can define the intensity of heat islands across the area of interest and is capable of determining the most affected regions in the study area. LSTs for core urban areas and dispersed or small urban pockets with low-density buildup were considered, and the NUHII was estimated using Equation 13.8.

$$NUHII = \frac{\left(LST_i - LST_{Rural}\right)}{\left(LST_{max} - LST_{min}\right)} \quad (13.8)$$

Here,
    $LST_i \rightarrow LST$ *of pixel under consideration.*

13.7　APPLYING THE METHODOLOGICAL FRAMEWORK

Kolkata Metropolitan Area (KMA) was chosen as the region of interest (Figure 13.4) to perform the entire analysis as shown in Figure 13.4. Table 13.2 elaborates information about the data used for analysis.

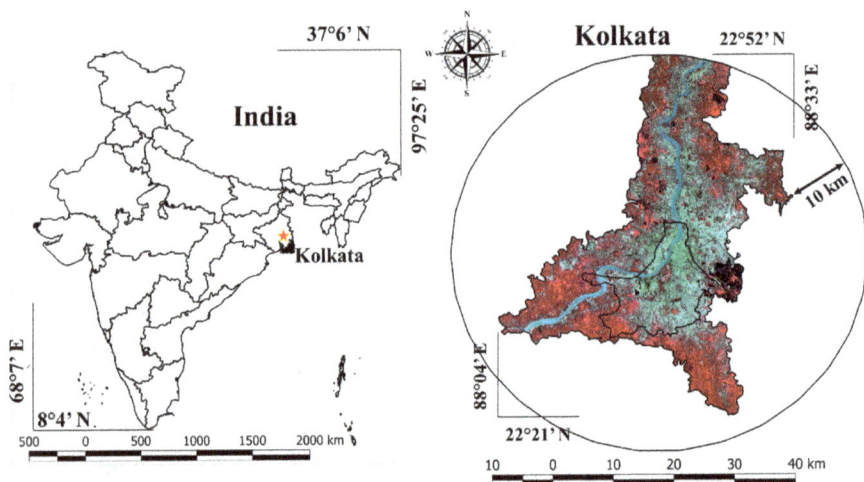

FIGURE 13.4　Region of interest: The KMA with 10-km buffer.

TABLE 13.2
Details of Satellite Data Used

Satellite	Sensor	Date
Landsat-5	Thematic Mapper (TM)	February 11, 2000
		March 25, 2004
		March 07, 2009
Landsat-8	Operational Land Imager/Thermal Infrared Sensors (OLI/TIRS)	March 21, 2014
		May 06, 2019

13.7.1 ANALYSIS OF LANDSCAPE—LAND USE

A comprehensive landscape analysis was carried out in the form of land use, which collectively reflects the study area with several categories (four for this analysis). The population increase and influx of migrants are often used to characterize the region's urban expansion (Subasinghe et al., 2016). The KMA region has seen both infill (core) and distributed (buffer) urban expansion. The metropolitan area has grown in all directions, particularly to the north. Impervious surface in the region has increased approximately threefold (from 235.11 km² to 704.03 km²) during the study period, and undulation has been observed in percentage of vegetation and others categories due to the presence of agricultural fields in the buffer region. Due to an increase in wetlands and aquaculture in the eastern part of the area of interest, water bodies in the region have increased from 3.09–4.37% of the entire study area. Figure 13.5 illustrates the land use of the study area during a span of 19 years, and Figure 13.6 represents the percentage contribution of each category to formulate the land use of the study area.

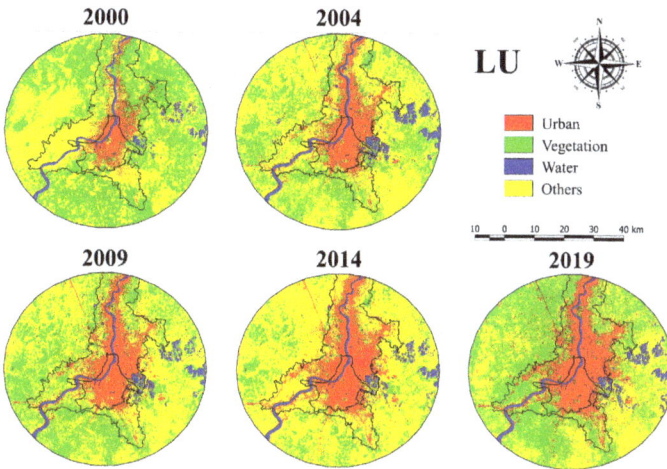

FIGURE 13.5 Land-use maps.

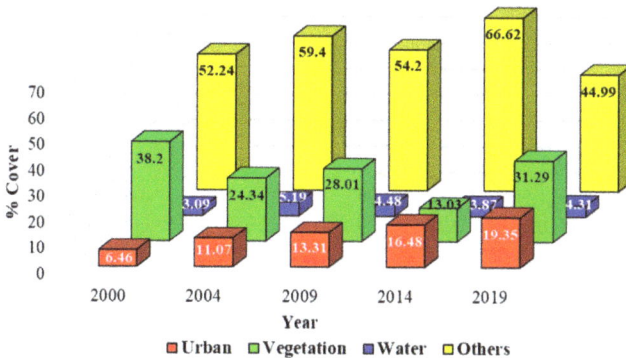

FIGURE 13.6 Percentage cover by each land-use category.

13.7.2 LST

A temporal LST analysis for 2000, 2004, 2009, 2014, and 2019 over the region of interest was performed, and the results are shown in Figure 13.7 and Table 13.2. It was observed that the mean surface temperature of the region has increased from 22.97°C to 29.74°C. It was observed that the minimum temperature of the region has increased from 12.50°C to 24.35°C, and the maximum temperature has also increased by 7.44°C. Class-wise analysis as illustrated in Table 13.3, showed that there has been a rise in the mean LST of the urban class by 6.37°C. The vegetation and others classes have witnessed a rise from 22.48°C to 29.43°C and 23.16°C to 29.44°C, respectively. The mean LST measured for water class has increased by 6.16°C from 2000 to 2019.

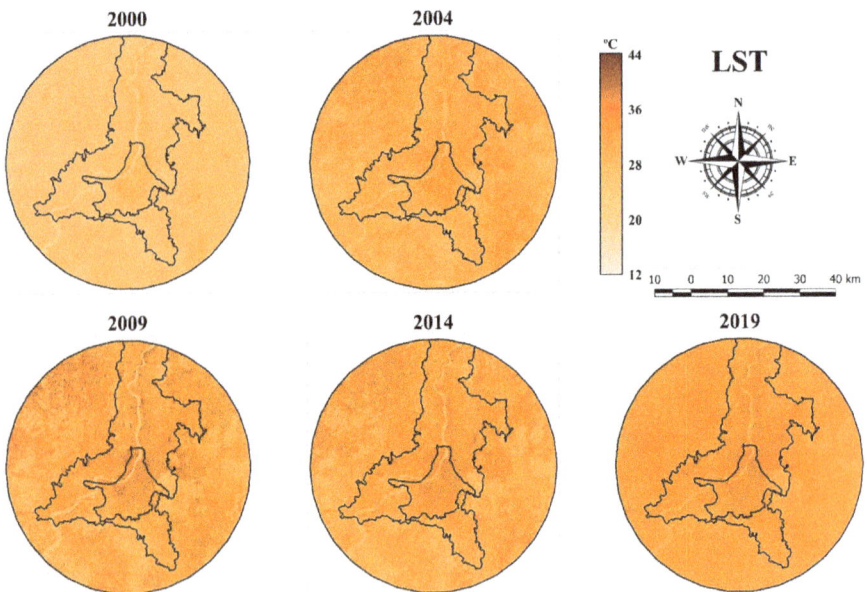

FIGURE 13.7   LST.

TABLE 13.3

Class-Wise LST

Landuse Class	Mean Land Surface Temperature (°C)				
	2000	2004	2009	2014	2019
Urban	25.16 ± 1.56	29.23 ± 1.39	31.89 ± 2.29	31.17 ± 1.93	31.53 ± 1.46
Vegetation	22.48 ± 0.70	26.34 ± 1.06	27.41 ± 1.82	26.18 ± 1.39	29.43 ± 0.73
Water	21.04 ± 1.24	25.49 ± 1.36	25.71 ± 2.29	25.32 ± 1.35	27.20 ± 0.80
Others	23.16 ± 1.04	28.22 ± 1.48	30.54 ± 2.19	29.25 ± 2.02	29.44 ± 0.98
Mean	22.97 ± 1.22	27.74 ± 1.73	29.63 ± 2.77	29.01 ± 2.46	29.74 ± 1.42

13.7.3 TEMPERATURE PROFILE GRAPH

A transact A–B across the study area was considered to illustrate the variation of surface temperature with each class. The LST across this transact was plotted and is shown in Figure 13.8. The relationship between land use and LST reveals high LST values (crest) corresponding to urban structures, and moderate to high surface temperatures for the others category in the temperature profile graph. On the other hand, depressions (troughs) in the graph were noticed for vegetated regions, and a radical concavity was observed across the water bodies, illustrating low and very low temperatures associated with these land-use classes.

13.7.4 SPATIAL INDICES

ARCLST: According to the World Population Review (2019), the population of the KMA has risen thrice during the last five decades. The population of the KMA was estimated to be approximately 13 million in 2000, which increased to 14.8 million in 2011 and is approximated to be around 16 million at present. The ARCLST was calculated to estimate the yearly rate of change in surface temperatures and was then subdivided into eight classes for easier comprehension (Figure 13.9), with the percentage area corresponding to each class listed in Table 13.4.

NUHII: The NUHII for 2019 depicted that most of the buildup in the urban class has exhibited positive heat islands. More than two-thirds of the impervious surface in urban areas have an NUHII equal to 0.1–0.3 normalized unit variation (affects more than 70% of the population; Figure 13.10). Kolkata city has positive heat islands for more than 80% area while the KMA has more than half a fraction of the total area being affected by positive heat islands. The maximum values of the NUHII were observed to be in the core region, a few urban structures in the western part (Jala Dhulagiri and Biparnna Para), and at Netaji Subhas Chandra Bose International Airport. A high value on the NUHII was mostly observed for the industries and factories near Tikiapara, Khidirpur, Ghusuri, and the ones adjacent to river Hooghly. This can be inferred to be the aluminum and tin sheets used as roofing material for most of them.

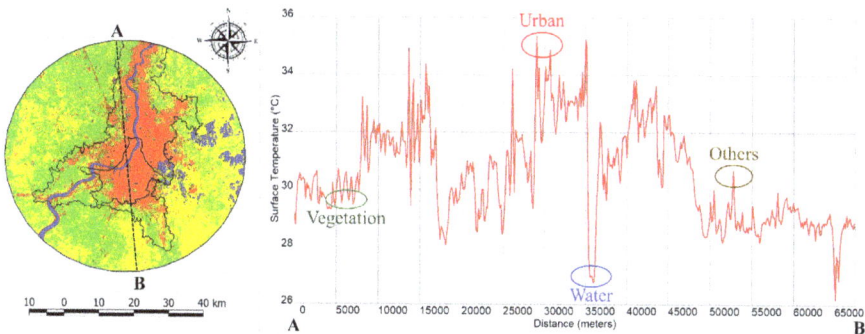

FIGURE 13.8 Temperature profile graph (2019) across the transact A–B.

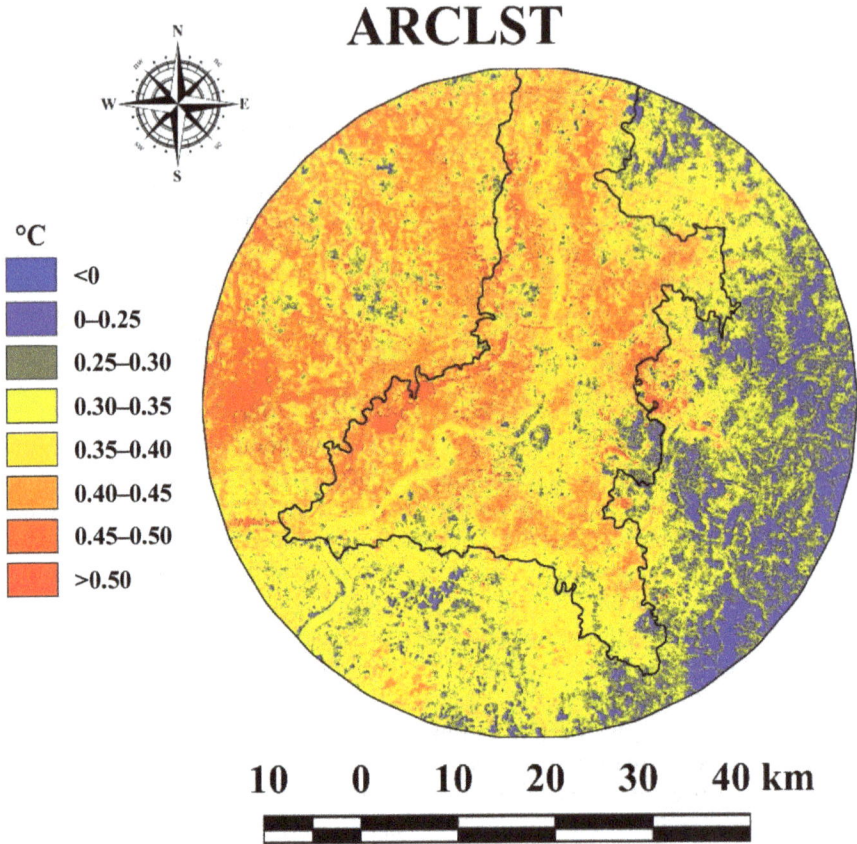

FIGURE 13.9   Annual rate of change in LST.

TABLE 13.4

Percentage Area under Each Category of ARCLST

Range (°C)	Area (%)
<0	0.02
0–0.25	7.55
0.25–0.3	12.56
0.3–0.35	23.71
0.35–0.4	27.74
0.4–0.45	20.59
0.45–0.5	6.51
>0.5	1.31

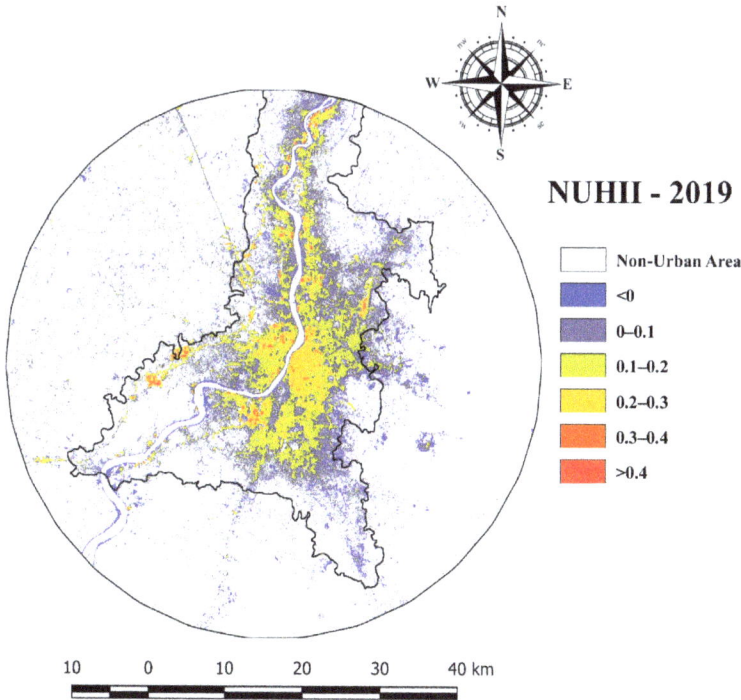

FIGURE 13.10   The NUHII.

13.8   BETTER PLANNING AND PROACTIVENESS LEADS TO COMFORTABLE URBAN ENVIRONMENT

13.8.1   FORECASTING LST

Forecasting land surface temperature for an urbanized area (LULC changes) using linear (or nonlinear) time-series analysis can be challenging and chaotic (Ranjan et al., 2018), but it can help mitigate the UHI effect by facilitating the development of new sustainable policies and aiding the development of new innovative urban design and planning strategies (Tran et al., 2017). It is crucial for the management of environmental and earth resources (Bisht and Dodamani, 2016). The LST forecast is associated with other studies, including urban growth modeling, the estimation of GHG, and the prediction for change in vegetation cover. This is one of the hot and stimulating topics among climate researchers, and numerous studies are being conducted throughout the globe. With the increasing urbanized areas, the climate, temperature, and atmospheric profiles are changing, making it important necessary to predict future urban growth using sophisticated models for sustainable development. With LST being the primary indicator of several natural phenomena such as climate, energy balance, greenhouse effect, the health of the region's ecology and atmospheric conditions is necessary to be forecasted. Future LST simulation plays

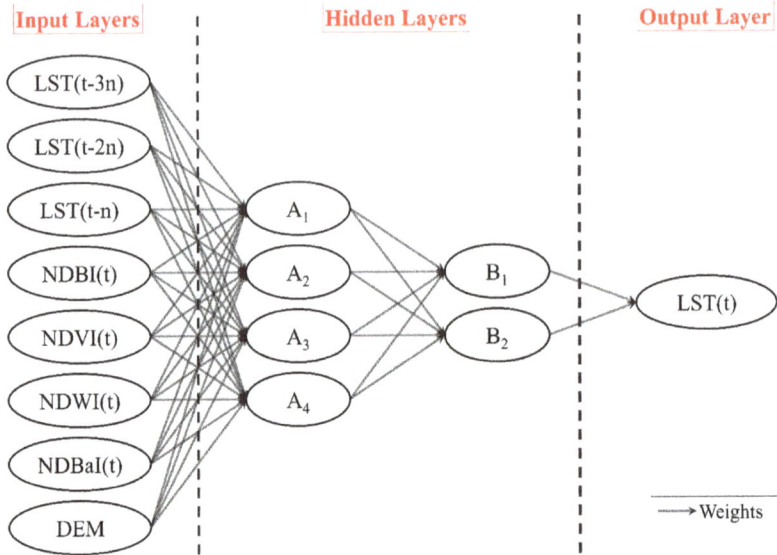

FIGURE 13.11   Neural network model for forecasting LST at time 't'.

a vital role in mitigating the negative impacts on the environment due to increasing urban growth. It serves as a valuable parameter for understanding and visualizing the areas that require attention at priority. It can help develop appropriate policies and frameworks to reduce or mitigate the dreadful effects of increasing LST and climate change on humans. Various statistical measures that can efficiently forecast the climatic parameters including LST include regression analysis (linear and nonlinear), time-series analysis (autoregressive moving average, autoregressive integrated moving average, seasonal autoregressive Integrated Moving Average, and many others). With technological advancements, numerous climatic models and machine learning algorithms have been developed to forecast LST with even greater efficiency than statistical measures while considering a variety of other climatic variables.

One of the most widely used methods for forecasting urban scenarios and climatic parameters, including LST, is the artificial neural network (ANN). The method derives a relationship between input and output via a series of iterations to estimate and assign the weights to each of the input variables. The model's hyperparameters, including activation function, type of neural network, architecture, learning rate, epochs, and many more, vary with the user's requirements and application. Considering a scenario wherein LST at time 't' is dependent on the temporal LSTs, the landscape (in the form of land-cover maps), and the elevation of the region (DEM). The example outputs of ANN applied over a sample area are shown in Figure 13.11, which also depicts the architecture used.

$$A_1 = -4.67 + 4.16 \times LST(t - n) + 3.77 \times LST(t - 2n) + 6.55 \times LST(t - 3n) + 0.81$$
$$\times NDBI(t) - 2.84 \times NDVI(t) + 3.32 \times NDWI(t) - 1.83 \times NDBaI(t) - 3.28$$
$$\times DEM$$

$A_2 = 1.33 + 2.63 \times LST(t - n) + 0.09 \times LST(t - 2n) + 4.73 \times LST(t - 3n) - 3.59 \times NDBI(t) - 0.83 \times NDVI(t) - 1.75 \times NDWI(t) + 2.72 \times NDBaI(t) - 2.57 \times DEM$

$A_3 = 22.72 + 1.81 \times LST(t - n) + 5.54 \times LST(t - 2n) + 3.59 \times LST(t - 3n) - 17.81 \times NDBI(t) - 21.94 \times NDVI(t) - 8.99 \times NDWI(t) - 4.11 \times NDBaI(t) - 5.10 \times DEM$

$A_4 = -1.23 + 1.25 \times LST(t - n) + 2.22 \times LST(t - 2n) - 2.41 \times LST(t - 3n) + 0.01 \times NDBI(t) - 1.43 \times NDVI(t) - 0.59 \times NDWI(t) - 1.26 \times NDBaI(t) + 1.42 \times DEM$

$B_1 = -1.44 + 2.39 \times fun(A_1) + 1.50 \times fun(A_2) - 0.97 \times fun(A_3) - 2.02 \times fun(A_4)$

$B_2 = -0.53 + 1.37 \times fun(A_1) - 0.84 \times fun(A_2) - 0.83 \times fun(A_3) - 1.79 \times fun(A_4)$

$\mathbf{LST(t)} = 0.54 + 0.50 \times fun(B_1) - 1.75 \times fun(B_2)$

Note: Here, $fun(x) = 1/(1 + e^{-x})$ = *sigmoid function.*

The relationship obtained should be compared with the original data and once satisfactory results are obtained, it can be applied to forecast LST. Providing inputs with respect to time 't + n' to the developed and verified ANN model, LST (t + n) can be estimated.

13.8.2 MITIGATION STRATEGIES

With the understanding of the urban LST, mitigation strategies in the form of policies and architectural interventions at meso-town level, city level, community level, neighborhood level, and in terms of building design can be efficiently developed to counteract the negative implications of the rising LST and heatwaves. A few of the already existing methods for cooling includes increasing the green and blue spaces within an urban area (Parida et al., 2021) as in the following:

- Shading the streets and public ventures via artificial structures or vegetation in the form of trees
- Dedicating some piece of land within urban pockets for native plantations that can serve as microclimate enhancers
- Developing community parks in every locality with a water body (artificial or natural) can improve the cooling of the entire locality

Besides these, a few other measures that can help in reducing the effect of high LST and UHI include the following:

- Designing new urban pockets considering the solar gains, wind direction, building orientation, and the ratio of building height to width of the adjacent street
- Introduction of cool and green roof/walls for the existing constructions or the upcoming ones
- Adapting architecture to the region's climate regime in order to minimize the energy consumption
- Replacing or renovating the urban constructions with highly reflective materials/applying a coat of reflective paints/insulating the sun-exposed

area, can essentially reduce the amount of heat absorbed and retained by the surfaces
- Development of water-sensitive or water-integrated urban designs—inclusion of ponds, pools, and fountains can increase the evaporative cooling
- Creation of pollution-free zones by restricting the entry of vehicles within some parts of the urban area, especially markets can help in reducing air pollution
- Government can start an initiative of incentivizing the usage of green energy at the individual or industrial level, which will help in reducing emissions
- Organizing various outreach programs and organizing public gatherings to make the residents environmentally conscious and educated, to minimize the environmental nuisance. These programs can include the promotion of car-pooling, Swachh Bharat, efficient usage of electricity, and many more.

13.9 CONCLUSION

The chapter provided a comprehensive overview of how alteration in the landscape of urban areas may disrupt environmental balance and affect the climate in terms of the land surface temperature of the region. The study sets out to understand how an increase in LST can lead to the development of the UHI phenomenon and creates uncomfortable living conditions for the dwellers. The last section of the research illustrates the KMA (with a buffer of 10 km) as an example to show how alterations in land-use patterns and unplanned urbanization are causing an escalation in the surface temperature of the region and its vicinities. The key purpose of this chapter was to showcase a methodological framework for monitoring and analyzing the urban land surface temperatures. One of the key aspects to gain from this chapter is to understand how important it is to have a continuous assessment of urban areas, since alterations in them can lead to large-term impacts on the residents in the form of micro-climatic variations and thermal discomforts. It can be stated that an unplanned and inadvertent urban growth and sprawl, with minimum areas dedicated to vegetation, open spaces and water bodies can lead to a significant increment in the warming of the region. The open and vegetated areas in the cities serve as breathing spaces, help in the dispersal of pollutants and dust particles for improved air quality, and further provide ventilation to the nearby localities. Water bodies, on the other hand, act as regional conditioners, cooling the wind that passes via them to the concretized areas. Altogether, it can be inferred that an adequate equilibrium in the landscape needs to be maintained in the urban areas for a better and healthy standard of living for the residents.

ACKNOWLEDGMENT

We are grateful to Indian Institute of Technology Kharagpur for the financial and infra-structure support. We thank (1) the U.S. Geological Survey and (2) the National Remote Sensing Centre (NRSC Hyderabad) for providing temporal remote-sensing data.

REFERENCES

AATSR, and SLSTR. "LST Portal: Welcome to the AATSR/SLSTR Land Surface Temperature Portal." Accessed March 10, 2018. http://lst.nilu.no/

BBC News. "Canada weather: Dozens dead as heatwave shatters records." Accessed June 30, 2021. www.bbc.com/news/world-us-canada-57654133

Bharath, H. A., M. C. Chandan, S. Vinay, and T. V. Ramachandra. 2017. "Intra and inter spatio-temporal patterns of urbanisation in Indian megacities." *International Journal of Imaging and Robotics* 17 (2): 28–39.

Bisht, K., and S. S. Dodamani. 2016. "Estimation and modelling of land surface temperature using landsat 7 ETM+ images and fuzzy system techniques." Paper presented at AGU Fall Meeting Abstracts, San Francisco, California, December 12–16, 2021. http://adsabs.harvard.edu/

Chandan, M. C., N. Aishwarya, and H. A. Bharath. 2017. "Multi temporal urban growth characterization using Geospatial technologies." Paper presented at 38th Asian Conference of Remote Sensing, Delhi, India, October 23–27, 2017.

Chen, M., H. Zhang, W. Liu, and W. Zhang. 2014. "The global pattern of urbanization and economic growth: evidence from the last three decades." *PLoS One* 9 (8): e103799. https://doi.org/10.1371/journal.pone.0103799

Chen, S., Z. Shan, and X. Chuangjie. 2000. "Remote sensing and GIS for urban growth analysis in China." *Photogrammetric Engineering and Remote Sensing* 66 (5): 593–598.

Copernicus. 2018. "Land surface temperature." Accessed January 18, 2021. https://land.copernicus.eu/global/products/lst

Dash, P., F-M. Göttsche, F-S. Olesen, and F. Herbert. 2010. "Land surface temperature and emissivity estimation from passive sensor data: Theory and practice-current trends." *International Journal of Remote Sensing* 23 (13): 2563–2594. https://doi.org/10.1080/01431160110115041

EPA. "Heat island impacts." Accessed January 11, 2019. www.epa.gov/heat-islands/heat-island-impacts

Gonzalez, R. C., and Woods, R. E. 2007. *Digital image processing* (3rd edition).

Henderson, V. 2003. "The urbanization process and economic growth: The so-what question." *Journal of Economic growth* 8 (1): 47–71. https://doi.org/10.1023/A:1022860800744

Ji, C. Y., Q. Liu, D. Sun, S. Wang, P. Lin, and X. Li. 2001. "Monitoring urban expansion with remote sensing in China." *International Journal of Remote Sensing* 22 (8): 1441–1455. https://doi.org/10.1080/01431160117207

Kandlikar, M., and A. Sagar. 1999. "Climate change research and analysis in India: An integrated assessment of a South–North divide." *Global Environmental Change* 9 (2): 119–138. https://doi.org/10.1016/S0959-3780(98)00033-8

Karoly, D. J., K. Braganza, P. A. Stott, J. M. Arblaster, G. A. Meehl, A. J. Broccoli, and K. W. Dixon. 2003. "Detection of a human influence on North American climate." *Science* 302 (5648): 1200–1203. https://doi.org/10.1126/science.1089159

Lal, D. S. 2017. *Climatology* (Revised edition: 2017). Allahabad, Sharda Pustak Bhawan.

Li, Z. L., B. H. Tang, H. Wu, H. Ren, G. Yan, Z. Wan, I. F. Trigo, and J. A. Sobrino. 2013. "Satellite-derived land surface temperature: Current status and perspectives." *Remote Sensing of Environment* 131 (April 2013): 14–37. https://doi.org/10.1016/j.rse.2012.12.008

Liang, S., and P. Shi. 2009. "Analysis of the relationship between urban heat island and vegetation cover through Landsat ETM+: A case study of Shenyang." Paper presented at 2009 Joint Urban Remote Sensing Event. Shanghai, China, May 20–22, 2009.

Lillesand, T. M., Kiefer, R. W., Chipman, J. W., and Lilles, T. M. 2003. *Remote sensing and image interpretation* (5th edition). New York, Wiley, John and Sons.

Mendelsohn, R., P. Kurukulasuriya, A. Basist, F. Kogan, and C. Williams. 2007. "Climate analysis with satellite versus weather station data." *Climatic Change* 81 (1): 71–83. https://doi.org/10.1007/s10584-006-9139-x

Ministry of Statistics and Programme Implementation, Government of India. "Statistics related to climate change—India 2015." Accessed September 28, 2021. www.mospi. gov.in/sites/default

Mojolaoluwa, T. D., E. O. Eresanya, and K. A. Ishola. 2018. "Assessment of the thermal response of variations in land surface around an urban area." *Modeling Earth Systems and Environment* 4 (2): 535–553. https://doi.org/10.1007/s40808-018-0463-8

National Geographic. "Climate change—5 ways it will affect you." Accessed January 11, 2020. www.nationalgeographic.com/climate-change/

Nuruzzaman, M. 2015. "Urban heat island: causes, effects and mitigation measures-a review." *International Journal of Environmental Monitoring and Analysis* 3 (2): 67–73. https:// doi.org/10.11648/j.ijema.20150302.15

Pachauri, R. K., and M. Chand. 2008. "A global perspective on climate change." In *Climate change and energy pathways for the mediterranean*, edited by Moniz E.J., 1–14. Dordrecht, Springer.

Pal, S., and S. K. Ziaul. 2017. "Detection of land use and land cover change and land surface temperature in English Bazar urban centre." *The Egyptian Journal of Remote Sensing and Space Science* 20 (1): 125–145. https://doi.org/10.1016/j.ejrs.2016.11.003

Parida, B. R., S. Bar, D. Kaskaoutis, A. C. Pandey, S. D. Polade, and S. Goswami. 2021. "Impact of COVID-19 induced lockdown on Land Surface Temperature, aerosol, and urban heat in Europe and North America." *Sustainable Cities and Society* 75: 103336. https://doi.org/10.1016/j.scs.2021.103336

Ramachandra, T. V., H. A. Bharath, and D. D. Sanna. 2012a. "Insights to urban dynamics through landscape spatial pattern analysis." *International Journal of Applied Earth Observation and Geoinformation* 18 (August 2012): 329–343. https://doi.org/10.1016/j.jag.2012.03.005

Ramachandra, T. V., H. A. Bharath, and M. V. Sowmyashree. 2014. "Urban footprint of Mumbai-the commercial capital of India." *Journal of Urban and Regional Analysis* 6 (1): 71–94. https://doi.org/10.37043/JURA.2020.6.1.5

Ramachandra, T. V., H. A. Bharath, and S. Sreekantha. 2012b. "Spatial metrics based landscape structure and dynamics assessment for an emerging Indian megalopolis." *International Journal of Advanced Research in Artificial Intelligence (IJARAI)* 1 (1). http://dx.doi.org/10.14569/IJARAI.2012.010109

Ranjan, A. K., A. Anand, P. B. S. Kumar, S. K. Verma, and L. Murmu. 2018. "Prediction of land surface temperature within Sun City Jodhpur (Rajasthan) in India using integration of artificial neural network and geoinformatics technology." *Asian Journal of Geoinformatics* 17 (3).

Reuters. "Scientists predict major increase in heatwave deaths as world warms." Accessed May 25, 2018. www.reuters.com/

Sanborn. "Land use/land cover mapping." Accessed Mar 29, 2021. www.sanborn.com/

Stott, P. A., J. F. Mitchell, M. R. Allen, T. L. Delworth, J. M. Gregory, G. A. Meehl, and B. D. Santer. 2006. "Observational constraints on past attributable warming and predictions of future global warming." *Journal of Climate* 19 (13): 3055–3069. https://doi. org/10.1175/JCLI3802.1

Subasinghe, S., R. C. Estoque, and Y. Murayama. 2016. "Spatiotemporal analysis of urban growth using GIS and remote sensing: A case study of the Colombo Metropolitan Area, Sri Lanka." *ISPRS International Journal of Geo-Information* 5 (11): 197. https://doi. org/10.3390/ijgi5110197

Tomlinson, C. J., L. Chapman, J. E. Thornes, and C. Baker. 2011. "Remote sensing land surface temperature for meteorology and climatology: A review." *Meteorological Applications* 18 (3): 296–306. https://doi.org/10.1002/met.287

Tran, D. X., F. Pla, P. Latorre-Carmona, S. W. Myint, M. Caetano, and H. V. Kieu. 2017. "Characterizing the relationship between land use land cover change and land surface temperature." *ISPRS Journal of Photogrammetry and Remote Sensing* 124 (February 2017): 119–132. https://doi.org/10.1016/j.isprsjprs.2017.01.001

UNDESA. 2018. *World urbanization prospects: The 2018 revision.* New York: United Nations Department of Economic and Social Affairs. https://population.un.org/wup/Publications/Files/WUP2018-KeyFacts.pdf

Weerakoon, K. 2017. "Analysis of spatio-temporal urban growth using GIS integrated urban gradient analysis; Colombo District, Sri Lanka." *American Journal of Geographic Information System* 6 (3): 83–89. http://dr.lib.sjp.ac.lk/handle/123456789/7183

Yang, L., F. Qian, D. Song, and K. Zheng. 2016. "Research on urban heat-island effect." *Procedia Engineering* 169 (2016): 11–18. https://doi.org/10.1016/j.proeng.2016.10.002

14 Effects of Land Use/ Land Cover Changes on Surface Temperature and Urban Heat Island over Kathmandu District in Nepal

Sourav Kumar[1],*, Aniket Prakash[2],
Sandeep Kumar[3], and Bikash Ranjan Parida[4]

1 Department Of Civil Engineering, Indian Institute of Technology Delhi, India

2 Department Of Civil Engineering, Indian Institute of Technology (ISM) Dhanbad, India

3 Centre of Studies in Resources Engineering, Indian Institute of Technology Bombay, India

4 Department of Geoinformatics, School of Natural Resource Management, Central University of Jharkhand, Ranchi–835222, India

CONTENTS

14.1 Introduction 276
14.2 Study Area 277
14.3 Materials and Method...278
    14.3.1 LULC...279
    14.3.2 LST Estimation.. 280
        14.3.2.1 Top of Atmosphere (TOA) Spectral Radiance (L_λ) 280
        14.3.2.2 Brightness Temperature 280
        14.3.2.3 NDVI.. 280
        14.3.2.4 Land Surface Emissivity (ε) Adopted From (Van De Griend and OwE 1993) 281
        14.3.2.5 LST 281
    14.3.3 Surface UHI..281

DOI: 10.1201/9781003265160-17

14.4 Results..281
    14.4.1 LULC Change Analysis between 2000 and 2020............................281
        14.4.1.2 LST and UHI Change Analysis between 2000
             and 2020.. 283
14.5 Discussion.. 286
14.6 Conclusion 287
Acknowledgment 287
Conflicts of Interest.. 287
References... 287

14.1 INTRODUCTION

Land use indicates the anthropogenic activity taking place on land, whereas land cover is the biophysical properties of the land surface (Lambin, Geist, and Rindfuss 2006). The terms *land use* and *land cover* (LULC) are interconnected and so studied together (Verburg et al. 2009). The rapid conversion of croplands, forests, water bodies, wasteland to settlements or impervious surfaces leads to urbanization (Ding and Shi 2013). Urbanization, causing a rapid transformation of LULC in a region, has become a serious concern for the environment (Choudhury, Das, and Das 2019). This leads to the development of urban heat islands (UHIs), causing unsustainable development with a reduction of green cover (Choudhury, Das, and Das 2019). Urbanization has altered the energy budgets of the land surface by converting natural green cover into impervious surfaces, changing the water cycle, and affecting physical properties of the land surface that influence global, regional, and local climate (Arnfield 2003; Changnon 1999; Du et al. 2007; Goodin, Harrington, and Rundquist 2002; Kalnay and Cai 2003; Kaufmann et al. 2007). Many researchers discovered that the long-term effect of the LULC change leads to the relative increase in land surface temperature (LST) (Abdullah-Al-Faisal et al. 2021; Maimaitiyiming et al. 2014; Pal and Ziaul 2017). LST is generally higher in cities by 2–4°C than in rural areas (Lai and Cheng 2010).

Although the urbanization process is a sign of economic progress and success, it has also negative consequences for city development in short and long-term terms (Maimaitiyiming et al. 2014). There are both positive as well as negative impacts of rapid urbanization. Work options and a higher quality of living are among the positive impacts, but pollution, health issues, and strain on city infrastructure are some of the negative consequences (Ullah et al. 2019). The city population exceeded the rural population in 2009, marking an important turning point in the history of human beings (Siu and Hart 2013) which will affect urban ecosystems and sustainability. In contrast to natural land cover, the growth of built-up regions alters the physical and geometric characteristics of the surface, resulting in changes in surface energy and radiation budgets (Choudhury, Das, and Das 2019). The presence of impervious surfaces, such as houses, roads, and industrial farms, increases short-wave radiation absorption and reduces energy loss owing to long-wave radiation emission (Oke 1976).

The development of a UHI effect is one of the more significant consequences of urbanization (Choudhury, Das, and Das 2019). The phenomenon of a UHI occurs when a city is substantially warmer than its surrounding rural regions due to a high proportion of impervious surfaces (Xie and Zhou 2015). UHI is traditionally identified by measuring the difference in air temperature between the urban and adjacent

rural areas. However, the availability of high spatial and temporal resolution satellite data leads to the extraction of LST and UHI for investigating the effects of urbanization on urban microclimate (Choudhury, Das, and Das 2019; Parida et al. 2021a). High–spatial resolution Landsat TM/ETM+ (Thematic Mapper/Enhanced Thematic Mapper) based thermal bands are widely used for the calculation of LST in urban areas for identifying UHI (Nichol 1994; Weng, Lu, and Liang 2006; Yuan and Bauer 2007).

Over the years, the use of combined remote-sensing (RS) and Geographic Information System (GIS) technologies to monitor LULC and LST changes in urbanized areas has grown rapidly (Balogun and Ishola 2017; Lilly Rose and Devadas 2009; Lal et al. 2019). Field-based measurements of LST and LULC change are time-consuming, laborious, and prone to mistakes (Hart and Sailor 2009; Lilly Rose and Devadas 2009). It's simpler to evaluate, monitor, and model LULC and LST changes when we use RS and GIS technologies together (Fu and Weng 2018; Niyogi 2019; R. Thapa and Murayama 2009). The spatiotemporal analysis of LULC and LST changes has brought substantial smart solutions to the temperature increasing concerns caused by unsystematic land-cover changes in urban ecosystems (Celik et al. 2019). The variation of LST is influenced by human–environment interactions, topography (i.e., elevation and slope) as well as underlying vegetation types (Parida et al. 2008; Peng et al. 2020).

Typically, mountain topography has a significant impact on LST, and mostly a negative correlation was seen with elevation and slope. A negative correlation was observed between vegetation and LST that suggests LST is greatly controlled by surface characteristics (Yuan et al. 2020). The relationships between vegetation and LST are widely studied with the help of satellite-based spectral indices (e.g., Normalized Difference Vegetation Index [NDVI]–LST relationships). For instance, a strong negative correlation was reported on LST with vegetation surface (r = 0.51), but a positive correlation with water bodies (0.45) and built-up area and bare land (0.14; Guha and Govil 2020). The NDVI has been utilized often to derive land surface emissivity and subsequently LST estimation (Sobrino, Jiménez-Muñoz, and Paolini 2004). The land conversion process also changes the LULC types and land surface properties (i.e., albedo, soil moisture, and roughness; Bonan 2008), and consequently, the LST variation in the urban ecosystem drives changes in UHI. The variation of LST and UHI in mountain topography is understudied. Therefore, the present study aims to (1) identify LULC change in the last 20 years in the Kathmandu district using satellite data, (2) estimate spatiotemporal LST variations, and (3) find impacts of LULC change over LST and UHI in urban ecosystems.

14.2 STUDY AREA

This study was conducted for Kathmandu district, located between 27° 30′ 11.8794″ N to 27° 49′ 2.6034″ N latitude and 85° 11′ 27.276″ E to 85° 34′ 5.1054″ E longitude and had an area of about 418.6 km² (Figure 14.1). The Kathmandu Valley consists of three districts of Nepal: Kathmandu, Lalitpur, and Bhktapur. Here, we selected only the Kathmandu district, where its elevation varies from 1300 m to 2800 m above mean sea level (Haack and Rafter 2006). Slopes of more than 20° comprise about 20% of the valley (Lilleso et al. 2005). It has both a subtropical and a temperate climate, which is mainly controlled by the South Asian monsoon (UNDESA 2015). The annual average rainfall is 1407 mm, but 80% of rainfall occurs in the monsoon season

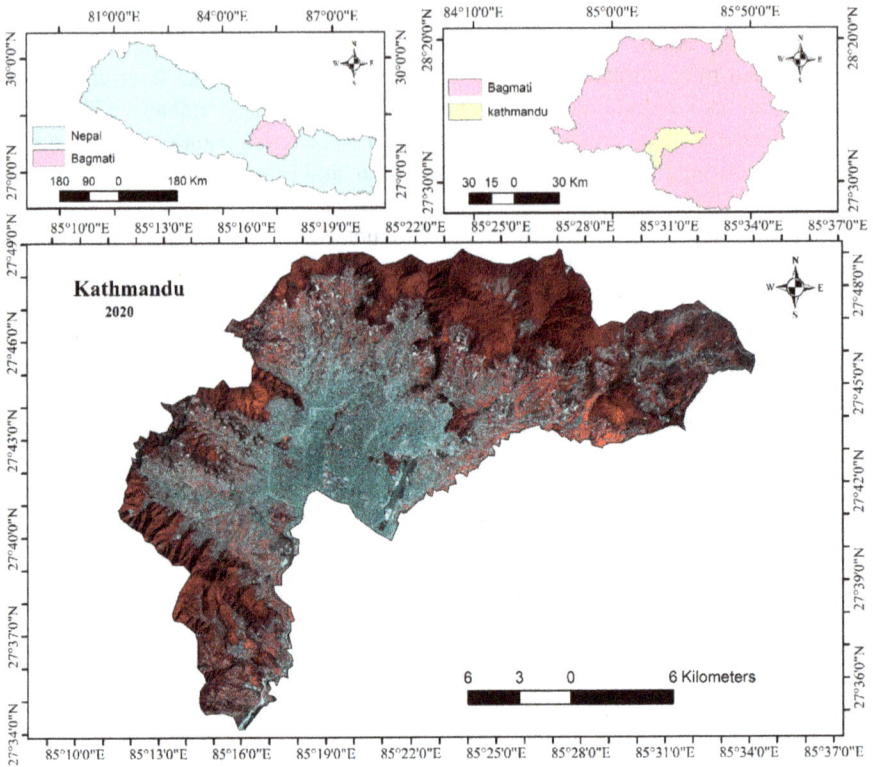

FIGURE 14.1 Study area map with false color composite (FCC) showing urban extent surrounded by high mountainous forest cover (data source: Sentinel-2 in 2020).

(June–September; Wang et al. 2020). The yearly average temperature in the valley is about 18.1°C, with some mountain peaks covered with seasonal snow (Ishtiaque, Shrestha, and Chhetri 2017). The predominant plant type in the area is a mix of conifer and broadleaved trees at lower altitudes, gradually changing to conifers and shrubland at higher elevations (Wang et al. 2020). Kathmandu city has a population of more than 1.42 million, which is exacerbated by significant rural–urban migration, and it is facing problems like unplanned and uncontrolled urban growth coupled with inadequate infrastructure. As a result, the International Institute for Environment and Development has marked Kathmandu as one of the world's 15 most vulnerable cities (IIED 2009).

14.3   MATERIALS AND METHOD

The present study has employed thermal bands and optical data sets for deriving LST and LULC maps, respectively. The thermal bands were acquired from the Landsat series (Landsat-5, Landsat-8) for determining LST. The optical data sets were acquired from the Landsat series and Sentinel-2 for LULC mapping for the years 2000 and 2020. The detailed sensor characteristics are given in Table 14.1 and the methodology is given in Figure 14.2. The LST and UHI change analysis was

TABLE 14.1

Data Used in This Study and Sensor Characteristics

Data Sets	Spatial/Temporal Resolution	Purpose	Source
Landsat-5	30 m, 16 days	LULC map for 2000	USGS
(a) optical bands	60 m, 16 days	LST estimation for 2000	
(b) thermal bands			
Landsat-8	30 m, 16 days	LST estimation for 2020	USGS
Sentinel-2	10 m, 5 days	LULC map for 2020	ESA

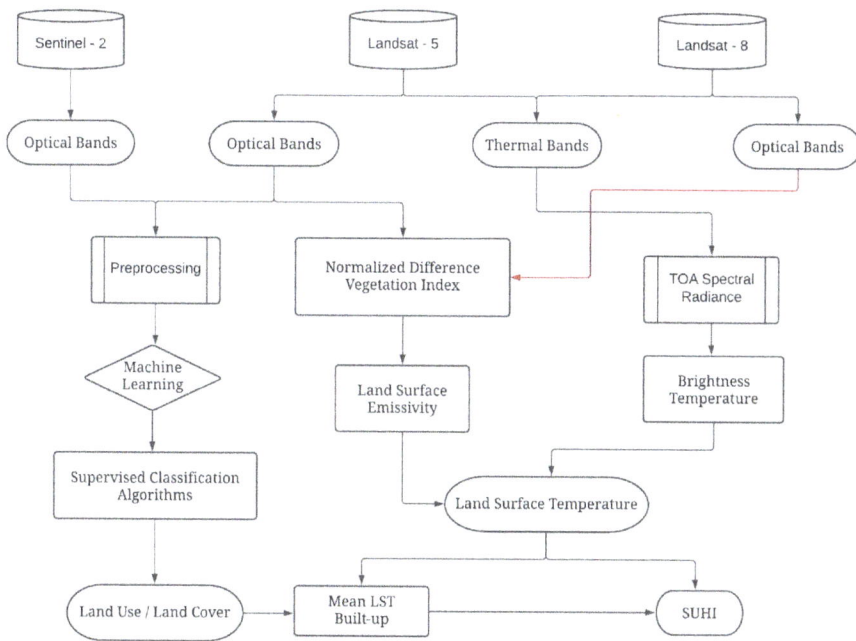

FIGURE 14.2 Flowchart of the methodology adopted in this study.

performed for three years: 2000, 2018, and 2020. As 2020 has undergone a series of lockdowns due to the COVID-19 pandemic (Parida et al. 2021b), we separately analyzed changes in LST and UHI for 2018. For LST estimation, all satellite data were acquired from April, such as 4 April 2000, 22 April 2018, and 11 April 2020.

14.3.1 LULC

The LULC classification maps for 2000 and 2020 were produced using the mean of 1 year of data to reduce the error of classification using Google Earth Engine (Yu et al. 2018; Zurqani et al. 2018). Landsat-5 and Sentinel-2 data have been used

for LULC classification. The impact of LULC change on both natural and human systems is key research precedence in multiple directions. The impact of LULC on LSTs is studied widely (Ishtiaque, Shrestha, and Chhetri 2017). The present study has used pixel-based supervised classification algorithms of the machine learning technique for LULC preparation (Sekertekin, Marangoz, and Akcin 2017; Shaharum et al. 2018; Vivekananda, Swathi, and Sujith 2021). In the technique, the user has to train the software for classification by developing the spectral signatures of known categories (i.e., built-up areas, forest, agriculture, barren land, water bodies), and then each pixel of the image is assigned to the land-cover types with which its signature is most similar (Zaidi et al. 2017).

14.3.2 LST ESTIMATION

For the LST study, we have used the thermal bands of Landsat-5 and Landsat-8 (Ermida et al. 2020; Lal et al. 2019; Mustafa et al. 2020). It is calculated using the following factors, which were performed in QGIS.

14.3.2.1 Top of Atmosphere (TOA) Spectral Radiance (L_λ)

$$L_\lambda = M_L . Q_{Cal} + A_L, \qquad (14.1)$$

where
  L_λ = TOA spectral radiance,
  M_L = band-specific multiplicative rescaling factor from the metadata,
  A_L = band-specific additive rescaling factor from the metadata, and
  Q_{Cal} = quantized and calibrated standard product pixel values.

14.3.2.2 Brightness Temperature

$$BT = \frac{K_2}{\ln\left(\dfrac{K_1}{L_\lambda} + 1\right)}, \qquad (14.2)$$

where
  BT = at-satellite brightness temperature (K),
  L_λ = TOA spectral radiance,
  K_1 = band-specific thermal conversion constant from the metadata, and
  K_2 = band-specific thermal conversion constant from the metadata.

14.3.2.3 NDVI

$$NDVI = \frac{(NIR - Red)}{(NIR + Red)} \qquad (14.3)$$

14.3.2.4 Land Surface Emissivity (ε) Adopted From (Van De Griend and OwE 1993)

$$\varepsilon = a + b \ln \mathbf{NDVI}, \tag{14.4}$$

where

ε = spetxltral surface emissivity, a = constant derived from regression analysis, and b = constant derived from regression analysis.

14.3.2.5 LST

$$LST = \frac{\mathbf{BT}}{\left\{1 + \left[\dfrac{\lambda.\mathbf{BT}}{\rho}.\ln \varepsilon\right]\right\}}, \tag{14.5}$$

where

LST = land surface temperature (K), λ = wavelength of emitted radiance (µm), and ρ = hc/σ.

14.3.3 Surface UHI

The difference between the surface temperature of an urban region and its rural environs is known as surface UHI (SUHI) and urban cold island (Rasul, Balzter, and Smith 2016). The spatiotemporal fluctuation of SUHI was computed using Equation 14.6. The reference urban LST (i.e., 27.98°C for 2000, 33.17°C for 2018, and 28.01°C for 2020) was initially created in order to quantify the SUHI, and the formulae were adopted from (O'Malley et al. 2015; Zhou et al. 2015).

$$SUHI = LST - LST_U, \tag{14.6}$$

where

SUHI = surface urban heat island and
LST_U = mean LST of the urban area.

14.4 RESULTS

14.4.1 LULC Change Analysis between 2000 and 2020

The LULC maps for 2000 and 2020, based on Landsat-5 and Sentinel-2A, respectively, are shown in Figures 14.3 and 14.4. The analysis of the LULC map suggests that Kathmandu has gone under a drastic change between 2000 and 2020. The LULC change assessment has been considered for five major LULC classes, namely, built-up, water body, forest/shrubland, agriculture, and barren land. In 2000, forest/shrubland and agriculture have covered an area of 166.03 km^2 and 153.6 km^2, respectively (Table 14.2). The minimum area was occupied by water bodies, which was 1.01 km^2, whereas the built-up and barren land covered an area of 54.08 km^2

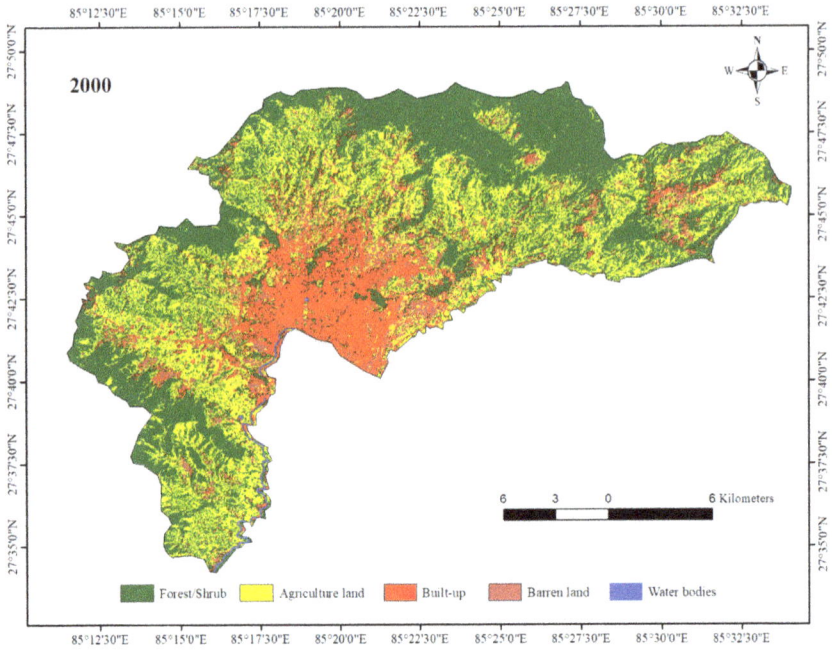

FIGURE 14.3   LULC map of 2000 for Kathmandu district.

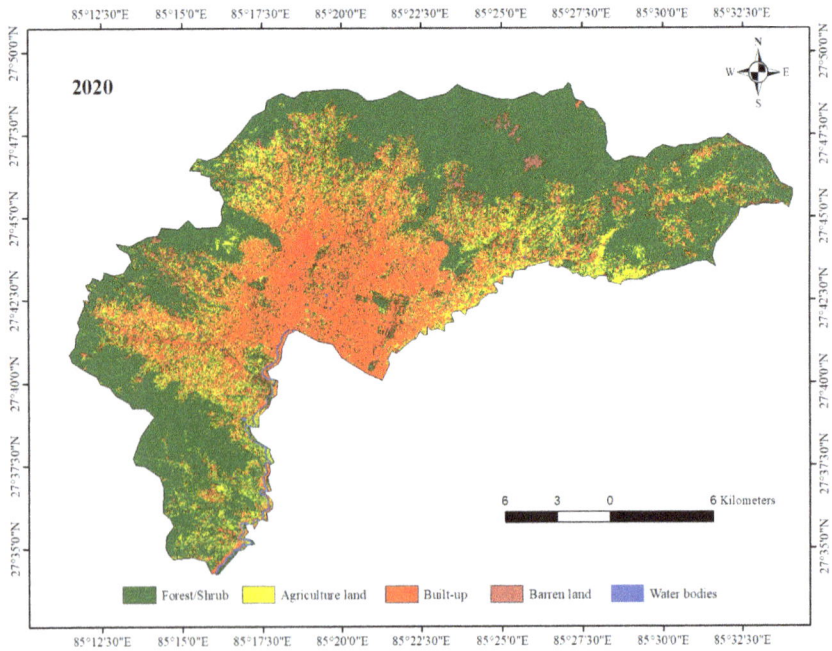

FIGURE 14.4   LULC map of 2020 for Kathmandu district.

TABLE 14.2

Area Statistics of LULC in km²

LULC Classes	2000	2020	Area Change
Built-up	54.08	111.4	+57.32
Water Bodies	1.01	1.1	+0.09
Forest/Shrub	166.03	224.92	+58.89
Agriculture	153.6	73.26	−80.34
Barren Land	44.16	7.56	−36.6

and 44.16 km², respectively. In 2020, the built-up, forest/shrubland, and waterbody areas increased to 111.4 km², 224.92 km², and 1.1 km², respectively. The area coverage by agriculture and barren land was found to be decreased to 73.26 km² and 7.56 km², respectively. Over the two decades, the urban expansion has been undergone rapidly at the cost of agricultural lands as well as barren land. Along with the migration of people toward urban areas and population rise, urban expansion grows by almost 100%. By contrast, increasing forest area by 35% can be attributed to reforestation activities including the policy taken up by the government of Nepal. The government has also emphasized local communities for protecting forests through community forestry.

14.4.1.2 LST and UHI Change Analysis between 2000 and 2020

The assessment of the LST variability was performed using Landsat-5 and Landsat-8, which revealed a shift in the pattern of LST in the study area from 2000 to 2020 (Figure 14.5). The spatial pattern of LST exhibited much higher in 2018 than in 2000. However, the pandemic year 2020 showed lower LST than 2000, which was simply due to the cessation of anthropogenic emissions. The zonal mean of LST was calculated for each LULC category for 2000, 2018, and 2020. In 2000, the mean LST was ranging between 25–30°C. In 2018, a major proportion of the study area was found to be having a mean LST between 30–35°C. A very different trend was observed in 2020, as the mean LST fell to the range of 25–30°C. As there was a nationwide declared lockdown due to COVID-19 restrictions, it contributed a lot to the rejuvenation of the atmosphere by discontinuing the emission of pollutants into the atmosphere (Bar et al. 2021). The zonal mean analysis is presented in Figure 14.6, which shows the mean LST of built-up areas was 27.98°C in 2000, 33.17°C in 2018, and 28.01°C in 2020. The mean LST for forest/shrubland was 25.25°C in 2000, 27.52°C in 2018, and 22.83°C in 2020, and for agriculture, the mean LST was 25.48°C in 2000, 30.37°C in 2018, and 26.41°C in 2020. The mean LST for barren land was 29.72°C in 2000, 30.04°C in 2018, and 26.1°C in 2020. Overall, it can be stated that the LST increased from 2000–2018, with increasing urbanization. Moreover, increasing infrastructure, industry, vehicles, and built-up surfaces, along with decreasing open surfaces such as agricultural land and barren land, were the major driving factors of increasing LST.

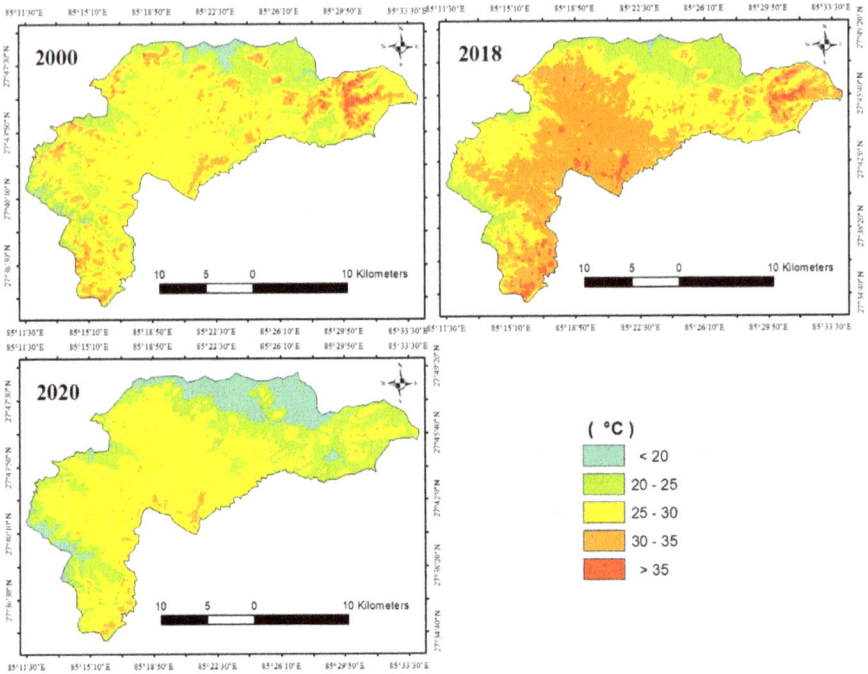

FIGURE 14.5   LST map of Kathmandu for 2000, 2018, and 2020.

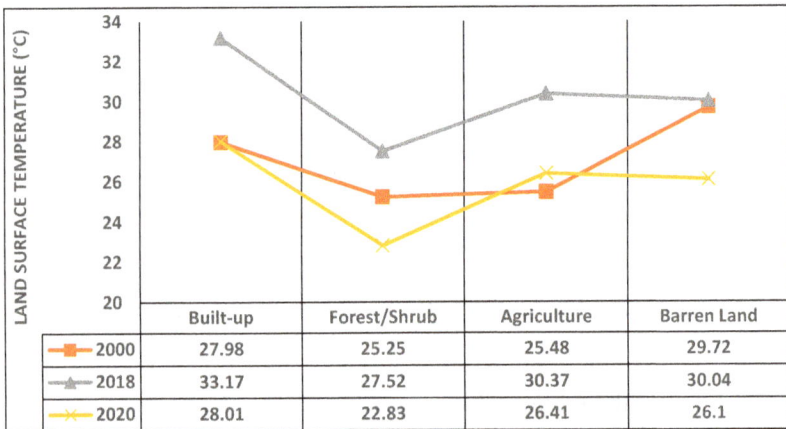

FIGURE 14.6   Mean LST variations for different LULC classes for 2000, 2018, and 2020.

Changes in SUHI magnitude for 2000, 2018, and 2020 have been shown in Figure 14.7, which shows that SUHI ranges from −10°C to 5°C. The higher magnitude of SUHI was mainly concentrated over built-up areas and magnitude generally changes with the LULC category. The LULC category-wise SUHI indicated that

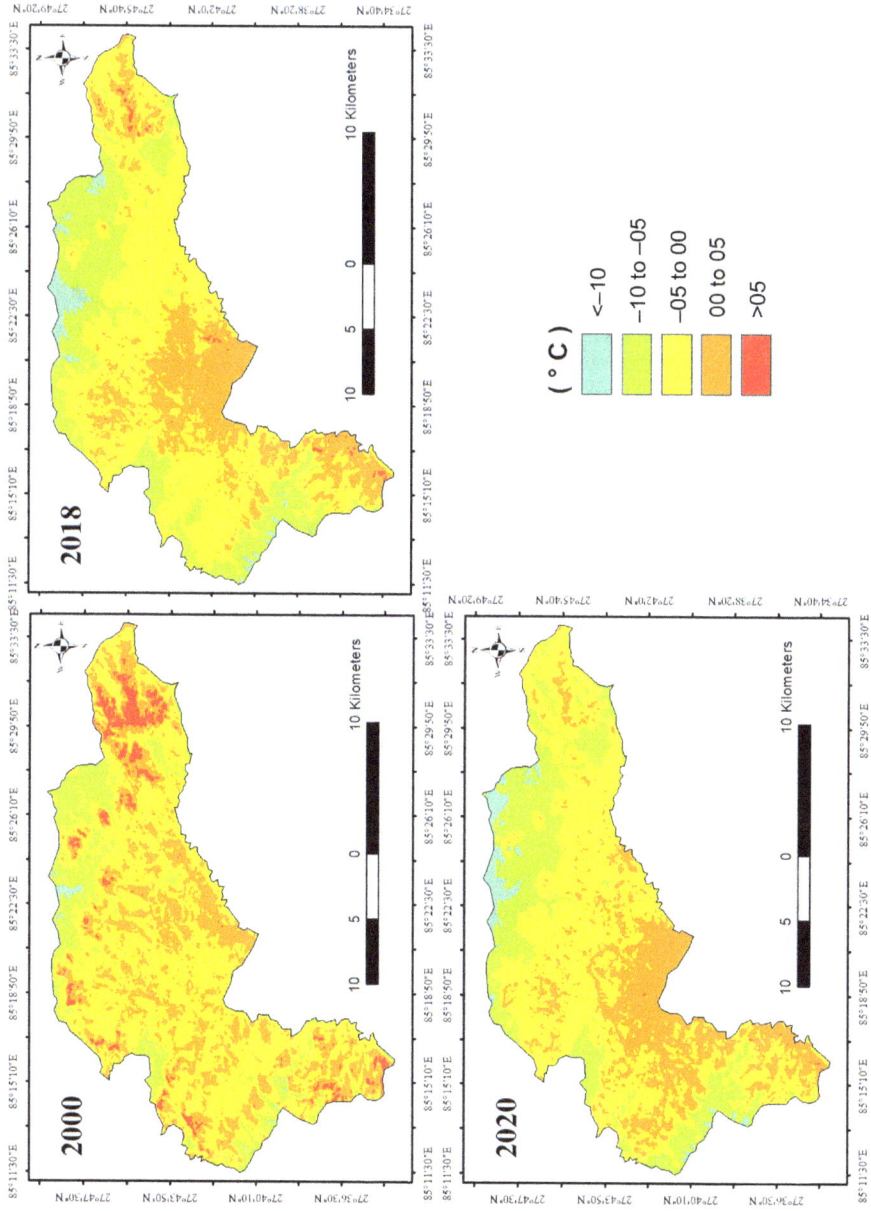

FIGURE 14.7 SUHI map of Kathmandu for 2000, 2018, and 2020.

urban areas were colocated with high SUHI ranges from 1–5°C, while the forest and agriculture regions were colocated with lower SUHI that ranges from −10°C to −5°C and agriculture −5°C to −0°C.

14.5 DISCUSSION

LULC study enables us to comprehend biophysical alterations such as the loss of vegetated areas, degradation of environmental quality, and the loss of productive ecosystems/biodiversity, all of which are crucial factors for the building of sustainable cities. The rapid conversion of croplands, forest, water bodies, and wasteland to settlements or impervious surfaces is the major drawback of urbanization, which leads to the formation of UHI. Over the years, the use of combined RS and GIS had widely used to evaluate, monitor, and model LULC and LST changes. This study is aimed to estimate LULC change in the last 20 years in Kathmandu, Nepal, and the spatiotemporal variation of LST with respect to LULC classes. This study showed a large-scale alteration of agricultural land and barren land toward urbanization. As a result, it has an impact on LST, especially over the built-up environment. The LULC category-wise LST analysis showed that 4.52°C and 2.3°C temperature was higher in built-up than the surrounding forest covers and agricultural lands, respectively. Consequently, the increasing LST has changed urban microclimate that was analyzed through the SUHI. According to the SUHI, about 2.16°C was higher in the built-up environment than surrounding conditions, which indicates the impacts of changes on LULC on both LST and UHIs.

Prior studies reported that LSTs of urban areas were consistently higher than surrounding forests and agricultural land over Kathmandu valley (Mishra, Sandifer, and Gyawali 2019) based on Moderate Resolution Imaging Spectroradiometer (MODIS) 8-day product (MOD11A2). The difference of LST of 2–4°C was reported within the built-up environment based on Landsat-8 satellite data, while a difference of about 5–10°C was found between built-up and vegetated land-cover types over Kathmandu Valley (Regmi and Chhetri 2021). Luintel et al. (2019) in their work hypothesized that the nighttime LST in Nepal increased more substantially than the daytime LST, with more pronounced warming in the pre-monsoon and monsoon seasons. The yearly nighttime LST increased at a rate of 0.05 K yr^{-1} (p < 0.01), but the daytime LST change was statistically negligible. Sharma, Tongkumchum, and Ueranantasun (2018) have analyzed the MODIS-based LST patterns along with altitudes around Kathmandu Valley and applied the second-degree polynomial regression model to study temperature patterns. It reported that 65.4% of the areas exhibited an 'accelerating' pattern (i.e., increasing) mostly over lower altitudes with dense population, whereas 34.6% of the areas showed a 'nonaccelerating' pattern (i.e., decreasing or no change) distributed over higher altitudes (e.g., northern and western areas).

The regression analysis between the spatial trend of LULC and the spatial pattern of LST (1988–2014) revealed the increase of urban area considerably by 259% (Thapa 2017). Surface temperatures were seen consistently to be higher in bare land and urban land-use types, with a positive relationship between urban expansion and LST (Thapa 2017). The built-up area significantly contributed to the rise in LST and especially, the built-up areas with dense settlement and population density

areas exhibited higher LSTs (Chidi, Magar, and Magar 2021). The city center of Kathmandu has a greater LST than less populated and adjacent areas. With rising urbanization, the LST has also risen strongly from 1999–2017. The expanding urbanization may result in a worse ecological condition in Kathmandu Valley in the future (Chidi, Magar, and Magar 2021). Using multispectral satellite data (Landsat), the UHIs pattern was studied in Kathmandu and Bhaktapur during the summer and winter seasons of 2014–2019 (Thapa 2021). It reported that the UHI levels displayed have increased significantly in 2019 compared to 2014 and suggested that these cities must think about urban heat and make people and decision-makers aware of the need to take steps to mitigate this effect.

14.6 CONCLUSION

The present study has utilized Landsat satellite data between 2000–2020 to evaluate LULC, LST, and SUHI. According to the analysis of thermal pattern, the LST was higher in built-up areas than the vegetated land types. It is evident that the higher urbanization is, the higher the temperatures, with an increasing intensity of the UHI effect as determined through SUHI. We found that urbanization is a key factor of influence SUHI during the 20 years of observation over Kathmandu, and the maximum magnitude of SUHI occurs mainly in low vegetation and the sun radiation–facing valley at lower altitudes. Kathmandu Valley is one of the most urbanization areas of Nepal where the built-up environment undergoing rapid changes. Therefore, to manage the microclimate of city like Kathmandu, the present study suggests that the systematic planning of open and green areas is vital to keep the city environment sustainable and resilient.

ACKNOWLEDGMENT

Authors thanks the U.S. Geological Survey and Copernicus, ESA for providing the Landsat and Sentinel-2 satellite data.

CONFLICTS OF INTEREST

The authors declare no conflicts of interest.

REFERENCES

Abdullah-Al-Faisal, Abdulla—Al Kafy, A.N.M. Foyezur Rahman, et al. 2021. "Assessment and Prediction of Seasonal Land Surface Temperature Change Using Multi-Temporal Landsat Images and Their Impacts on Agricultural Yields in Rajshahi, Bangladesh." *Environmental Challenges* 4: 100147. doi:10.1016/j.envc.2021.100147

Arnfield, A. John. 2003. "Two Decades of Urban Climate Research: A Review of Turbulence, Exchanges of Energy and Water, and the Urban Heat Island." *International Journal of Climatology* 23 (1): 1–26. doi:10.1002/joc.859

Balogun, I.A., and K.A. Ishola. 2017. "Projection of Future Changes in Landuse/Landcover Using Cellular Automata/Markov Model over Akure City, Nigeria." *Journal of Remote Sensing Technology* 5 (1): 22–31. doi:10.18005/JRST0501003

Bar, Somnath, B.R. Parida, S.P. Mandal, A.C. Pandey, N. Kumar, and B. Mishra. 2021. "Impacts of Partial to Complete COVID-19 Lockdown on NO_2 and PM2.5 Levels in Major Urban Cities of Europe and USA." *Cities* 107: 103308. https://doi.org/10.1016/j.cities.2021.103308

Bonan, Gordon B. 2008. "Forests and Climate Change: Forcings, Feedbacks, and the Climate Benefits of Forests." *Science* 320 (5882): 1444–1449. doi:10.1126/science.1155121

Celik, Bahadir, Sinasi Kaya, Ugur Alganci, Dursun Zafer Seker, and others. 2019. "Assessment of the Relationship between Land Use/Cover Changes and Land Surface Temperatures: A Case Study of Thermal Remote Sensing." *FEB-Fresenius Environmental Bulletin* 3: 541.

Changnon, Stanley A. 1999. "A Rare Long Record of Deep Soil Temperatures Defines Temporal Temperature Changes and an Urban Heat Island." *Climatic Change* 42 (3): 531–538. doi:10.1023/A:1005453217967

Chidi, Chhabi Lal, Ramesh Kumar Salami Magar, and Dipendra Salami Magar. 2021. "Assessment of Urban Heat Island in Kathmandu Valley (1999–2017)." *Geographical Journal of Nepal* 14: 1–20. doi:10.3126/gjn.v14i0.35544

Choudhury, Deblina, Kalikinkar Das, and Arijit Das. 2019. "Assessment of Land Use Land Cover Changes and Its Impact on Variations of Land Surface Temperature in Asansol-Durgapur Development Region." *The Egyptian Journal of Remote Sensing and Space Science* 22 (2): 203–218. doi:10.1016/j.ejrs.2018.05.004

Ding, Haiyong, and Wenzhong Shi. 2013. "Land-Use/Land-Cover Change and Its Influence on Surface Temperature: A Case Study in Beijing City." *International Journal of Remote Sensing* 34 (15): 5503–5517. doi:10.1080/01431161.2013.792966

Du, Yin, Zhiqing Xie, Yan Zeng, Yafeng Shi, and Jingang Wu. 2007. "Impact of Urban Expansion on Regional Temperature Change in the Yangtze River Delta." *Journal of Geographical Sciences* 17 (4): 387–398. doi:10.1007/s11442-007-0387-0

Ermida, Sofia L., Patrícia Soares, Vasco Mantas, Frank-M. Göttsche, and Isabel F. Trigo. 2020. "Google Earth Engine Open-Source Code for Land Surface Temperature Estimation from the Landsat Series." *Remote Sensing* 12 (9): 1471. doi:10.3390/rs12091471

Fu, Peng, and Qihao Weng. 2018. "Responses of Urban Heat Island in Atlanta to Different Land-Use Scenarios." *Theoretical and Applied Climatology* 133 (1–2): 123–135. doi:10.1007/s00704-017-2160-3

Goodin, Douglas G., John A. Harrington, and Bradley C. Rundquist. 2002. "Land Cover Change and Associated Trends in Surface Reflectivity and Vegetation Index in Southwest Kansas: 1972–1992." *Geocarto International* 17 (1): 45–52. doi:10.1080/10106040208542224

Guha, S., and H. Govil. 2020. "Land Surface Temperature and Normalized Difference Vegetation Index Relationship: A Seasonal Study on a Tropical City." *SN Applied Sciences* 2: 1661. https://doi.org/10.1007/s42452-020-03458-8

Haack, Barry N., and Ann Rafter. 2006. "Urban Growth Analysis and Modeling in the Kathmandu Valley, Nepal." *Habitat International* 30 (4): 1056–1065. doi:10.1016/j.habitatint.2005.12.001

Hart, Melissa A., and David J. Sailor. 2009. "Quantifying the Influence of Land-Use and Surface Characteristics on Spatial Variability in the Urban Heat Island." *Theoretical and Applied Climatology* 95 (3–4): 397–406. doi:10.1007/s00704-008-0017-5

IIED. 2009. "Climate Change and the Urban Poor: Risk and Resilience in 15 of the World's Most Vulnerable Cities." *London: International Institute for Environment and Development.* https://pubs.iied.org/g02597

Ishtiaque, Asif, Milan Shrestha, and Netra Chhetri. 2017. "Rapid Urban Growth in the Kathmandu Valley, Nepal: Monitoring Land Use Land Cover Dynamics of a Himalayan City with Landsat Imageries." *Environments* 4 (4): 72. doi:10.3390/environments4040072.

Kalnay, Eugenia, and Ming Cai. 2003. "Impact of Urbanization and Land-Use Change on Climate." *Nature* 423 (6939): 528–531. doi:10.1038/nature01675

Kaufmann, Robert K., Karen C. Seto, Annemarie Schneider, Zouting Liu, Liming Zhou, and Weile Wang. 2007. "Climate Response to Rapid Urban Growth: Evidence of a Human-Induced Precipitation Deficit." *Journal of Climate* 20 (10): 2299–2306. doi:10.1175/JCLI4109.1

Lai, Li-Wei, and Wan-Li Cheng. 2010. "Urban Heat Island and Air Pollution—An Emerging Role for Hospital Respiratory Admissions in an Urban Area." *Journal of Environmental Health* 72 (6): 32–36.

Lal, Preet, Sandeep Kumar, Sheetal Kumari, Aniket Prakash, and Sourav Kumar. 2019. "Spatio-Temporal Analysis of Land Surface Tem-Perature of Sikkim, India." *International Journal for Research in Applied Science and Engineering Technology* 7 (8): 267–271. doi:10.22214/ijraset.2019.8037

Lambin, Eric F., Helmut Geist, and Ronald R. Rindfuss. 2006. "Introduction: Local Processes with Global Impacts." In *Land-Use and Land-Cover Change*, edited by Eric F. Lambin and Helmut Geist, 1–8. Global Change—The IGBP Series. Berlin, Heidelberg: Springer Berlin Heidelberg. doi:10.1007/3-540-32202-7_1

Lilleso, J-P., T. Shrestha, L. Dhakal, R. Nayaju, and Shrestha. 2005. *The Map of Potential Vegetation of Nepal: A Forestry/Agro-Ecological/Biodiversity Classification Syste.* Development and Environment Serie 2-2005/CFC-TIS document series no.11. Horsholm, Denmark: Forest and Landscape Denmark.

Lilly Rose, A., and Monsingh D. Devadas. 2009. "Analysis of Land Surface Temperature and Land Use/Land Cover Types Using Remote Sensing Imagery-a Case in Chennai City, India." In Proceedings of the 7th International Conference on Urban Climate (ICUC-7), Yokohama, Japan. Vol. 29.

Luintel, Nirajan, Weiqiang Ma, Yaoming Ma, Binbin Wang, and Sunil Subba. 2019. "Spatial and Temporal Variation of Daytime and Nighttime MODIS Land Surface Temperature across Nepal." *Atmospheric and Oceanic Science Letters* 12 (5): 305–312. doi:10.1080/16742834.2019.1625701

Maimaitiyiming, Matthew, Abduwasit Ghulam, Tashpolat Tiyip, Filiberto Pla, Pedro Latorre-Carmona, Ümüt Halik, Mamat Sawut, and Mario Caetano. 2014. "Effects of Green Space Spatial Pattern on Land Surface Temperature: Implications for Sustainable Urban Planning and Climate Change Adaptation." *ISPRS Journal of Photogrammetry and Remote Sensing* 89: 59–66. doi:10.1016/j.isprsjprs.2013.12.010

Mishra, Bijesh, Jeremy Sandifer, and Buddhi Raj Gyawali. 2019. "Urban Heat Island in Kathmandu, Nepal: Evaluating Relationship between NDVI and LST from 2000 to 2018." *International Journal of Environment* 8 (1): 17–29. doi:10.3126/ije.v8i1.22546

Mustafa, Elhadi K., Yungang Co, Guoxiang Liu, Mosbeh R. Kaloop, Ashraf A. Beshr, Fawzi Zarzoura, and Mohammed Sadek. 2020. "Study for Predicting Land Surface Temperature (LST) Using Landsat Data: A Comparison of Four Algorithms." *Advances in Civil Engineering* 2020: 1–16. doi:10.1155/2020/7363546

Nichol, Janet Elizabeth. 1994. "A GIS-Based Approach to Microclimate Monitoring in Singapore's High- Rise Housing Estates." *Photogrammetric Engineering & Remote Sensing* 60 (10): 1225–1232.

Niyogi, Dev. 2019. "Land Surface Processes." In *Current Trends in the Representation of Physical Processes in Weather and Climate Models*, edited by David A. Randall, J. Srinivasan, Ravi S. Nanjundiah, and Parthasarathi Mukhopadhyay, 349–370. Springer Atmospheric Sciences. Singapore: Springer Singapore. doi:10.1007/978-981-13-3396-5_17

Oke, T.R. 1976. "The Distinction between Canopy and Boundary-layer Urban Heat Islands." *Atmosphere* 14 (4): 268–277. doi:10.1080/00046973.1976.9648422

O'Malley, Christopher, Poorang Piroozfar, Eric R.P. Farr, and Francesco Pomponi. 2015. "Urban Heat Island (UHI) Mitigating Strategies: A Case-Based Comparative Analysis." *Sustainable Cities and Society* 19: 222–235. doi:10.1016/j.scs.2015.05.009

Pal, Swades, and Sk. Ziaul. 2017. "Detection of Land Use and Land Cover Change and Land Surface Temperature in English Bazar Urban Centre." *The Egyptian Journal of Remote Sensing and Space Science* 20 (1): 125–145. doi:10.1016/j.ejrs.2016.11.003

Parida, B.R., B. Oinam, N.R. Patel, N. Sharma, R. Kandwal, and M.K. Hazarika. 2008. "Land Surface Temperature Variation in Relation to Vegetation Type Using MODIS Satellite Data in Gujarat State of India." *International Journal of Remote Sensing* 29 (14): 4219–4235. doi:10.1080/01431160701871096

Parida, B.R., Somnath Bar, Dimitris Kaskaoutis, A.C. Pandey, S.D. Polade, and S. Goswami. 2021a. "Impact of COVID-19 Induced Lockdown on Land Surface Temperature, Aerosol, and Urban Heat in Europe and North America." *Sustainable Cities and Society* 75: 103336. doi:10.1016/j.scs.2021.103336

Parida, B.R., Somnath Bar, Gareth Roberts, S.P. Mandal, A.C. Pandey, M. Kumar, and J. Dash. 2021b. "Improvement in Air Quality and Its Impact on Land Surface Temperature in Major Urban Areas across India during the First Lockdown of the Pandemic." *Environmental Research* 199: 111280. doi:10.1016/j.envres.2021.111280

Peng, Xiaoxue, Wenyuan Wu, Yaoyao Zheng, Jingyi Sun, Tangao Hu, and Pin Wang. 2020. "Correlation Analysis of Land Surface Temperature and Topographic Elements in Hangzhou, China." *Scientific Reports* 10 (1): 10451. doi:10.1038/s41598-020-67423-6

Rasul, Azad, Heiko Balzter, and Claire Smith. 2016. "Diurnal and Seasonal Variation of Surface Urban Cool and Heat Islands in the Semi-Arid City of Erbil, Iraq." *Climate* 4 (3): 42. doi:10.3390/cli4030042

Regmi, Pukar, and D.B.T Chhetri. 2021. "A Study of Urban Heat Island Relating "Local Climate Zones" Using Landsat Images -The Case of Kathmandu Valley." KEC Conference 2021, 96–100. ISBN 978-9937-0-9019-3.

Sekertekin, A., A.M. Marangoz, and H. Akcin. 2017. "Pixel-Based Classification Analysis of Land Use Land Cover Using Sentinel-2 and Landsat-8 Data." *The International Archives of the Photogrammetry, Remote Sensing and Spatial Information Sciences* XLII-4/W6: 91–93. doi:10.5194/isprs-archives-XLII-4-W6-91-2017

Shaharum, Nur Shafira Nisa, Helmi Zulhaidi Mohd Shafri, Jibrin Gambo, and Fauzul Azim Zainal Abidin. 2018. "Mapping of Krau Wildlife Reserve (KWR) Protected Area Using Landsat 8 and Supervised Classification Algorithms." *Remote Sensing Applications: Society and Environment* 10: 24–35. doi:10.1016/j.rsase.2018.01.002

Sharma, Ira, Phattrawan Tongkumchum, and Attachai Ueranantasun. 2018. "Modeling of Land Surface Temperatures to Determine Temperature Patterns and Detect Their Association with Altitude in the Kathmandu Valley of Nepal." *Chiang Mai University Journal of Natural Sciences* 17 (4). doi:10.12982/CMUJNS.2018.0020

Siu, Leong Wai, and Melissa A. Hart. 2013. "Quantifying Urban Heat Island Intensity in Hong Kong SAR, China." *Environmental Monitoring and Assessment* 185 (5): 4383–4398. doi:10.1007/s10661-012-2876-6

Sobrino, José A., Juan C. Jiménez-Muñoz, and Leonardo Paolini. 2004. "Land Surface Temperature Retrieval from LANDSAT TM 5." *Remote Sensing of Environment* 90 (4): 434–440. doi:10.1016/j.rse.2004.02.003

Thapa, P. 2021. "Urban Heat Island Analysis Using Landsat 8 Satellite Data." Conference: 11th International Geographic Information Science, 1–16. Poland.

Thapa, Rajesh, and Yuji Murayama. 2009. "Examining Spatiotemporal Urbanization Patterns in Kathmandu Valley, Nepal: Remote Sensing and Spatial Metrics Approaches." *Remote Sensing* 1 (3): 534–556. doi:10.3390/rs1030534

Thapa, Sushil. 2017. "Exploring the Impact of Urban Growth on Land Surface Temperature of Kathmandu Valley, Nepal," 77. NIMS—MSc Dissertations Geospatial Technologies (Erasmus-Mundus). Available online: http://hdl.handle.net/10362/34223

Ullah, Siddique, Adnan Ahmad Tahir, Tahir Ali Akbar, Quazi K. Hassan, Ashraf Dewan, Asim Jahangir Khan, and Mudassir Khan. 2019. "Remote Sensing-Based Quantification of the Relationships between Land Use Land Cover Changes and Surface Temperature over the Lower Himalayan Region." *Sustainability* 11 (19): 5492. doi:10.3390/su11195492

UNDESA. 2015. "World Urbanization Prospects: The 2014 Revision." *United Nations Department of Economics and Social Affairs, Population Division: New York, NY, USA* 41.

Van De Griend, A.A., and M. Owe. 1993. "On the Relationship between Thermal Emissivity and the Normalized Difference Vegetation Index for Natural Surfaces." *International Journal of Remote Sensing* 14 (6): 1119–1131. doi:10.1080/01431169308904400

Verburg, Peter H., Jeannette van de Steeg, A. Veldkamp, and Louise Willemen. 2009. "From Land Cover Change to Land Function Dynamics: A Major Challenge to Improve Land Characterization." *Journal of Environmental Management* 90 (3): 1327–1335. doi:10.1016/j.jenvman.2008.08.005

Vivekananda, Gn, R Swathi, and Avln Sujith. 2021. "Multi-Temporal Image Analysis for LULC Classification and Change Detection." *European Journal of Remote Sensing* 54 (sup2): 189–199. doi:10.1080/22797254.2020.1771215

Wang, Sonam Wangyel, Belay Manjur Gebru, Munkhnasan Lamchin, Rijan Bhakta Kayastha, and Woo-Kyun Lee. 2020. "Land Use and Land Cover Change Detection and Prediction in the Kathmandu District of Nepal Using Remote Sensing and GIS." *Sustainability* 12 (9): 3925. doi:10.3390/su12093925

Weng, Qihao, Dengsheng Lu, and Bingqing Liang. 2006. "Urban Surface Biophysical Descriptors and Land Surface Temperature Variations." *Photogrammetric Engineering & Remote Sensing* 72 (11): 1275–1286. doi:10.14358/PERS.72.11.1275

Xie, Qijiao, and Zhixiang Zhou. 2015. "Impact of Urbanization on Urban Heat Island Effect Based on Tm Imagery in Wuhan, China." *Environmental Engineering and Management Journal* 14 (3): 647–655. doi:10.30638/eemj.2015.072

Yuan, Fei, and Marvin E. Bauer. 2007. "Comparison of Impervious Surface Area and Normalized Difference Vegetation Index as Indicators of Surface Urban Heat Island Effects in Landsat Imagery." *Remote Sensing of Environment* 106 (3): 375–386. doi:10.1016/j.rse.2006.09.003

Yuan, Moxi, Lunche Wang, Aiwen Lin, Zhengjia Liu, Qingjun Li, and Sai Qu. 2020. "Vegetation Green up under the Influence of Daily Minimum Temperature and Urbanization in the Yellow River Basin, China." *Ecological Indicators* 108: 105760. doi:10.1016/j.ecolind.2019.105760

Yu, Zhiqi, Liping Di, Junmei Tang, Chen Zhang, Li Lin, Eugene Genong Yu, Md. Shahinoor Rahman, Juozas Gaigalas, and Ziheng Sun. 2018. "Land Use and Land Cover Classification for Bangladesh 2005 on Google Earth Engine." In *2018 7th International Conference on Agro-Geoinformatics*, 1–5. Hangzhou: IEEE. doi:10.1109/Agro-Geoinformatics.2018.8475976

Zaidi, Syeda Maria, Abolghasem Akbari, Azizan Abu Samah, Ngien Kong, and Jacqueline Gisen. 2017. "Landsat-5 Time Series Analysis for Land Use/Land Cover Change Detection Using NDVI and Semi-Supervised Classification Techniques." *Polish Journal of Environmental Studies* 26 (6): 2833–2840. doi:10.15244/pjoes/68878

Zhou, Decheng, Shuqing Zhao, Liangxia Zhang, Ge Sun, and Yongqiang Liu. 2015. "The Footprint of Urban Heat Island Effect in China." *Scientific Reports* 5 (1): 11160. doi:10.1038/srep11160

Zurqani, Hamdi A., Christopher J. Post, Elena A. Mikhailova, Mark A. Schlautman, and Julia L. Sharp. 2018. "Geospatial Analysis of Land Use Change in the Savannah River Basin Using Google Earth Engine." *International Journal of Applied Earth Observation and Geoinformation* 69: 175–185. doi:10.1016/j.jag.2017.12.006

15 Assessing Aerosol and Nitrogen Dioxide Concentration in Major Urban Cities over the Himalayan Region during the COVID-19 Lockdown Phases

Shyama Prasad Mandal[1], Avinash Kumar Ranjan[2],*, Bikash Ranjan Parida[1], and Sailesh Narayan Behera[3]

1 Department of Geoinformatics, School of Natural Resource Management, Central University of Jharkhand, Ranchi-835205, India.

2 Department of Mining Engineering, National Institute of Technology Rourkela, Odisha-769008, India

3 Department of Civil Engineering, Centre for Environmental Sciences & Engineering (CESE), Shiv Nadar University, Greater Noida 201314, Uttar Pradesh, India

CONTENTS

15.1 Introduction 294
15.2 Materials and Method.. 297
    15.2.1 AOD Dataset... 297
    15.2.2 NO_2 Dataset 297
    15.2.3 Gridded Population Density 298
15.3 Methods 298
15.4 Results..299
    15.4.1 Phase-Wise AOD Concentration over Great Himalayan Region 299
    15.4.2 Phase-Wise AOD Anomaly Concentrations over Great Himalayan Region 299

DOI: 10.1201/9781003265160-18

        15.4.3 Phase-Wise AOD Anomaly Concentration over 14 Cities............... 303
        15.4.4 Phase-Wise NO$_2$ Concentration over Great Himalayan
               Region.. 304
        15.4.5 Phase-Wise NO$_2$ Anomaly Concentration over Great
               Himalayan Region 305
        15.4.6 Phase-Wise NO$_2$ Anomaly Concentration over 14 Cities 306
        15.4.7 Relationship between AOD and NO$_2$ Concentrations with
               Population Density.. 307
15.5 Discussion 309
15.6 Conclusion and Recommendations..310
Acknowledgments.. 311
Conflicts of Interest... 311
References... 311

15.1 INTRODUCTION

SARS-CoV-2 (COVID-19) has resulted in high mortality rates and huge economic losses around the world (Bukhari and Jameel 2020). Since the first COVID-19 case outbreak on 31 December 2019, it has spread to more than 216 countries, and consequently, the World Health Organization (WHO) declared a health emergency (WHO 2020). The pandemic has caused approximately 5.1 million deaths worldwide as of 15 November 2021, and approximately 0.42 million deaths in India (MoHFW 2021; WHO 2020). India has a large geographical boundary with a total population of more than 1.35 billion, and the vast diversity of demography carries a momentous risk factor. Higher population density and air pollution due to the extensive urbanization with high consumption of fossil fuel in different sectors is the major concern of human health risk (Bhadra et al. 2021). The poor air quality index across the globe has caused 8.8 (7.11–10.41) million deaths per year (Lelieveld et al. 2018). According to Deep et al. (2021), the Indian transport sector is the primary contributor of NO$_x$ emission (32%), followed by the power generation sector (28%), industries (27%), and biomass burning (19%). In India during 2019, air pollution has caused about 1.67 million deaths (Pandey et al. 2021), accounting for 17.8% of the entire deaths within the country. The majority of deaths (0.98 million) were caused by ambient particulate matter (PM) and household air pollution (0.61 million).

More than 100 countries took preventive measures (e.g., social distancing, lockdowns, travel bans, and quarantines, among others) to stop or reduce the COVID-19 transmission risk across the world. The impact had led to slow economic growth and global sociopolitical relationships (Barbate et al. 2021). As a result, the use of transportation has been reduced overall by 50% (surface transportation) and 75% (air transportation) compared to the 2019 average (IEA 2020). The COVID-19-induced lockdowns compelled stopping or reducing such activities against the environment; subsequently, the terrestrial environment got a brake to heal itself in terms of air quality, land surface temperatures, and urban heat islands, among others (Ranjan et al. 2020; Lal et al. 2020; Parida et al. 2021a).

Several studies have reported a substantial drop in atmospheric pollutants, namely, CO, nitrogen dioxide (NO$_2$), aerosol optical depth (AOD), sulfur dioxide (SO$_2$), and

PM across the globe during the lockdown periods (Kanniah et al. 2020; Lal et al. 2020; Ranjan et al. 2020; Acharya et al. 2021). A noteworthy drop in NO$_2$ concentrations (up to 20–30%) was also detected in China, India, Malaysia, Europe, and the US (Shrestha et al. 2020; Tobías et al. 2020; Zhang et al. 2020). The concentration of NO$_2$ decreased by about 30% and 70% over China and India, respectively, while in Europe, NO$_2$ concentration decreased by 25% in Spain, and 30% in France and Italy during the lockdown period (Gautam 2020). In Barcelona (Spain), PM$_{10}$, NO$_2$, SO$_2$, and black carbon reduction were reported to be 45%, 51%, 3%, and 19%, respectively (Tobías et al. 2020). Similarly, in Ecuador (South America), the capital city of Quito, NO$_2$ (68%) and PM$_{2.5}$ (29%) levels significantly declined (Zalakeviciute et al. 2020). In India, Ranjan et al. (2020) have also reported a considerable decline in AOD concentration by 45% over the whole Indian Territory and by approximately 6–37% over the four metropolitan cities (i.e., National Capital Region, Mumbai, Kolkata, Bengaluru). The drop in atmospheric CO$_2$ concentration was also reported up to 18–39% over Kolkata and its nearest island Sundarban, India (Parida et al. 2020). The previously mentioned studies have shown a substantial drop in all atmospheric pollutants during the COVID-19 lockdown phases. Some more studies are illustrated in the Table 15.1, which shows the significant improvement in the air quality status across the world. Hence, it is perceived that the decline was due to the partial to complete shutdown of major pollution sources like fossil fuel burning, industrial emission, heavy transportation emission, human mobility, etc.

TABLE 15.1

Summary of Some Recent Studies on COVID-19 and Air Quality Conducted across the Globe

Study Area (City, Country)	Key Findings	Source
India (41 cities in different regions)	Reported significant reduction in NO$_2$ concentration over the study sites during the 2020 lockdown periods as compared to 2019. A remarkable reduction in NO$_2$ concentration was noted over New Delhi (~62%) followed by Delhi (60%), Bangalore (48%), Ahmedabad (46%), Nagpur (46%), Gandhinagar (46%), and Mumbai (43%).	(Vadrevu et al. 2020)
India (21cities in different regions)	The study assessed the impacts of lockdown on air pollutants concentration. During the lockdown period, air pollution is reduced up to 43% (PM$_{2.5}$), 31% (PM$_{10}$), 10% (CO), and 18% (NO$_2$) as compared to a three years' mean condition (2017–2019), while O$_3$ is increased up to 17% with trivial changes in SO$_2$. Air Quality Index (AQI) values were reduced up to 44% (North), 33% (South), 29% (East), 15% (Central), and 32% (West) over the continent.	(Sharma et al. 2020)

(Continued)

TABLE 15.1
(Continued)

Study Area (City, Country)	Key Findings	Source
India (15 Major cities in different regions)	Study analyzed the air quality (i.e., AOD, CO, $PM_{2.5}$, NO_2, O_3, and SO_2) during the pre-lockdown period (9 February 2020–23 March 2020) and during the lockdown period (24 March 2020–4 May 2020) using two different data source. The study results indicated that the concentration of $PM_{2.5}$ and NO_2 has declined up to 14% and 30%, respectively, over the study cities. On the other hand, ground-recorded data (Central Pollution Control Board/State Pollution Control Boards) showed more than 40% and 47% declination in $PM_{2.5}$, and PM_{10} concentrations, respectively. In Chennai and Nagpur, SO_2 concentration was decreased by approximately 85%, while O_3 concentration over five cities was increased by about 17% during the 43 days of the lockdown period.	(Rahaman et al. 2021)
India (Delhi, Mumbai, Chennai, Kolkata, and Bangalore)	Evaluated the influence of lockdowns on air pollution over the five megacities of India (i.e., Delhi, Mumbai, Chennai, Kolkata, and Bangalore). Study reported a significant reduction in air pollutants ($PM_{2.5}$ by 41%, PM_{10} by 52%, NO_2 by 51%, and CO by 28%) in Delhi. Similarly, other megacities have also shown declination in air pollutant concentration during the lockdown period.	(Jain and Sharma 2020)
South Asia	Explored the co-benefits of the COVID-19 lockdowns (January–October 2020) on the air pollution and water pollution as compared to the last five years' (2015–2019) mean conditions. Study results showed a remarkable reduction in air and water pollution (~30–40% drop in NO_2, 45% in AOD, and 50% decline in coastal Chl-a concentration.	(Shafeeque et al. 2021)
India	Study investigated the NO_2, CO, and AOD concentrations during the lockdown. Approximately 17% decline in mean NO_2 concentration. With respect to the last 5 years' mean, the AOD level was reduced approximately 25% over the country, specifically over the Indo-Gangetic plains region.	(Pathakoti et al. 2020)
New York, Los Angeles, Zaragoza, Rome, Dubai, Mumbai, Delhi, Beijing, and Shanghai	Assessed the $PM_{2.5}$ concentration during the COVID-19 outbreak period and reported an approximate 11% (Dubai), 35% (Delhi), 14% (Mumbai), 50% (Beijing), 50% (Shanghai), 32% (New York) and 4% (Los Angeles) reduction in $PM_{2.5}$ during March 2020 compared to last 3 years' condition (2017–2019).	(Chauhan and Singh 2020)
China, Spain, France, Italy, USA	Study reported a significant reduction in NO_2 concentration (~ 20–30%) over Wuhan (China), Spain, France, Italy, and the US.	(Muhammad et al. 2020)

It is well-known from previous studies that aerosol and NO$_2$ have a significant role in climate change events on a regional to global scale (Qian and Giorgi 1999; Suddick et al. 2013; Bar et al. 2021). The AOD and NO$_2$ are directly or indirectly responsible for unstable the atmosphere by absorbing solar radiation (Superczynski and Christopher 2011; Ranjan et al. 2021). Thus, monitoring air pollutants (AOD, NO$_x$) is significant for monitoring and understanding the terrestrial environment and climate interactions. So that environmentalists, scientists, government bodies, and policymakers can formulate health and sustainable development-related schemes.

So far, rarely any study has been conducted over the Himalayan region for understanding the effect of lockdown on atmospheric pollution. The Himalayan region is typically known for its rich and fragile vegetation ecosystem. Most of the region is predominantly covered with vegetation. Although this region is less polluted (air) owing to its geography as compared to Central India, the Indo-Gangetic plain, and other locations (Unnithan and Gnanappazham 2020; Maheshwarkar and Sunder Raman 2021), we believe that there shall be some deviations in the air quality parameters due to the COVID lockdowns. Thus, the present investigation focuses on analyzing the AOD and NO$_2$ concentration during lockdown periods over selected cities in the Himalayan region, which can help to understand the influence of anthropogenic activity on the tropospheric concentrations. The main objectives of the study are (1) to examine the effect of lockdown from March to May 2020 due to COVID-19 on AOD and NO$_2$ levels over the Greater Himalayan Region with special emphasis on major urban cities and (2) to examine the correlation between AOD and NO$_2$ with a population density of the specific cities.

15.2 MATERIALS AND METHOD

15.2.1 AOD Dataset

The Moderate Resolution Imaging Spectroradiometer (MODIS)–based aerosol product MCD19A2 was acquired for the study period (25 March–31 May 2017–2020) shown in Table 15.2. The MCD19A2 provides Multi-Angle Implementation of Atmospheric Correction (MAIAC) land AOD-gridded product (level 2) with a spatial and temporal resolution of 1 km and daily. MCD19A2 is available in two wavelength bands, i.e., 470 nm (blue) and 550 nm (green). The AOD at 550 nm was employed in this study due to its better consistency (Lyapustin et al. 2018). The precision of the MODIS-derived AOD in comparison to a set of ground-based sun photometer (SP), and the Aerosol Robotic Network (AERONET) was estimated to have an expected error ($\pm 0.05 + 15\%$) over the land (Remer et al. 2005; Remer et al. 2008; Levy et al. 2010).

15.2.2 NO$_2$ Dataset

Sentinel–5P-based NO$_2$ dataset was used from March to May into a phase-wise national lockdown implemented by the Indian government over the past 2 years (25 March–31 May, 2019–2020) shown in Table 15.2. The Copernicus Sentinel–5 satellite carries onboard the Tropospheric Monitoring Instrument (TROPOMI) that was

TABLE 15.2
Satellite/Gridded Dataset Used in the Present Investigation

Datasets	Time Period	Band Used	Spatial/Temporal Resolution	Citation
MCD19A2 (AOD)	25 March to 31 May (2017–2020)	550 nm (green band)	1 km, daily	(Lyapustin et al. 2018)
Sentinel–5P/ TROPOMI (NO_2)	26 March to 31 May (2019–2020)	NO_2 (column number density)	3.5 × 7 km, daily	(Veefkind et al. 2012)
Gridded Population density (GPWv4.11)	2020	Population Density	30 arc-second	(CIESIN 2017)

launched in October 2017, which acquired daily observation at a spatial resolution 7 × 3.5 km (along-track × across-track) in a near-polar sun-synchronous orbit. The NO_2 column density data are available in mol/m² unit, and for conversion, the unit in μmol/m² is used to multiply by 1.0E-6 (1000000). The TROPOMI has separate four spectrometers that cover a spectral range in the ultraviolet and visible (0.27–0.5 μm), near-infrared (0.675–0.775 μm), and short-wave infrared (2.305–2.385 μm) use to monitor top atmospheric pollutants gases, namely, ozone (O_3), methane (CH_4), formaldehyde (HCHO), CO, NO_2, and sulfur dioxide (SO_2; Griffin et al. 2019; Veefkind et al. 2012).

15.2.3 GRIDDED POPULATION DENSITY

The global gridded population data (GPWv4) was used in this study for 2020 to link with the AOD and NO_2 concentrations. The GPW data are developed by NASA's Socioeconomic Data and Application Center (SEDAC) and hosted by the CIESIN (Center for International Earth Science Information Network at Columbia University). GPWv4 provides gridded population density data for 2000–2020 (at 5-year intervals) based on population registers in the number of persons per square kilometer in a spatial resolution of 30 arc-second (~1 km at the equator; CIESIN 2017).

15.3 METHODS

The present study aims to analyze the impact of partial to complete lockdown events (during 25 March–31 May 2020) on the atmospheric condition (AOD and NO_2) over the Great Himalayan Region as compared to the last 3 years' (2017–2019) mean condition, wherein NO_2 was compared with the previous year's normal condition (2019). The spatiotemporal pattern of mean AOD and NO_2 is characterized in four periods, such as phase 1 (25 March–14 April 2020), phase 2 (15 April–3 May 2020), phase 3 (4 May–17 May 2020), and phase 4 (18 May–31 May 2020). Simultaneously, the AOD (mean over 2017–2019) and NO_2 (2019) for the same periods are computed to comprehend the variance in both concentration-level patterns during the lockdown periods. Equation 15.1 is used to generate the phase-wise AOD and NO_2 anomaly maps.

Furthermore, attention was paid to investigating the variation of both pollutants' concentrations within the 3-km buffer (from midpoint) of 14 cities, namely, Jammu, Shimla, Dehradun, Kathmandu, Gangtok, Darjeeling, Thimpu, Itanagar, Kohima, Imphal, Aizawl, Agartala, Shillong, and Dishpur, over that region. The details of the study region, including 14 cities, land use/land cover pattern, and the population density of that area are presented in Figure 15.1. Furthermore, a relationship was built to understand the influence of both concentrations (AOD and NO$_2$) on highly populous cities over the study region.

$$\text{Anomaly}\left(\%\right) = \frac{\left(X\right) - \overline{(x)}}{\overline{(x)}} \times 100, \tag{15.1}$$

where x is the phase-wise mean for AOD and NO$_2$ during 2020; $\bar{x}$ is the phase-wise mean over 2017–2019) for AOD while the mean over 2019 for NO$_2$.

15.4   RESULTS

15.4.1   PHASE-WISE AOD CONCENTRATION OVER GREAT HIMALAYAN REGION

The lockdown phase-wise AOD distribution map of 2020 with respect to the mean of 2017–2019 over the Great Himalayan Region is presented in Figure 15.2. A significant reduction of AOD concentration was noticed across all states over the Great Himalayan region during the lockdown period (Figure 15.2a′–b′). In some areas of West Bengal, Assam, Tripura, and Mizoram states showed higher AOD concentrations than other states during the first lockdown period (phase 1, 25 March–14 April 2020; Figure 15.2a′). During the second and third lockdown periods (phase 2, 5 April–3 May; phase 3, 4 May–17 May), a substantial reduction in AOD in all states was noted. While some parts of West Bengal and Assam still showed slightly higher AOD than the other states (Figure 15.2b′–c′). During Phase 4 (2020), a noteworthy reduction in AOD concentration was observed as compared to the mean AOD (2017–2019) in all states during the same period (Figure 15.2d′). Higher AOD levels in some places during Phases 3 and 4 of the lockdown periods (2020) may be accredited to regional emission sources and higher population density. Nevertheless, it can be concluded that lockdown has driven the enormous decline in AOD value across the study area. However, countries/states like Jammu and Kashmir, Ladakh, Himachal Pradesh, Bhutan, Arunachal Pradesh, and a major part of Uttarakhand and Sikkim showed lower AOD.

15.4.2   PHASE-WISE AOD ANOMALY CONCENTRATIONS OVER GREAT HIMALAYAN REGION

Phase-wise AOD anomalies were computed as shown in Figure 15.3. The AOD anomaly map has shown a sharp decline in the AOD concentration during the lockdown periods compared to the mean AOD. During Phase 1, a significant

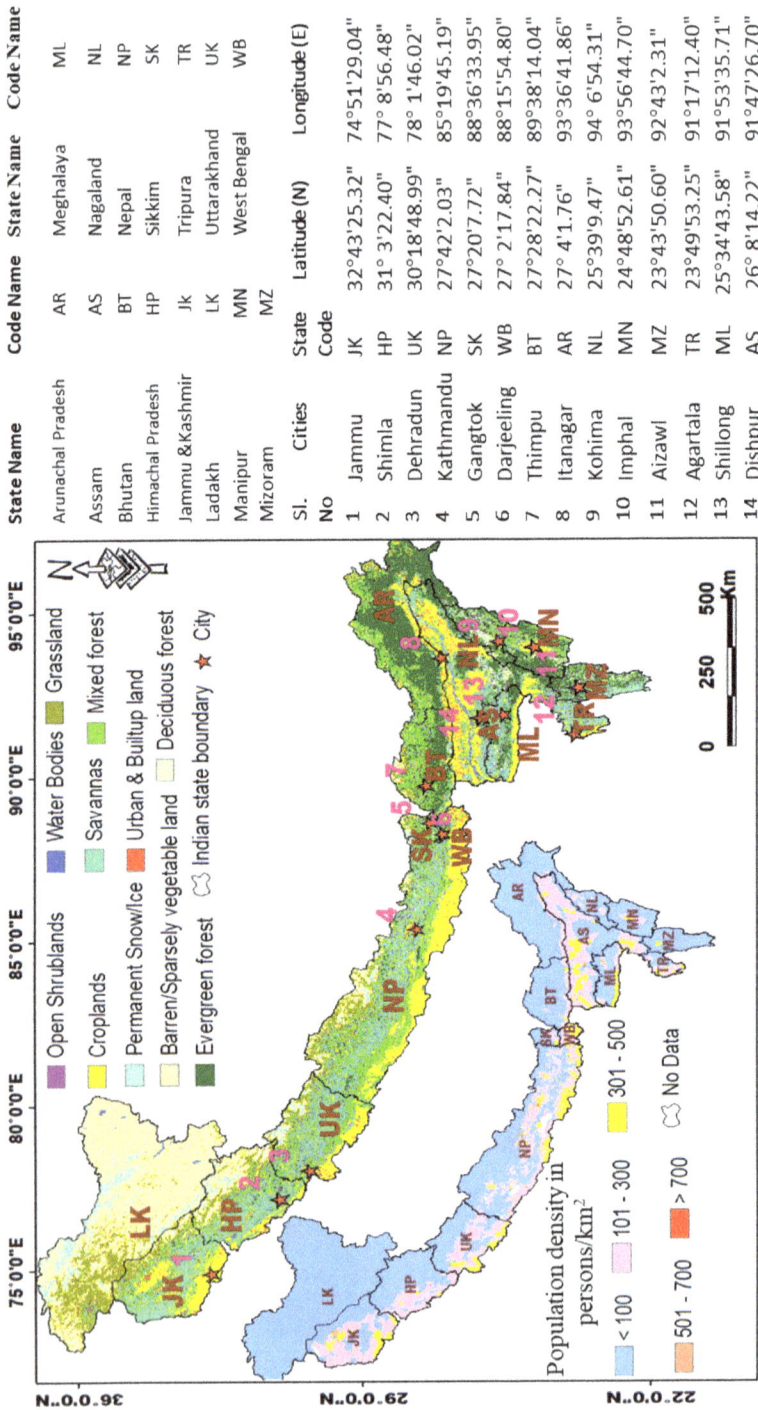

State Name	Code Name
Arunachal Pradesh	AR
Assam	AS
Bhutan	BT
Himachal Pradesh	HP
Jammu &kashmir	Jk
Ladakh	LK
Manipur	MN
Mizoram	MZ
Meghalaya	ML
Nagaland	NL
Nepal	NP
Sikkim	SK
Tripura	TR
Uttarakhand	UK
West Bengal	WB

Sl. No	Cities	State Code	Latitude (N)	Longitude (E)
1	Jammu	JK	32°43'25.32"	74°51'29.04"
2	Shimla	HP	31°3'22.40"	77°8'56.48"
3	Dehradun	UK	30°18'48.99"	78°1'46.02"
4	Kathmandu	NP	27°42'2.03"	85°19'45.19"
5	Gangtok	SK	27°20'7.72"	88°36'33.95"
6	Darjeeling	WB	27°2'17.84"	88°15'54.80"
7	Thimpu	BT	27°28'22.27"	89°38'14.04"
8	Itanagar	AR	27°4'1.76"	93°36'41.86"
9	Kohima	NL	25°39'9.47"	94°6'54.31"
10	Imphal	MN	24°48'52.61"	93°56'44.70"
11	Aizawl	MZ	23°43'50.60"	92°43'2.31"
12	Agartala	TR	23°49'53.25"	91°17'12.40"
13	Shillong	ML	25°34'43.58"	91°53'35.71"
14	Dishpur	AS	26°8'14.22"	91°47'26.70"

FIGURE 15.1 The map is showing the Great Himalayan Region with state administrative boundaries. The upper map shows land cover (LULU) of the Great Himalayan Region as per the International Geosphere–Biosphere Programme classification scheme of MCD12Q1 data. The lower map represents the population density. The 14 cities in that region are also represented by the red star mark.

Note: Graticule is given for LULC map.

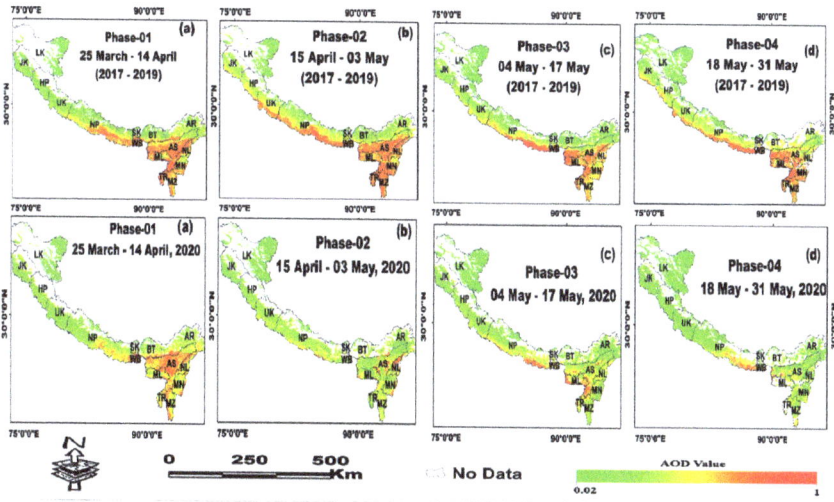

FIGURE 15.2   Phase-wise AOD-level map of the Great Himalayan region: (a–d) show the mean of AOD during 2017–2019, and (a′–d′) show the mean AOD for the same periods in 2020 for the lockdown period.

reduction in AOD was observed in most of the states (except a few portions of West Bengal and Assam that reported higher AOD values; Figure 15.3a). During phase 2, almost all states were observed with negative AOD anomaly (Figure 15.3b). However, in phase 3, slightly higher positive anomaly patches were observed over West Bengal, Nepal, and Assam (Figure 15.3c). Furthermore, during phase 4, all states showed negative anomalies. It could be because of the strict lockdown impact, as the majority of the unfavorable anthropogenic-based activities were stopped or reduced. Table 15.3 shows the phase-wise AOD anomaly of different states. West Bengal (11%) and Nagaland (10.5%) recorded a maximum positive AOD anomaly during phase 2 and phase 4, respectively. The maximum negative AOD anomaly during phase 1 was observed over Ladakh (57%), and the rest of all states showed a negative AOD anomaly by 1–56%. Similar to the current study, few other studies have already reported the consequential decline in air pollutants concentrations over the different parts of the globe. Ranjan et al. (2020) has reported a substantial drop in AOD level (up to ~45%) during the lockdown period as compared to the long-term mean AOD (2000–2019) across the Indian territory. Also, another study by Rahaman et al. (2021) reported a steady decreasing pattern of AOD moving from South Indian to North Indian cities. An increasing value was observed in Bengaluru (+13.58%), Chennai (+7.02%), Hyderabad (+3.54%). In contrast, a decreasing trend was observed in Nagpur (−11.56%), Mumbai (−9.96%), Delhi (−15%), and Ahmadabad (−15%). The highest decrease in AOD concentration was observed in Lucknow (−22.5%), Patna (−21.95%), and Siliguri (−19.05%).

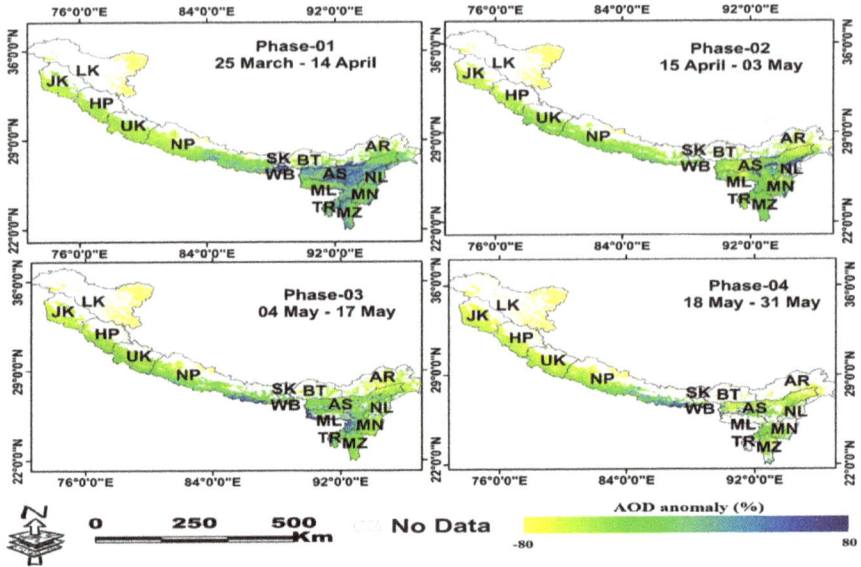

FIGURE 15.3 Phase-wise AOD anomaly maps of the Great Himalayan Region estimated using phase-wise AOD during the lockdown period (2020) against the mean AOD for the same period (2017–2019).

TABLE 15.3

AOD Anomaly Percentage (mean) during the Lockdown Phases by States

States	AOD Anomaly (%)			
	Phase 1	Phase 2	Phase 3	Phase 4
Arunachal Pradesh	−7.8	−13.8	−37.5	−1.17
Assam	−35.0	−49.4	−44.2	−21.7
Bhutan	−16.7	−20.8	−32.7	−2.8
Himachal Pradesh	−18.4	−4.5	−34.1	−18.5
Jammu & Kashmir	−41.3	−37.2	−56.8	−28.3
Ladakh	−57.1	−41.9	−56.0	−37.8
Manipur	−41.0	−32.5	−41.8	−2.4
Mizoram	−45.3	−40.6	−39.4	−10.7
Meghalaya	−41.6	−46.1	−26.2	−1.4
Nagaland	−22.9	10.3	−25.4	5.4
Nepal	−44.5	−34.6	−40.5	−7.3
Sikkim	−9.0	−13.0	−13.2	−34.8
Tripura	−46.5	−51.9	−41.1	−14.6
Uttarakhand	−49.6	−20.4	−46.3	−38.3
West Bengal	−15.8	−45.8	−40.2	11.5

15.4.3　Phase-Wise AOD Anomaly Concentration over 14 Cities

Phase-wise AOD anomaly (mean) with the 3-km buffer zone (from midpoint) of 14 cities (i.e., Jammu, Shimla, Dehradun, Kathmandu, Gangtok, Darjeeling, Thimpu, Itanagar, Kohima, Imphal, Aizawl, Agartala, Shillong, and Dishpur) are shown in Table 15.4. The majority of the cities have shown a negative AOD anomaly within the 3-km buffer (from midpoint) zone area. A negative anomaly was observed from 0.04–36% over most of the cities during phase 1. However, Gangtok (28.6%), Itanagar (37.1%), and Kohima (7.5%) showed slightly higher positive anomalies during phase 1. During phase 2, Shimla (11.9%), Thimpu (40.8%), Itanagar (79.3%), Imphal (8.7%), and Agartala (1.4%) were observed with a higher positive AOD anomaly. All cities have shown negative anomaly during phase 2 that ranges up to 45%. During phase 3, most of the cities have shown negative AOD anomalies that range from 1–77%, but Gangtok has shown a positive anomaly of 10.4%. During phase 4, Kathmandu (28.6%) and Kohima (8.4%) cities also showed positive anomalies, and the rest of the cities showed negative anomalies (2–59%). Pixel values within the 3-km buffer zone of Gangtok, Darjeeling, Thimpu, and Agartala were missing; therefore, no trend was depicted for these regions.

Moreover, a few cities (i.e., Gangtok, Darjeeling, Itanagar, Kathmandu, Kohima, and Dishpur) also showed higher AOD anomalies during different lockdown phases. It is very unlikely during these strict lockdown phases, wherein most of the aerosol sources and anthropogenic activities (i.e., traffic, industrial operations, etc.) were almost closed. The discrepancy in the number of AOD pixels within the boundary

TABLE 15.4

Phase-Wise AOD Anomaly over 14 Cities within a 3-km Buffer Zone

Cities	3-km Buffer AOD Anomaly (%)			
	Phase 1	Phase 2	Phase 3	Phase 4
Jammu	−34.7	−17.7	−30.9	−34.1
Shimla	−2.8	12.0	−8.3	−26.1
Dehradun	−38.5	−0.2	−37.1	−41.3
Kathmandu	−32.7	−18.3	−77.8	28.6
Gangtok	28.6	No data	10.4	No data
Darjeeling	−20.2	No data	−10.1	No data
Thimpu	−7.2	40.9	−25.6	No data
Itanagar	37.1	79.3	−47.7	−9.5
Kohima	7.6	−45.1	−14.6	8.5
Imphal	−32.5	8.7	−54.2	−41.9
Aizawl	−34.5	−17.1	−38.2	−2.2
Agartala	−35.1	1.5	−1.1	No data
Shillong	−29.6	−43.4	−29.8	−15.2
Dishpur	−0.5	−13.9	−35.5	−59.2

Note: "No data" indicates no pixel within the 3-km buffer zone.

can be a possible reason for the increased AOD value during these phases. Another possible cause could be the transportation of aerosols from the nearby boundary states like the Tibetan Plateau or others (Moore and Semple 2021; Zhang et al. 2017; Saikawa et al. 2019).

15.4.4 Phase-Wise NO_2 Concentration over Great Himalayan Region

The phase-wise NO_2 distribution map of the Great Himalayan Region is presented in (Figure 15.4 a–d) for 2019. Similar to AOD, a drastic reduction was observed in mean NO_2 concentration level across all of the states of the Great Himalayan Region during the lockdown periods (Figure 15.4 a′–d′). During phase 1 (2019), a relatively higher NO_2 concentration (mean) was observed over Tripura, Mizoram, Manipur, and Assam. In phase 2 (2019), Meghalaya, Assam (74.89 µmol/m²), Tripura, and West Bengal were observed with maximum (mean) NO_2 concentration. In phase 3, West Bengal, Nepal, Uttarakhand, and Meghalaya, whereas in phase 4, Assam, Meghalaya, Tripura, and Uttarakhand were noted to have higher NO_2 concentrations. But in 2020, comparatively lower mean NO_2 concentration was recorded over the states as mentioned above (except West Bengal in phase 2) (**Figure 15.4 a′-d′**). In phase 1, Tripura, Mizoram, Manipur, and Assam; in phase 2, Meghalaya, Assam, Tripura, and West Bengal; in phase 3, West Bengal, Nepal, Uttarakhand, and Meghalaya, and in phase 4, Assam, Meghalaya, Tripura, and Uttarakhand were found to have a lower NO_2 concentration in 2020 lockdown periods compared to 2019.

The finding of the current study shows the same trend as reported by the other studies over different study regions. Sharma et al. (2020) reported a reduction of NO_2 18%, and AQI reduction was up to 44% (North), 33% (South), 29% (East), 15%

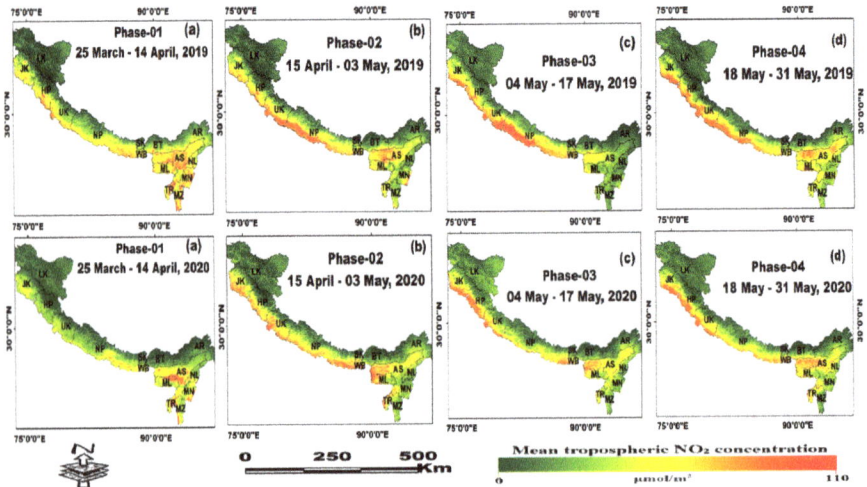

FIGURE 15.4   Phase-wise NO_2-level map of the Great Himalayan region. (a–d) shows the mean AOD during 2019, and (a′–d′) shows the mean AOD for the same periods in 2020.

(Central), and 32% (West) over the 22 Indian cities. An another study by Rahaman et al. (2021) and Vadrevu et al. (2020) reported a sharp decrease in NO$_2$ concentration over Delhi (−37.40%), Ghaziabad (−35.32%), Mumbai (−32.96%), Chennai (−30.21%) Nagpur (−23%), Kolkata (−26.29%), Asansol (−25.06%), Jaipur (−10%), Patna (−9.99%) and Hyderabad (−8.99%). Some other studies a publicized reduction in NO$_2$ anomaly over the Indian states namely, Arunachal Pradesh (−10.2%), Himachal Pradesh (−37.6%), Manipur (−9.7%), Mizoram (−33.3%), Nagaland (−16.5%), Tripura (−9.7%), Uttarakhand (−22%), and West Bengal (−15.4%) (Biswal et al. 2020; Targino et al. 2013; Sidhu et al. 2017; Chowdhury and Karim 2018; Val Martin et al. 2008).

15.4.5 PHASE-WISE NO$_2$ ANOMALY CONCENTRATION OVER GREAT HIMALAYAN REGION

Phase-wise NO$_2$ anomalies are presented in Figure 15.5, which helps to understand the spatial variation in NO$_2$ level during the lockdown period (2020) in the Great Himalayan region compared to 2019. The NO$_2$ anomaly map shows a sharp decline in NO$_2$ concentration (mean) during the lockdown period. During phase 1, a significant reduction was observed in all states as compared to 2019, wherein the highest reduction was observed over Bhutan (19.6%) and the lowest over Tripura (6.4%; Figure 15.5a). During phase 2, almost all states showed a negative anomaly, while West Bengal (3.6 %) and Sikkim (1.7%) showed slightly positive anomalies (Figure 15.5b). Notably,

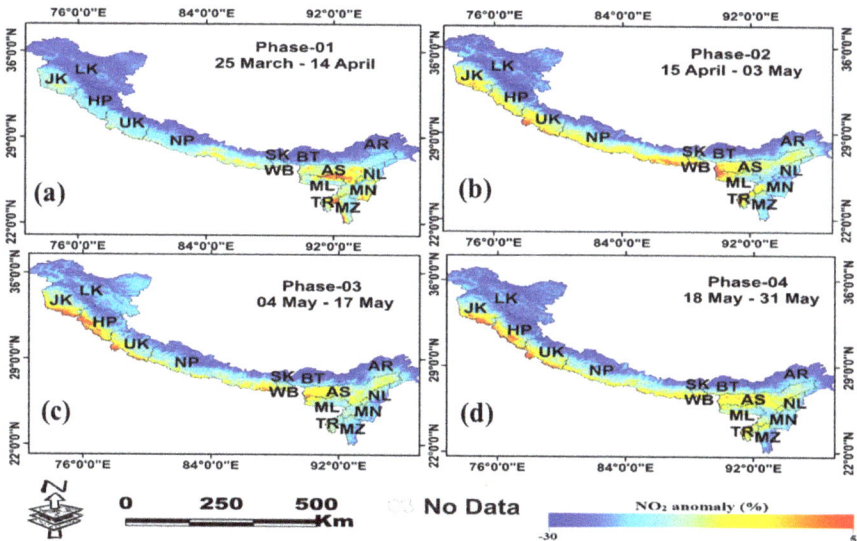

FIGURE 15.5 Phase-wise NO$_2$ anomaly map of the Great Himalayan Region estimated using phase-wise NO$_2$ during the lockdown period (2020) as compared to the mean NO$_2$ for the same period in 2019.

during phases 3 and 4, most of the states were observed with sharp negative anomalies and range between (2–20.2%; Figure 15.5c–d). The maximum and minimum negative anomalies (means) were observed over Nepal (20.2%) and Arunachal Pradesh (2%) during phase 3 (Table 15.5). In contrast, the maximum and minimum positive anomaly was observed over West Bengal (3.6%) and Sikkim (1.7%) during phase 2. The rest of all cities, including West Bengal and Sikkim, also showed negative anomalies during all lockdown phases, ranging from 2–20%.

15.4.6 PHASE-WISE NO$_2$ ANOMALY CONCENTRATION OVER 14 CITIES

Phase-wise NO$_2$ anomaly (mean) within the 3-km buffer zone (from the midpoint), shown in Table 15.6, showed negative anomalies over most cities during phase 1. The maximum and minimum negative anomaly were observed over Shimla and Itanagar (26.4%) and Shillong (0.3%), respectively. Unexpectedly, Gangtok (2.7%) and Darjeeling (4.1%) were found with a positive anomaly during phase 2, which is quite surprising because the major sources of NO$_2$ pollutants over the region were closed due to lockdown. The higher NO$_2$ mean anomaly in this specific period could be attributed to inconsistency in the total number of NO$_2$ pixels within the specific buffer zone. Or possibly pollutants might be carried from the nearby boundary states (Moore and Semple 2021; Zhang et al. 2017). Besides, during phases 3 and 4, all cities have shown negative anomalies.

TABLE 15.5

Phase-Wise NO$_2$ Anomaly Percentage (mean) during Lockdown Periods

States	NO$_2$ Anomaly (%)			
	Phase 1	Phase 2	Phase 3	Phase 4
Arunachal Pradesh	−18.5	−2.4	−2.0	−16.2
Assam	−11.7	−8.5	−4.4	−15.1
Bhutan	−19.6	−5.4	−10.7	−11.7
Himachal Pradesh	−15.3	−5.0	−5.1	−9.2
Jammu & Kashmir	−9.2	−2.5	−8.0	−10.5
Ladakh	−8.2	−3.6	−6.6	−12.0
Manipur	−13.2	−11.3	−8.3	−15.9
Mizoram	−14.5	−9.1	−7.6	−18.2
Meghalaya	−7.8	−7.7	−7.7	−15.8
Nagaland	−17.0	−8.1	−4.5	−14.6
Nepal	−10.8	−8.5	−20.2	−15.4
Sikkim	−15.3	1.7	−10.1	−12.4
Tripura	−6.4	−3.1	−4.7	−8.7
Uttarakhand	−12.2	−5.5	−14.1	−14.6
West Bengal	−10.9	3.6	−15.0	−4.7

TABLE 15.6

Phase-Wise NO$_2$ Anomaly Concentration over 14 Cities in 3-km Buffer Zone

Cities	3-km Buffer NO$_2$ Anomaly (%)			
	Phase 1	Phase 2	Phase 3	Phase 4
Jammu	−21.6	−8.2	−4.2	−15.7
Shimla	−26.4	−6.7	−10.4	−11.1
Dehradun	−25.2	−6.4	−17.2	−14.5
Kathmandu	−22.3	−1.1	−17.1	−15.9
Gangtok	−17.6	2.7	−14.5	−9.5
Darjeeling	−13.3	4.1	−11.8	−9.2
Thimpu	−21.7	−3.0	−1.0	−12.1
Itanagar	−26.4	−5.2	−2.3	−17.9
Kohima	−0.7	−10.1	−7.9	−3.3
Imphal	−19.7	−9.4	−5.1	−17.9
Aizawl	−13.0	−10.0	−10.1	−18.1
Agartala	−11.2	−8.5	−9.5	−0.5
Shillong	−0.3	−25.3	−10.7	−18.7
Dishpur	−5.9	−1.4	−11.3	−11.3

15.4.7 RELATIONSHIP BETWEEN AOD AND NO$_2$ CONCENTRATIONS WITH POPULATION DENSITY

Ten cities (i.e., Jammu, Shimla, Dehradun, Gangtok, Darjeeling, Thimpu, Imphal, Agartala, Shillong, and Dishpur) were selected to analyze the impact of population density on NO$_2$ and AOD concentrations. The highest and lowest population densities were observed over Dishpur (4200 person/km^2), and Shimla (190 person/km^2), whereas the rest of the cities ranged from 200–3500 person/km^2. Visually, variation in air pollutants (AOD and NO$_2$) can be perceived along with population density. The correlation coefficient I between AOD concentration and population density was found moderate positive relationship during the lockdown phase, and the highest r value during phase 1 was observed as 0.63 ($p < 0.10$; n = 10; Figure 15.6a–b). But in the case of NO$_2$ concentration, the correlation coefficient with population density was found slightly higher than the AOD and during phase 1 (r = 0.68, $p < 0.10$; Figure 15.6c–d). It can be interpreted that a larger population in a region leads to a major role in decreasing NO$_2$ and AOD concentration during the lockdown period over the study area. Similar to the present investigation, few studies (Lamsal et al. 2013; Wang 2017) have also accounted good correlation between AOD and NO$_2$ concentrations with population density. They found that higher population density led to higher concentration and vice versa. However, the spatial heterogeneity, that is, emission pattern, local climate, and different topographic conditions, led to substantial urban-to-rural variations in AOD and NO$_2$ concentration values (Cyrys et al. 2012; Jerrett et al. 2009; Novotny et al. 2011).

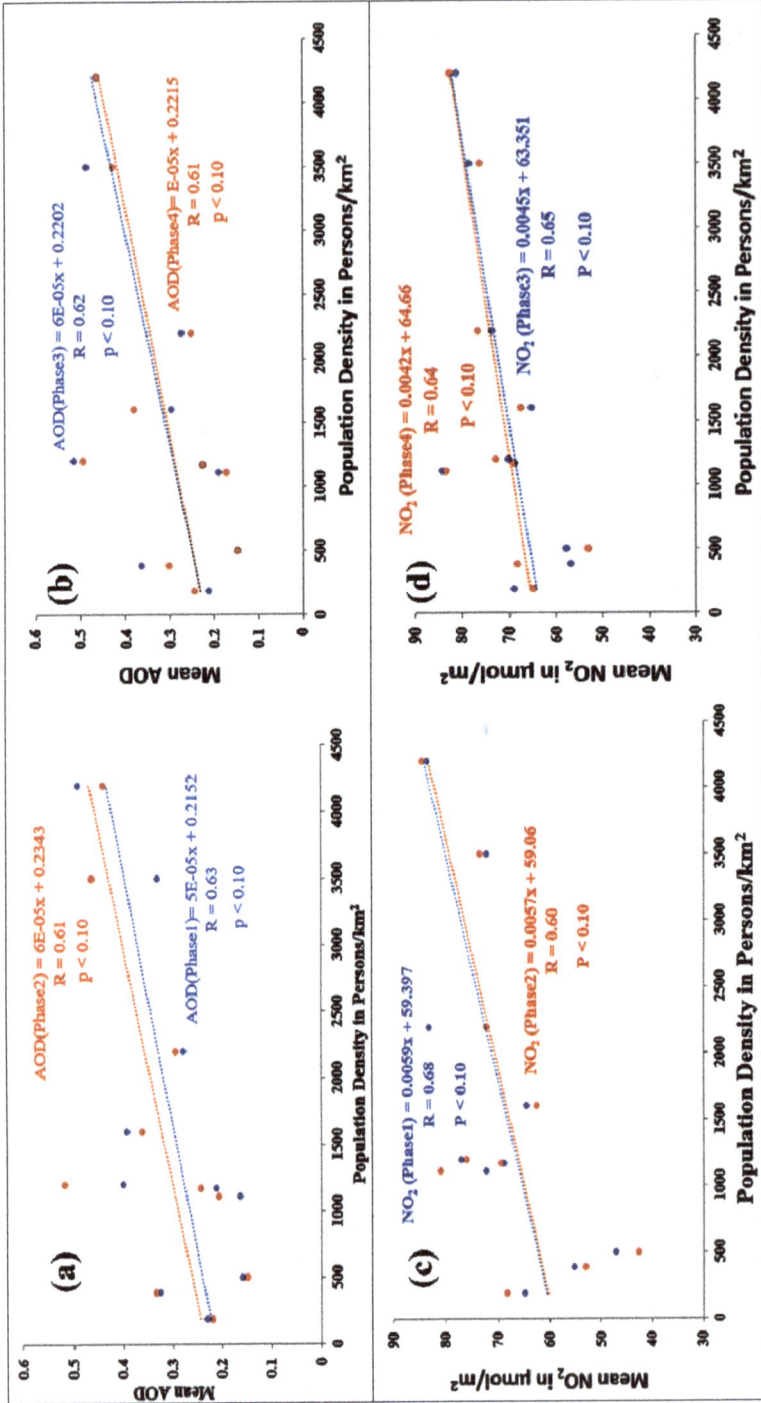

FIGURE 15.6 The relationship between population density with AOD (a–b) and NO$_2$ (c–d) as represented by correlation coefficient (R).

15.5 DISCUSSION

The study has evaluated the changes in atmospheric NO$_2$ and aerosol (AOD) concentration during the lockdown phases (2020) compared to the same period in 2017–2019 (for AOD) and in 2019 (for NO$_2$) over the Greater Himalayan Region. The satellite-based analysis demonstrated that the AOD concentration was primarily reduced by 30–60% over different states (Himalayan region) during lockdown phases. At the city scale, the AOD anomalies were reduced by 1.8–78.4% across 14 selected cities during most lockdown phases. The reduction of the air pollutants and AOD observed in these states are comparable with the results from similar studies that have reported a reduction in AOD by 40–60% during COVID-19 lockdown periods (Chauhan and Singh 2020; Kanniah et al. 2020; Ranjan et al. 2020). The NO$_2$ concentration was also reduced by 10–20% during most of the lockdown phases in most of the states. Phase-wise NO$_2$ anomalies within a 3-km buffer of 14 cities over the study area were observed with a negative mean anomaly by 1–26%. Our finding is consistent with several studies, wherein they have also accounted for a significant reduction in NO$_2$ emission over the Indian states/cities during the lockdown periods (Acharya et al. 2021; Biswal et al. 2020). Vadrevu et al. (2020) analyzed the air pollution across 41 cities over the Indian subcontinent and found a reduction in tropospheric NO$_2$ concentration by 20% (Aizawl), 14% (Itanagar, Gangtok, Dishpur), and 3% (Imphal, Agartala, Shimla, Shillong).

The inconsistency in the total number of pixels in a particular zone (missing data and pixel values) was perceived as one of the drawbacks of the current study, which might have somewhat influenced the mean values. It was very surprising that during the strict lockdown period when all major sources of AOD were completely or partially closed, but a few states showed a slightly positive anomaly. One of the possible reasons behind increased AOD could be discrepancies in the number of AOD pixels and within the state's administrative boundary during the corresponding phases. Although having some minor inconsistencies in the outcome, we believe that the main findings of the current study will remain valid. Similarly, in West Bengal and Sikkim (during phase 2), a higher NO$_2$ concentration (mean) was observed in 2020 compared to 2019, when all the major sources of air pollutants were expected to be closed. However, it was not clear why only this phase has shown higher NO$_2$ concentration. It could be due to inconsistent pixel values between two specific periods, as discussed for AOD also. It is also possible that pollutants could have been transported from nearby boundary states (Moore and Semple 2021; Zhang et al. 2017; Saikawa et al. 2019).

Furthermore, the study revealed that most populated and anthropogenically modified regions (e.g., Dehradun, Shimla, Kathmandu, and Dishpur, among others) are associated with a huge decline in AOD and NO$_2$ concentrations during lockdown phases. The partial to total restriction of transportation and industrial mode of production with a pattern of decrease in west and increase in east aided to enhance the air pollutants. As a result, the Himalayan peaks were also visible from India's National Capital Region for the first time in recent history (Moore and Semple 2021). Apart from this, meteorological parameters and geographical

conditions (LULC) might have also contributed to reducing air pollution (Sarkar and Kafatos 2004), which needs to be evaluated in future studies. For instance, Parida et al. (2021b) have found an increasing trend in relative humidity by 15–25% over India during the pre-monsoon season (March–May), which could have enabled the aerosol formation in the atmosphere. The wind speed and direction at 1000 hPa pressure level (mostly originating from the Arabian Sea) could play an important role in transporting the air pollutant concentration from one place to another place. In this line, Ranjan et al. (2020) have also found a higher positive AOD anomaly over Central India, which could be the reason for coming to the dust particles from the Thar desert toward the interior landmass. Precipitation and air pollutants also have a direct relationship; the higher precipitation over the Himalayan region might be responsible for reducing the air pollutant concentration (Parida et al. 2021b).

15.6 CONCLUSION AND RECOMMENDATIONS

The devastating COVID-19 epidemic has unwittingly caused a critical issue for human health and economic damage across the globe. On the other hand, lockdown-triggered events demonstrated that nature can cure itself if given a chance by humans. The current investigation revealed that AOD and NO_2 levels across the Great Himalayan Region were significantly decreased due to COVID-19 lockdown events. Based on the present investigation, a few conclusions can be drawn:

- Among the various study locations, Ladakh state had the highest positive impact of lockdown, whereas the AOD level reduced up to about 57% during phase 1 (25 March–14 April).
- NO_2 concentrations were also significantly decreased (up to ~20%) over the study regions during lockdown phases (2020) compared to the same period of 2019.
- Most cities within the 3-km buffer zone (from midpoint) were found to have a negative AOD anomaly up to 54.2% (except Shimla, Gangtok, Thimpu, Itanagar, Imphal, and Agartala). The NO_2 anomalies in all cities were also found with negative anomalies up to 26.4% (except Gangtok and Darjeeling).
- A good correlation was observed between air pollutant (AOD, NO_2) and population density during the lockdown phase (r = 0.63, p < 0.10; r = 0.68, p < 0.10, respectively).
- The anomaly described in this paper has the drawback of being unable to extract some pixel values or missing datasets (specifically in AOD) for a few locations. As a result, when all pixel values or datasets are taken into account, the results may vary slightly. Nonetheless, we believe that the investigation's major findings will stand.
- It can be deduced that in near future, policymakers may implement a strategically tailored lockdown at air pollution hot-spots areas to control the air pollution to meet SDG–11, which targets resilient metropolitan areas.

ACKNOWLEDGMENTS

Authors sincerely thanks to European Union Copernicus Services, NASA LPDAAC, and SEDAC for providing access to satellite-based NO$_2$ data (Sentinel–5P TROPOMI), MODIS AOD, and population density datasets. Google Earth Engine cloud platform is sincerely acknowledged for providing the data processing and analyzing platform.

CONFLICTS OF INTEREST

The authors declare no conflicts of interest.

REFERENCES

Acharya, Prasenjit, Gunadhar Barik, Bijoy Krishna Gayen, Somnath Bar, Arabinda Maiti, Ashis Sarkar, Surajit Ghosh, Sikhendra Kisor De, and S. Sreekesh. 2021. "Revisiting the Levels of Aerosol Optical Depth in South-Southeast Asia, Europe and USA amid the COVID-19 Pandemic Using Satellite Observations." *Environmental Research* 193: 110514. https://doi.org/10.1016/j.envres.2020.110514

Bar, Somnath, B.R. Parida, S.P. Mandal, A.C. Pandey, Navneet Kumar, and Bibhudatta Mishra. 2021. "Impacts of Partial to Complete COVID-19 Lockdown on NO2 and PM2.5 Levels in Major Urban Cities of Europe and USA." *Cities* 117: 103308. https://doi.org/10.1016/j.cities.2021.103308

Barbate et al., Vikas. 2021. "COVID-19 and Its Impact on the Indian Economy." *Vision: The Journal of Business Perspective* 25 (1): 23–35. https://doi.org/10.1177/0972262921989126

Bhadra et al., Arunava. 2021. "Impact of Population Density on Covid-19 Infected and Mortality Rate in India." *Modeling Earth Systems and Environment* 7 (1): 623–629. https://doi.org/10.1007/s40808-020-00984-7

Biswal, Akash, Tanbir Singh, Vikas Singh, Khaiwal Ravindra, and Suman Mor. 2020. "COVID-19 Lockdown and Its Impact on Tropospheric NO2 Concentrations over India Using Satellite-Based Data." *Heliyon* 6 (9): e04764. https://doi.org/10.1016/j.heliyon.2020.e04764

Bukhari, Qasim, and Yusuf Jameel. 2020. "Will Coronavirus Pandemic Diminish by Summer?" *SSRN Electronic Journal.* https://doi.org/10.2139/ssrn.3556998

Chauhan, Akshansha, and Ramesh P. Singh. 2020. "Decline in PM2.5 Concentrations Over Major Cities around the World Associated with COVID-19." *Environmental Research* 187: 109634. https://doi.org/10.1016/j.envres.2020.109634

Chowdhury, Shirin Sultana, and Fahd A.A. Karim. 2018. "Biodentin- A Bioactive Dentin Substitute in Operative Dentistry." *Update Dental College Journal* 8 (2): 1. https://doi.org/10.3329/updcj.v8i2.40376

CIESIN. 2017. "(Center for International Earth Science Information Network) Gridded Population of the World, Version 4 (GPWv4): Population Density, Revision 11." Palisades, NY: Socioeconomic Data and Applications Center (SEDAC). https://doi.org/10.7927/H49C6VHW

Cyrys, Josef, Marloes Eeftens, Joachim Heinrich, Christophe Ampe, Alexandre Armengaud, Rob Beelen, Tom Bellander, et al. 2012. "Variation of NO2 and NOx Concentrations between and within 36 European Study Areas: Results from the ESCAPE Study." *Atmospheric Environment* 62 (December): 374–390. https://doi.org/10.1016/j.atmosenv.2012.07.080

Deep, Amar, Chhavi Pant Pandey, Hemwati Nandan, Narendra Singh, Garima Yadav, P.C. Joshi, K.D. Purohit, and S.C. Bhatt. 2021. "Aerosols Optical Depth and Ångström Exponent over Different Regions in Garhwal Himalaya, India." *Environmental Monitoring and Assessment* 193 (6): 324. https://doi.org/10.1007/s10661-021-09048-4

Gautam, Sneha. 2020. "COVID-19: Air Pollution Remains Low as People Stay at Home." *Air Quality, Atmosphere & Health* 13 (7): 853–857. https://doi.org/10.1007/s11869-020-00842-6

Griffin, Debora, Xiaoyi Zhao, Chris A. McLinden, Folkert Boersma, Adam Bourassa, Enrico Dammers, Doug Degenstein, et al. 2019. "High-Resolution Mapping of Nitrogen Dioxide with TROPOMI: First Results and Validation Over the Canadian Oil Sands." *Geophysical Research Letters* 46 (2): 1049–1060. https://doi.org/10.1029/2018GL081095

IEA. 2020. "Changes in Transport Behaviour during the Covid-19 Crisis, *IEA, Paris*. www.Iea.Org/Articles/Changes-in-Transport-Behaviour-during-the-Covid-19-Crisis

Jain, Suresh, and Tanya Sharma. 2020. "Social and Travel Lockdown Impact Considering Coronavirus Disease (COVID-19) on Air Quality in Megacities of India: Present Benefits, Future Challenges and Way Forward." *Aerosol and Air Quality Research* 20: 1222–1236. https://doi.org/10.4209/aaqr.2020.04.0171

Jerrett, Michael, Murray M. Finkelstein, Jeffrey R. Brook, M. Altaf Arain, Palvos Kanaroglou, Dave M. Stieb, Nicolas L. Gilbert, et al. 2009. "A Cohort Study of Traffic-Related Air Pollution and Mortality in Toronto, Ontario, Canada." *Environmental Health Perspectives* 117 (5): 772–777. https://doi.org/10.1289/ehp.11533

Kanniah, Kasturi Devi, Nurul Amalin Fatihah Kamarul Zaman, Dimitris G. Kaskaoutis, and Mohd Talib Latif. 2020. "COVID-19's Impact on the Atmospheric Environment in the Southeast Asia Region." *Science of The Total Environment* 736: 139658. https://doi.org/10.1016/j.scitotenv.2020.139658

Lal, Preet, Amit Kumar, Shubham Kumar, Sheetal Kumari, Purabi Saikia, Arun Dayanandan, Dibyendu Adhikari, and M.L. Khan. 2020. "The Dark Cloud with a Silver Lining: Assessing the Impact of the SARS COVID-19 Pandemic on the Global Environment." *Science of The Total Environment* 732 (August): 139297. https://doi.org/10.1016/j.scitotenv.2020.139297

Lamsal, L.N., R.V. Martin, D.D. Parrish, and N.A. Krotkov. 2013. "Scaling Relationship for NO$_2$ Pollution and Urban Population Size: A Satellite Perspective." *Environmental Science & Technology* 47 (14): 7855–7861. https://doi.org/10.1021/es400744g

Lelieveld et al., Jos Lelieveld, ndy Haines, Andrea Pozzer. 2018. "Air Pollution—A Neglected Cause of Death," June 29, 2018. www.mpg.de/12118117/air-pollution-cause-of-death

Levy, R.C., L.A. Remer, R.G. Kleidman, S. Mattoo, C. Ichoku, R. Kahn, and T.F. Eck. 2010. "Global Evaluation of the Collection 5 MODIS Dark-Target Aerosol Products over Land." *Atmospheric Chemistry and Physics* 10 (21): 10399–10420. https://doi.org/10.5194/acp-10-10399-2010

Lyapustin et al. 2018. "MCD19A2 MODIS/Terra+Aqua Land Aerosol Optical Depth Daily L2G Global 1km SIN Grid V006." *NASA EOSDIS Land Processes DAAC*. https://doi.org/10.5067/MODIS/MCD19A2.006

Maheshwarkar, Prem, and Ramya Sunder Raman. 2021. "Population Exposure across Central India to PM2.5 Derived Using Remotely Sensed Products in a Three-Stage Statistical Model." *Scientific Reports* 11 (1): 544. https://doi.org/10.1038/s41598-020-79229-7

MoHFW. 2021. www.Mohfw.Gov.In/

Moore, G.W.K., and J.L. Semple. 2021. "Himalaya Air Quality Impacts From the COVID-19 Lockdown Across the Indo-Gangetic Plain." *GeoHealth* 5 (6). https://doi.org/10.1029/2020GH000351

Muhammad, Sulaman, Xingle Long, and Muhammad Salman. 2020. "COVID-19 Pandemic and Environmental Pollution: A Blessing in Disguise?" *Science of The Total Environment* 728 (August): 138820. https://doi.org/10.1016/j.scitotenv.2020.138820

Novotny, Eric V., Matthew J. Bechle, Dylan B. Millet, and Julian D. Marshall. 2011. "National Satellite-Based Land-Use Regression: NO $_2$ in the United States." *Environmental Science & Technology* 45 (10): 4407–4414. https://doi.org/10.1021/es103578x

Pandey, Anamika, Michael Brauer, Maureen L. Cropper, Kalpana Balakrishnan, Prashant Mathur, Sagnik Dey, Burak Turkgulu, et al. 2021. "Health and Economic Impact of Air Pollution in the States of India: The Global Burden of Disease Study 2019." *The Lancet Planetary Health* 5 (1): e25–e38. https://doi.org/10.1016/S2542-5196(20)30298-9

Parida, Bikash Ranjan, Somnath Bar, Dimitris Kaskaoutis, A.C. Pandey, Suraj D. Polade, and Santonu Goswami. 2021a. "Impact of COVID-19 Induced Lockdown on Land Surface Temperature, Aerosol, and Urban Heat in Europe and North America." *Sustainable Cities and Society* 75: 103336. https://doi.org/10.1016/j.scs.2021.103336

Parida, Bikash Ranjan, Somnath Bar, Gareth Roberts, S.P. Mandal, A.C. Pandey, Manoj Kumar, and Jadunandan Dash. 2021b. "Improvement in Air Quality and Its Impact on Land Surface Temperature in Major Urban Areas across India during the First Lockdown of the Pandemic." *Environmental Research* 111280. https://doi.org/10.1016/j.envres.2021.111280

Parida, Bikash Ranjan, Somnath Bar, Nilendu Singh, Bakimchandra Oinam, A.C. Pandey, and Manoj Kumar. 2020. "A Short-Term Decline in Anthropogenic Emission of CO$_2$ in India Due to COVID-19 Confinement." *Progress in Physical Geography: Earth and Environment* 030913332096674. https://doi.org/10.1177/0309133320966741

Pathakoti, Mahesh, Aarathi Muppalla, Sayan Hazra, Mahalakshmi Dangeti, Raja Shekhar, Srinivasulu Jella, Sesha Sai Mullapudi, Prasad Andugulapati, and Uma Vijayasundaram. 2020. "An Assessment of the Impact of a Nation-Wide Lockdown on Air Pollution—a Remote Sensing Perspective over India." Preprint. Gases/Remote Sensing/Troposphere/Chemistry (chemical composition and reactions). https://doi.org/10.5194/acp-2020-621

Qian, Yun, and Filippo Giorgi. 1999. "Interactive Coupling of Regional Climate and Sulfate Aerosol Models over Eastern Asia." *Journal of Geophysical Research: Atmospheres* 104 (D6): 6477–6499. https://doi.org/10.1029/98JD02347

Rahaman, Saidur, Selim Jahangir, Ruishan Chen, Pankaj Kumar, and Swati Thakur. 2021. "COVID-19's Lockdown Effect on Air Quality in Indian Cities Using Air Quality Zonal Modeling." *Urban Climate* 36 (March): 100802. https://doi.org/10.1016/j.uclim.2021.100802

Ranjan, Avinash Kumar, A.K. Patra, and A.K. Gorai. 2020. "Effect of Lockdown Due to SARS COVID-19 on Aerosol Optical Depth (AOD) over Urban and Mining Regions in India." *Science of The Total Environment* 745: 141024. https://doi.org/10.1016/j.scitotenv.2020.141024

Ranjan, A.K., A. Patra, and A.K. Gorai. 2021. A Review on Estimation of Particulate Matter from Satellite-Based Aerosol Optical Depth: Data, Methods, and Challenges. *Asia-Pacific Journal of Atmospheric Sciences* 57: 679–699. https://doi.org/10.1007/s13143-020-00215-0

Remer, L.A., Y.J. Kaufman, D. Tanré, S. Mattoo, D.A. Chu, J.V. Martins, R.-R. Li, et al. 2005. "The MODIS Aerosol Algorithm, Products, and Validation." *Journal of the Atmospheric Sciences* 62 (4): 947–973. https://doi.org/10.1175/JAS3385.1

Remer, Lorraine A., Richard G. Kleidman, Robert C. Levy, Yoram J. Kaufman, Didier Tanré, Shana Mattoo, J. Vanderlei Martins, et al. 2008. "Global Aerosol Climatology from the MODIS Satellite Sensors." *Journal of Geophysical Research* 113 (D14): D14S07. https://doi.org/10.1029/2007JD009661

Saikawa, Eri, Arnico Panday, Shichang Kang, Ritesh Gautam, Eric Zusman, Zhiyuan Cong, E. Somanathan, and Bhupesh Adhikary. 2019. "Air Pollution in the Hindu Kush Himalaya." In *The Hindu Kush Himalaya Assessment*, edited by Philippus Wester, Arabinda Mishra, Aditi Mukherji, and Arun Bhakta Shrestha, 339–387. Cham: Springer International Publishing. https://doi.org/10.1007/978-3-319-92288-1_10

Sarkar, Sudipta, and Menas Kafatos. 2004. "Interannual Variability of Vegetation over the Indian Sub-Continent and Its Relation to the Different Meteorological Parameters." *Remote Sensing of Environment* 90 (2): 268–280. https://doi.org/10.1016/j.rse.2004.01.003

Shafeeque, Muhammad, Arfan Arshad, Ahmed Elbeltagi, Abid Sarwar, Quoc Bao Pham, Shahbaz Nasir Khan, Adil Dilawar, and Nadhir Al-Ansari. 2021. "Understanding Temporary Reduction in Atmospheric Pollution and Its Impacts on Coastal Aquatic System during COVID-19 Lockdown: A Case Study of South Asia." *Geomatics, Natural Hazards and Risk* 12 (1): 560–580. https://doi.org/10.1080/19475705.2021.1885503

Sharma, Shubham, Mengyuan Zhang, Anshika, Jingsi Gao, Hongliang Zhang, and Sri Harsha Kota. 2020. "Effect of Restricted Emissions during COVID-19 on Air Quality in India." *Science of The Total Environment* 728: 138878. https://doi.org/10.1016/j.scitotenv.2020.138878

Shrestha, Asheshwor, Uttam Shrestha, Roshan Sharma, Suraj Bhattarai, Hanh Tran, and Maheswar Rupakheti. 2020. "Lockdown Caused by COVID-19 Pandemic Reduces Air Pollution in Cities Worldwide." Preprint. *Life Sciences*. https://doi.org/10.31223/OSF.IO/EDT4J

Sidhu, Maninder Kaur, Khaiwal Ravindra, Suman Mor, and Siby John. 2017. "Household Air Pollution from Various Types of Rural Kitchens and Its Exposure Assessment." *Science of The Total Environment* 586 (May): 419–429. https://doi.org/10.1016/j.scitotenv.2017.01.051

Suddick, Emma C., Penelope Whitney, Alan R. Townsend, and Eric A. Davidson. 2013. "The Role of Nitrogen in Climate Change and the Impacts of Nitrogen–Climate Interactions in the United States: Foreword to Thematic Issue." *Biogeochemistry* 114 (1–3): 1–10. https://doi.org/10.1007/s10533-012-9795-z

Superczynski, Stephen D., and Sundar A. Christopher. 2011. "Exploring Land Use and Land Cover Effects on Air Quality in Central Alabama Using GIS and Remote Sensing." *Remote Sensing* 3 (12): 2552–2567. https://doi.org/10.3390/rs3122552

Targino et al., Admir C. 2013. "Deterioration of Air Quality across Sweden Due to Transboundary Agricultural Burning Emissions." http://hdl.handle.net/10138/229277

Tobías, Aurelio, Cristina Carnerero, Cristina Reche, Jordi Massagué, Marta Via, María Cruz Minguillón, Andrés Alastuey, and Xavier Querol. 2020. "Changes in Air Quality during the Lockdown in Barcelona (Spain) One Month into the SARS-CoV-2 Epidemic." *Science of The Total Environment* 726: 138540. https://doi.org/10.1016/j.scitotenv.2020.138540

Unnithan, S.L. Kesav, and L. Gnanappazham. 2020. "Spatiotemporal Mixed Effects Modeling for the Estimation of $PM_{2.5}$ from MODIS AOD over the Indian Subcontinent." *GIScience & Remote Sensing* 57 (2): 159–173. https://doi.org/10.1080/15481603.2020.1712101

Vadrevu, Krishna Prasad, Aditya Eaturu, Sumalika Biswas, Kristofer Lasko, Saroj Sahu, J.K. Garg, and Chris Justice. 2020. "Spatial and Temporal Variations of Air Pollution over 41 Cities of India during the COVID-19 Lockdown Period." *Scientific Reports* 10 (1): 16574. https://doi.org/10.1038/s41598-020-72271-5

Val Martin, M., R.E. Honrath, R.C. Owen, and K. Lapina. 2008. "Large-Scale Impacts of Anthropogenic Pollution and Boreal Wildfires on the Nitrogen Oxides over the Central North Atlantic Region." *Journal of Geophysical Research* 113 (D17): D17308. https://doi.org/10.1029/2007JD009689

Veefkind, J.P., I. Aben, K. McMullan, H. Förster, J. de Vries, G. Otter, J. Claas, et al. 2012. "TROPOMI on the ESA Sentinel-5 Precursor: A GMES Mission for Global Observations of the Atmospheric Composition for Climate, Air Quality and Ozone Layer Applications." *Remote Sensing of Environment* 120 (May): 70–83. https://doi.org/10.1016/j.rse.2011.09.027

Wang. 2017. "Relationship between Aerosol Optical Depth and Population Density in Typical Area of East China." *Applied Ecology and Environmental Research* 15 (3): 1041–1056. https://doi.org/10.15666/aeer/1503_10411056

WHO. 2020. Www.Who.Int/Emergencies/Diseases/Novel-Coronavirus-2019

Zalakeviciute, Rasa, Renne Vasquez, Daniel Bayas, Adrian Buenano, Danilo Mejia, Rafael Zegarra, Valeria Diaz, and Brian Lamb. 2020. "Drastic Improvements in Air Quality in Ecuador during the COVID-19 Outbreak." *Aerosol and Air Quality Research* 20 (8): 1783–1792. https://doi.org/10.4209/aaqr.2020.05.0254

Zhang, Ruixiong, Yuhang Wang, Qiusheng He, Laiguo Chen, Yuzhong Zhang, Hang Qu, Charles Smeltzer, et al. 2017. "Enhanced Trans-Himalaya Pollution Transport to the Tibetan Plateau by Cut-off Low Systems." *Atmospheric Chemistry and Physics* 17 (4): 3083–3095. https://doi.org/10.5194/acp-17-3083-2017

Zhang, Ruixiong, Yuzhong Zhang, Haipeng Lin, Xu Feng, Tzung-May Fu, and Yuhang Wang. 2020. "NOx Emission Reduction and Recovery during COVID-19 in East China." *Atmosphere* 11 (4): 433. https://doi.org/10.3390/atmos11040433

16 Development of Spatially Distributed GIS-based Emission Inventory of Particulate Matter from Anthropogenic Sources over India and Assessment of Trends of Pollution

Sailesh N. Behera[1,2,*], Bikash R. Parida[3], Jitendra K. Tripathi[1], and Mukesh Sharma[4]

1 Air Quality Laboratory, Department of Civil Engineering, Shiv Nadar University, Delhi-NCR, Greater Noida, Gautam Buddha Nagar, Uttar Pradesh, PIN: 201314, India.

2 Centre for Environmental Sciences and Engineering, Shiv Nadar University, Delhi-NCR, Greater Noida, Gautam Buddha Nagar, Uttar Pradesh, PIN: 201314, India.

3 Department of Geoinformatics, School of Natural Resource Management, Central University of Jharkhand, Ranchi, PIN: 835222, India.

4 Department of Civil Engineering, Indian Institute of Technology Kanpur, Kanpur, Uttar Pradesh, PIN: 208016, India.

CONTENTS

16.1 Introduction 318
16.2 Materials and Methods 320
    16.2.1 Overall Methodology Considered in This Study 320
    16.2.2 Characteristics of Study Domain...321
    16.2.3 Identification of Sources 322

DOI: 10.1201/9781003265160-19

        16.2.4 Collection and Compilation of Activity Data 323
        16.2.5 Emission Estimation and Selection of Emission Factors..................325
        16.2.6 GIS-Based Spatially Distributed Emission Maps............................327
        16.2.7 Trend Analysis of India's Emission and a Himalayan City 328
16.3 Results and Discussion 329
        16.3.1 Overall Emission Inventory of PM_{10}..329
        16.3.2 Source-Wise Emission Inventories of PM_{10}.......................................331
        16.3.3 Trends in Emission Loads of PM_{10} in India..334
        16.3.4 Trends in Emission Loads and Ambient Concentration of PM_{10}
                 in a Himalayan City...336
16.4 Conclusion 337
References.. 338

16.1 INTRODUCTION

The issues of air pollution have received significant attention from various sections across the globe during the last two decades (2000–2020) due to its association with potential human health implications and environmental impacts. The environmental impacts include a reduction in visibility through haze formation and affecting the climate and several other ecosystems through long-range transport under favorable synoptic meteorological conditions (*Von-Schneidemesser et al., 2015; Al-Thani et al., 2018; Madadi et al., 2021*). As a result, numerous studies on pollution issues are being conducted from a research point of view and from policy implementation perspectives in both developed and developing countries. This is to be noted that synoptic meteorological condition is defined as the scale in meteorology that can spread in a horizontal length to 1000 km or even more, and the influence of meteorological factors through this phenomenon helps in the transboundary transport of air pollutants. The developed countries are several steps ahead of developing countries with systematic regulations in place, being followed by responsible stakeholders for the reduction of ambient concentration levels of air pollutants (*Rafaj et al., 2018; Anjum et al., 2020*). Due to rapid urbanization, fast growth in transportation sectors, and unbridled industrial growth, a developing country like India is facing many challenges with higher levels of air pollutants in most of its urban regions.

The geographical location of half of the Indian subcontinent falls within the tropical region with higher levels of solar radiation and these conditions create a typical topography that induces an extremely complex structure of wind flow (*Lau et al., 2000; Umar et al., 2021*). Such unique structures of wind flow and meteorology have significant effects on the transportation of air pollutants after being emitted from respective sources. Once pollutants are emitted from sources, their behaviors are capable enough to produce very unhealthy situations those are having several environmental implications at a regional scale of observations (*Garaga et al., 2018*). For example, thick haze and smoke originating from the burning of biomass and municipal solid wastes and the combustion of fossil fuels in industries in northern and western India normally concentrate inside or central parts of the Indo-Gangetic Plain (IGP) region (*Singh et al., 2020*). However, prevailing westerlies wind (moving from the west toward the east) and favorable synoptic meteorological conditions push forward particulate matter (PM) or aerosols from these central parts of the IGP region to various parts of the Himalayan region, eastern India,

and the Bay of Bengal (*Aswini et al., 2019; Romshoo et al., 2021*). This is noteworthy that the IGP region is famously known as the largest river basin area in India (*Behera and Sharma, 2015; Romshoo et al., 2021*), with more than 40% of its population and 26% of its total landmass (*Behera et al., 2016; Yadav et al., 2020*).

Many complex behaviors of such atmospheric turbulence in the Indian subcontinent region lead to an in-depth and rigorous analysis of the dynamics of pollutants that would be helpful in finding economical solutions to various air pollution issues that cover the entire country including rural areas (*Chakraborty et al., 2017; Gautam and Bolia, 2020*). To achieve such objectives, it is necessary to develop a systematic, digitized inventorization (i.e., records of the quantity of pollution activities and their emission loads into a digital form that can be easily read and processed by a computer) of pollutants from various sources with respect to different regions in finer resolutions. In this context, the Geographic Information System (GIS) is an important informatics tool that is capable of keeping records of relevant activity-level data, fitting a mathematical emission model into it, doing calculations with a desired estimate, and finally producing maps of different perspectives known as thematic layers (*Aneja et al., 2012; Singh et al., 2016; Hakkim et al., 2021*). Therefore, a GIS-based emission inventory is useful in conducting various options of iterations in air quality modeling (*Ma et al., 2019*). These air quality models need input parameters such as topographic features, surface roughness, surface temperatures, and spatially distributed pollution sources, such as point sources with large industrial stacks, area sources with small/medium-scale industries having shorter stacks, and area sources like open burning and line source like vehicles. The GIS is a useful tool to process emission activities into preprocessing files that are compatible with air quality models. Additionally, many air quality models are coupled with the GIS to produce various environmental processes through computational simulations (*Lam et al., 2021*).

The rise in concentration levels of PM with aerodynamic diameter ≤10 μm (PM_{10}; the most significant air pollutant) in India is caused by a steady increase in pollution activities at source levels. These sources include primary sources of point industrial and area industrial sources, various vehicle categories, off-road stationary engines, different domestic cooking activities, agricultural activities, open biomass burning, garbage burning, and many others. Most of the air pollution studies (*Sharma et al., 2019; Vohra et al., 2021*) in India cover experimental or field studies, and the objectives of air quality management consider the development of monitoring networks that comprises several cities and regions, followed by collection of experimental data for further processes and studies. For a huge country like India with population of over 1.38 billion and an area of 3.287 million km^2, the current approach of air pollution monitoring at every single location is an unfathomable task, both economically and for managing the information on multiple pollutants. In other words, such objectives are more cost-effective and cumbersome in managing large set of resources required for the maintenance of such a vast air pollution–monitoring network. Therefore, the use of atmospheric emission models in informatics systems like GIS is becoming more popular in an inventorization of pollution loads from various perspectives.

An atmospheric emission model can be defined as a combination of mathematical sub-models of different degrees of complexity in an integrated manner. These sub-models are useful in estimating emission loads of pollutants from various responsible sources located in a portrayed geographical area in a designated period of time

(Singh et al., 2016; Ma et al., 2019). In the perspective of Indian emission inventory work, many studies in the past *(Garg et al., 2006; Hakkim et al., 2021)* have developed emission maps and provided emission estimates using techniques of partially or fully top-down approaches. However, the bottom-up approach gives more accurate and reliable results compared to the top-down approach, as activity level data are compiled at a finer resolution. This is to be noted that the emission activities data in a top-down approach are converted from a lower resolution to a higher resolution with reference to a standard parameter, whereas the emission activities data in a bottom-up approach are converted from a higher resolution to a lower resolution.

The present study is a first of its kind that used a distinct methodology in the GIS to develop a grid-based emission map of India at a resolution of 40 km × 40 km using a bottom-up approach from pollution activity data from 593 districts in India. Such grid-based emission inventory outcomes can be used as preprocessing inputs for chemical transport modeling to study various scenarios of pollution through transport. For example, the scope of the study can be formulated as the effects of emissions of PM_{10} from India on deposition in the Himalayan regions, where ecological sensitive peaks and glaciers are located. The methodology described in this study would be helpful for researchers to develop spatially distributed informatics-based emission loads and explore various scenarios that would be able to reduce atmospheric levels of PM_{10} at the national level. The specific objectives of this study are described as follows: (1) identifying and categorizing sources responsible for the emission of PM_{10} and collecting and compiling activity-level data from 593 districts of India followed by considering a method of estimating emissions, (2) developing a spatially distributed GIS-based emission inventory of PM_{10} at a resolution of 40 km × 40 km and examining contributing sources for the base year as 2010, (3) assessing growth trends of pollution loads at the national level for past 10 years and the contemporary future 10 years, and (4) examining trends of pollution loads and ambient concentration levels of PM_{10} at an ecologically sensitive Himalayan region, the city of Shimla.

16.2 MATERIALS AND METHODS

16.2.1 Overall Methodology Considered in This Study

Various steps adopted in this study to develop grid-based emission inventory are described in Figure 16.1. Under the first step is identifying responsible individual sources of PM_{10} for major categories of domestic, industries, open burning, and vehicles was done based on district-wise information. During this exercise, inputs from relevant literature were considered in finalizing sources under these four distinct major sources. With the next step of emission inventory development, the collection and compilation of district-wise source-specific activity-level data from various reports and literature (e.g., human population, vehicle population, vehicle kilometer traveled, details of fuel usage for domestic cooking, the production capacity of industries, stack height of industries) were done.

ArcGIS platform was used for digitization of base map of India followed by extraction of boundaries of states and nation as a whole to district levels in polygonal forms. The selection of source-specific emission factors of PM_{10} and the decision of emission estimation methods in finding emissions from individual sources were done to take forward these techniques to the platform of ArcGIS. Various thematic layers with different purposes were formed from the vector data for computing emissions and extracting source-specific

Identification of responsible individual sources of PM_{10} under major categories of domestic, industries, open burning and vehicles based on district-wise information

⬇

Collection and compilation of district-wise source-specific activity level data from various reports and literature (e.g., population, vehicle population, fuel usage)

⬇

Digitization of base map of India with ArcGIS to extract state and national boundaries in the form of polygons to district levels

⬇

Selection of source specific emission factors of PM_{10} and decision of emission estimation methods in finding emissions from individual sources

⬇

Formation of thematic layers from the vector data for computing emissions and extracting source-specific inventories using attribute tables in ArcGIS

⬇

Development of district-wise emission inventories of PM_{10} considering activity data of 593 districts of four distinct sources (domestic, industries, open burning and vehicles) and suitable emission factors using emission load estimation method

⬇

Development of spatially-resolved grid-wise emission map at a resolution of 40 km × 40 km with emission inventory estimated values from distinct sources adopting nearest neighborhood technique in ArcGIS

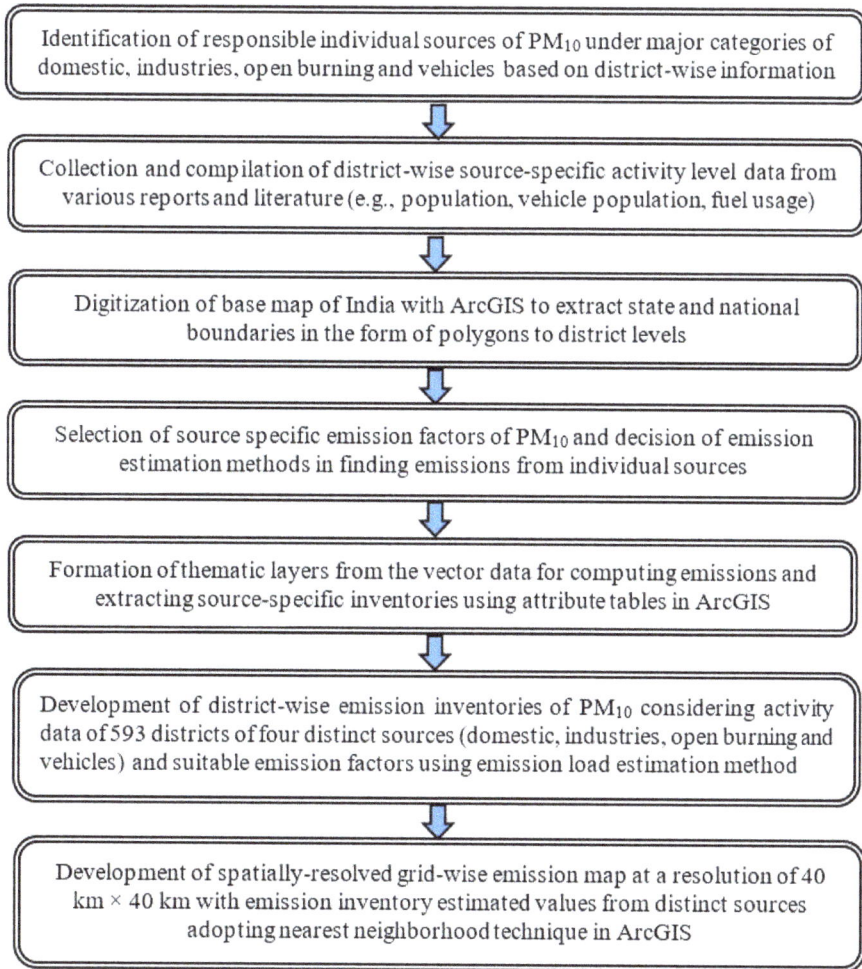

FIGURE 16.1 Sequence of steps adopted in this study in developing grid-based emission inventory.

inventories using attribute tables in ArcGIS. With this capability, a map of distinct purpose was generated. At the end, a spatially resolved grid-wise emission map was developed at a resolution of 40-km × 40-km grids with emission inventory estimated values from distinct sources adopting the nearest neighborhood technique in ArcGIS.

16.2.2 CHARACTERISTICS OF STUDY DOMAIN

The whole of India (6° 44′–35° 30′ N and 68° 7′–97° 25′ N) comprises 593 districts, 28 states, and 9 union territories was considered with the scope of research in this study. The population of India was estimated to be 1217 million in 2010 using the logistics growth method from 2001 population data. Figure 16.2 shows India on the world map with corresponding districts and state boundaries. In the first 50 years of the post-independence

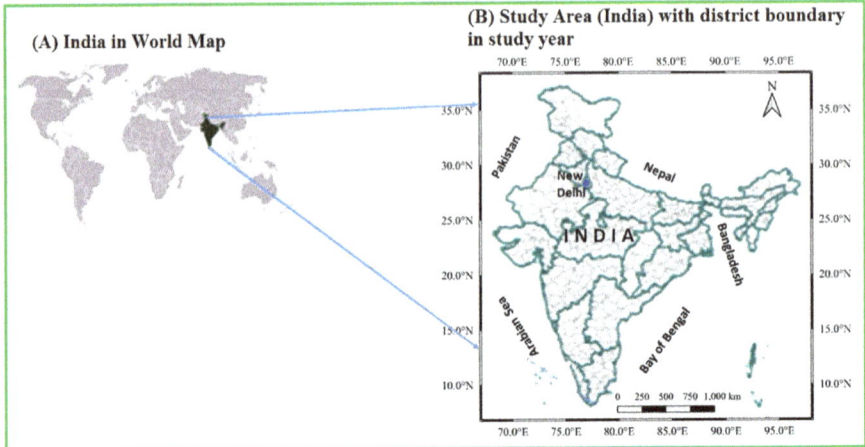

FIGURE 16.2   The study area India: (a) Map of India on the globe, and (b) map of study area with district and state boundaries during the study year.

era in India (1950–2000), the installed capacity of power plants increased almost by 50 times, indicating that the usage of fossil fuels has been increasing rapidly. Such an increase in usage of fuel for power and energy production is a matter of concern, as production processes involve large emissions of PM_{10} to the atmosphere. The desire for becoming higher income for citizens has been soaring ownership in multiple ways, causing a fast increase in the population of motor vehicles. During the last two decades (2000–2020), it has been seen that the vehicular population has increased by three times (*Pucher et al., 2007; MoRTH, 2011; Chakraborty et al., 2017*).

   With a rise in the activities of urbanization, the generation of solid waste is also increasing in India. This is noteworthy that garbage burning in India is one of the responsible sources of PM_{10}. The agricultural residue burning in India is on the rise; as farmers have a tendency to burn waste on the field after harvesting is done, such activities increase emissions of PM_{10} to the atmosphere. For domestic cooking and heating, people use fuel as liquefied petroleum gas (LPG), coal, fuelwood, kerosene, and cow-dung cake in both rural and urban areas. As population grows substantially, fuel for cooking and heating purposes for domestic uses is also increasing. To have a better gross domestic product growth rate, an increase in manufacturing industries is happening in India. Overall, the demand for power, motorization, industrial products, a rise in usage of domestic fuels, garbage burning, and the burning of agricultural residues cause a significant rise in levels of atmospheric PM_{10} in India.

16.2.3 IDENTIFICATION OF SOURCES

We have categorized all related sources responsible for emission of PM_{10} to the atmosphere into four major sectors, that is, domestic cooking and heating (namely, domestic), industrial activities (known as industries), vehicular movement (known as vehicles), and burning activities occurring openly (known as open burning; *Reddy*

and Venkataraman, 2002a, b; Aneja et al., 2012). The pollution activities in the domestic sector included cooking and heating usages through fuels of LPG, coal, fuelwood, cow-dung, and kerosene. The pollution activities under industries divided into sub-major sources, such as iron and steel plants, cement industries, sugar industries, power plants, and brickmaking kilns (known as brick kilns). The pollution activities under vehicles were categorized based on various types of vehicles, that is, buses, trucks, tractors and trailers, light motor vehicles (LMVs), light commercial vehicles (LCVs), two-wheeler motor bikes (two-wheelers), and four-wheeler passenger cars and jeeps (cars). The source activities under open burning were subdivided into open garbage burning occurring in the streets and agricultural burning occurring on the agricultural fields after harvesting crops in India.

In this study year of 2010, we considered 593 districts that had relevant pollution activity information in development of emission inventory meant for PM_{10}. As the database was larger to handle, an emission inventory of PM_{10} for a specific source and location was provided, with a specific code using a short text description to be associated them with pertinent emission factors under certain codes of source characterization falling in that particular district. For example, PM_LPG_A_1 symbolizes the load of PM_{10} emitted from domestic cooking use of LPG for cooking and heating purpose from the district of Anantapur in the first state, that is, Andhra Pradesh. We considered specific codes meant for sources that were helpful for the purposes of data compiling, processing, and reporting in estimates of PM_{10} emission loads in a large database. During various steps of execution for development of emission inventory of PM_{10} in the platform of ArcGIS, the previously mentioned assigned codes were utilized in identifying, categorizing, and distributing emission loads of PM_{10} in desired formats. We collected and compiled population data in each of the districts of India during 2001 from the Census database, Government of India (http://censusindia.gov. in/). The population in the break of urban and rural in all districts of India were collected from this Census database. Subsequently, the database was projected for 2010 (our study year) from a given database of 2001 in the means of district-wise information. The population forecasting method through logistics method approach was used to predict the unknown population during 2010 and 2020. The separate growth rates meant for both urban and rural of specific districts were considered in the population forecasting process in this study. We used a logistics growth method for prediction of population during 2010 and 2020 using different growth rates for urban and rural populations for their respective districts.

16.2.4 Collection and Compilation of Activity Data

The activity-level data for the domestic sector considered for collection were domestic heating source, number of inhabitants, fuel-type use, fuel consumption rates, and population data, having various classes of people based on their economic standards. The data on state-wise average consumption of kerosene per head were utilized in estimating consumption at district levels both for urban and rural areas considering values of relative fraction of population, that is, district/state (*NCAER, 2010; MoPNG, 2010; MoSPI, 2010*). Similarly, the activity-level data meant for cooking purposes were generated at district levels for LPG consumption from the *MoPNG*

(2010), coal usage from *MoC (2010)*, and fuelwood and cow-dung cake uses from *MoSPI (2010)*. In production of electric power in India, the majority of thermal power plants were coal-based and small amounts were gas- and oil-based plants. We considered more than 800 plant units for collection and compilation of information that included boiler size in terms of capacity, category of fuel, amount of electricity generation, data on specific fuel consumption during plant operation, time of operation for a specified purpose, and ultimately the geographical location of that particular industry. The review reports published by Central Electricity Authority (CEA) and Ministry of Power (MoP) of India were used in derivation of relevant information required for compilation of activity level data meant for thermal power plants *(CEA, 2010; MoP, 2010)*. In case of unavailability of data on load factors of power or specific oil consumption at district levels meant for any specific plant or industry, we considered respective state average values. As an overall assessment, it was seen that large-scale industrial units achieved their power requirement partially or fully through installing their own power-generating appliances or units in their respective industrial premises. During the compilation of activity-level data meant for every fuel type and specific fuel consumption in the unit of expression of kg/kWh for various industrial units, the fuel consumption rate in related power plants was estimated using the data on installed generating capacity and load factors on an average basis *(CMIE, 2010)*.

Several categories of iron and steel industries, including steel plants, secondary plants producing iron and steel, sponge iron, and alloy steel plants, were considered in this study for the inventorization of PM_{10} loads in India. The compiled activity-level data revealed that coal was mostly used as the predominant fuel in the production of iron and steel followed by other fuels like oil and gases in smaller quantities. The collection of primary activity-level data was done in the unit of consumption of specific fuel per ton of hot metal production, applicable to all possible forms of steel and iron plants *(CMIE, 2010)*. The yield rates of specific integrated steel plants and their respective coke over were considered to estimate generation of derived fuels that included coke, coke oven gas, and blast furnace gas. This is to be noted that we assumed consumption of derived fuels in the sector of iron and steel industries only, and the average yield rates of an integrated steel plant were considered in an estimation exercise *(MoSPI, 2010)*.

The predominant fuel was coal in manufacturing bricks in small brick kilns, and low-grade fuels locally available were used in brick-kiln units in some places. This trend varied with geographical locations of that particular region. We used information available on the web (http://practicalaction.org/) on the production of bricks at the state level to estimate respective coal consumptions considering variations in the mix of fuels in that particular region.

Activity level data were collected and compiled for vehicles, and it was seen during the study year (i.e., 2010) that petrol and diesel were the mostly used fuels for road transportation system of India, and compressed natural gas (CNG) was used in fewer places like Delhi. The activity level data for vehicles included in this study were categories of fuel use, quantity of consumption of fuel in respective vehicle categories, road lengths under various types, and the number of vehicles under different categories. The vehicle population at the state and district levels was used as

a surrogate reference in the estimation of the consumption of fuel usage by respective vehicles in urban areas (*MoRHT, 2011*). In other words, fuel consumption was estimated in proportion to the respective vehicular population at every state in urban areas. Moreover, road lengths of various types of roads from the database of Ministry of Road Transport and Highways (MoRHT) and digitized road map of India were used in estimation of fuel consumption in road transportation at the district level for rural areas. The vehicle population in rural areas was found to be lower than in urban areas in every state in India.

In the agricultural sector, India is known to be one of the leading countries in the world to produce varieties of crops. A common practice happens to burn crop residues on the field in India after each harvesting. This practice of burning waste in open areas is known as agricultural open burning. Such a practice by the farming community is being done to avoid retaining crop residues on the field for longer times, which need to be cleared for the next crop through rotation of cultivation. The production rate or yield of a specific crop is normally a key player in deciding quantities of its residues on the field after harvesting. The value for quantity of residue leftover on the field was considered as 12% of total rice straw (*Gadde et al., 2009*). For wheat straw cultivation, 7.5% of total wheat straw generated was considered as leftover on the field those are subjected to on-site open burning (*Sahai et al., 2007*). For other crops, the assumption was made that as much as 10% of the total residue is burned in open burning on the field after harvesting. The report published by the Department of Agriculture and Cooperation was used for retrieving activity-level information district-wise crop production (*MoA, 2010*).

Burning of waste spreading in streets, known as garbage burning, has been practiced in most of urban areas in India during the evening period. We used reports published by Central Pollution Control Board (CPCB), Delhi, and respective State Pollution Control Boards (SPCBs) in collecting and compiling state-wise activity-level data meant for garbage generation (*CPCB, 2010*). The proportion of the urban population at the district level was used in estimating activity-level data required at district level.

16.2.5 EMISSION ESTIMATION AND SELECTION OF EMISSION FACTORS

As described in the aforementioned section, two approaches, including top-down and bottom-up approaches, are normally considered for emission estimates of air pollutants in a region. In a top-down approach, total emissions are estimated for a designated study area, and the emissions in different cells are distributed through spatial disintegration using reference parameters that include population, density of industrial development, and traffic volume. However, such approach of top-down is inappropriate for emission inventory with a larger geographical area due to more assumptions and less accuracy accumulated in the database and limitation on opting for better resolution in final outcomes. We used a bottom-up approach on the GIS platform, where emissions for all cells were estimated, in which geographical area was divided by means of establishing every parameter for each cell in particular (*Aneja et al., 2012*). Then total emission value was obtained by aggregation of estimation carried out for every single cell. Such approach of bottom-up in development

of spatially resolved emission inventory is preferred over the top-down approach in producing outcomes for better resolution and model performance (*Wang et al., 2009; Gao et al., 2019*).

The standard method for emission estimation from activity-level data was adopted, in which suitable emission factor was multiplied with activity data to find out the desired emission. The emission at the district level was estimated as the sum of emissions from all responsible source sectors that were using fuels for combustion. PM_{10} emission from the sources of industries or domestic was estimated by using Equation 16.1:

$$E_{Ind\ or\ Dom} = \sum_{j}\sum_{k}\sum_{l} A_{j,k,l} \times EF_{j,k,l} \times (1-\eta_{j,k,l}) \qquad (16.1)$$

where $E_{Ind\ or\ Dom}$ is the emission of PM_{10} from fuels of industries or domestic cooking or heating, either in a district or in the nation/state. The term, A is considered as the activity data in terms of consumption of fuel, η is considered as the removal efficiency, and EF is the respective emission factor. The subscripts j, k, and l expressed in Equation 16.1 represent region (e.g., district or state), sector, and fuel type, respectively.

PM_{10} emission from area source like open burning was calculated using Equation 16.2:

$$E_{OB} = \sum_{m}\sum_{n}\sum_{o} A_{m,n,o} \times B_{m,n,o} \times C_{m,n,o} \times EF_{m,n,o} \qquad (16.2)$$

where E_{OB} is emission of PM_{10} from area source like agricultural residue burning or garbage burning. A represents the burning area of that particular region, B represents the fuel load per unit area, C represents burning efficiency, EF represents the respective emission factor, and subscripts m, n and o represent the region (e.g., district or state), sector, and fuel type, respectively.

PM_{10} emission from vehicles was estimated using Equation 16.3:

$$E_V = \sum_{s}\sum_{t}\sum_{u} Veh_{s,t,u} \times D_{s,t,u} \times EF_{s,t,u} \qquad (16.3)$$

where E_V is the emission of PM_{10} from vehicles. *Veh* represents the number of vehicles of a particular category, D represents the distance traveled in km in 1 year for the respective vehicle category, EF is the mass emission factor of one vehicle for 1 km of travel. The subscripts s, t and u represent the region (e.g., district), sector, and vehicle type, respectively.

In the task of development of emission inventory, the selection of emission factors has a pivotal role, as it decides the extent of uncertainty in emission estimation. In this study, emission factors for each of the sources were taken from studies having experimental results suitable to Indian region specifically. In the case of data, absent of Indian studies, we have chosen the international reports (e.g., USEPA AP42, IPCC reports) and research studies to decide the emission factors suitable for Indian

activity processes. The values of considered emissions fractions (EFs) of PM_{10} are provided as follows: LPG at 2.1 g/kg, kerosene at 1.9 g/kg, coal at 9.6 g/kg, fuelwood at 11.7 g/kg, dung cake at 10.8 g/kg, brick kiln at 2.2 g/kg, sugar industry at 7.8 g/kg, cement plant at 1.3 g/kg, steel industry at 6.4 g/kg, power plant at 1.15 g/kg, bus at 1.1 g/km, truck at 1.25 g/km, tractor and trailer at 1.8 g/km, LMV and LCV at 0.8 g/km, car at 0.72 g/km, two-wheeler at 0.23 g/km, crop-residue burning at 11.2 g/kg, and garbage burning at 8.1 g/kg.

16.2.6 GIS-Based Spatially Distributed Emission Maps

The primary objective of the methodology of this study with various steps was to develop a variable GIS-based grid PM_{10} emission database, which can be utilized in the scope of the present study, testing the efficacy of pollution control policies and for future use in predicting concentrations at various locations in India. Additionally, this developed grid-based database on loads of PM_{10} emission can be utilized to predict the dry and wet deposition rates of PM_{10} in various eco-sensitive regions of the Himalayas through the platform of any chemical transport model (CTM). Moreover, the purpose for improving the accuracy of emission inventory can be achieved through GIS-based emission inventorization exercise, which possess a systematic process of an emission inventory, including the development of thematic layers of maps, producing a database of those that can be in the meteorology-air quality/climate modeling system.

A standard U.S. Environmental Protection Agency (USEPA)-based GIS methodology was adopted for developing a national-level emission inventory of PM_{10} for the base year of 2010. the Geostatistical Analyst extension of GIS (ArcGIS) was selected in this study due to its relative user-friendliness, and its frequent use in studies of air quality management by local, regional, and federal authorities and research institutes (*Singh et al., 2016; Ma et al., 2019*). To be specific, various tools in ArcGIS were used to view, edit, create, analyze, and explore geospatial data with framework of modeling framework. We divided digitized India's map into 40-km × 40-km size grids, and each grid was assigned an identification number for better handling of data on the database (as described in an earlier section). The nearest neighborhood interpolation technique in ArcGIS was used in the development of an emission inventory.

The population densities in all districts were estimated after incorporating areas of respective districts with irregular boundaries, using Equation 16.4. The area of grids falling inside respective districts were calculated for estimation of population in the grids using Equation 16.5.

$$Population\ density\ \left(person\,/\,km^2\right) = \frac{Population\ of\ a\ district\ \left(person\right)}{district\ area\ \left(km^2\right)} \quad (16.4)$$

$$Grid\ population = \sum_{i=1}^{N}\left(inter\ sected\ district\ area_i \times density\ of\ district_i\right) \quad (16.5)$$

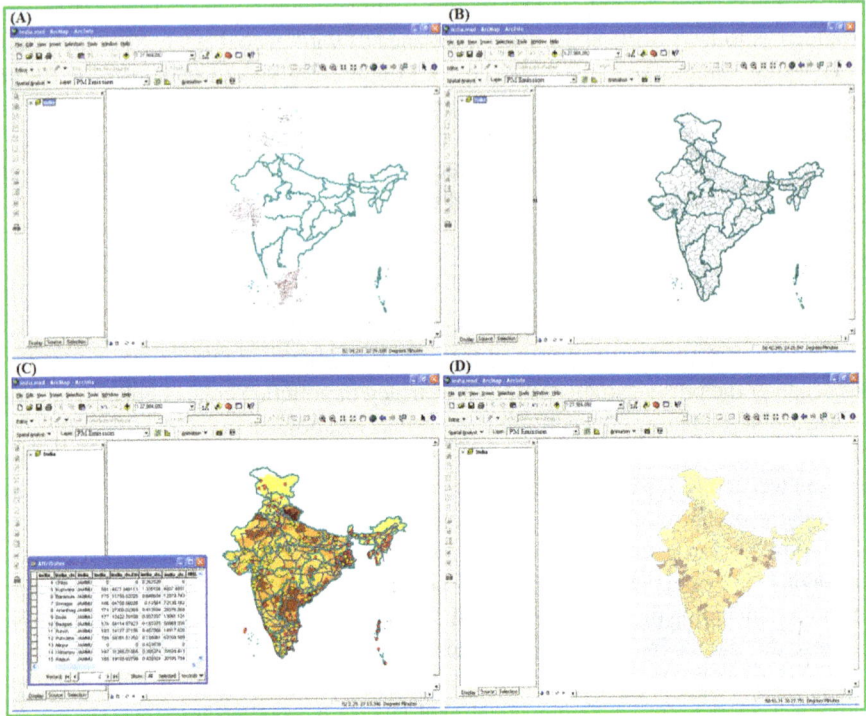

FIGURE 16.3 Schematic diagrams for development of GIS-based emission inventory: (a) Map digitization, (b) district-wise source database, (c) district-wise pollutant emissions, and (d) grid-wise (40-km × 40-km) emission inventory.

where N represents the number of districts or fraction of full district in that particular grid and i represents a particular district. Thus, the population of each grid was estimated and stored in the database of ArcGIS.

During emission inventory development, the same approach was applied to other scopes of work that included vehicular activities, industrial activities and emissions, open-burning matters, and domestic cooking activities. The vector data of the emission inventorization process were converted to different thematic layers to interact and to determine the locations of specific sources as industries, areas of interest, and for editing factors. Various thematic layers were prepared that included maps of industrial activity, population, transportation layers, area boundaries, and many others. The attribute values in the map table can be edited allowing users to tweak values, such as a feature's population, agricultural field area, and the number of vehicles on a road in each district.

16.2.7 TREND ANALYSIS OF INDIA'S EMISSION AND A HIMALAYAN CITY

To assess India's emission trends for past and future years for PM_{10} emission loads at the national level, we considered four major sources, namely domestic, industries, vehicles, and open burning. We assumed the same EFs of all responsible sources

during these years and included growth rates of respective activity level data. The reports from the National Survey of Household Income and Expenditure (NCAER), the Ministry of Petroleum and Natural Gas (MoPNG), the Ministry of Coal (MoC), and the Ministry of Statistics and Program Implementation (MoSPI) were considered for finding the growth rates of activity data of various sources under domestic sector. The reports of the CEA, the MoP, the Centre for Monitoring Indian Economy (CMIE), and the MoSPI were taken into account in arriving at various growth rates of activity data of sources of industrial sector. The reports of the Ministry of Road Transport and Highways (MoRTH) and data from the Regional Transport Office (RTO) of various states helped in getting growth rates of activity data of various vehicle categories. Reports from the CPCB and the Ministry of Agriculture (MoA) were considered for finding growth rates of activity data of garbage burning and agricultural residue burning. The trend analysis of the growth of emission loads of PM_{10} in India at the national level for individual years from 2000–2020 was done with the base year as 2010. In a similar way, we have also quantified the emission load of PM_{10} at a Himalayan eco-sensitive city (Shimla city) for the base year (2010) and assessed the growth trends for past and future 10 years with respect to the base year. Additionally, we compiled ambient air concentration levels of PM_{10} from the website of the SPCBs and the CPCB for these 20 years to assess the growth trends of atmospheric levels of PM_{10} concentration and to compare the concentration levels with trends of pollution loads of PM_{10}.

16.3 RESULTS AND DISCUSSION

16.3.1 OVERALL EMISSION INVENTORY OF PM_{10}

The spatially distributed map of PM_{10} emissions in India with a grid resolution of 40 km × 40 km is shown in Figure 16.4. Figure 16.4b also presents the breakup of emission estimates from four major polluting sources for the study year of 2010. The estimation revealed that 11,218 Gg/yr of total PM_{10} emission load were generated from four major sources of domestic, industry, vehicles, and open burning with a breakup of 39%, 28%, 19%, and 14%, respectively. The highest contribution (39%) from domestic cooking and heating process suggested that the old practice of fuel usage in rural areas prevailed. The old practice of the usage of fuels for this purpose included cow-dung cake, fuelwood, kerosene oil, and fuel coal (*Behera et al., 2015; Choudhury and Desai, 2020*). The source of open burning included crop-residue burning and street garbage burning, and this particular source of emission with significant contribution (10%) is a matter of concern, as it mostly adds to the pollution load during the winter season (*Yadav et al., 2020; Nagar and Sharma, 2022*). The crop-harvesting period and the timing of severe cold cause an increase in activity-level data for open-burning sources. Solid fuels, including varieties of fuel coal, were mostly used in manufacturing and energy industries in India. This might be the reason for a larger contribution of pollution loads (>25%) to the total load of PM_{10}. The contribution of vehicles at 19% to total PM_{10} emission load was due to the presence of large number of vehicles in urban areas compared to rural areas.

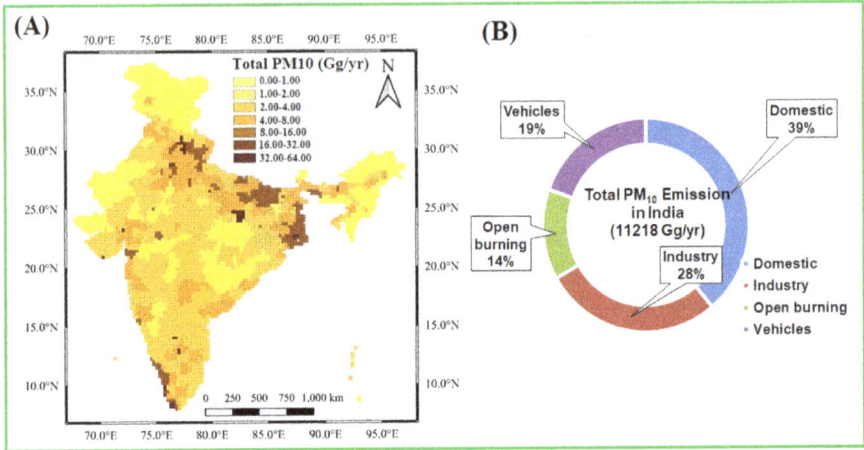

FIGURE 16.4 Emission inventory of total PM_{10} in India: (a) Spatial distribution in grid overlay of (40 km × 40 km) and (b) emission estimates from major sources.

Although this study, being a first of its kind that was conducted for development of emission inventory with higher resolution through application of a unique methodology from existing responsible activity-level data in India, we attempted to compare and contrast our results with those reported in the literature (*Kurokawa et al., 2013; Li et al., 2017; Janssens-Maenhout et al., 2013*). The results from previous studies conducted in the global context with India as a domain may not be directly compared with the results of our study, as the methodology used for emission estimates was different in each study. The activity-level data collection method, the pattern of application, and their quantities were different from each other. The consideration of EFs of all responsible sources were different from each other. However, we tried our best to provide an understanding of similarity in the range of our results. The Indian total emission load of PM_{10} from the study by *Kurokawa et al. (2013)* was 6651 Gg/yr in the study year of 2008. *Li et al. (2017)* provided the estimation of 5874 Gg/yr in the study year of 2006, and *Janssens-Maenhout et al. (2013)* provided 8268 Gg/yr in the study year of 2010. Our value of 11,218 Gg/yr for the study year of 2010 was at the upper side compared to previous studies. The reason for such results was due to selection of higher EFs and larger activity-level data compared to previous studies. Overall, we believe that our results were in the reasonable side of estimation, as we compiled activity-level data at district-level information and with a bottom-up approach of the emission inventory. Comparing the results of our previous study of *Paliwal et al. (2016)*, we got activity-level data from all responsible sources from the whole of India as a domain in a similar range of values. Additionally, we estimated uncertainty in our approach as explained subsequently.

To quantify the overall uncertainty in estimates of emission inventory computed for PM_{10} emission inventory for India, we used Monte Carlo simulations and

considered uncertainties in estimating EFs and activity-level data collected from district-level information compiled from various government reports during the simulation process (*Wang et al., 2008; Lu et al., 2011; Yang et al., 2015*). Based on the methodology reported by *Zhang et al. (2008)*, the statistical distributions of activity-level data were determined, and the uncertainty of distribution of activity level was estimated with the database obtained from the reports, as explained earlier. To quantify the uncertainty in the estimation, variations in two methods (bottom-up and top-down) were considered. The uncertainty estimation can be affected by the daily, monthly, and yearly fluctuations in activity-level data due to the methodology explained in the section "Materials and Methods." Similarly, uncertainty in the consideration of EFs from various sources also causes uncertainty in the overall estimation of the emissions of PM_{10}. The overall uncertainties at 95% confidence intervals (CIs) of the emission inventory developed for PM_{10} in this study during the base year (2010) were estimated as −41.5% to 38.7%. In inference, note that it has been found that emission speciation is an ongoing topic for additional study, and the results presented here include the caveat of this uncertainty (*Aneja et al., 2012; Thompson et al., 2017*). The fact remains that despite caveats in emission estimates, the presented method of estimation of PM_{10} emission loads from responsible sources provided results of parameters in a better range of significance.

The spatial distribution map shows that the IGP region comprising the states of Delhi, Uttarakhand, Uttar Pradesh, Bihar, West Bengal, Punjab, Haryana, and Himachal Pradesh generated higher emission loads in the range between 4.00–8.00 Gg/yr and 16.00–32.00 Gg/yr compared to other regions of India. Moreover, the urban areas such as Delhi, Ahmedabad, Chennai, Hyderabad, Kochi, Bangalore, Chandigarh, Lucknow, Patna, Bhopal, and Jaipur showed the highest pollution loads of PM_{10} in the range of 32.00–64.00 Gg/yr. The fact of the matter is that the IGP region contains fast growing urban areas and localities with larger population, industrial establishments, vehicular population, and more agricultural crop productions. The fast-growing cities showing the highest range of pollution loads showed maximum activities of vehicles, industries, densely populated with domestic cooking, and open burning during winter was occurring in these regions. Next to the IGP region, developed states like Kerala, Gujarat, Maharashtra, Andhra Pradesh, and Tamil Nadu showed emission loads in the range between 2.00–4.00 Gg/yr and 4.00–8.00 Gg/yr. The states of Jammu and Kashmir, Madhya Pradesh, Chhattisgarh, Orissa, Assam, and northeastern states showed lower emission loads in the range between 0.00–1.00 Gg/yr and 1.00–2.00 Gg/yr. The states of Rajasthan and Gujarat with regions of desert showed the least emission loads in the range of Gg/grid to 0.00–1.00 Gg/yr.

16.3.2   SOURCE-WISE EMISSION INVENTORIES OF PM_{10}

Figure 16.5 shows the spatial distribution of emission loads of PM_{10} in India from four major sectors of domestic, industries, vehicles, and open burning. The spatial distribution under the source category of domestic cooking varied in the range from 0.00–0.50 Gg/yr to 8.00–16.00 Gg/yr. The states under IGP region showed higher emission loads of PM_{10}, in the range from 2.00 to 8.00 Gg/yr. The cities in the

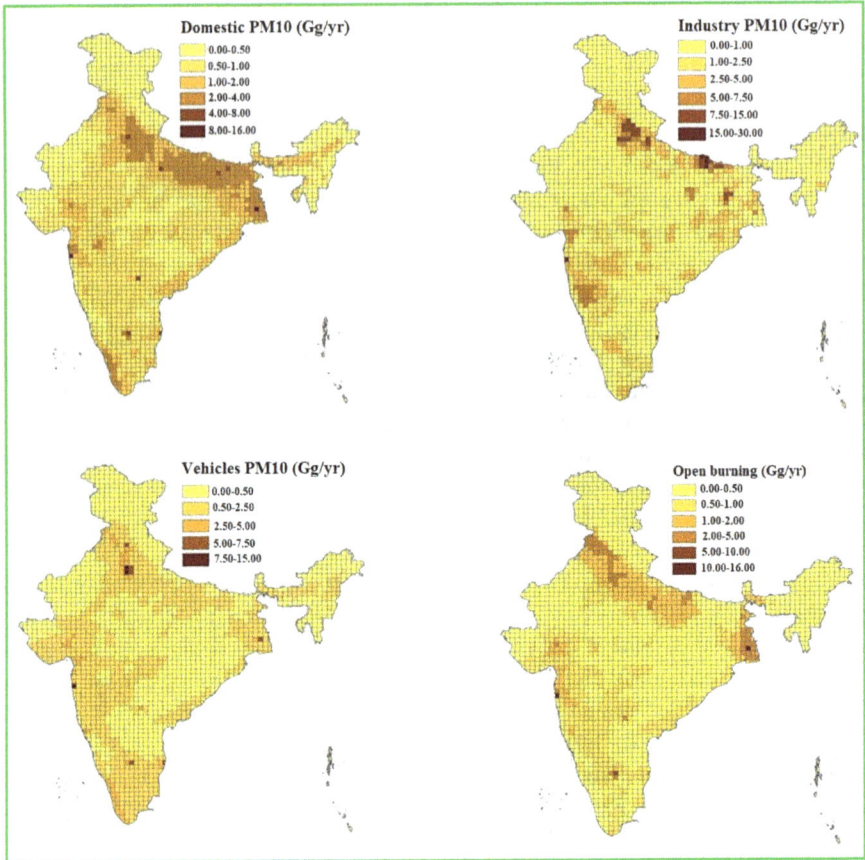

FIGURE 16.5 Spatial distribution of PM_{10} in India from four major sectors of domestic, industries, vehicles, and open burning.

metropolitan region and with higher population showed the highest emission loads in the range of 8.00–16.00 Gg/yr. The reason may be due to the fact that these regions had higher population densities (number/km2). Additionally, fuel uses of coal, fuel-wood, and other traditional fuel happened in semiurban regions (*Behera et al., 2015; Choudhury and Desai, 2020*). The regions under deserts, forested areas, and mountainous or hilly regions with lower populations experienced lower emission loads in the range from 0.00–0.50 to 1.00–2.00 Gg/yr. Under the source category of industries, the spatial variations of PM_{10} emission loads varied in the range from 0.00–1.00 to 15.00–30.00 Gg/yr. The industrial areas in the states of Punjab, Haryana, Gujarat, Tamil Nadu, Maharashtra, Uttar Pradesh, and other parts of India showed higher emission loads of PM_{10} in the range from 2.00–5.00 to 7.50–15.00 Gg/yr. The regions in the metropolitan coverages and their surroundings showed the maximum range PM_{10} pollution loads in the range from 15.00–30.00 Gg/yr. The regions of less developed areas in India showed lower PM_{10} emission loads in the range from 1.00–2.50 Gg/yr (*Paliwal et al., 2016*).

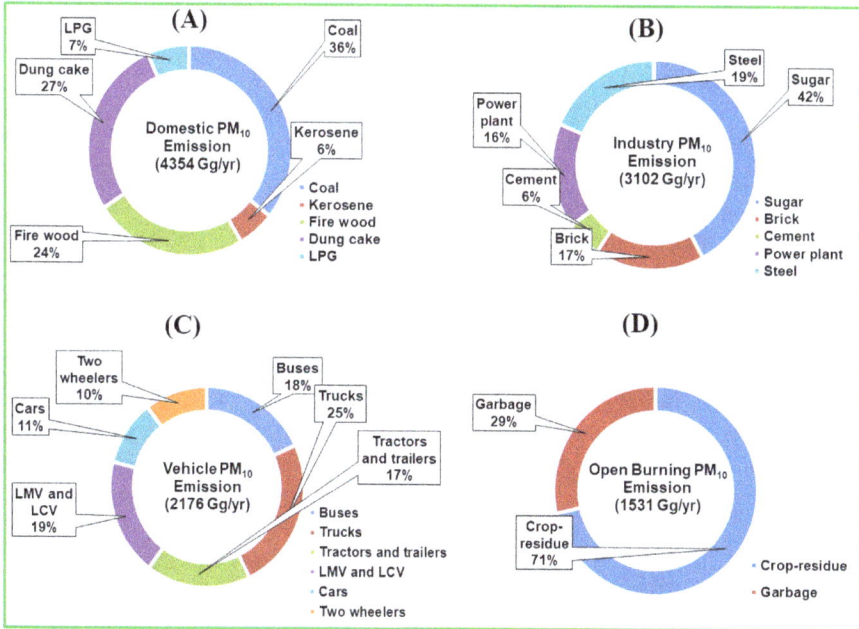

FIGURE 16.6   Emission estimates of PM_{10} in India's four major sectors: (a) domestic, (b) industries, (c) vehicles, and (d) open burning.

Under source categories of vehicles, the spatial variations of PM_{10} emission loads varied from 1.00–15.00 Gg/yr. The regions under metropolitan cities, urban and sub-urban areas showed higher PM_{10} pollution loads in the range from 2.50 to 15.00 Gg/yr. The regions coming under rural, deserted, forested, mountainous, and hilly areas in India showed the lowest PM_{10} emission loads in the range from 0.00–0.50 Gg/yr. Under source category of open burning, PM_{10} emission loads varied in the range from 0.00–1.500 to 10.00–16.00 Gg/yr. The regions with more crop harvesting (developed in terms of agricultural activities) and urban areas generating more garbage showed larger PM_{10} emission loads in the range from 2.00–16.00 Gg/yr. Similar to other sources, the regions with less frequency of crops, forested, mountainous and hilly, and deserted areas showed the least PM_{10} emission loads in the range from 0.00–1.00 Gg/yr. This is noteworthy that agricultural residue burning after crop harvesting and garbage burning in urban and semi-urban regions are very common in India (*Yadav et al., 2020; Nagar and Sharma, 2022*).

The major source-wise emission estimates with the breakup of sub-major sources of PM_{10} emissions at the national level are presented in Figure 16.6. The highest major contributor to PM_{10} emission, domestic sources had a breakdown of sub-major sources as follows: the use of coal at 1543 Gg/yr (36%), the use of dung cake at 1195 Gg/yr (27%), the use of firewood at 1056 Gg/yr (24%), use of LPG 286 Gg/yr (7%), and use of kerosene at 255 Gg/yr (6%). The results showed that uses of coal, dung cake, and firewood prevailed in large quantities in rural areas and semiurban regions in India causing higher contributions under the major source category of domestic cooking (*Rana*

et al., 2019; Choudhury and Desai, 2020). Next to domestic cooking, a major source in the industry sector, the following estimate values of PM_{10} emission were estimated as the sugar industry at 1307 Gg/yr (42%), the steelmaking industry at 581 Gg/yr (8%), brick manufacturing kilns at 535 Gg/yr (17%), power plants at 502 Gg/yr (16%), and the cement manufacturing industry at 177 Gg/yr (6%). The fuel, bagasse, is being burned for energy purposes in manufacturing of sugar. With higher contents of biomass matter, bagasse has more affinity to emit larger PM_{10} to the atmosphere (Mitchell et al., 2016). In India, most of the steelmaking and metallurgical industries use coal as their primary fuel during processes of manufacturing (*Shanmugam et al., 2021*).

Under the major source category of vehicles, the contributions from sub-major sources are estimated as follows: trucks at 536 Gg/yr (25%), LMV and LCV at 418 Gg/yr (19%), buses at 396 Gg/yr (18%), tractors and trailers at 380 Gg/yr (17%), cars at 231 Gg/yr (11%), and two-wheelers with 228 Gg/yr (10%). The results showed that engines with higher size and power (trucks, buses, tractors, trailers) emits more PM_{10} due to higher EF compared to other vehicles (cars and two-wheelers; *Wang et al., 2008; Lu et al., 2011; Yang et al., 2015*). The major source, open burning, emitted PM_{10} to the atmosphere from the following sub-major sources: agricultural crop-residue burning contributed 1090 Gg/yr at 71%, and garbage burning contributed 436 Gg/yr at 29% to emission estimates of open burning. Such a significant contribution from agricultural crop-residue burning was quite clear in our emission estimate, indicating that the maximum area under agricultural fields in India was the responsible player. Agricultural residue burning in India is very common after each harvesting period of crops (*Negi, 2006; Singh et al., 2020*).

16.3.3 TRENDS IN EMISSION LOADS OF PM_{10} IN INDIA

Figure 16.7 shows the trends in emission loads of PM_{10} in India at the national level during the last 10 years and the next 10 years. It was observed that the overall growth rate of PM_{10} emission load was equivalent to 3.41% per year. The emission load of PM_{10} was estimated at 8279 Gg/year in 2000, 11,218 Gg/year in 2010, and 16,604 Gg/year in 2020. The growth rate of PM_{10} emission load during 2000–2001 was found to be 2.28%, during 2009–2010 as 3.71%, and 2019–2020 as 4.51%. The overall trends showed an equivalent linear increase in growth of PM_{10} emission load. On assessing the growth in activity-level data, it was revealed that the average annual population growth rate in India is equivalent to 1.0%, an average annual growth rate in vehicle numbers equivalent to 8.5%, and an average annual growth in industries equivalent to 3.4%. The demand for agricultural production has increased due to the rise in population and a reduction in agricultural land area. This is noteworthy that the agricultural land area in India is being converted to residential or institutional zones. As it was clear that the agricultural residue burning is directly proportional to the field area, the rise in population caused the production of more agricultural residue followed by open burning on the field. Therefore, the overall annual growth rate for all responsible sources during these two decades fell in the range of growth rates of all individual activity-level data.

We compared our estimates on growth trends in PM_{10} emission with similar studies conducted in India. *Sahu et al. (2008)* reported particulate black carbon (BC) loads in India and provided the decadal growth rate as 61%. They provided the

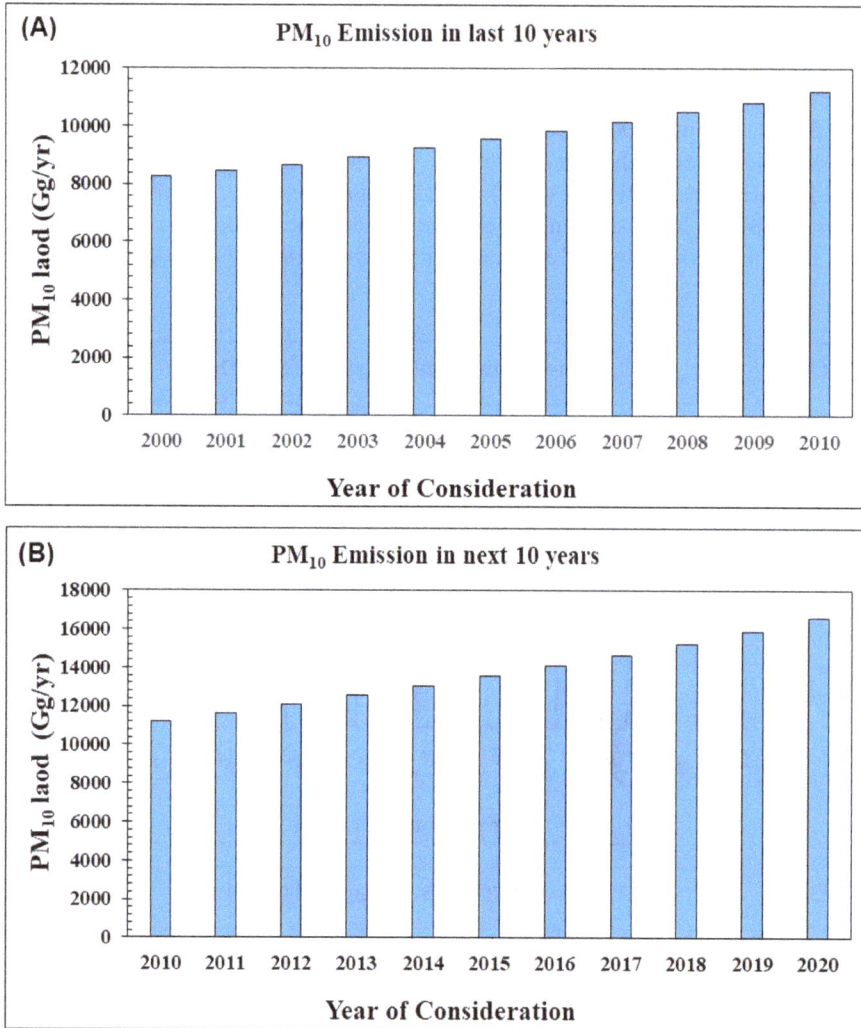

FIGURE 16.7   Trends in emission loads of PM_{10} in India at the national level during (a) last 10 years and (b) next 10 years.

estimation of BC emission at 1343.78 Gg and 835.50 Gg for the base years 2001 and 1991, respectively, with a decadal growth of about 61%, which was highly significant. *Kurokawa and Ohara (2020)* conducted a long-term historical emission inventory of air and climate pollutants in East, Southeast, and South Asia during 1950–2015 through the database developed as the Regional Emission inventory in ASia version 3 (REASv3). They have also provided the prediction of about 4.2% annual growth rate in PM_{10} emission loads in Asia. Hence, the trend analysis in the annual growth rate of PM_{10} in India in this study falls in the range of the previous study.

16.3.4 TRENDS IN EMISSION LOADS AND AMBIENT CONCENTRATION OF PM$_{10}$ IN A HIMALAYAN CITY

Figure 16.8 shows the trends in emissions of PM$_{10}$ in the Himalayan Eco-sensitive city, Shimla during 2000–2020. Both PM$_{10}$ emission load in the unit of Gg/yr, and concentration levels in the unit of μg/m^3 are depicted in the figure. The emission load of PM$_{10}$ during the study year (2010) was 1.49 Gg/yr, and the city with a population of about 1.8 lakh had a substantial generation of emission, which is a matter of concern from the perspectives of eco-sensitivity of the region. The estimated values in

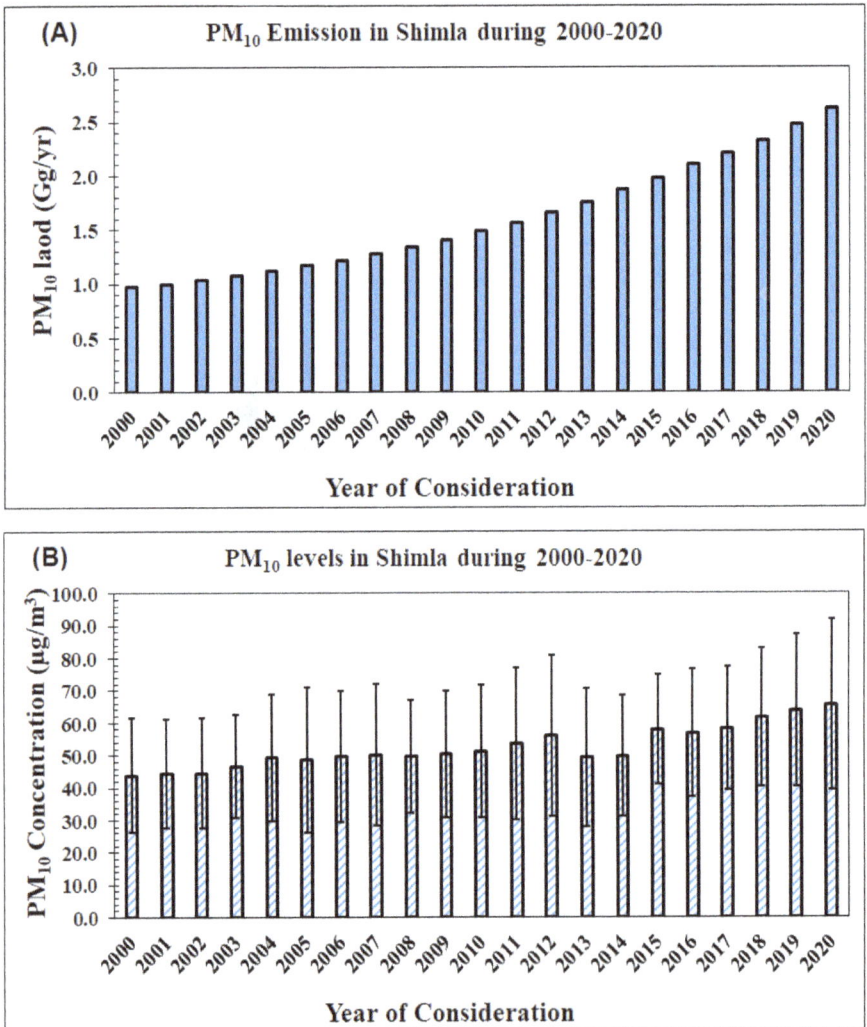

FIGURE 16.8 Trends in emissions of PM$_{10}$ in the Himalayan eco-sensitive city Shimla during 2000–2020: (a) PM$_{10}$ emission load (Gg/yr) and (b) PM$_{10}$ concentration levels (μg/m^3).

the last 10 years showed an increasing trend in emission load of PM_{10} with 0.97 Gg/yr during 2000. The annual average growth rate during the last decade was 4.39%, which was higher than the national annual average growth of 3.1%. The emission load of PM_{10} was estimated at 2.62 Gg/yr during 2020. The annual average growth rate of PM_{10} emission load during the next decade was estimated to be 5.71%, which was higher than the national annual average of 4.05%. The growth rate of PM_{10} emission load in Shimla was at 3.01% during 2000–2001, 5.37% during 2009–2010, and 6.05% during 2019–2020. With an overall assessment, the Indian national growth rate of PM_{10} emission load during these two decades (2000–2020) 3.41%, whereas the average growth of PM_{10} emission load in Shimla was 5.04% during these two decades. The results and trends showed that the growth of PM_{10} emission load was higher in Shimla compared to the growth rate of the nation. The fact of the matter is that the population and vehicular growth rates of Shimla city were much higher than the national average growth rates.

The monthly average data on the concentration levels of PM_{10} in the unit of $\mu g/m^3$ at monitoring sites of Shimla city were averaged in finding the annual average with a standard deviation. During the last decade, the concentration level of PM_{10} was observed as 43.9 ± 17.6 $\mu g/m^3$ in GIS-based emission inventory 2000 and 51.2 ± 20.4 $\mu g/m^3$ in 2010. There was clear increase in concentration level of PM_{10} during the last decade (2000–2010) at 1.57%. Similarly in the next decade (2010–2020), the concentration level of PM_{10} 65.6 ± 26.3 $\mu g/m^3$ with a decadal growth rate of 2.65%. Although there were some years, the fall in concentration levels of PM_{10} occurred during these two decades (2000–2020), the average annual growth rate was at 2.15%. The growth rates in both the emission load and the concentration level of PM_{10} in Shimla city indicated that pollution control policies should be strategized to reduce levels of emission loads in this eco-sensitive city.

16.4   CONCLUSION

Through the systematic GIS-based methodology described and adopted in this study, we developed a GIS-based emission inventory of PM_{10} from four major sources (domestic, industries, vehicles, open burning) in grid resolution of 40 km × 40 km with 2010 as the base year. The results showed that 11,218 Gg/yr of total PM_{10} was emitted during 2010 in India from four major pollution sources of domestic, industry, vehicles, and open burning with a breakup of 39%, 28%, 19%, and 14%, respectively, to total PM_{10} emission load. From the spatial distribution of emission load of PM_{10}, it was seen that the urban areas such as Delhi, Ahmedabad, Chennai, Hyderabad, Kochi, Bangalore, Chandigarh, Lucknow, Patna, Bhopal, and Jaipur showed the highest pollution loads of PM_{10} in the range of 32.00–64.00 Gg/yr. Next to these urban regions, the IGP region comprising the states of Uttarakhand, Uttar Pradesh, Bihar, West Bengal, Punjab, Haryana, and Himachal Pradesh generated higher emission loads in the range of 4.00–8.00 Gg/yr and 16.00–32.00 Gg/yr compared to remaining regions of India. The overall annual growth rate of PM_{10} emission load during these two decades (2000–2020) was equivalent to 3.41% per year at the national level in India. Specifically, the emission load of PM_{10} was estimated at 8279 Gg/year during 2000, 11,218 Gg/year during 2010, and 16,604 Gg/year during 2020. The average

growth of PM_{10} emission load in the Himalayan eco-sensitive region at Shimla city was 5.04% during these two decades (2000–2020). The average annual growth rate in ambient concentration levels of PM_{10} in Shimla city was found to be 2.15% during these two decades (2000–2020). Overall, the growth rates of both emission load and ambient concentration level of PM_{10} in Shimla city suggested strategizing stringent pollution control policies is required to reduce levels of emission loads in this eco-sensitive city with a future perspective.

REFERENCES

Al-Thani, H., M. Koç, and R.J. Isaifan. 2018. "A review on the direct effect of particulate atmospheric pollution on materials and its mitigation for sustainable cities and societies." *Environmental Science and Pollution Research* 25 (28): 27839–27857. https://doi.org/10.1007/s11356-018-2952-8

Aneja, V.P., W.H. Schlesinger, J.W. Erisman, S.N. Behera, M. Sharma, and W. Battye. 2012. "Reactive nitrogen emissions from crop and livestock farming in India." *Atmospheric Environment* 47: 92–103. https://doi.org/10.1016/j.atmosenv.2011.11.026

Anjum, M.S., S.M. Ali, M.A. Subhani, M.N. Anwar, A.S. Nizami, U. Ashraf, and M.F. Khokhar. 2020. "An emerged challenge of air pollution and ever-increasing particulate matter in Pakistan; a critical review." *Journal of Hazardous Materials* 402: 123943. https://doi.org/10.1016/j.jhazmat.2020.123943

Aswini, A.R., P. Hegde, P.R. Nair, and S. Aryasree. 2019. "Seasonal changes in carbonaceous aerosols over a tropical coastal location in response to meteorological processes." *Science of the Total Environment* 656: 1261–1279. https://doi.org/10.1016/j.scitotenv.2018.11.366

Behera, B., A. Jeetendra, and A. Ali. 2015. "Household collection and use of biomass energy sources in South Asia." *Energy* 85: 468–480. https://doi.org/10.1016/j.energy.2015.03.059

Behera, S.N. and M. Sharma. 2015. "Spatial and seasonal variations of atmospheric particulate carbon fractions and identification of secondary sources at urban sites in North India." *Environmental Science and Pollution Research* 22 (17): 13464–13476. https://doi.org/10.1007/s11356-015-4603-7

Behera, S.N., M. Sharma, and S.P. Shukla. 2016. "Characterization of gaseous pollutants and water-soluble inorganic ions in PM2.5 during summer time at an urban site of North India." *Journal of Hazardous, Toxic, and Radioactive Waste* 20 (4): A4015002. https://doi.org/10.1061/(ASCE)HZ.2153-5515.0000299

CEA. 2010. "Central Electricity Authority, Ministry of Power, India. Annual Report 2009–10." www.cea.nic.in/reports/annual/annualreports/annual_report-2010.pdf. Last accessed on 20 April 2011.

Chakraborty, R., U. Saha, A.K. Singh, and A. Maitra. 2017. "Association of atmospheric pollution and instability indices: A detailed investigation over an Indian urban metropolis." *Atmospheric Research* 196: 83–96. https://doi.org/10.1016/j.atmosres.2017.04.033

Choudhury, P., and S. Desai. 2020. "Gender inequalities and household fuel choice in India." *Journal of Cleaner Production* 265: 121487. https://doi.org/10.1016/j.jclepro.2020.121487

CMIE. 2010. "Centre for Monitoring Indian Economy. Reports during 2009–2010." www.cmie.com/. Last accessed on 24 March 2011.

CPCB. 2010. "Central Pollution Control Board. Air quality monitoring, emission inventory and source apportionment study for Indian cities." https://cpcb.nic.in/source-apportionment-studies. Last accessed on 22 February 2011.

Gadde, B., S. Bonnet, C. Menke, and S. Garivait. 2009. "Air pollutant emissions from rice straw open field burning in India, Thailand and the Philippines." *Environmental Pollution* 157 (5): 1554–1558. https://doi.org/10.1016/j.envpol.2009.01.004

Gao, C., W. Gao, K. Song, H. Na, F. Tian, and S. Zhang. 2019. "Spatial and temporal dynamics of air-pollutant emission inventory of steel industry in China: A bottom-up approach." *Resources, Conservation and Recycling* 143: 184–200. https://doi.org/10.1016/j.resconrec.2018.12.032

Garaga, R., S.K. Sahu, and S.H. Kota. 2018. "A review of air quality modeling studies in India: Local and regional scale." *Current Pollution Reports* 4 (2): 59–73. https://doi.org/10.1007/s40726-018-0081-0

Garg, A., P.A. Shukla, and M. Kapshe. 2006. "The sectoral trends of multigas emissions inventory of India." *Atmospheric Environment* 40 (24): 4608–4620. https://doi.org/10.1016/j.atmosenv.2006.03.045

Gautam, D., and N.B. Bolia. 2020. "Air pollution: impact and interventions." *Air Quality, Atmosphere & Health* 13 (2): 209–223. https://doi.org/10.1007/s11869-019-00784-8

Hakkim, H., A. Kumar, S. Annadate, B. Sinha, and V. Sinha. 2021. "RTEII: A new high-resolution (0.1°× 0.1°) road transport emission inventory for India of 74 speciated NMVOCs, CO, NOx, NH_3, CH_4, CO_2, $PM_{2.5}$ reveals massive overestimation of NOx and CO and missing nitromethane emissions by existing inventories." *Atmospheric Environment: X* 11: 100118. https://doi.org/10.1016/j.aeaoa.2021.100118

Janssens-Maenhout, G., V. Pagliari, D. Guizzardi, and M. Muntean. 2013. "Global emission inventories in the Emission Database for Global Atmospheric Research (EDGAR)–Manual (I). Gridding: EDGAR emissions distribution on global gridmaps." *Publications Office of the European Union, Luxembourg* 775. https://publications.jrc.ec.europa.eu/repository/bitstream/JRC78261/edgarv4_manual_i_gridding_pubsy_final.pdf

Kurokawa, J., and T. Ohara. 2020. "Long-term historical trends in air pollutant emissions in Asia: Regional Emission inventory in ASia (REAS) version 3. *Atmospheric Chemistry and Physics* 20 (21): 12761–12793. https://doi.org/10.5194/acp-20-12761-2020

Kurokawa, J., T. Ohara, T. Morikawa, S. Hanayama, G. Janssens-Maenhout, T. Fukui, K. Kawashima, and H. Akimoto. 2013. "Emissions of air pollutants and greenhouse gases over Asian regions during 2000–2008: Regional emission inventory in ASia (REAS) version 2" *Atmospheric Chemistry and Physics* 13: 11019–11058. https://doi.org/10.5194/acp-13-11019-2013

Lam, Y.F., C.C. Cheung, X. Zhang, J.S. Fu, and J.C.H. Fung. 2021. "Development of a new emission reallocation method for industrial sources in China." *Atmospheric Chemistry and Physics* 21 (17): 12895–12908. https://doi.org/10.5194/acp-21-12895-2021

Lau, K.M., K.M. Kim, and S. Yang. 2000. "Dynamical and boundary forcing characteristics of regional components of the Asian summer monsoon." *Journal of Climate* 13 (14): 2461–2482. https://doi.org/10.1175/1520-0442(2000)013<2461:DABFCO>2.0.CO;2

Li, M., Q. Zhang, J.I. Kurokawa, J.H. Woo, K. He, Z. Lu et al. 2017. "MIX: A mosaic Asian anthropogenic emission inventory under the international collaboration framework of the MICS-Asia and HTAP." *Atmospheric Chemistry and Physics* 17 (2): 935–963. https://doi.org/10.5194/acp-17-935-2017

Lu, Z., Q. Zhang, and D.G. Streets. 2011. "Sulfur dioxide and primary carbonaceous aerosol emissions in China and India, 1996–2010." *Atmospheric Chemistry and Physics* 11 (18): 9839–9864. https://doi.org/10.5194/acp-11-9839-2011

Madadi, A., A.H. Sadr, A. Kashani, A.G. Gilandeh, V. Safarianzengir, and M. Kianian. 2021. "Monitoring of aerosols and studying its effects on the environment and humans health in Iran." *Environmental Geochemistry and Health* 43 (1): 317–331. https://doi.org/10.1007/s10653-020-00709-w

Ma, X., I. Longley, J. Gao, A. Kachhara, and J. Salmond. 2019. "A site-optimised multi-scale GIS based land use regression model for simulating local scale patterns in air pollution." *Science of the Total Environment* 685: 134–149. https://doi.org/10.1016/j.scitotenv.2019.05.408

Mitchell, E.J.S., A.R. Lea-Langton, J.M. Jones, A. Williams, P. Layden, and R. Johnson. 2016. "The impact of fuel properties on the emissions from the combustion of biomass and other solid fuels in a fixed bed domestic stove." *Fuel Processing Technology* 142: 115–123. https://doi.org/10.1016/j.fuproc.2015.09.031

MoA. 2010. "Ministry of Agriculture, India. Annual Report 2009–2010." https://icar.org.in/content/dareicar-annual-report-2009–2010. Last accessed on 20 March 2011.

MoC. 2010. "Ministry of Coal, India. Annual Report 2009–2010." https://coal.nic.in/content/annual-report-2009–10. Last accessed on 17 March 2011.

MoP. 2010. "Ministry of Power, India. Annual Report 2009–2010." https://powermin.gov.in/sites/default/files/uploads/Annual_Report_2009-10_English.pdf. Last accessed on 20 March 2011.

MoPNG. 2010. "Ministry of Petroleum and Natural Gas, India. Annual Report 2009–2010." http://petroleum.nic.in/sites/default/files/AR09-10.pdf. Last accessed on 20 March 2011.

MoRTH. 2011. "Ministry of Road Transport and Highways, India. Annual Report 2010–2011." https://morth.nic.in/sites/default/files/Annual-Report-2010-2011.pdf. Last accessed on 24 March 2012.

MoSPI. 2010. "Ministry of Statistics and Program Implementation, India. Household Consumption of Various Goods and Services in India. Reports of 2009–2010." http://mospi.nic.in/download-reports. Last accessed on 24 February 2011.

Nagar, P.K., and M. Sharma. 2022. "A hybrid model to improve WRF-Chem performance for crop burning emissions of $PM_{2.5}$ and secondary aerosols in North India." *Urban Climate* 41: 101084. https://doi.org/10.1016/j.uclim.2022.101084

NCAER. 2010. "National Survey of Household Income and Expenditure (2009–10), NCAER New Delhi.". www.ncaer.org/data.php#data27. Last accessed on 14 March 2011.

Negi, G.C.S. 2006. "Leaf and bud demography and shoot growth in evergreen and deciduous trees of central Himalaya, India." *Trees* 20 (4): 416–429. https://doi.org/10.1007/s00468-006-0056-4

Paliwal, U., M. Sharma, and J.F. Burkhart. 2016. "Monthly and spatially resolved black carbon emission inventory of India: uncertainty analysis." *Atmospheric Chemistry and Physics* 16 (19): 12457–12476. https://doi.org/10.5194/acp-16-12457-2016

Pucher, J., Z.R. Peng, N. Mittal, Y. Zhu, and N. Korattyswaroopam. 2007. "Urban transport trends and policies in China and India: impacts of rapid economic growth." *Transport Reviews* 27(4): 379–410. https://doi.org/10.1080/01441640601089988

Rafaj, P., G. Kiesewetter, T. Gül, W. Schöpp, J. Cofala, Z. Klimont, P. Purohit, C. Heyes, M. Amann, J. Borken-Kleefeld, and L. Cozzi. 2018. "Outlook for clean air in the context of sustainable development goals." *Global Environmental Change* 53: 1–11. https://doi.org/10.1016/j.gloenvcha.2018.08.008

Rana, A., S. Jia, and S. Sarkar. 2019. "Black carbon aerosol in India: A comprehensive review of current status and future prospects." *Atmospheric Research* 218: 207–230. https://doi.org/10.1016/j.atmosres.2018.12.002

Reddy, M.S., and C. Venkataraman. 2002a. "Inventory of aerosol and sulphur dioxide emissions from India: I—Fossil fuel combustion." *Atmospheric Environment* 36 (4): 677–697. https://doi.org/10.1016/S1352-2310(01)00463-0

Reddy, M.S., and C. Venkataraman. 2002b. "Inventory of aerosol and sulphur dioxide emissions from India. Part II—biomass combustion." *Atmospheric Environment* 36 (4): 699–712. https://doi.org/10.1016/S1352-2310(01)00464-2

Romshoo, S.A., M.A. Bhat, and G. Beig. 2021. "Particulate pollution over an urban Himalayan site: Temporal variability, impact of meteorology and potential source regions." *Science of The Total Environment* 799: 149364. https://doi.org/10.1016/j.scitotenv.2021.149364

Sahai, S., C. Sharma, D.P. Singh, C.K. Dixit, N. Singh, P. Sharma, K. Singh, S. Bhatt, S. Ghude, V. Gupta, R.K. Gupta, M.K. Tiwari, S.C. Garg, A.P. Mitra, and Gupta, P.K. 2007. "A study for development of emission factors for trace gases and carbonaceous particulate species from in situ burning of wheat straw in agricultural fields in India." *Atmospheric Environment* 41 (39): 9173–9186. https://doi.org/10.1016/j.atmosenv.2007.07.054

Sahu, S.K., G. Beig, and C. Sharma. 2008. "Decadal growth of black carbon emissions in India." *Geophysical Research Letters* 35 (2): L02807. https://doi.org/10.1029/2007GL032333

Shanmugam, S.P., V.N. Nurni, S. Manjini, S. Chandra, and L.E. Holappa. 2021. Challenges and Outlines of Steelmaking toward the Year 2030 and beyond—Indian Perspective. *Metals* 11 (10): 1654. https://doi.org/10.3390/met11101654

Sharma, R., R. Kumar, D.K. Sharma, I. Priyadarshini, B.T. Pham, D.T. Bui, and S. Rai. 2019. "Inferring air pollution from air quality index by different geographical areas: Case study in India." *Air Quality, Atmosphere & Health* 12 (11): 1347–1357. https://doi.org/10.1007/s11869-019-00749-x

Singh, D., S.P. Shukla, M. Sharma, S.N. Behera, D. Mohan, N.B. Singh, and G. Pandey. 2016. "GIS-based on-road vehicular emission inventory for Lucknow, India." *Journal of Hazardous, Toxic, and Radioactive Waste* 20 (4): A4014006. https://doi.org/10.1061/(ASCE)HZ.2153-5515.0000244

Singh, R., D.B. Yadav, N. Ravisankar, A. Yadav, and H. Singh. 2020. "Crop residue management in rice–wheat cropping system for resource conservation and environmental protection in north-western India." *Environment, Development and Sustainability* 22 (5): 3871–3896. https://doi.org/10.1007/s10668-019-00370-z

Thompson, T.M., D. Shepherd, A. Stacy, M.G. Barna, and B.A. Schichtel. 2017. "Modeling to evaluate contribution of oil and gas emissions to air pollution." *Journal of the Air & Waste Management Association* 67 (4): 445–461. https://doi.org/10.1080/10962247.2016.1251508

Umar, S.I.U., D. Konwar, A. Khan, M.A. Bhat, F. Javid, R. Jeelani, B. Nabi, A.A. Najar, D. Kumar, and B. Brahma. 2021. "Delineation of temperature-humidity index (THI) as indicator of heat stress in riverine buffaloes (Bubalus bubalis) of a sub-tropical Indian region." *Cell Stress and Chaperones* 26: 657–669: https://doi.org/10.1007/s12192-021-01209-1

Vohra, K., E.A., Marais, S. Suckra, L. Kramer, W.J. Bloss, R. Sahu, A. Gaur, S.N. Tripathi, M. Van Damme, L. Clarisse, and P.F. Coheur. 2021. "Long-term trends in air quality in major cities in the UK and India: A view from space." *Atmospheric Chemistry and Physics* 21 (8): 6275–6296. https://doi.org/10.5194/acp-21-6275-2021

Von-Schneidemesser, E., P.S. Monks, J.D. Allan, L. Bruhwiler, P. Forster, D. Fowler, A. Lauer, W.T. Morgan, P. Paasonen, M. Righi, K. Sindelarova, and M.A. Sutton. 2015. "Chemistry and the linkages between air quality and climate change." *Chemical Reviews* 115 (10): 3856–3897. https://doi.org/10.1021/acs.chemrev.5b00089

Wang, H., C. Chen, C. Huang, and L. Fu. 2008. "On-road vehicle emission inventory and its uncertainty analysis for Shanghai, China." *Science of The Total Environment* 398 (1–3): 60–67. https://doi.org/10.1016/j.scitotenv.2008.01.038

Wang, H., L. Fu, X. Lin, Y. Zhou, and J. Chen. 2009. "A bottom-up methodology to estimate vehicle emissions for the Beijing urban area." *Science of The Total Environment* 407 (6): 1947–1953. https://doi.org/10.1016/j.scitotenv.2008.11.008

Yadav, A., S.N. Behera, P.K. Nagar, and M. Sharma. 2020. "Spatio-seasonal concentrations, source apportionment and assessment of associated human health risks of $PM_{2.5}$-bound polycyclic aromatic hydrocarbons in Delhi, India." *Aerosol and Air Quality Research* 20: 2805–2825. https://doi.org/10.4209/aaqr.2020.04.0182

Yang, X.F., H. Liu, Y.H. Man, and K.B. He. 2015. "Characterization of road freight trans-
    portation and its impact on the national emission inventory in China." *Atmospheric
    Chemistry and Physics* 15 (4): 2105–2118. https://doi.org/10.5194/acp-15-2105-2015
Zhang, Q., Y. Wei, W. Tian, and K. Yang. 2008. GIS-based emission inventories of urban
    scale: A case study of Hangzhou, China. *Atmospheric Environment* 42 (20): 5150–5165.
    https://doi.org/10.1016/j.atmosenv.2008.02.012

17 Spatial Distribution of Particulate Organic Carbon over India and the Prediction of Its Deposition in the Himalayas through the GIS-WRF-CAMx Modeling System

Ajay K. Singh[1], Sailesh N. Behera[2,3,*],
Mukesh Sharma[1], and Bikash Ranjan Parida[4]

1 Department of Civil Engineering, Indian
  Institute of Technology Kanpur, Kanpur,
  Uttar Pradesh, PIN: 208016, India.

2 Air Quality Laboratory, Department of Civil Engineering,
  Shiv Nadar University Delhi-NCR, Greater Noida, Gautam
  Buddha Nagar, Uttar Pradesh, PIN: 201314, India.

3 Centre for Environmental Sciences and Engineering, Shiv
  Nadar University Delhi-NCR, Greater Noida, Gautam
  Buddha Nagar, Uttar Pradesh, PIN: 201314, India.

4 Department of Geoinformatics, School of Natural
  Resource Management, Central University of
  Jharkhand, Ranchi, PIN: 835222, India.

CONTENTS

17.1 Introduction 344
17.2 Materials and Methods 346
    17.2.1 Characteristics of the Study Domain.................................... 346
    17.2.2 Configuration of Study Aspects... 348
    17.2.3 Simulation Scenarios 348

DOI: 10.1201/9781003265160-20

    17.2.4 Development of the GIS-Based Emission Inventory 350
    17.2.5 Postprocessing GIS-Based EI for the CAMx Model........................352
    17.2.6 WRF Meteorological Modeling 352
    17.2.7 Chemical Transport Modeling Using the CAMx Model..................353
17.3 Results and Discussion 356
    17.3.1 Overall Emission Inventory of POC...356
    17.3.2 Sector-Wise Emissions of POC at the National Level.....................358
    17.3.3 Backward Trajectory Analysis...359
    17.3.4 Model Evaluation and Validation Exercise......................................361
    17.3.5 Deposition Rates and Ambient Concentration of POC
             at the Himalayan Range.. 364
    17.3.6 Diurnal Variations of Deposition Rates and Ambient
             Concentration of POC.. 366
    17.3.7 Deposition Rates and Ambient Concentration of POC at Peaks
             and Glaciers 367
17.4 Conclusion 368
References.. 369

17.1 INTRODUCTION

Airborne particulate matter (PM) or aerosol influences the planet Earth's radiation budget directly by scattering of sunlight and indirectly by the formation of cloud condensation nuclei in the atmosphere. As an abundant component of PM, carbonaceous aerosol has significant contributions to many environmental impacts including reduction in atmospheric visibility, climate change, and associated human health effects (*Carmichael et al., 2009; Behera and Sharma, 2015*). The particulate organic carbon (OC) is the nonabsorptive fraction of carbonaceous aerosol, which has a lower molecular weight than elemental carbon (EC; *Szidat et al., 2009; Bauer and Menon, 2012*). The particulate OC may contain reactive compounds, which take part in atmospheric chemical transformations and support the enhancing condensation of clouds or act as toxins and allergens (*Szidat et al., 2009; Yadav et al., 2020*). The particulate OC is emitted as primary organic carbon particles and can be formed as secondary organic carbon particles from condensation of organic gases of low-volatile and semi-volatile origins (*Jiang et al., 2011; Singh et al., 2021*).

Primary organic carbon (POC) has been considered as a larger contributor to total OC, which is a complex mixture of chemical compounds containing carbon–carbon bonds produced mainly from the combustion of fossil fuels and biofuels and the burning of biomass present in the relevant matter (*Behera and Sharma, 2015; Bhowmik et al., 2021; Izhar et al., 2021*). The major impacts of particulate OC emissions on the environment can be compiled as changes in a decrease in atmospheric visibility, temperature, precipitation, melting of snow, global warming, and many other effects (*Armitage et al., 2011; Schmale et al., 2021*). Although particulate OC has a net cooling effect, then too much of its emission is considered as bad because EC is co-emitted from the same sources of origin (*Hallquist et al., 2009; Moran et al., 2012*). The particulate EC is held responsible substantially for global warming, and hence, emissions of particulate OC have indirect impacts on climate through the melting of snow and global warming (*Marcq et al., 2010; Bauer and Menon, 2012*).

To meet the growth potential in urbanization, industrialization. and transportation, a developing country like India requires a great amount of energy from conventional means. As a result, a steady increase in emissions of greenhouse gases (GHGs) and PM happens in the entire country (*Yadav et al., 2020; Dhanya et al., 2021; Mele and Magazzino, 2021*). This is noteworthy that the prevailing westerlies wind and favorable synoptic meteorological conditions have the higher potential to take emitted PM from the Indian mainland to various parts of the Himalayan region through long-range transport followed by its deposition through dry and wet ways on the ecologically sensitive locations of peaks and glaciers (*Dumka et al., 2019; Romshoo et al., 2021*). Overall, this atmospheric phenomenon causes impacts on the Himalayan region in melting glaciers and increasing temperature (through co-emission of particulate EC) in that ecosystem. The emissions of particulate OC from the activities like fossil fuel, biomass and biofuel burning, transportation, and land clearing have major impacts on climate change, particularly at regional scales (*Liu et al., 2020; Price-Allison et al., 2021*). Nevertheless, the precise role of particulate OC emissions in any climate-change mitigation strategy remains uncertain because of the coexistence of the emitted particulate EC. The co-emission of OC and EC varies with fuel type, combustion efficiency in engines or industries, and the extent of emission control adopted. When fossil fuels, such as oil and coal are incompletely combusted (not completely oxidized to carbon dioxide [CO_2]), particulate EC tends to form in much larger amounts compared to particulate OC (*Bond et al., 2004; Mancilla et al., 2015*). When biomass fuel, such as fuel wood is incompletely combusted, particulate OC forms in larger amounts than that of particulate EC (*Streets et al., 2004; Williams et al., 2012; Mitchell et al., 2016*).

Like other particulate compositions, EC and OC have indirect radiative effects on clouds by altering their albedo and lifetime. Some researchers have postulated that reduction in particulate OC may be a relatively inexpensive and achievable method to reduce future temperature (*Jacobson, 2002; Hansen et al., 2000; Shindell et al., 2012*). Different sectors generate different net positive or negative forcing based on the differences in relative amounts of EC versus OC, implying that targeting OC reductions as a means of climate change mitigation may be a viable option for certain sectors. Regardless of the role of particulate OC emissions in climate change, a reduction in those emissions expects to have significant health benefits, as particulate OC can create severe breathing problems leading to asthmatic conditions (*He et al., 2021; Wu et al., 2021*).

In addition to several known environmental impacts of particulate OC, the recent thinning of glaciers over the Himalayas (sometimes referred to as the third polar cap) due to the co-emitted particulate EC is a matter of concern for all sections of responsible stakeholders across the globe. Such a consequence has raised concern for future water supplies since these glaciers supply water to large river systems that support millions of people inhibiting the surrounding areas, especially in India (*Menon et al., 2010; Qian et al., 2011*). A past study has suggested that 915 km² of Himalayan glaciers in Spiti/Lahaul (30–33°N, 76–79°E) thinned by an annual average of 0.85 m from 1999 to 2004 (*Berthier et al., 2007*). The assessment of the deposition of these aerosols (combined effect of EC and OC) on snow over the Himalayas reveals an increase in the thinning rate of glaciers, which in turn exposes land and open water. Hence, an even greater portion of solar radiation is absorbed due to its relatively darker surfaces (low albedo; *Niu et al., 2020; Nie et al., 2021*). This creates positive feedback, which further accelerates the melting of snow. For the past decade, the Himalayan glaciers have decreased by 1% due

to the deposition of these aerosols. Some Indian past studies (*Ojha et al., 2016; Guo et al., 2017; Sharma et al., 2017*) have addressed the scope of work on emission inventory development and its application in air quality modeling exercise at the national level. However, the fate of ambient PM or its composition like OC from the source of origin to their deposition over the Himalayan regions has not been studied so far.

This study is a first of its kind that predicted hourly average deposition rate (dry and wet), and concentration of POC (OC from the primary source of origin) on the Himalayas, especially at the locations of seven major peaks and seven major glaciers during three seasons (summer, monsoon and winter) in 2008. The predicted deposition rates and ambient concentrations of POC were assessed due to transport of emissions of POC over the study region with India and some of its neighboring countries including Pakistan, Nepal, Sri Lanka, Bangladesh, and Myanmar to the Himalayan regions. The emission inventory results for India developed from this study were used in the Comprehensive Air Quality Model (CAMx) modeling. For emission results from neighboring countries, we considered emission data from the REAS (Regional Emission Inventory in Asia) database (*Ohara et al., 2007*). The scopes of research in this study demonstrated a thorough modeling concept of a chemical transport model (CTM), that is, the CAMx model. This technical method is famously known as a complete third generation modeling approach that combines various capabilities of Geographic Information System (GIS), meteorological model (Weather Research Forecasting [WRF]), and the CAMx model. The emission, dispersion, chemical reactions, and deposition rate of POC can be simulated by solving the pollutant continuity on a system of nested three-dimensional grids in this integrated modeling platform.

The specific objectives of this study are described as follows: (1) to develop a detailed national-level GIS-based emission inventory of POC emitted from major sources, such as open burning, domestic cooking, industries and vehicles on a finer resolution of 40 km × 40 km grids; (2) to configure and run the models with different scenarios in checking the sensitivity of the models in terms predictions of emissions, meteorology and deposition rates; and (3) to formulate a three-tier modeling system using the applications of GIS-WRF-CAMx to assess emission rates of POC from India, and neighboring countries, its atmospheric transport and finally its deposition rates (dry and wet) over the regions of Himalayan Range.

17.2 MATERIALS AND METHODS

17.2.1 CHARACTERISTICS OF THE STUDY DOMAIN

The Indian climatic conditions are much varied due to its unique geography and geology. This can be seen such as large variations from the colder Himalayas in the northern direction to the hotter Thar Desert in the northwest direction (*Aneja et al., 2012; Sen et al., 2017*). The specialty of the Indian region is that the designated regions have different microclimatic conditions and other features compared to other regions. Hence, it can be seen that rainforests in the southwest and southern Indian peninsular region situated under humid tropical conditions, alpine tundra and glaciers located in the north, and deserts existing in the west (*Negi, 2006; Saraf et al., 2011*). Although the Tropic of Cancer (line dividing the tropics and the subtropics) passes through the middle of the country, overall, India is considered a tropical

country. The rise in activities of open burning of biomass waste, agricultural residue, and garbage burning in India is a matter of concern that helps in increase in levels of POC emissions to the atmosphere (*Behera et al., 2016; Singh et al., 2021; Yadav et al., 2020*). Moreover, the fast increase in power requirement, industrialization, domestic cooking fuel, and motorization in India are responsible for large emissions of POC to the atmosphere (*Behera and Sharma, 2015; Reddy et al., 2021*).

Figure 17.1 shows the study region as a map of India, which was considered for development of emission inventory of POC, and some areas of its neighboring countries, including Pakistan, Nepal, Sri Lanka, Bangladesh, and Myanmar. For emission estimates of POC from our neighboring countries, the database from REAS was considered. The Himalayan range considered for the prediction of deposition rates of POC and sketches of locations of seven peaks and seven glaciers in six regions of the Himalayas are also depicted in Figure 17.1. The name of peaks and glaciers

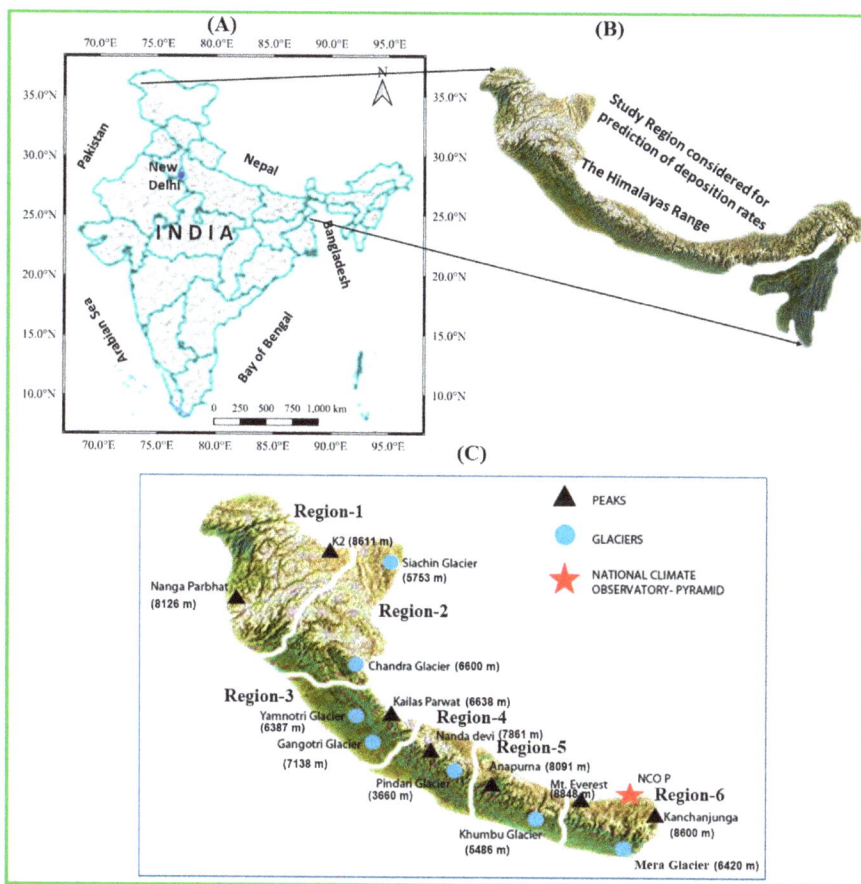

FIGURE 17.1 (a) Study region as map of India for emission inventory development. (b) The Himalayan range considered for prediction of deposition rates of primary OC and its ambient concentration. (c) Sketches of locations of seven peaks and seven glaciers in six regions of the Himalayas.

with respective elevation levels (m) in these six regions are provided as follows: (1) Region 1: K2 peak (8611 m) and Nanga Parvat Peak (8126 m); (2) Region 2: Siachen Glacier (5753 m) and Chandra Glacier (6600 m); (3) Region 3: Kailash Parvat Peak (6638 m), Yamnotri Glacier (6387 m), and Gangotri Glacier (7138 m); (4) Region 4: Nanda Devi Peak (7816 m) and Pindari Glacier (3660 m); (5) Region 5: Annapurna Peak (8091 m) and Khumbu Glacier (5486 m); and (6) Mount Everest Peak (8848 m), Kangchenjunga Peak (8600 m), and Mera Glacier (6420 m).

From its diverse geographical locations, India possesses a varied meteorology, and the meteorological conditions strongly influence pollutant transport (in our case POC reaching the Himalayas). The CTM model used in this study, that is, CAMx easily accounts for such varying meteorology, terrain, and geographical features like snow, desert, and mountains. With the scope of work in this study, we considered the emissions of POC from the Indian subcontinent and some of the neighboring countries, their deposition rates (both dry and wet), and ambient concentrations of POC over the Himalayan range, which serves as our complete domain. Overall, Figure 17.1 serves as our complete study domain.

17.2.2 CONFIGURATION OF STUDY ASPECTS

The domain of the simulation covering a 3280-km × 3280-km area with an outer bigger square grid containing 82 × 82 cells with each cell at a resolution of 40 km × 40 km in both x and y directions is shown in Figure 17.2a. The output of the WRF model provided meteorological conditions in 31 vertical layers, and successively, these layers were collapsed to eight vertical layers as inputs in the CAMx model using a preprocessing tool. The predicted values of deposition rates of POC and ambient concentration in the grids overlaying the Himalayan region were separated, and the average value of all separated-out grids was considered for final reporting of data. Both 1-h and 24-h average temporal resolutions were fitted into the simulation exercises for three distinct seasons including summer (15 April–31 May 2008), monsoon (1 July–31 July 2008), and winter (1 November–15 December 2008). The four different major polluting sectors considered for this study were open burning, domestic, industries, and vehicles. The selected time resolution helped in understanding the diurnal, seasonal and source-wise variability of deposition rates, and ambient concentrations of POC.

A considerable amount of computing power was required for this study, where modeling of such a high resolution was required. An Intel Core i3 processor coupled with 2 GB RAM was used for modeling POC deposition in the Himalayas. Using these configurations, the simulations give comparably good results and in a reasonable amount of time with a faster simulation speed. Fedora 13 was used as the operating system with a software model configured as ArcGIS, WRF3.2 model, CAMx 5.2 model, visualization tool as VERDI 3.2, and compilers as PGI Fortran and CC, GCC and Intel Fortran.

17.2.3 SIMULATION SCENARIOS

We performed the simulations for 45 days in summer and winter, and 30 days in monsoon during the study year, 2008. The consideration of fewer number of days for simulation during monsoon compared to summer and winter was done because

(A)

(B)

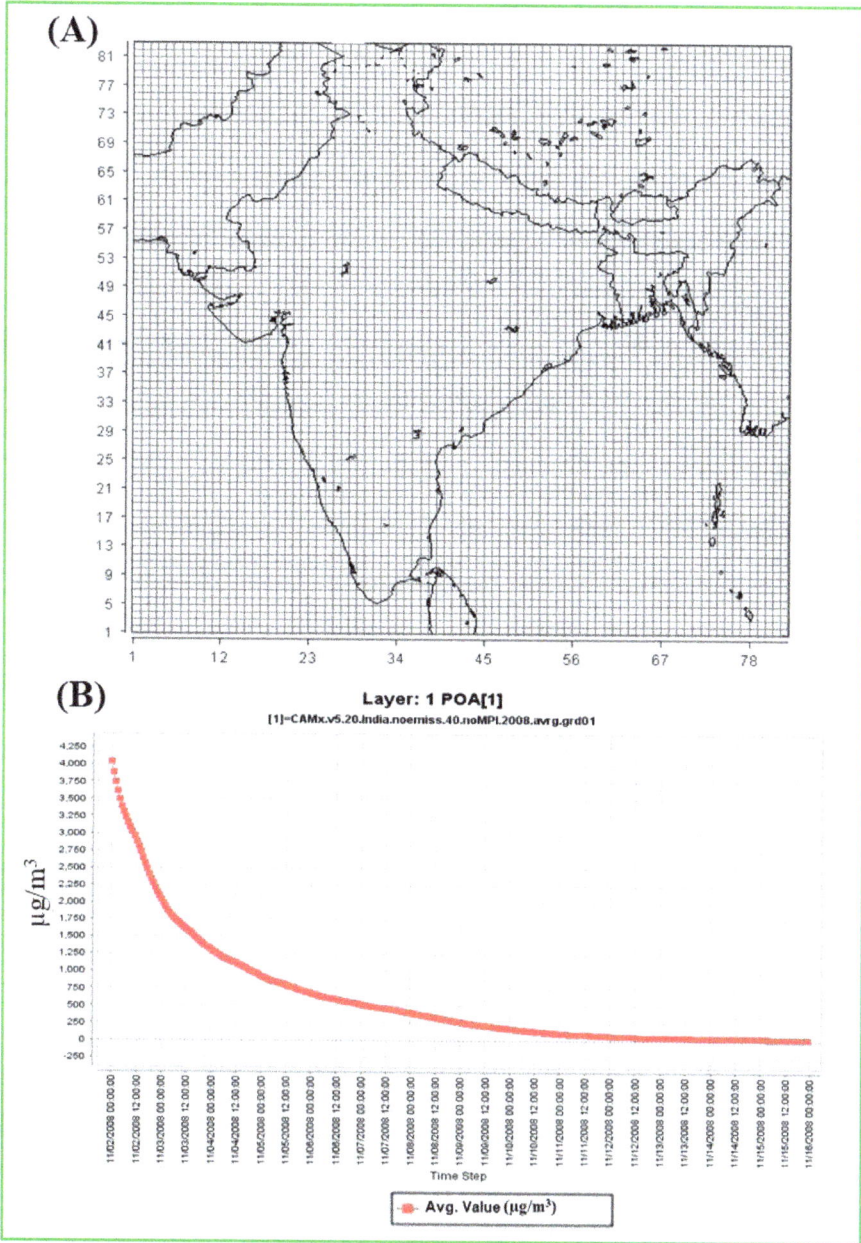

FIGURE 17.2   (a) The study region considered for WRF modeling and CTM exercise in grid overlay of 82 × 82 grids with the size of each grid as 40 km × 40 m. (b) Representative simulated values of POC with time for no emission input.

this season is not critical as pollutants are normally washed out due to precipitation and removed from the atmosphere. Simulations were performed for the first 15 days of study period for the emissions from each of these source sectors (open burning, domestic, industries, vehicles). These 15 days of simulation for each of the sources were used as a spin-up time for initialization of the model, that is, to relax initial and boundary conditions. A simulation with no emission input was also performed to demonstrate the effect of initial conditions and the purpose of spin-up time. The database from the REAS in Asia (*Ohara et al., 2007*) was used in spin-up simulation as the emission input to consider the emission impact of other neighboring countries.

The deposition of POC due to spin-up simulation was added to the deposition due to the emissions from India to account for the impact of neighboring countries. To configure the model for a better simulation for three seasons, input data sensitivity analysis was performed with no emission value of POC. In other words, to assess the impact of initial conditions or emissions on modeled results, a simulation performed with a zero-emission value of POC. The retention time of any initial emission value of POC lasted fewer than 15 days due to higher meteorological turbulence and solar radiation–driven photochemical reaction removal. This clearly explained the role of spin-up time and ensured that the impact of initial conditions was minimal due to removal processes such as dispersion, deposition, and chemical transformation in the atmosphere (*Berge et al., 2001; Baker and Bash, 2012; Rafael et al., 2021*). Figure 17.2b shows representative results of simulation of POC with time for no emission input.

17.2.4 DEVELOPMENT OF THE GIS-BASED EMISSION INVENTORY

The primary model for emission estimation was used for emission load in this study, in which the product of two variables was utilized, that is, activity-level data for a particular source and emission factor (EF) of POC from that source (*Aneja et al., 2012; Gao et al., 2019*). The estimates of emissions were put together into databases or into inventories those contained all supporting data including location of sources, measurements of emission characteristics, capacity/production, emission factors, activity rates for all sources, and method of estimation (*Garg et al., 2006; Singh et al., 2020*). The main objective of GIS-based emission inventory was to generate a variable grid POC emission database to provide inputs in further modeling exercise using CAMx model. The distribution of time-invariant emissions from district-level inventory to finely gridded values by interpolating from local-level data, emission sources, emission factors, and activity data using relational data structure was developed in the GIS. ArcGIS10.0 application developed by ESRI (Environmental Systems Research Institute, USA) was used because of its relative user-friendliness, reliability in data management in the attributes, and generation of thematic layer-based emission maps (*Singh et al., 2016; Gao et al., 2019*). Particular display of different shape files with reference to various emitting sources and geographical locations was possible due to thematic maps/themes and these themes could be selected/viewed in any order. This is noteworthy that GIS is not only a map

viewer in the coupled system of GIS-WRF-CAMx but more of an integrated tool to handle data from many sources.

During the emission inventory exercise, four major source sectors were divided into several subsectors as follows: (1) open burning into crop-residue burning and garbage burning; (2) domestic into liquefied petroleum gas (LPG), kerosene, coal, fuel wood, and dung cake; (3) industries into brick kilns, sugar industry (bagasse burning), cement plants, steel industry, and power plants; and (4) vehicles into buses, trucks, tractors and trailers, light motor vehicles (LMVs) and light commercial vehicles (LCVs), cars, and two-wheelers. The district-level activity-level data were collected from various government reports, market research reports, country-level statistical survey reports, and previous emission inventory studies (*Gurjar, 2004, Garg et al., 2006; CEA, 2010; MoA, 2010; CMIE, 2010; MoC, 2010; MoPNG, 2010; MoSPI, 2009; MoRHT, 2009; CPCB, 2010, 2011*). The Indian Census for 2001 was considered for collection of district-wise population, and later these data were projected for the base year of this study (2008) using a logistics prediction method for both rural and urban areas of a district. Suitable source-specific EFs for sources of open burning, domestic, and industry in the unit of g/kg were selected from Indian and other relevant literature. Similarly, the unit of expression of selected EFs for vehicles was g/km. The values of considered EFs are provided as follows: crop-residue burning at 6.92, garbage burning at 5.27, LPG at 0.05, kerosene at 0.8, coal at 1.8, fuel wood at 5.4, dung cake at 4.9, brick kiln at 1.1, sugar industry at 5.3, cement plant at 0.37, steel industry at 1.8, power plant at 0.48, bus at 0.46, truck at 0.64, tractor and trailer at 0.94, LMV and LCV at 0.6, car at 0.32, and two-wheeler at 0.11.

The vector data from a digitized map of India and its surroundings were converted to thematic layers to interact and to determine the locations of industries, areas of specific interest. Different prepared thematic maps included population, domestic activities, industrial activities, area boundaries and vehicles. These thematic maps or layers were used to identify the sources and represent the data visually. A spatially resolved source-wise emission inventory was prepared on a resolution of 40-km × 40-km grid for the emission estimation. Four stages were involved in development of GIS-based emission inventory including digitization of maps, calculation of district level emissions, generation of thematic layers and attribution of emission values to thematic layers. Figure 17.3 presents basic concepts of developing the GIS-based gridded emission map of POC in India.

For emission calculation, the following basic expressions were used for development of GIS-based emission inventory, as

$$Raw\ activity\ data_{projected\ year} = f(Raw\ activity\ data_{known\ year},$$
$$Growth\ rate) \qquad (17.1)$$

$$POC_{i,j} = f(Raw\ activity\ data_{i,j}, Emission\ factor_i) \qquad (17.2)$$

where, i = particular source, j = specific location (district), and $POC_{i,j}$ = primary organic carbon of i source at j location.

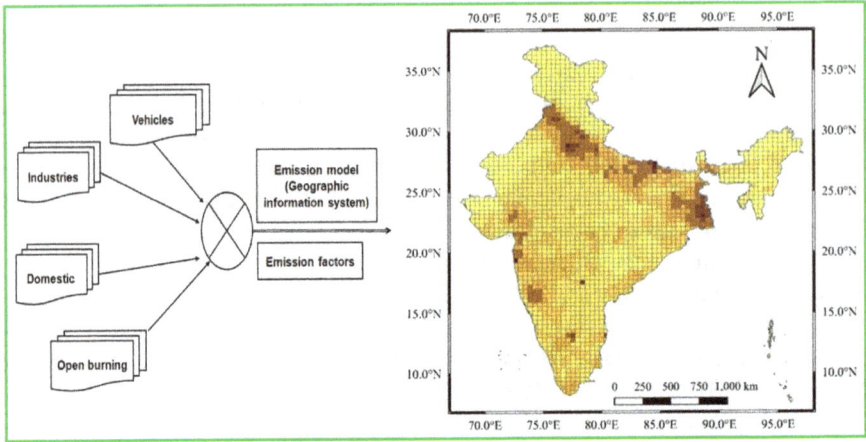

FIGURE 17.3 Basic concepts on development of GIS-based gridded emission map of POC.

17.2.5 Postprocessing GIS-Based EI for the CAMx Model

As input information in CAMx modeling was required with grid-wise emission profiles, the district-wise emission data were converted to a grid-wise map for visualization using nearest neighborhood technique in GIS. Specifically, the CAMx model platform required information about the quantity and kind of pollutant in different grid cells over the entire study domain, and all information was provided by GIS-based emission inventory outputs. The formats of output files from the GIS were in dBASE format and shape file; however, inputs in CAMx were required in the formats of text and ASCII. The text emissions files were prepared manually using FORTRAN program. The main components of these text input files were described as follows: the header, containing information about the domain of study, time of simulation, and so on. The grid-wise emission data was arranged as specified in UAM-V for the application in CAMx (*Baker and Bash, 2012; Rafael et al., 2021*).

The row, column, and layer (grids) in XY plane were arranged with the following relations, as $i + 1 = ROW$, $j + 1 = COL$, $k + 1 = LAY$, with ROW, COL, and LAY being the number of rows, column, and layers that define the location of a grid in the study domain. For each of the grids, the corresponding hourly emission data were estimated from the developed emission inventory for 1 year by dividing 366×24. In each layer, the hourly emission data were arranged according to the rows and columns. At the end, for its implementation to CAMx framework, the previously mentioned text file was converted to the binary format using ASC2BIN preprocessing tool.

17.2.6 WRF Meteorological Modeling

The WRF model is a next-generation mesoscale numerical weather prediction system designed for operational forecasting based on further inputs in different modeling systems. In this study, the WRF simulation system configured to provide inputs

with meteorological dynamic parameters in the CAMx modeling platform. The version used for modeling purposes was WRF version 3.2 (WRF3.2). The WRF model simulated the meteorological parameters in the desired format for the CAMx modeling platform for the study domain. In the modeling practice, interpolation scheme used known meteorological data with variations in temporal and spatial scales and predicted dynamic data in the desired formats (*Skamarock et al., 2008; Coelho et al., 2014*). The past studies on rigorous performance testing of WRF model fields (*Deb et al., 2008; Kumar et al., 2008; Guttikunda and Jawahar, 2018*) demonstrated that dynamic and thermodynamic fields generated by WRF were quite compatible in predicting variable meteorological parameters in the study domain that represented a regional modeling system. During the simulation process, the control file contained the details of the study domain, and the desired inputs were provided in the WRF modeling system. To get further detailed information on fundamental concepts for numerical expressions, the principle of physics in the derivation of the driving equations, and their dynamics and procedures to conduct simulations in this model, the relevant literature can be refereed (*Skamarock et al., 2008; Cassano et al., 2011*). With a broader overview about the model, the respective equations (Advanced Research WRF [ARW] solver) are formulated in terms of terrain, following hydrostatic pressure vertical coordinate (η), expressed as follows:

$$\eta = \frac{(P_h - P_{ht})}{\mu}, \quad \text{where, } \mu = (P_{hs} - P_{ht}) \tag{17.3}$$

where P_h represents the hydrostatic component of the pressure and P_{hs} and P_{ht} represent the values along the surface and top boundaries, respectively.

Since $\mu(x, y)$ represents the mass per unit area within the column in the model domain, the appropriate flux form variables are expressed, as follows:

$$V = \mu v = (U, V, W), \quad \Omega = \mu\dot{\eta}, \quad \Theta = \mu\theta \tag{17.4}$$

where $V = (U, V, W)$ represent the covariant velocities in two horizontal and vertical directions, respectively, while $\omega = \dot{\eta}$ represents contravariant 'vertical' velocity. θ represents potential temperature. The non-conserved variables $\varphi = gz$ (the geopotential), p (pressure), and $\alpha = 1/\rho$ (the inverse density) also appear in the governing equations of ARW.

The WRF3.2 preprocessing system (WPS), the inputs files were prepared for the domain using detailed information about the topography, land use, albedo, soil temperature, and other related parameters. The initial and boundary conditions for this mesoscale modeling exercise were introduced with the forecast data of the Global Forecast System (GFS) of the NCEP (National Centers for Environmental Prediction), USA (*Kumar et al., 2012; Heming et al., 2019*).

17.2.7 CHEMICAL TRANSPORT MODELING USING THE CAMx MODEL

The three-dimensional Eulerian photochemical transport model, known as the chemical transport model, applies relevant scientific concepts to model photochemical-based

reactions and their removal processes (*Baker and Bash, 2012; Rafael et al., 2021*). In this study, the CAMx modeling system with the extension version of 5.2 was used to simulate pollutant emission, its dispersion and dilution, chemical reactions responsible for transformation, transport of pollutants, and finally its removal in the study domain (*Akritidis et al., 2014; Ma et al., 2019*). The average pollutant concentration (C_l) in every grid cell volume is the sum of all physical and chemical processes operating in that volume. The governing numerical expression is an Eulerian continuity equation in the terrain following height (z) coordinates. The mathematical expression is described, as follows Equation 17.5:

$$\frac{\partial C_l}{dt} = -\nabla_H \bullet V_H C_l + \left[\frac{\partial(C_l \eta)}{\partial z} - C_l \frac{\partial}{\partial z}\left(\frac{\partial h}{\partial t}\right) \right] + \nabla \bullet \rho K \nabla \left(\frac{C_l}{\rho}\right)$$
$$+ \frac{\partial C_l}{\partial t_{Emission}} + \frac{\partial C_l}{\partial t_{Chemistry}} + \frac{\partial C_l}{\partial t_{Removal}} \tag{17.5}$$

where V_H represents horizontal wind vector, η represents net vertical entrainment rate, h represents layer interface height, ρ represents atmospheric density, and K represents turbulent exchange or diffusion coefficient. In the previous governing equation, the first term on right-hand side represents the horizontal advection in the atmosphere. The second term represents the net resolved vertical transport across an arbitrary space- and time-varying height grid. The third term of the equation represents the sub-grid-scale turbulent diffusion. A set of reaction equations defined from specific chemical mechanisms and transformation are solved in parallel to consider the relevant concept of chemistry.

After successful simulation, the output from WRF model was processed further using a tool of meteorological chemical interface preprocessor of WRF-CAMx. That specific tool collapsed 31 vertical sigma levels of the WRF output to equivalent eight vertical layers required for inputs into the CAMx model (*Cassano et al., 2011; Coelho et al., 2014; Evans et al., 2012; Brasseur et al., 2019*). With a grid resolution of 40 km × 40 km, the configured WRF model performed three sets of simulations in three different seasons to produce outputs that were utilized in CAMx modeling. The configurations of parameters for modeling exercises both in WRF and CAMx models are summarized in Tables 17.1a and 17.1b, respectively.

TABLE 17.1A

Summary of Configurations and Components in WRF Modeling Exercises

Particulars	Parameters Configured
Model version	Non-hydrostatic version WRF version 3.2
Domain details	Single domain: Indian subcontinent 5–38°N latitude and 65–98°E longitude
Simulation period	15 April–31 May 2008 (summer case)
	1 July–31 July 2008 (monsoon case)
	1 November–15 December 2008 (winter case)

Particulars	Parameters Configured
Horizontal resolution	40 km × 40 km
Vertical resolution	31 sigma levels
Timestamp	90 s
Data preprocessing	Real data using WRF preprocessing system, conversion from Grib files
Boundary condition	6 hourly boundary condition
Dynamics	Eulerian mass coordinates
Microphysics	Cloud microphysics (*Morrison et al., 2005, 2009*)
Cumulus parameterization	Kain–Fritsch new eta scheme (*Kain and Fritsch, 1990, 1993*)
Planetary boundary layer	Yonsei University, South Korea (*Noh et al., 2003; Kwun et al., 2009*)
Surface layer	Monin-Obukhov scheme
Land-surface model	Unified Noah land-surface model
Slab soil model	5-layer thermal diffusion
Long-wave radiation	Rapid radiative transfer model (RRTM).
Short-wave radiation	Dudhia scheme
Sub-grid turbulence	Horizontal Smagorinsky first-order closure
Topographic and land use	30-s data from U.S. Geological Survey (USGC ver 2.0).
Soil type	30-s data from U.S. State Soil Geographic (STATSGO).
Terrestrial details	10-min 12 monthly vegetation fraction data with 1-degree deep soil temperature data
Postprocessing	NCAR Command Language (NCL) graphics utilities, and MET—tool for statistical analysis

TABLE 17.1B
Summary of Configurations and Components in CAMx Modeling Exercises

Particulars	Parameters Configured
Model version	CAMx version 5.2
Horizontal resolution	40 km × 40 km
Number of grid points	82 × 82
Number of vertical layers	8 (up to 12 km)
Simulation period	15 April–31 May 2008 (summer case) 01 July–31 July 2008 (monsoon case) 01 November–15 December 2008 (winter case)
Projection system	Lambert Projection system LAMBERT_Center_Longitude = 80.18 LAMBERT_Center_Latitude = 21 LAMBERT_True_Latitude1 = 21.96
Chemical mechanism	CB05 (Mechanism 6), CMC chemistry solver with full aerosol chemistry (SOAP, RADM, ISORROPIA) with CF approach
Advection solver	PPM
Vertical layers height (m)	25, 100, 500, 1000, 2000, 5000, 8000, 12000
Emissions inventory	Year 2008. For India, own GIS inventory, and for neighboring countries, REAS database.

17.3 RESULTS AND DISCUSSION

17.3.1 OVERALL EMISSION INVENTORY OF POC

Figure 17.4 shows spatially distributed maps of emission load of total POC and population density with a grid resolution of 40 km × 40 km in India. The census data of India for 2001 provided population data at the district level. These data were used in predicting population during the base year of this study (2008) using the logistics growth method. The district-level data were then converted to the gridded population using GIS (detailed procedures are described in a previous section). The legend of the population density map shows different ranges of population density in the unit of no of persons/km². It was observed that population of most of the regions in India ranged from 0 to 500 persons/km², and urban population regions varied in the ranges between 1000–2500, and 2500–5000 persons/km². The population density of metropolitan cities like Delhi, Chennai, Kolkata, Mumbai, Bangalore, and Hyderabad ranged from 5000–10,000 persons/km². The IGP region comprised the states of Punjab, Haryana, Uttarakhand, Uttar Pradesh, Bihar, and West Bengal showed higher population density ranged from 500–1000 to 1000–2500 persons/km². Moreover, some parts of the western states, like Gujarat and Maharashtra, and the southern states, like Kerala, Tamil Nadu, Karnataka, and Andhra Pradesh, showed a higher population density.

The estimate on total POC emission in India revealed the total pollution load as 2368 Gg/yr during the study year from four major sources of open burning, industries, domestic cooking and heating, and vehicles. On assessing the spatial distribution of POC pollution load, it was observed that emission load in each of the grids ranged from 0.00–30.00 Gg/yr. The desert areas in states of Gujarat and Rajasthan, some areas under states of Jammu and Kashmir and Himachal Pradesh showed lower emission loads in the range from 0.00–0.40 Gg/yr. Similarly, parts of northeastern states, Chhattisgarh and Orissa showed the lowest emission loads. Apart from these regions, most of Indian regions showed moderate emission loads in the range between 1.00–1.60 Gg/yr, and 1.60–2.50 Gg/yr. Substantial areas coming under Indo-Gangetic Plain (IGP) region comprising of states of Punjab, Haryana, Uttarakhand, Uttar Pradesh, Bihar and West Bengal showed higher values of POC emission load in the range from 2.5 to 5.00 Gg/yr. As expected in the perspective of the extent of activity level data responsible for POC emission, urbanized areas showed larger emission loads in the ranges between 2.50–5.00 Gg/yr, and 5.00–10.00 Gg/yr. Moreover, some urban areas and metropolitan areas had the potential to generate the highest ranges of emission loads from 10.00–30.00 Gg/yr.

Comparing spatially distributed maps of population and emission loads of POC in India, it was observed that higher-population-density region showed larger pollution loads. In addition to this observation, some areas with lower population density had the potential in emitting more POC to the atmosphere. Such trends indicated that sources like agricultural crop-residue burning would be a key player in influencing emission of POC. This reason for the explanation of the trend was matched well with land-use pattern under areas of agricultural fields, and the region with higher emission loads had more frequency of crop production in a year. We compared our results

of POC estimate with previous studies done by *Streets et al. (2004)* and *Ohara et al. (2007)* and found quite similar trends in the results. The total POC emission in our study had a lower side of estimate, which may be due to the consideration of a smaller number of source categories in this study.

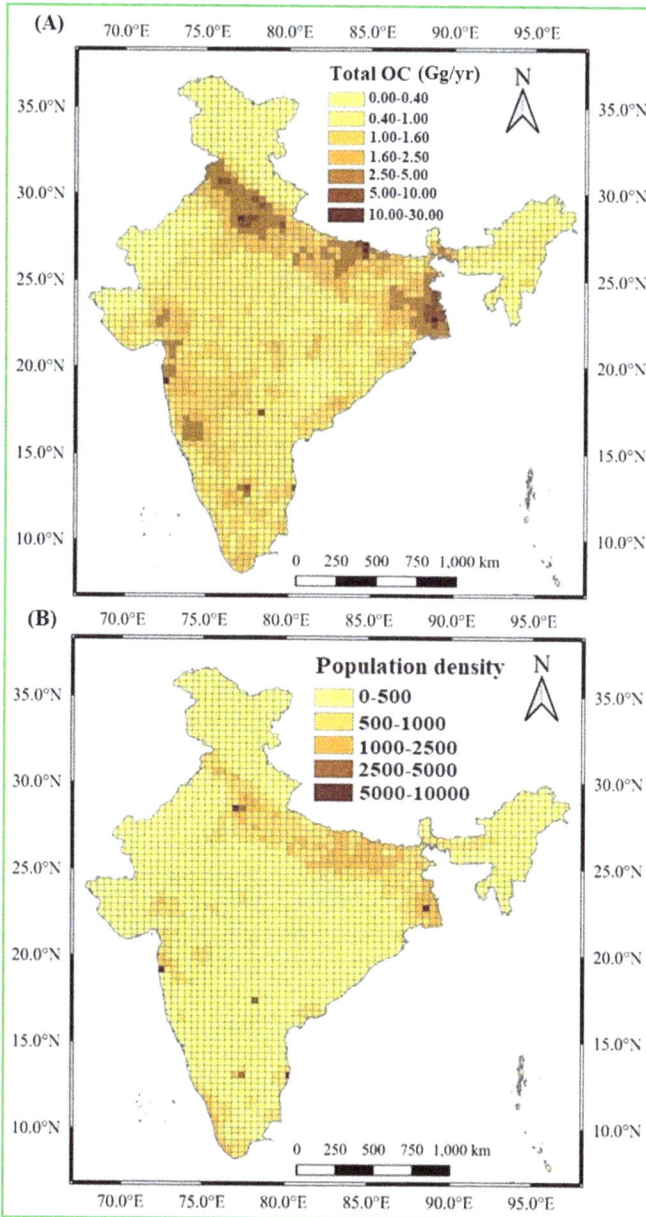

FIGURE 17.4 Spatial distributed maps with 40-km × 40-km grid resolutions of during 2008: (a) Emission of primary OC. (b) Population density (no of persons/km²) of India.

17.3.2 SECTOR-WISE EMISSIONS OF POC AT THE NATIONAL LEVEL

The sector-wise estimated values of emission of POC at the national level are depicted in Figure 17.5. Open burning contributed the highest emission with 1133 Gg/yr (48%) to total emission of POC, followed by industry with 503 Gg/yr (21%). Domestic cooking contributed 454 Gg/yr with 19%, and vehicles at 278 Gg/yr with 12% to the total emission of POC in India. Such trends in the emission estimate showed open burning composed of agriculture crop-residue burning and garbage burning had higher potential to emit more emission of POC due to higher EF

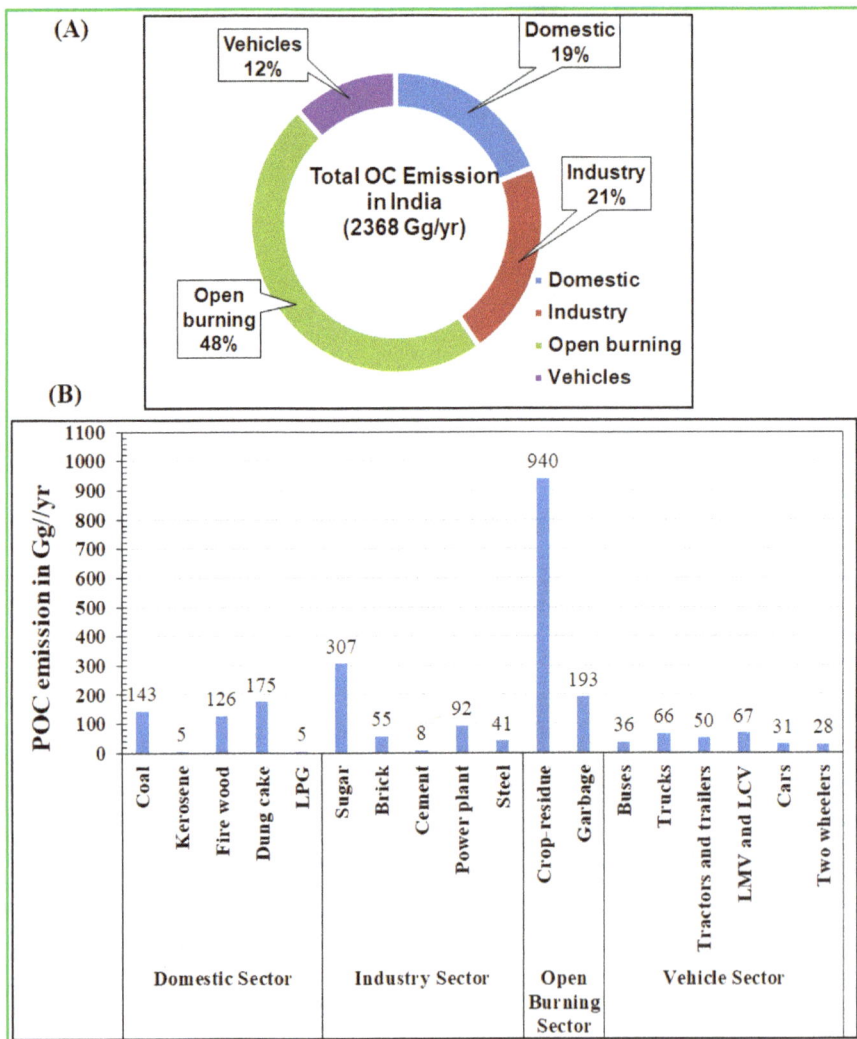

FIGURE 17.5   Emission estimates of POC in India: (a) Emissions from major source sectors with their contributions. (b) Emissions from significant sources in Gg/yr.

per unit of activity data compared to other sources (*Szidat et al., 2009; Singh et al., 2021*). The subsector source contributions to POC emission of that particular source sector were also assessed for each of the major sources. Under the open-burning sector, agricultural crop-residue burning contributed 940 Gg/yr, with 83%, and garbage burning contributed 193 Gg/yr, with 17% to emission estimate of open burning. Such a higher contribution from agricultural crop-residue burning was quite clear, as India has a maximum area under agricultural fields. The fact of agricultural residue burning after each harvesting period is very common in India (*Negi, 2006; Singh et al., 2020*).

Under the industry sector, the following estimate values were generated: sugar industry, with 307 Gg/yr (61%); power plant, with 92 Gg/yr (18%); brick manufacturing kilns, with 55 Gg/yr (11%); steelmaking industry, with 41 Gg/yr (8%); and cement manufacturing industry, with 8 Gg/yr (2%). The fuel in the combustion process for sugar manufacturing industry is normally from bagasse burning, which is of biomass origin. This estimate suggested that biomass burning has an upper edge compared to fossil fuel combustion in terms of contribution to POC emission (*Williams et al., 2012; Mitchell et al., 2016*). Under the domestic source sector, the contributions from subsector sources are provided as follows: use of dung cake with 39 Gg/yr (39%), use of coal with 143 Gg/yr (31%), use of firewood as 126 (28%), use of LPG as 5 Gg/yr (1%), and kerosene with LPG as 5 Gg/yr (1%). Under vehicle sectors, the contributions from subsector sources are estimated as follows: LMVs and LCVs, with 67 Gg/yr (24%); trucks, with 66 Gg/yr (24%); tractors and trailers, with 50 Gg/yr (18%); buses, with 36 Gg/yr (13%); cars, with 31 Gg/yr (11%); and two-wheelers with 28 Gg/yr (10%).

17.3.3 Backward Trajectory Analysis

Backward trajectory analysis is normally conducted to study the advection pattern of air pollutants and the coherency of origin of air masses and their respective pollutant concentrations at a receiver location. The prediction of backward trajectory is a common method for identification of origin and path of air masses for estimation of boundary conditions. We acquired air-mass back trajectories reaching at one of the Himalayan Peak points, that is, Kailash Parvat Peak at 31.14°N 81.18°E. For this purpose, we used archived data from Global Data Assimilation System of the National Oceanic and Atmospheric Administration (NOAA), USA in the platform of the Air Resource Laboratory (ARL) Hybrid Single-Particle Lagrangian Integrated Trajectory (HYSPLIT) model (*Escudero et al., 2006; Draxler and Rolph, 2013*). The simulations were performed with a 24-h interval based on 168-h backward trajectory, illustrating typical air-mass flow to the Kailash Parvat at 500, 1000, and 1500 m above the ground level or at arrival height. These air masses picked up POC pollutant from different parts of Indian subcontinent during their travel and deposited them at the receptor site of the Kailash Paravt. Figure 17.6 shows representative backward trajectories during summer, monsoon, and winter. Table 17.2 presents results of the HYSPLIT backward trajectory analysis during three seasons in terms of the percentage of trajectories with overall wind direction to the arrival site at heights of 500, 1000 and 1500 m. These backward trajectories clearly demonstrated that the effect

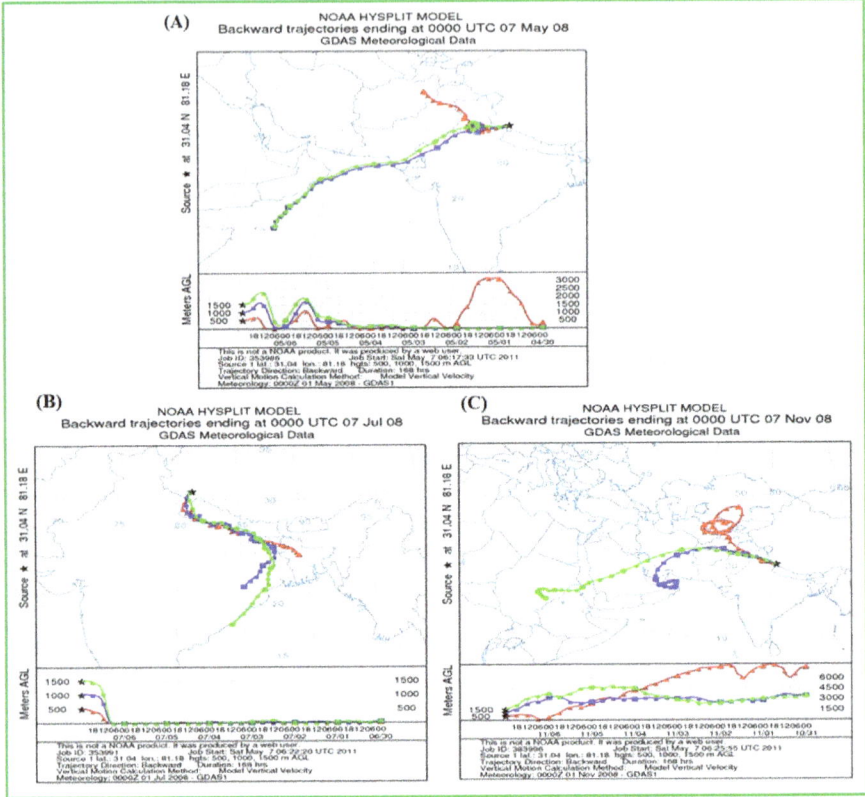

FIGURE 17.6   Representative backward trajectories from the outputs of HYSPLIT model (NOAA, USA) to show air-mass flow to the Kailash Parvat in the Himalayas during different seasons: (a) summer, (b) monsoon, and (c) winter.

TABLE 17.2
Results of HYSPLIT Backward Trajectory Analysis

Season	Arrival Height (m)	Percentage of Trajectories with Overall Wind Direction to Arrival Site (%)							
		N	NW	W	SW	S	SE	E	NE
Summer	500	40	37	17	6	NA	NA	NA	NA
	1000	14	16	49	21	NA	NA	NA	NA
	1500	12	18	47	23	NA	NA	NA	NA
Monsoon	500	NA	NA	NA	14	24	42	16	4
	1000	NA	NA	NA	17	26	39	13	5
	1500	NA	NA	NA	14	24	42	16	4
Winter	500	22	62	9	4	3	NA	NA	NA
	1000	24	61	8	3	4	NA	NA	NA
	1500	23	64	7	2	4	NA	NA	NA

Note: NA = not applicable.

of westerly winds over the Indian plateau in transporting air masses with pollutants to arrive at the Himalayan region. As most parts of India fall in a tropical region, it gets more solar radiation compared to its northwestern Eurasian plate, creating a low-pressure zone and causes the northwesterly wind in this region (*Pathirana et al., 2014; Umar et al., 2021*). This effect was seen clearly during winter. However, inconsistent local-level heating and seasonal wind from the Arabian Sea become deciding factors for the air masses' advection during summer. The results revealed that the model generated outcomes, which are more realistic during summer and winter than during monsoon, perhaps due to the effect of seasonal turbulence occurring during monsoon (*Pathirana et al., 2014; Umar et al., 2021*). Hence, the west advection of air masses was important on the boundary of this region and that played a major role when deciding the boundary conditions for the CAMx model (*Baker and Bash, 2012; Rafael et al., 2021*).

17.3.4 MODEL EVALUATION AND VALIDATION EXERCISE

The predicted results from modeling exercises from the WRF and CAMx models were evaluated for their performance by comparing predicted results with respective observed results using a set of statistical parameters. The statistical metrics used in evaluation performances included two bias metrics (normalized mean bias [NMB] and fractional bias [FB]), two error metrics (normalized mean error [NME] and fractional error [FE]), correlation coefficient (r), and root mean square error (RMSE; *Baker and Bash, 2012;Nopmongcol et al., 2017;Rafael et al., 2021*). These parameters are described in Table 17.3.

The predicted results from meteorological modeling from WRF model were evaluated with observations from ground meteorological station data at Kanpur city belonging to Central Pollution Control Board (CPCB). The performance evaluation of the WRF modeling included wind speed and ambient air temperature. The performance evaluation of predicted results from the CAMx modeling was conducted with observed data during 2008 from reported studies by *Bonasoni et al. (2008, 2010)*. The location of the observed data was the Nepal Climate Observatory–Pyramid (NCO-P) (5079 m) on southern slope of the Himalayan range. The NCO-P is the highest aerosol observatory managed within the Ev-K2-CNR Stations at High Altitude for Research on the Environment (SHARE) and the United Nations Environmental Programme (UNEP), Atmospheric Brown Cloud (ABC) projects (*Bonasoni et al., 2008, 2010*). As POC is co-emitted with EC from same set of source activities and their ratio remains the same in ambient air, the evaluation based on the observed ambient EC would be applicable for the ambient concentration of POC *(Hallquist et al., 2009; Moran et al., 2012; Behera and Sharma, 2015)*.

For the purpose of model evaluation validation, we used measurement data from the site of the NCO-P assuming that the transported air mass from remote regions had more influence on the ambient air concentration of primary EC. Although local emissions might have some influence on measurement site results, previous studies inferred that the grid cells, especially affecting the atmospheric concentration of anthropogenic pollutants at NCO-P, are those pertinent to the northern Indian subcontinent. The atmospheric circulation characterizing the NCO-P site

TABLE 17.3

Model Performance Metrics Used in the Evaluation of Prediction Exercise in the WRF and CAMx Models

Equation Number	Particular of Parameter	Expression	Potential Range	Significance in Air Quality Modeling[a]
6	Normalized mean bias (NMB) (%)	$NMB = \dfrac{\sum\limits_{i=1}^{N}(C_m - C_o)}{\sum\limits_{i=1}^{N} C_o}$	-100% to $+\infty$	$\pm 15\%$
7	Normalized mean error (NME) (%)	$NME = \dfrac{\sum\limits_{i=1}^{N} |C_m - C_o|}{\sum\limits_{i=1}^{N} C_o}$	0% to $+\infty$	
8	Fractional bias (FE) (%)	$FE = \dfrac{1}{N}\sum\limits_{i=1}^{N}\dfrac{|C_m - C_o|}{\left(\dfrac{C_o + C_m}{2}\right)}$	0 to 200%	$\pm 35\%$
9	Fractional bias (FB) (%)	$FB = \dfrac{1}{N}\sum\limits_{i=1}^{N}\dfrac{(C_m - C_o)}{\left(\dfrac{C_o + C_m}{2}\right)}$	-200 to 200%	
10	Correlation coefficient (r)	$r = \dfrac{1}{N-1}\sum\limits_{i=1}^{N}\left(\dfrac{C_m - \overline{C_m}}{\sigma_m}\right)\left(\dfrac{C_o - \overline{C_o}}{\sigma_o}\right)$	0 to 1.0	Close to 1 is better and precise value depends on number of data points
11	Root mean square error (RMSE)	$RMSE = \sqrt{\dfrac{\sum\limits_{i=1}^{N}(C_m - C_o)^2}{N}}$	Starts from 0	The lesser value is always preferable

[a]For better prediction as per *USEPA (1991)*, where Co represents individual observed value and Cm represents the respective individual predicted value. $\overline{C_o}$ represents the mean of observed data points, and $\overline{C_m}$ represents the mean of respective predicted data points. σ_o represents the standard deviation of observed data points, and σ_m represents the respective standard deviation of predicted data points. N represents number data points, and C represents the parameter of interest, that is, temperature, wind speed, and ambient concentration.

was dominated by westerly, southwesterly, and northwesterly directions (*Bonasoni et al., 2010; Maione et al., 2011*). Thus, pollution characteristics at NCO-P was predominately dependent on transported air mass coming from the Indian subcontinent to this remote site. Moreover, this particular site is an ecologically sensitive location in the Himalayas range, which has negligible local emission compared to transboundary-transported pollution to the site. The change in terrain and meteorology was accounted for in the setup of the CAMx model. The research scope of this study was limited to the prediction of dry and wet deposition through the transport of pollutants due to emissions from different parts of India and its neighboring countries. The changes in topography of various locations in the study domain were taken into account during simulation exercise, as described in the configuration table of models.

Table 17.4 presents the model performance metrics with values of statistical parameters of wind speed and temperature predicted from WRF model and ambient EC concentration predicted from CAMx model. The model biases (NMB, FB) were found to be less than 11% for predictions of wind speed, temperature, and EC concentration through modeling exercises in the WRF and CAMx models. Similarly, model prediction errors (NME, FE) were found to be less than 14% for the WRF and CAMx models. These performance parameters were well within bias and error goals of less than ±15% and 35% recommended by U.S. Environmental Protection Agency (USEPA), indicating these models provided better results in prediction for the study area. Moreover, correlation coefficient values for both these values were significant and RMSE values were well within the range of validation.

Modeled primary EC concentration at NCO-P was based on four distinct sources coming from India and its neighboring countries. The emission inventory consisted of estimation from India with limited sources and data from REAS for neighboring countries. This approach generates uncertainty in emission inventory causing errors in CTM modeling, as described in Tables 17.3 and 17.4. There is uncertainty associated with the speciation profiles used in individual source categories to assign primary EC emissions. In some areas, profiles and inventory assumptions may be updated and region-specific, while others may not be. Emissions speciation is an ongoing topic for additional study, and

---

TABLE 17.4

Model Performance Metrics for the WRF and CAMx Modeling Exercise

Parameter	Unit of Expression	No of Data Points	Average Observed Value	NMB (%)	NME (%)	FB (%)	FE (%)	r	RMSE
Wind speed	m/s	180	2.7	8.6	9.8	10.6	11.9	0.79	7.6
Temperature	°C	180	31.4	7.9	11.6	9.6	13.4	0.76	6.4
EC concentration	μg/m^3	140	1.6	9.3	12.4	10.2	13.8	0.68	8.3

the results presented here include the caveat of this uncertainty (*Aneja et al., 2012; Nopmongcol et al., 2017; Thompson et al., 2017)*. The fact remains that despite caveats in emission estimates, the modeling performance provided results of parameters in a better range of significance.

We compared our model performance results from WRF-CAMs modeling exercises with previous meteorological and air quality modeling studies. We have considered *Fountoukis et al. (2011), Baker and Bash (2012), Yu et al. (2014),* and *Nopmongcol et al. (2017)* to compare our results of validation. *Fountoukis et al. (2011)* conducted air quality study to simulate the mass concentration and chemical composition of PM during May 2008 in Europe using the platform of PMCAMx-2008, a detailed three-dimensional CTM model in which meteorology was simulated through the WRF model. In predictions of four species of PM, the range of values of parameters were reported, as follows: NMB, from −11% to 67%; NME, from 30% to 105%; FB, from −0.1% to 0.4%; and FE, from 0.3% to 0.8%. In the prediction of ambient mercury using two CTM models (CAMx and CMAQ), *Baker and Bash (2012)* reported the range of values of evaluating parameters, as follows: FB from −60% to −7% for the CAMx model and −45% to 110% for the CMAQ model and FE from 69% to 91% for the CAMx model and 72% to 121% for the CMAQ model. *Yu et al. (2014)* through their modeling exercise in the platform of WRF-CMAQ reported the range of values of evaluating parameters, as follows: NMB, from −0.4 to 37.8%; NME, from 14.8 to 42.8%; and RMSE from 11.3 to 25.1 ppbv. *Nopmongcol et al. (2017)* through their modeling exercise in the platform of WRF-CAMx reported the range of values of evaluating parameters, as follows: NMB from −10.6% to 5.6%, NME from 10.3% to 14.0%; FB, from −11.8% to 5.6%; FE, from 10.2% to 13.1%, correlation coefficient (r) from 0.30 to 0.71, and RMSE from 6.67 ppb to 9.53 ppb. Overall, our estimates on model evaluating parameters were in the range of previous studies with similar scope of objectives.

17.3.5 DEPOSITION RATES AND AMBIENT CONCENTRATION OF POC AT THE HIMALAYAN RANGE

Figure 17.7 shows the overall results of modeling in the Himalayan range for dry deposition, wet deposition and total deposition rates, and ambient concentration of POC during summer, monsoon, and winter. These results represent average of predicted values at seven peaks and seven glaciers in the region. Dry deposition rate varied from 1.27 µg/m²/day during monsoon to 4.27 µg/m²/day during summer. Dry deposition rate during winter was predicted as 2.06 µg/m²/day. Higher values of dry deposition rate during summer and winter compared to monsoon could be explained through favorable synoptic meteorology in Indian landmass with occurrence of less precipitation that helped in transportation of air masses containing pollutants to remote regions of this mountainous region (*Kulshrestha and Kumar, 2014; Singh et al., 2020*). The wet deposition rate varied from 2.83 µg/m²/day during winter to 65.29 µg/m²/day during monsoon. Wet deposition rate during summer was predicted as 22.85 µg/m²/day. Higher values for the wet deposition rate during monsoon and summer compared to winter were observed due to the occurrence of relatively higher

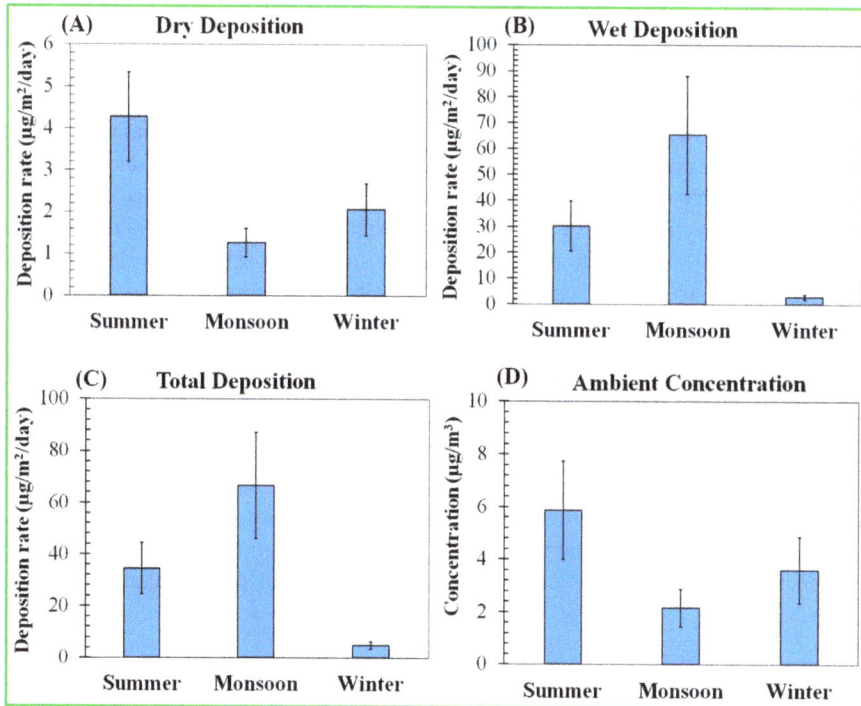

FIGURE 17.7 Overall values of the modeled results of POC in the Himalayas range: (a) dry deposition, (b) wet deposition, (c) total deposition, and (d) ambient concentration.

precipitation during those two seasons in this mountainous region (*Deb et al., 2008; Pathirana et al., 2014*).

Total deposition rate (sum of dry and wet) of POC varied from 4.89 ± 1.47 µg/m²/day during winter to 66.56 ± 20.63 µg/m²/day during monsoon. Total deposition rate during summer was predicted as 34.45 ± 9.82 µg/m²/day. The higher values of total deposition during monsoon were explained through the reason of dominance of wet deposition value over dry deposition value for all three seasons. The ambient concentration of POC varied from 2.16 ± 0.71 µg/m³ during monsoon to 5.85 ± 1.87 µg/m³ during summer, and it was predicted as 3.59 ± 1.25 µg/m³ during winter. Higher values of predicted ambient concentration during summer compared to winter were understood through mechanisms of transport of pollutants along with air masses under favorable synoptic meteorological conditions with higher wind speed from the Indian landmass to the Himalayan region. The lower predicted values of the ambient concentration of POC during monsoon could be explained through the occurrence of local-level washout of pollutants in areas of the Indian landmass, leaving a lower quantity of pollutants for their transport to the remote mountainous region (*Deb et al., 2008; Qian et al., 2011; Ojha et al., 2016*).

17.3.6   DIURNAL VARIATIONS OF DEPOSITION RATES AND AMBIENT CONCENTRATION OF POC

Diurnal variations of predicted values of dry deposition rate, wet deposition rate, total deposition rate, and ambient concentrations of POC in the Himalayan region are shown in Figure 17.8, considering average results of peaks and glaciers decided in the region. Dry deposition rate varied from 0.20 $\mu g/m^2$ at 1:00 to 0.39 $\mu g/m^2$ at 14:00. During daytime from 9:00 to 18:00, predicted dry depositions were observed to be higher than night-time periods and early morning periods. Similar observations of diurnal variations were also predicted for wet deposition and total deposition rates. Specifically, wet deposition rate varied from 0.95 $\mu g/m^2$ at 1:00 to 1.84 $\mu g/m^2$ at 14:00, and total deposition rate varied from 1.08 $\mu g/m^2$ at 1:00 to 2.09 $\mu g/m^2$ at 14:00. The pattern of diurnal variations of ambient concentrations of POC was slightly different than that of deposition rates, with predicted concentration values varieying from 2.96 $\mu g/m^3$ at 0:00 to 5.14 $\mu g/m^3$ at 12:00. Similar trends in diurnal variations of deposition rates and ambient concentration of carbonaceous aerosols were reported in previous studies (*Decesari et al., 2010; Yasunari et al., 2010; Zhao et al., 2013*). The reasons for such a trend could be explained through residence time of pollutants coupled with local levels of pollution in the region.

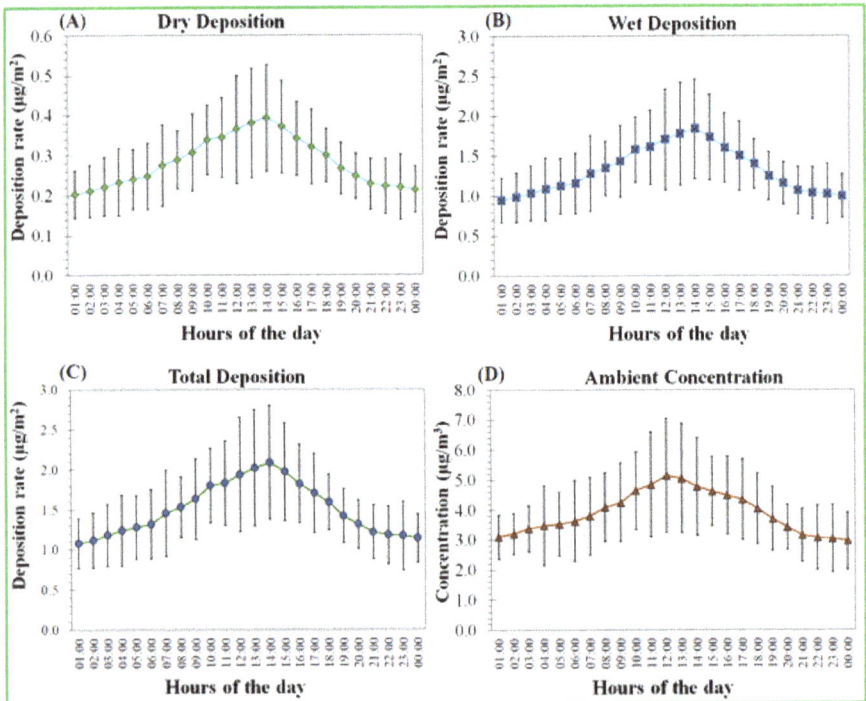

FIGURE 17.8   Diurnal variations of modeled results of POC in the Himalayas range: (a) dry deposition, (b) wet deposition, (c) total deposition, and (d) ambient concentration.

The cycle of residence time came into effect through transportation. For example, air masses with pollutants were transported from locations of source generation to remote regions, and the peak time for higher concentration at source generation location might be midday. The cycle of continuous transport of air masses with pollutants to receiver sites in this mountainous region caused peak ambient concentration during midday timing (*Berthier et al., 2007; Menon et al., 2010; Qian et al., 2011*). Additionally, local pollutions were added to the transported pollutants. As a result, the peak time for deposition through dry and wet means shifted just after the peak of ambient concentration at midday (12:00).

17.3.7 DEPOSITION RATES AND AMBIENT CONCENTRATION OF POC AT PEAKS AND GLACIERS

The average values of predicted deposition rates and ambient concentrations of POC at different region-specific peaks and glaciers of the Himalayan range are shown in Figure 17.9. Predicted values of deposition rates and ambient concentration of POC at various peaks are explained in this paragraph. The dry deposition rate varied from 0.81 µg/m²/day at K2 Peak in Region 1 to 5.83 µg/m²/day at Kangchenjunga Peak in Region 6. Similarly, wet deposition rate varied from 8.28 µg/m²/day at K2 Peak to 81.92 µg/m²/day at Kangchenjunga Peak. Total deposition rate varied from 9.09 µg/m²/day at K2 Peak to 87.75 µg/m²/day at Kangchenjunga Peak. Ambient concentration varied from 1.85 µg/m³ at K2 Peak to 8.01 µg/m³ at Kangchenjunga Peak. The pattern in variations of deposition rates and ambient concentration at different regions and peaks revealed that predicted results decreased while moving from upper positions in Region 1 to lower positions in Region 6 in the Himalayan range. This indicated that northwesterly and northern winds prevailing in the Indian landmass played a major role in deciding the extent of pollutant transportation to the Himalayan region (*Baker and Bash, 2012; Pathirana et al., 2014*). In other words, the air masses covering from less area of the Indian landmass hit Region 1 and gradual increase in covering areas from the Indian landmasses happened while going from Region 1 to Region 2 and so on. As a result, the quantity of transported POC emission increased with a rise in the covered area of the Indian landmass (*Umar et al., 2021; Rafael et al., 2021*).

The dry deposition rate varied from 0.12 µg/m²/day at Siachen Glacier in Region 2 to 2.91 µg/m²/day at Mera Glacier in Region 6. Similarly, the wet deposition rate varied from 2.01 µg/m²/day at Siachen Glacier to 42.82 µg/m²/day at Mera Glacier. The total deposition rate varied from 2.13 µg/m²/day at Siachen Glacier to 42.82 µg/m²/day at Mera Glacier. The ambient concentration varied from 1.5 µg/m³ at Siachen Glacier to 10.51 µg/m³ at Mera Glacier.

Comparing deposition rates and ambient concentrations of POC at peaks and glaciers region-wise, it was found that for all regions (Region 1–6), individual region-specific dry deposition, wet deposition and total deposition rates, and the ambient concentration at glaciers were significantly higher than those at peaks with 95% level of confidence. Overall trends suggested that deposition rates and ambient concentrations of POC at glaciers were higher than at peaks. Such trend in results could be explained through the elevation levels of those particular locations. This is noteworthy

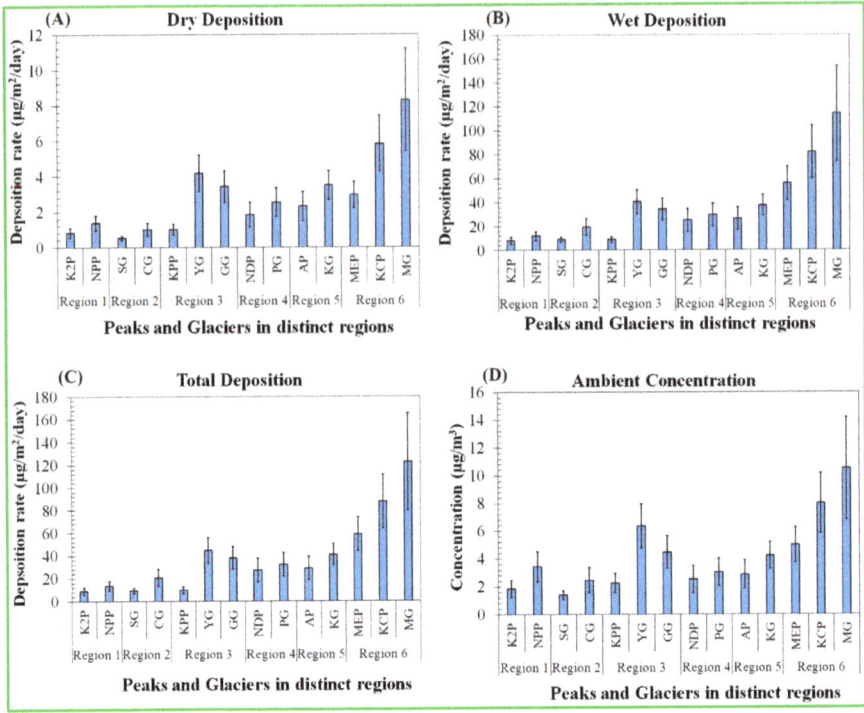

FIGURE 17.9   Overall values of modeled results of POC at different Peaks and Glaciers in distinct regions of the Himalayas range: (a) dry deposition, (b) wet deposition, (c) total deposition, and (d) ambient concentration. Abbreviations used are explained as follows: K2P: K2 Peak, NPP: Nanga Parvat Peak, SG: Siachen Glacier, CG: Chandra Glacier, KPP: Kailas Parvat Peak, YG: Yamnotri Glacier, GG: Gangotri Glacier, NDP: Nanda Devi Peak, PG: Pindari Glacier, AP: Annapurna Peak, KG: Khumbu Glacier, MEP: Mt. Everest Peak, KCP: Kangchenjunga Peak, and MG: Mera Glacier.

that the elevation level of a glacier in particular region is at lower height compared to that of a peak in that region. Therefore, the transported air mass containing pollutant would first settle at the glacier, and the remaining air mass successively would settle at the peak in that region.

17.4   CONCLUSION

With the research scopes of this study, we predicted ambient concentration and deposition rate of POC in peaks and glaciers of the Himalayan range using a modeling system that coupled outputs from the GIS and the WRF models in the CAMx simulation system. A national-level emission inventory revealed that open burning contributed the highest emission, with 1133 Gg/yr (48%), to the total emission of POC, followed by industry, with 503 Gg/yr (21%); domestic cooking, with 454 Gg/yr (19%); and vehicles, with 278 Gg/yr (12%) to total emission of POC in

India. Backward-trajectory analysis inferred that air masses along with pollutants from the Indian landmass areas were transported through northwest and west winds to remote regions to the Himalayan range, followed by deposition through dry and wet means. The model evaluation performance was conducted for the WRF and CAMx using statistical metrics with parameters of biases, errors, correlation coefficients, and RMSEs. The results were well within bias and error goals of less than ±15% and 35% recommended by USEPA, indicating these models provided better results in prediction in the study area.

The dry deposition rate during summer, monsoon, and winter was 4.27, 1.27, and 2.06 $\mu g/m^2$/day, respectively. The wet deposition rate varied from 2.83 $\mu g/m^2$/day during winter to 65.29 $\mu g/m^2$/day during winter. The wet deposition rate during summer was predicted as 22.85 $\mu g/m^2$/day. Total deposition rate (sum of dry and wet) of POC during summer, monsoon, and winter were 34.45 ± 9.82, 66.56 ± 20.63, and 4.89 ± 1.47 $\mu g/m^2$/day, respectively. The ambient concentration of POC varied from 2.16 ± 0.71 $\mu g/m^3$ during monsoon to 5.85 ± 1.87 $\mu g/m^3$ during summer, and it was predicted as 3.59 ± 1.25 $\mu g/m^3$ during winter. Dry deposition rate varied from 0.20 $\mu g/m^2$ at 1:00 to 0.39 $\mu g/m^2$ at 14:00. During daytime from 9:00 to 18:00, the predicted dry depositions were observed to be higher than nighttime periods and early morning periods. The wet deposition and total deposition rates also showed similar trends in diurnal variations like the dry deposition rate. The pattern of diurnal variations of ambient concentrations of POC was slightly different than that of deposition rates, with predicted concentration values varied from 2.96 $\mu g/m^3$ at 0:00 to 5.14 $\mu g/m^3$ at 12:00. Comparing region-wise deposition rates and ambient concentrations, it was revealed that the air masses covering less area of the Indian landmass hit Region 1 and gradual increase in covering areas from the Indian landmass happened while going from Region 1 to Region 2 and so on. As a result, the quantity of transported POC emissions increased with a rise in the covered area of the Indian landmass.

REFERENCES

Akritidis, D., P. Zanis, I. Pytharoulis, and T. Karacostas. 2014. "Near-surface ozone trends over Europe in RegCM3/CAMx simulations for the time period 1996–2006." *Atmospheric Environment* 97: 6–18. https://doi.org/10.1016/j.atmosenv.2014.08.002

Aneja, V.P., W.H. Schlesinger, J.W. Erisman, S.N. Behera, M. Sharma, and W. Battye. 2012. "Reactive nitrogen emissions from crop and livestock farming in India." *Atmospheric Environment* 47: 92–103. https://doi.org/10.1016/j.atmosenv.2011.11.026

Armitage, J.M., C.L. Quinn, and F. Wania. 2011. "Global climate change and contaminants—an overview of opportunities and priorities for modelling the potential implications for long-term human exposure to organic compounds in the Arctic." *Journal of Environmental Monitoring* 13 (6): 1532–1546. https://doi.org/10.1039/C1EM10131E

Baker, K.R., and J.O. Bash. 2012. "Regional scale photochemical model evaluation of total mercury wet deposition and speciated ambient mercury." *Atmospheric Environment* 49: 151–162. https://doi.org/10.1016/j.atmosenv.2011.12.006Bauer, S.E., and S. Menon. 2012. "Aerosol direct, indirect, semidirect and surface albedo effects from sector contributions based on the IPCC AR5 emissions for preindustrial and present-day conditions." *Journal of Geophysical Research: Atmospheres* 117 (D1): D01206. https://doi.org/10.1029/2011JD016816

Behera, S.N., and M. Sharma. 2015. "Spatial and seasonal variations of atmospheric particulate carbon fractions and identification of secondary sources at urban sites in North India." *Environmental Science and Pollution Research* 22 (17): 13464–13476. https://doi.org/10.1007/s11356-015-4603-7

Behera, S.N., M. Sharma, and S.P. Shukla. 2016. "Characterization of gaseous pollutants and water-soluble inorganic ions in PM2.5 during summer time at an urban site of North India." *Journal of Hazardous, Toxic, and Radioactive Waste* 20 (4): A4015002. https://doi.org/10.1061/(ASCE)HZ.2153-5515.0000299

Berge, E., H.C. Huang, J. Chang, and T.H. Liu. 2001. "A study of the importance of initial conditions for photochemical oxidant modeling." *Journal of Geophysical Research: Atmospheres* 106 (D1): 1347–1363. https://doi.org/10.1029/2000JD900227

Berthier, E., Y. Arnaud, R. Kumar, S. Ahmad, P. Wagnon, and P. Chevallier. 2007. "Remote sensing estimates of glacier mass balances in the Himachal Pradesh (Western Himalaya, India)." *Remote Sensing of Environment* 108 (3): 327–338. https://doi.org/10.1016/j.rse.2006.11.017

Bhowmik, H.S., S. Naresh, D. Bhattu, N. Rastogi, A.S. Prévôt, S.N. Tripathi. 2021. "Temporal and spatial variability of carbonaceous species (EC; OC; WSOC and SOA) in PM2.5 aerosol over five sites of Indo-Gangetic Plain." *Atmospheric Pollution Research* 12 (1): 375–390. https://doi.org/10.1016/j.apr.2020.09.019

Bonasoni, P., P. Laj, F. Angelini, J. Arduini, U. Bonafe, F. Calzolari et al. 2008. "The ABC-pyramid atmospheric research observatory in Himalaya for aerosol, ozone and halocarbon measurements." *Science of the Total Environment* 391 (2–3): 252–261. https://doi.org/10.1016/j.scitotenv.2007.10.024

Bonasoni, P., P. Laj, A. Marinoni, M. Sprenger, F. Angelini, J. Arduini, U. Bonafè, F. Calzolari, T. et al. 2010. "Atmospheric brown clouds in the Himalayas: First two years of continuous observations at the Nepal-Climate Observatory at Pyramid (5079 m)." *Atmospheric Chemistry and Physics* 10: 1–15. https://doi.org/10.5194/acp-10-7515-2010

Bond, T.C., D.G. Streets, K.F. Yarber, S.M. Nelson, J.H. Woo, and Z. Klimont. 2004. "A technology-based global inventory of black and organic carbon emissions from combustion." *Journal of Geophysical Research: Atmospheres* 109 (D14): D14203. https://doi.org/10.1029/2003JD003697

Brasseur, G.P., Y. Xie, A.K. Petersen, I. Bouarar, J. Flemming, M. Gauss, F. Jiang, R. Kouznetsov, et al. 2019. "Ensemble forecasts of air quality in eastern China–Part 1: Model description and implementation of the MarcoPolo–Panda prediction system, version 1." *Geoscientific Model Development* 12 (1): 33–67. https://doi.org/10.5194/gmd-12-1241-2019

Carmichael, G.R., B. Adhikary, S. Kulkarni, A. D'Allura, Y. Tang, D. Streets, Q. Zhang, T.C. Bond, V. Ramanathan, A. Jamroensan, and P. Marrapu. 2009. "Asian aerosols: Current and year 2030 distributions and implications to human health and regional climate change." *Environmental Science & Technology* 43 (15): 5811–5817. https://doi.org/10.1021/es8036803

Cassano, J.J., M.E. Higgins, and M.W. Seefeldt. 2011. "Performance of the weather research and forecasting model for month-long pan-arctic simulations." *Monthly Weather Review* 139 (11): 3469–3488. https://doi.org/10.1175/MWR-D-10-05065.1

CEA. 2010. "Central Electricity Authority, Ministry of Power, India. Annual Report 2009–10." www.cea.nic.in/reports/annual/annualreports/annual_report-2010.pdf. Last accessed on 20 March 2011.

CMIE. 2010. "Centre for Monitoring Indian Economy. Reports during 2009–2010." www.cmie.com/. Last accessed on 26 February 2011.

Coelho, M.C., T. Fontes, J.M. Bandeira, S.R. Pereira, O. Tchepel, D. Dias, E. Sá, J.H. Amorim, and Borrego, C. 2014. "Assessment of potential improvements on regional air quality

modelling related with implementation of a detailed methodology for traffic emission estimation." *Science of the Total Environment* 470: 127–137. https://doi.org/10.1016/j.scitotenv.2013.09.042

CPCB. 2010. "Central Pollution Control Board. Air quality monitoring, emission inventory and source apportionment study for Indian cities." https://cpcb.nic.in/source-apportionment-studies. Last accessed on 22 February 2011.

CPCB. 2011. "Central Pollution Control Board. Vehicular source emission profiles. Microsoft excel spreadsheet." www.cpcb.nic.in/Vehicular_Sources_Emission_Profiles.xls. Last accessed on 15 April 2011.

Decesari, S., M.C. Facchini, C. Carbone, L. Giulianelli, M. Rinaldi, E. Finessi, S. Fuzzi, A. et al. 2010. "Chemical composition of PM10 and PM1 at the high-altitude Himalayan station Nepal Climate Observatory-Pyramid (NCO-P) (5079 m a.s.l.)." *Atmospheric Chemistry and Physics* 10: 4583–4596, https://doi.org/10.5194/acp-10-4583-2010

Deb, S.K., T.P. Srivastava, and C.M. Kishtawal. 2008. "The WRF model performance for the simulation of heavy precipitating events over Ahmedabad during August 2006." *Journal of Earth System Science* 117 (5): 589–602. https://doi.org/10.1007/s12040-008-0055-5

Dhanya, G., T.S. Pranesha, K. Nagaraja, D.M. Chate, and G. Beig. 2021. "Variability of ozone and oxides of nitrogen in the tropical city, Bengaluru, India." *Environmental Monitoring and Assessment* 193 (12): 1–15. https://doi.org/10.1007/s40890-021-00136-1

Draxler, R.R., and G.D. Rolph. 2013. "HYSPLIT (HYbrid Single-Particle Lagrangian Integrated Trajectory) Model access via NOAA ARL READY Website." www.arl.noaa.gov/HYSPLIT.php. NOAA Air Resources Laboratory, College Park, MD.

Dumka, U.C., D.G. Kaskaoutis, D. Francis, J.P. Chaboureau, A. Rashki, S. Tiwari, S. Singh, E. Liakakou, and N. Mihalopoulos. 2019. "The role of the intertropical discontinuity region and the heat low in dust emission and transport over the Thar Desert, India: A Premonsoon case study." *Journal of Geophysical Research: Atmospheres* 124 (23): 13197–13219. https://doi.org/10.1029/2019JD030836

Escudero, M., A. Stein, R.R. Draxler, X. Querol, A. Alastuey, S. Castillo, and A. Avila. 2006. "Determination of the contribution of northern Africa dust source areas to PM_{10} concentrations over the central Iberian Peninsula using the Hybrid Single-Particle Lagrangian Integrated Trajectory model (HYSPLIT) model." *Journal of Geophysical Research: Atmospheres* 111: D06210. https://doi.org/10.1029/2005JD006395

Evans, J.P., M. Ekström, and F. Ji. 2012. "Evaluating the performance of a WRF physics ensemble over South-East Australia." *Climate Dynamics* 39 (6): 1241–1258. https://doi.org/10.1007/s00382-011-1244-5

Fountoukis, C., P.N. Racherla, H.A.C. DeniervanderGon, P. Polymeneas, P.E. Charalampidis, et al. 2011. "Evaluation of a three-dimensional chemical transport model (PMCAMx) in the European domain during the EUCAARI May 2008 campaign." *Atmospheric Chemistry and Physics* 11: 10331–10347. https://doi.org/10.5194/acp-11-10331-2011

Gao, C., W. Gao, K. Song, H. Na, F. Tian, and S. Zhang. 2019. "Spatial and temporal dynamics of air-pollutant emission inventory of steel industry in China: A bottom-up approach." *Resources, Conservation and Recycling* 143: 184–200. https://doi.org/10.1016/j.resconrec.2018.12.032

Garg, A., P.A. Shukla, and Kapshe, M. 2006. "The sectoral trends of multigas emissions inventory of India." *Atmospheric Environment* 40 (24): 4608–4620. https://doi.org/10.1016/j.atmosenv.2006.03.045

Guo, H., S.H. Kota, S.K. Sahu, J. Hu, Q. Ying, A. Gao, and H. Zhang. 2017. "Source apportionment of $PM_{2.5}$ in North India using source-oriented air quality models." *Environmental Pollution* 231: 426–436. https://doi.org/10.1016/j.envpol.2017.08.016

Gurjar, B.R., J.A. Van Aardenne, J. Lelieveld, and M. Mohan. 2004. "Emission estimates and trends (1990–2000) for megacity Delhi and implications. *Atmospheric Environment*. 38 (33): 5663–5681. https://doi.org/10.1016/j.atmosenv.2004.05.057

Guttikunda, S.K., and P. Jawahar, P. 2018. "Evaluation of particulate pollution and health impacts from planned expansion of coal-fired thermal power plants in India using WRF-CAMx modeling system." *Aerosol and Air Quality Research* 18 (12): 3187–3202. https://doi.org/10.4209/aaqr.2018.04.0134

Hallquist, M., J.C. Wenger, U. Baltensperger, Y. Rudich, D. Simpson, M. Claeys, J. Dommen, N.M. et al. 2009. "The formation, properties and impact of secondary organic aerosol: current and emerging issues." *Atmospheric Chemistry and Physics* 9 (14): 5155–5236. https://doi.org/10.5194/acp-9-5155-2009

Hansen, J., M. Sato, R. Ruedy, A. Lacis, and V. Oinas. 2000. "Global warming in the twenty-first century: An alternative scenario." *Proceedings of the National Academy of Sciences* 97 (18): 9875–9880. https://doi.org/10.1073/pnas.170278997

He, K., H. Xu, R. Feng, Z. Shen, Y. Li, Y. Zhang, and J. Cao. 2021. "Characteristics of indoor and personal exposure to particulate organic compounds emitted from domestic solid fuel combustion in rural areas of northwest China." *Atmospheric Research* 248: 105181. https://doi.org/10.1016/j.atmosres.2020.105181

Heming, J.T., F. Prates, M.A. Bender, R. Bowyer, J. Cangialosi, P. Caroff et al. 2019. "Review of recent progress in tropical cyclone track forecasting and expression of uncertainties." *Tropical Cyclone Research and Review* 8 (4): 181–218. https://doi.org/10.1016/j.tcrr.2020.01.001

Izhar, S., T. Gupta, A.M. Qadri, and A.K. Panday. 2021. "Wintertime chemical characteristics of aerosol and their role in light extinction during clear and polluted days in rural Indo Gangetic plain." *Environmental Pollution* 282: 117034. https://doi.org/10.1016/j.envpol.2021.117034

Jacobson, M.Z. 2002. "Control of fossil-fuel particulate black carbon and organic matter, possibly the most effective method of slowing global warming." *Journal of Geophysical Research: Atmospheres* 107 (D19): 4410. https://doi.org/10.1029/2001JD001376

Jiang, M., Y. Wu, G. Lin, L. Xu, Z. Chen, and F. Fu. 2011. "Pyrolysis and thermal-oxidation characterization of organic carbon and black carbon aerosols." *Science of The Total Environment* 409 (20): 4449–4455. https://doi.org/10.1016/j.scitotenv.2011.07.016

Kain, J.S., and J.M. Fritsch. 1990. "A one-dimensional entraining/detraining plume model and its application in convective parameterization." *Journal of Atmospheric Sciences* 47 (23): 2784–2802. https://doi.org/10.1175/1520-0469(1990)047<2784:AODEPM>2.0.CO;2

Kain, J.S., and J.M. Fritsch. 1993. "Convective parameterization for mesoscale models: The kain-fritsch scheme." In: Emanuel K.A., Raymond D.J. (eds) *The Representation of Cumulus Convection in Numerical Models. Meteorological Monographs.* American Meteorological Society, Boston, MA. https://doi.org/10.1007/978-1-935704-13-3_16

Kulshrestha, U., and B. Kumar. 2014. "Airmass trajectories and long range transport of pollutants: review of wet deposition scenario in South Asia." *Advances in Meteorology* 2014: 596041 https://doi.org/10.1155/2014/596041

Kumar, A., J. Dudhia, R. Rotunno, D. Niyogi, and U.C. Mohanty. 2008. "Analysis of the 26 July 2005 heavy rain event over Mumbai, India using the Weather Research and Forecasting (WRF) model." *Quarterly Journal of the Royal Meteorological Society* 134 (636): 1897–1910. https://doi.org/10.1002/qj.325

Kumar, A., S.R. Bhowmik, and A.K. Das. 2012. "Implementation of Polar WRF for short range prediction of weather over Maitri region in Antarctica." *Journal of Earth System Science* 121 (5): 1125–1143. https://doi.org/10.1007/s12040-012-0217-3

Kwun, J.H., Y.K. Kim, J.W. Seo, J.H. Jeong, and S.H. You. 2009. Sensitivity of MM5 and WRF mesoscale model predictions of surface winds in a typhoon to planetary boundary layer parameterizations. *Natural Hazards* 51 (1): 63–77.

Liu, H., Y. Wang, S. Zhao, H. Hu, C. Cao, A. Li, Y. Yu, and H. Yao. 2020. "Review on the current status of the co-combustion technology of organic solid waste (OSW) and coal in China." *Energy & Fuels* 34 (12): 15448–15487. https://doi.org/10.1021/acs.energyfuels.0c02177

Ma, S., X. Zhang, C. Gao, D.Q. Tong, A. Xiu, G. Wu, C. Xinyuan, L. Huang, H. Zhao, S. Zhang, S. Ibarra-Espinosa, X. Wang, X. Li, and M. Dan. 2019. "Multimodel simulations of a springtime dust storm over northeastern China: implications of an evaluation of four commonly used air quality models (CMAQ v5. 2.1, CAMx v6. 50, CHIMERE v2017r4, and WRF-Chem v3. 9.1)." *Geoscientific Model Development* 12 (11): 4603–4625. https://doi.org/10.5194/gmd-12-4603-2019

Maione, M., M. Maione, U. Giostra, J. Arduini, F. Furlani, P. Bonasoni, P. Cristofanelli, P. Laj, and E. Vuillermoz. 2011. "Three-year observations of halocarbons at the Nepal Climate Observatory at Pyramid (NCO-P, 5079 m asl) on the Himalayan range." *Atmospheric Chemistry and Physics* 11 (7): 3431–3441.

Mancilla, Y., P. Herckes, M.P. Fraser, and A. Mendoza. 2015. "Secondary organic aerosol contributions to $PM_{2.5}$ in Monterrey, Mexico: Temporal and seasonal variation." *Atmospheric Research* 153: 348–359. https://doi.org/10.1016/j.atmosres.2014.09.009

Marcq, S., P., Laj, J.C., Roger, P., Villani, K., Sellegri, P., Bonasoni, A. Marinoni, P. Cristofanelli, G.P. Verza, and M. Bergin. 2010. "Aerosol optical properties and radiative forcing in the high Himalaya based on measurements at the Nepal Climate Observatory-Pyramid site (5079 m asl)." *Atmospheric Chemistry and Physics* 10 (13): 5859–5872. https://doi.org/10.5194/acp-10-5859-2010

Mele, M., and C. Magazzino. 2021. "Pollution, economic growth, and COVID-19 deaths in India: a machine learning evidence." *Environmental Science and Pollution Research* 28 (3): 2669–2677. https://doi.org/10.1007/s11356-020-10689-0

Menon, S., D. Koch, G. Beig, S. Sahu, J. Fasullo, and D. Orlikowski. 2010. "Black carbon aerosols and the third polar ice cap." *Atmospheric Chemistry and Physics* 10 (10): 4559–4571. https://doi.org/10.5194/acp-10-4559-2010

Mitchell, E.J.S., A.R. Lea-Langton, J.M. Jones, A. Williams, P. Layden, and R. Johnson. 2016. "The impact of fuel properties on the emissions from the combustion of biomass and other solid fuels in a fixed bed domestic stove." *Fuel Processing Technology* 142: 115–123. https://doi.org/10.1016/j.fuproc.2015.09.031

MoA. 2010. "Ministry of Agriculture, India. Annual Report 2009–2010." https://icar.org.in/content/dareicar-annual-report-2009-2010. Last accessed on 20 March 2011.

MoC. 2010. "Ministry of Coal, India. Annual Report 2009–2010." https://coal.nic.in/content/annual-report-2009-10. Last accessed on 27 February 2011.

MoPNG. 2010. "Ministry of Petroleum and Natural Gas, India. Annual Report 2009–2010." http://petroleum.nic.in/sites/default/files/AR09-10.pdf. Last accessed on 19 February 2011.

Moran, S.B., M.W. Lomas, R.P. Kelly, R. Gradinger, K. Iken, and Mathis, J.T. 2012. "Seasonal succession of net primary productivity, particulate organic carbon export, and autotrophic community composition in the eastern Bering Sea." *Deep Sea Research Part II: Topical Studies in Oceanography* 65: 84–97. https://doi.org/10.1016/j.dsr2.2012.02.011

Morrison, H.C., J.A. Curry, and V.I. Khvorostyanov. 2005. "A new double-moment microphysics parameterization for application in cloud and climate models. Part I: Description." *Journal of the Atmospheric Sciences* 62 (6): 1665–1677. https://doi.org/10.1175/JAS3446.1

Morrison, H., G., Thompson, and V. Tatarskii. 2009. "Impact of cloud microphysics on the development of trailing stratiform precipitation in a simulated squall line: Comparison of one-and two-moment schemes." *Monthly Weather Review* 137 (3): 991–1007. https://doi.org/10.1175/2008MWR2556.1

MoRTH. 2009. "Ministry of Road Transport and Highways, India. Annual Report 2008–2009." https://morth.nic.in/sites/default/files/Road_Transport_Year_Book_2009_11.pdf. Last accessed on 20 February 2011.

MoSPI. 2010. "Ministry of Statistics and Program Implementation, India. Household Consumption of Various Goods and Services in India. Reports of 2009–2010." http://mospi.nic.in/download-reports. Last accessed on 24 February 2011.

Negi, G.C.S. 2006. "Leaf and bud demography and shoot growth in evergreen and deciduous trees of central Himalaya, India." *Trees* 20 (4): 416–429. https://doi.org/10.1007/s00468-006-0056-4

Nie, Y., H.D. Pritchard, Q. Liu, T. Hennig, W. Wang, X. Wang, S. Liu, S. Nepal, D. Samyn, K. Hewitt, and Chen, X. 2021. "Glacial change and hydrological implications in the Himalaya and Karakoram." *Nature Reviews Earth & Environment* 2 (2): 91–106. https://doi.org/10.1038/s43017-020-00124-w

Niu, H., S. Kang, H. Wang, J. Du, T. Pu, G. Zhang, X. Lu, X. Yan, S. Wang, and Shi, X. 2020. "Light-absorbing impurities accelerating glacial melting in southeastern Tibetan Plateau." *Environmental Pollution* 257: 113541. https://doi.org/10.1016/j.envpol.2019.113541

Noh, Y., W.G. Cheon, S.Y. Hong, and S. Raasch. 2003. "Improvement of the K-profile model for the planetary boundary layer based on large eddy simulation data. *Boundary-Layer Meteorology* 107 (2): 401–427. https://doi.org/10.1023/A:1022146015946

Nopmongcol, U., Z. Liu, T. Stoeckenius, and G. Yarwood. 2017. "Modeling intercontinental transport of ozone in North America with CAMx for the Air Quality Model Evaluation International Initiative (AQMEII) Phase 3." *Atmospheric Chemistry and Physics* 17 (16): 9931–9943. https://doi.org/10.5194/acp-17-9931-2017

Ohara, T., H. Akimoto, J. Kurokawa, N. Horii, K. Yamaji, X. Yan, and T. Hayasaka. 2007. "An Asian emission inventory of anthropogenic emission sources for the period 1980–2020." *Atmospheric Chemistry and Physics* 7: 4419–4444. https://doi.org/10.5194/acp-7-4419-2007

Ojha, N., A. Pozzer, A. Rauthe-Schöch, A.K. Baker, J. Yoon, C.A. Brenninkmeijer, and J. Lelieveld. 2016. "Ozone and carbon monoxide over India during the summer monsoon: regional emissions and transport." *Atmospheric Chemistry and Physics* 16 (5): 3013–3032. https://doi.org/10.5194/acp-16-3013-2016

Pathirana, A., H.B. Denekew, W. Veerbeek, C. Zevenbergen, and A.T. Banda. 2014. "Impact of urban growth-driven landuse change on microclimate and extreme precipitation—A sensitivity study." *Atmospheric Research* 138: 59–72. https://doi.org/10.1016/j.atmosres.2013.10.005

Price-Allison, A., P.E. Mason, J.M. Jones, E.K. Barimah, G. Jose, G., A.E. Brown, AB Ross, and A. Williams. 2021. "The Impact of Fuelwood Moisture Content on the Emission of Gaseous and Particulate Pollutants from a Wood Stove." *Combustion Science and Technology*: 1–20. https://doi.org/10.1080/00102202.2021.1938559

Qian, Y., M.G. Flanner, L.R. Leung, and W. Wang. 2011. "Sensitivity studies on the impacts of Tibetan Plateau snowpack pollution on the Asian hydrological cycle and monsoon climate." *Atmospheric Chemistry and Physics* 11 (5): 1929–1948. https://doi.org/10.5194/acp-11-1929-2011

Rafael, S., V. Rodrigues, K. Oliveira, S. Coelho, and M. Lopes. 2021. "How to compute long-term averages for air quality assessment at urban areas?." *Science of The Total Environment* 795: 148603. https://doi.org/10.1016/j.scitotenv.2021.148603

Reddy, S.K.K., H. Gupta, U. Badimela, D.V. Reddy, R.M. Kurakalva, and D. Kumar. 2021. "Export of particulate organic carbon by the mountainous tropical rivers of Western Ghats, India: Variations and controls." *Science of the Total Environment* 751: 142115. https://doi.org/10.1016/j.scitotenv.2020.142115

Romshoo, S.A., M.A. Bhat, and G. Beig. 2021. "Particulate pollution over an urban Himalayan site: Temporal variability, impact of meteorology and potential source regions." *Science of The Total Environment* 799: 149364. https://doi.org/10.1016/j.scitotenv.2021.149364

Saraf, A.K., A.K., Bora, J., Das, V., Rawat, K. Sharma, and S.K. Jain. 2011. "Winter fog over the Indo-Gangetic Plains: mapping and modelling using remote sensing and GIS." *Natural Hazards* 58 (1): 199–220. https://doi.org/10.1007/s11069-010-9660-0

Schmale, J., P. Zieger, and A.M. Ekman. 2021. "Aerosols in current and future Arctic climate." *Nature Climate Change* 11 (2): 95–105. https://doi.org/10.1038/s41558-020-00969-5

Sen, A., A.S. Abdelmaksoud, Y.N. Ahammed, T. Banerjee, M.A. Bhat, A. Chatterjee et al. 2017. "Variations in particulate matter over Indo-Gangetic Plains and Indo-Himalayan Range during four field campaigns in winter monsoon and summer monsoon: role of pollution pathways." *Atmospheric Environment* 154: 200–224. https://doi.org/10.1016/j.atmosenv.2016.12.054

Sharma, A., N. Ojha, A. Pozzer, K.A. Mar, G. Beig, J. Lelieveld, and S.S. Gunthe. 2017. "WRF-Chem simulated surface ozone over south Asia during the pre-monsoon: effects of emission inventories and chemical mechanisms." *Atmospheric Chemistry and Physics* 17 (23): 14393–14413. https://doi.org/10.5194/acp-17-14393-2017

Shindell, D., J.C. Kuylenstierna, E. Vignati, R. vanDingenen, M. Amann, Z. Klimont, S.C. Anenberg, et al. 2012. "Simultaneously mitigating near-term climate change and improving human health and food security." *Science* 335 (6065): 183–189. https://doi:10.1126/science.1210026

Singh, D., S.P. Shukla, M. Sharma, S.N. Behera, D. Mohan, N.B. Singh, and G. Pandey. 2016. "GIS-based on-road vehicular emission inventory for Lucknow, India." *Journal of Hazardous, Toxic, and Radioactive Waste* 20 (4): A4014006. https://doi.org/10.1061/(ASCE)HZ.2153-5515.0000244

Singh, G.K., V. Choudhary, P. Rajeev, D. Paul, and T. Gupta. 2021. "Understanding the origin of carbonaceous aerosols during periods of extensive biomass burning in northern India." *Environmental Pollution* 270: 116082. https://doi.org/10.1016/j.envpol.2020.116082

Singh, T., A. Biswal, S. Mor, K. Ravindra, V. Singh, and S. Mor. 2020. A high-resolution emission inventory of air pollutants from primary crop residue burning over Northern India based on VIIRS thermal anomalies. *Environmental Pollution* 266: 115132. https://doi.org/10.1016/j.envpol.2020.115132

Skamarock, W.C., J.B. Klemp, J. Dudhia, D.O. Gill, D. Barker, M.G. Duda, X. Huang, W. Wang, and J.G. Powers. 2008. "A Description of the Advanced Research WRF Version 3 (No. NCAR/TN-475+STR)." University Corporation for Atmospheric Research. http://dx.doi.org/10.5065/D68S4MVH

Streets, D.G., T.C. Bond, T. Lee, and C. Jang. 2004. "On the future of carbonaceous aerosol emissions." *Journal of Geophysical Research: Atmospheres* 109 (D24): D24212. https://doi.org/10.1029/2004JD004902

Szidat, S., M. Ruff, N. Perron, L. Wacker, H.A. Synal, M. Hallquist, Shannigrahi, K.E. Yttri, C. Dye, and D. Simpson. 2009. "Fossil and non-fossil sources of organic carbon (OC) and elemental carbon (EC) in Göteborg, Sweden." *Atmospheric Chemistry and Physics* 9 (5): 1521–1535. https://doi.org/10.5194/acp-9-1521-2009

Thompson, T.M., D. Shepherd, A. Stacy, M.G. Barna, and B.A. Schichtel. 2017. "Modeling to evaluate contribution of oil and gas emissions to air pollution." *Journal of the Air & Waste Management Association* 67 (4): 445–461.

Umar, S.I.U., D. Konwar, A. Khan, M.A. Bhat, F. Javid, R. Jeelani, B. Nabi, A.A. Najar, D. Kumar, and B. Brahma. 2021. "Delineation of temperature-humidity index (THI) as indicator of heat stress in riverine buffaloes (Bubalus bubalis) of a sub-tropical Indian region." *Cell Stress and Chaperones* 26: 657–669: https://doi.org/10.1007/s12192-021-01209-1

USEPA. 1991. "United States Environmental Protection Agency: Guidance for Regulatory Application of the Urban Airshed Model (UAM)." Office of Air Quality Planning and Standards, Research Triangle Park, NC.

Williams, A., J.M. Jones, L. Ma, and M. Pourkashanian. 2012. "Pollutants from the combustion of solid biomass fuels." *Progress in Energy and Combustion Science* 38 (2): 113–137. https://doi.org/10.1016/j.pecs.2011.10.001

Wu, Y., H. Li, D. Xu, H. Li, Z. Chen, Y. Cheng, G. Yin, Y. Niu, C. Liu, H. Kan, D. Yu, and R. Chen. 2021. "Associations of fine particulate matter and its constituents with airway inflammation, lung function, and buccal mucosa microbiota in children." *Science of The Total Environment* 773: 145619. https://doi.org/10.1016/j.scitotenv.2021.145619

Yadav, A., S.N. Behera, P.K. Nagar, and M. Sharma. 2020. "Spatio-seasonal concentrations, source apportionment and assessment of associated human health risks of $PM_{2.5}$-bound polycyclic aromatic hydrocarbons in Delhi, India." *Aerosol and Air Quality Research* 20: 2805–2825. https://doi.org/10.4209/aaqr.2020.04.0182

Yasunari, T.J., P. Bonasoni, P. Laj, K. Fujita, E. Vuillermoz, A. Marinoni, P. Cristofanelli, R. Duchi, G. Tartari, and K.-M. Lau. 2010. "Estimated impact of black carbon deposition during pre-monsoon season from Nepal Climate Observatory—Pyramid data and snow albedo changes over Himalayan glaciers, *Atmospheric Chemistry and Physics* 10: 6603–6615. https://doi.org/10.5194/acp-10-6603-2010

Yu, S., R. Mathur, J. Pleim, D. Wong, R. Gilliam, K. Alapaty, C. Zhao, and X. Liu. 2014. Aerosol indirect effect on the grid-scale clouds in the two-way coupled WRF–CMAQ: model description, development, evaluation and regional analysis, *Atmospheric Chemistry and Physics* 14: 11247–11285, https://doi.org/10.5194/acp-14-11247-2014

Zhao, S., J. Ming, J. Sun, and C. Xiao. 2013. "Observation of carbonaceous aerosols during 2006–2009 in Nyainqêntanglha Mountains and the implications for glaciers." *Environmental Science and Pollution Research* 20 (8): 5827–5838. https://doi.org/10.1007/s11356-013-1548-6

Index

A

ablation zone, 24, 25, 26, 27
accumulation zone, 24, 25, 26
agriculture, 39, 70, 71, 72, 82, 84–86, 92, 100, 105, 115, 117, 158, 166, 191, 197, 199, 201, 206, 215, 217, 220, 226, 243, 280, 281, 283, 284, 286, 325, 329, 340, 358, 373
air temperature, 5, 7, 60, 117, 206, 211, 276, 361
ALOS PALSAR, 6, 8, 21, 33, 94, 194
ALOS-PALSAR-based, 94
ALOS W30D, 22
aqua, 209, 256, 312
area of glaciers, 46

B

backscatter coefficient, 96, 105
Bara Sigri Glacier, 21, 25–29
base flow, 8, 118, 133
built-up, 69, 75, 80, 82, 92, 100, 191, 192, 197, 199–201, 226, 227, 230, 231, 234, 237, 244, 246, 276, 277, 280, 281, 283, 284, 286, 287

C

catchment(s), 8–10, 14, 17, 32, 67, 69, 71, 73–77, 81, 83, 84, 86, 87, 92, 97, 99, 102, 113, 114, 134, 139, 181, 193, 201, 203, 206, 226, 248
central Himalayas, 24, 25, 30, 36, 164
climate change, 3, 4, 14, 17, 20, 30, 32, 37, 46, 69, 84, 85, 109, 111, 112, 117, 123, 126, 130, 134–136, 138, 139, 141, 157, 158, 161–164, 202–204, 206, 218, 220–222, 226, 250–252, 254, 268, 271, 272, 288, 289, 297, 314, 341, 344, 345, 369, 370, 375
climatology, 13, 161, 162, 164, 203, 204, 252, 256, 271, 272, 287, 288, 313
cloud, 5, 21, 27, 31, 39, 40, 73, 209, 311, 312, 344, 355, 361, 373, 374
cloud coverage, 31
CMIP6, 112, 113, 116, 117, 123, 130, 134, 135, 137–139, 141–143, 146, 150, 151, 158, 161–164
coal, 322–324, 327, 329, 332–334, 340, 345, 351, 359, 373
combustion, 318, 326, 340, 344, 345, 359, 370, 372–374, 376

D

decision-making, 55
deforestation, 201, 246
deglaciated, 58, 63, 66
degree-day factor, 7, 11
DEM, 5, 6, 8, 20, 22, 23, 31, 39, 58, 89, 91, 93–97, 104, 115, 193, 194, 209, 229, 231, 268, 269
demography, 294, 340, 374
deposition, 175, 320, 327, 343–348, 350, 363–369, 372, 376
de-seasonality, 210
dielectric constant, 30
digitization, 39, 231, 320, 328, 351
discharge, 3, 4, 8, 10, 13, 14, 31, 46, 83, 112, 115, 119, 121, 123, 126, 127, 130, 133, 135, 136, 164, 167, 175, 176, 193, 194, 202, 231, 251
district-level, 330, 331, 350, 351, 356
downscaling, 5, 161–164
downstream, 36, 37, 46, 97, 112, 113, 168, 170, 172, 173, 175, 176, 178–180, 183, 192, 208
drainage, 49, 70, 83, 84, 225, 226, 228, 229, 231, 236, 237, 241–243
droughts, 90, 158, 250, 255
dry-season, 182
dual-polarization, 91, 93, 94, 105

E

earthquake, 38, 49, 50, 54
eastern Himalaya(s), 24, 25, 31, 35–38, 47, 48, 50, 192
elevation zones, 9, 10, 77, 208, 209, 213, 214, 217
environment, 14, 17, 31–33, 48, 49, 51, 70, 84, 86, 134, 135, 141, 161–164, 170, 182–185, 194, 201–204, 219, 220, 222, 226, 227, 247, 249, 251, 258, 267, 268, 271, 272, 276, 278, 286–291, 294, 297, 311–315, 338–342, 344, 361, 369–372, 374–376
evapotranspiration, 4, 118, 124, 190, 196, 199–201, 205–208, 210, 215–217, 220, 221, 252
extremes, 70, 83, 85, 111–113, 116, 119, 126, 127, 130, 134, 136, 138, 139, 161, 162

F

feature tracking, 21, 29, 32
flood-affected, 75–78, 96
flooding, 70, 71, 73, 75–77, 79, 82, 83, 90–92, 97, 98, 100, 104, 112, 168, 175, 180, 182, 201
floodplain(s), 58, 83, 90, 91, 104, 106, 166–168, 170, 175, 176, 180, 181, 183
flood-prone, 70, 83, 84, 104
forecasting, 54, 68, 83, 84, 104, 105, 135, 164, 190, 203, 204, 227, 249, 267, 268, 323, 346, 352, 370, 372
Friedman test, 41
fuelwood, 322–324, 327, 329, 374

G

Gangotri Glacier, 14, 17, 20, 21, 25–30, 32, 33, 49, 55, 68, 71, 348, 368
gauge-satellite, 74
geographic information system, 70, 91, 226, 255, 273, 277, 319, 346
geoinformatics, 247, 272
geoinformation, 32, 50, 85, 105, 272, 291
geomorph, 47, 48, 182, 186
geomorphic, 33, 86, 96, 97, 106, 165–170, 172–174, 179–181, 185, 186
geomorphological, 20, 32, 35, 53–55, 58, 61, 63, 66, 70, 92, 239, 240
geomorphology, 31, 47, 48, 50, 67, 168, 182, 184, 186, 225, 227, 229, 231, 237, 241–243
geophysical, 17, 31, 86, 163, 164, 202, 312–314, 341, 369–372, 375
geosciences, 67, 68, 202, 203
geospatial, 1, 14, 16–17, 32, 50, 67, 84, 94, 105, 109, 115, 202, 223, 227, 247, 271, 290, 291, 327
geospatial techniques, 14
geotechnical, 49
GIS, 47, 55, 63, 66, 68, 70, 91, 186, 196, 225–228, 231, 237, 241, 247, 248, 255, 256, 271–273, 277, 286, 291, 314, 319, 320, 325, 327, 339, 346, 350, 352, 355, 356, 368, 375
glacier boundary, 20, 22, 23, 26
glacier melt, 3–5, 9, 10, 12–17, 130, 150, 206, 221
glacier retreat, 3, 4
glacier velocity, 19, 21, 22, 25, 26, 27–29, 32, 33
glaciological, 4, 14
GLDAS, 210, 211, 220
GLOVIS, 257

H

heterogeneous, 41, 43, 47, 117, 166, 221, 258
high-elevation, 209
high-mountain, 49, 50
high-resolution, 30, 31, 36, 39, 90, 91, 94, 229, 256, 312, 339, 375
high-velocity, 25, 26, 27
Himalaya–Karakoram, 4, 14
Himalayan glaciers, 3, 20, 33, 35, 47, 49, 157, 345, 376
Himalaya(s), 3, 14, 17, 19, 20, 24, 25, 30–33, 35–38, 40, 44, 46–51, 60, 68, 70, 72, 85, 87, 92, 97, 106, 112, 113, 138, 139, 143, 145, 146, 148–150, 152, 154–158, 160, 162–164, 184, 189, 191, 192, 201, 202–205, 218, 220, 222, 227, 248, 312, 314, 327, 340, 343, 345–348, 360, 363, 365, 366, 368, 370, 373, 374
Hindukush-Himalaya, 220
hydro-geomorphic, 84, 165, 167, 171, 181, 185

I

indices, 90, 196, 249, 261, 265, 277, 338
Indo-China, 53
Indo-Gangetic, 296, 297, 312, 318, 356, 370, 375
Indo-Gangetic-Brahmaputra, 85
Indus, 3–5, 14, 17, 32, 49, 138, 163, 207, 218, 220, 221
inferential statistics, 38, 40, 46, 47
InSAR, 20
intra-annual, 213
inundated, 75, 77, 79–82, 91, 96–98, 100, 102, 103, 105
inundation, 71, 73, 78, 79, 82, 84, 86, 89–92, 94–98, 100, 102–106, 167, 180
inventorization, 318, 319, 324, 327, 328
inventory, 14, 36, 37, 47–51, 221, 317, 319–321, 323, 325–331, 335, 337–342, 344–347, 350–352, 355, 356, 363, 368, 370, 371, 374, 375

J

JAXA, 31, 94, 104, 115

K

Kabul River Basin (KRB), 205, 207
Kappa coefficient, 227, 230, 231, 258
Karakoram, 4, 21, 31–33, 36, 37, 47, 49, 220, 222, 374
Kohonen's map, 91
Kolkata, 85, 179, 251, 262, 265, 295, 296, 305, 356
Koshi, 50, 85, 134, 137–139, 161–163, 171, 182
Kosi-Gandak-Karnali, 84

L

land-change, 203
land-cover, 94, 166, 202, 212, 229, 251, 268, 275, 277, 280, 286, 287, 288, 289, 299
landforms, 53–55, 57, 58, 61–64, 66, 72, 114, 182, 185, 240
landsat, 5–8, 17, 21, 31, 33, 39, 46, 50, 90, 170, 174, 177, 180, 183, 191, 192, 194, 204, 247, 256, 257, 261, 271, 277, 278, 287–291
Landsat-8, 33, 194, 256, 261, 290
Landsat imagery, 291
landscape, 48, 165–167, 184, 185, 192, 204, 221, 230, 237, 249, 251, 252, 256, 257, 261, 263, 268, 270, 272, 289
landslide(s), 50, 55, 66, 67, 71, 112, 138, 139, 158, 177, 191, 201, 204, 250
land-use, 83, 89, 94, 114, 115, 117, 120, 167, 176, 202, 229, 231, 246–249, 257, 258, 261, 263, 264, 265, 270, 275, 286, 287, 288, 289, 299, 313, 356, 374
Langtang Glacier, 21, 25–30, 33
leeward, 60

livelihood(s), 3, 13, 48, 157, 158, 201
lockdown, 272, 283, 288, 290, 293–299, 301–307, 309–314
low-dispersion, 234
LSM, 5
LST(s), 211, 249, 251, 252, 254–262, 264–271, 275–281, 283, 284, 286, 287, 289
LULC, 69, 73–75, 80, 89, 92–95, 100, 189–201, 203, 225, 226, 229, 231, 232, 234, 241–243, 246, 248, 267, 275–284, 286, 287, 291, 300, 310

M

mass exchange, 3
melt discharge, 3, 13, 14
melt runoff, 3–5, 9, 10, 12–17, 206, 221
meteorological, 4–6, 8, 14, 17, 55, 92, 106, 115, 136, 139, 141, 193, 194, 202, 210, 211, 221, 255–257, 272, 309, 314, 318, 338, 344, 345, 346, 348, 350, 352–354, 361, 364, 365, 372
meteorological data, 4, 5, 8, 106, 115, 139, 141, 193, 194, 210, 353
model, 4, 5, 9, 14, 20, 39, 53, 55, 58, 63, 66, 67, 82, 86, 91, 96, 105, 106, 111–113, 115, 117, 118, 123–125, 133–136, 138, 139, 142, 143, 150, 162–164, 166, 189, 190, 193–196, 199–204, 209–211, 220, 221, 226, 227, 230, 231, 241, 247, 248, 268, 269, 277, 286, 287, 312, 319, 326, 327, 339, 340, 343, 346, 348, 350, 352–355, 359–364, 369–374, 376
modeling, 5, 46, 50, 53, 85, 104, 105, 117, 134, 189, 190, 194, 196, 202, 203, 225, 227, 231, 241, 252, 267, 272, 288, 290, 311, 313, 314, 319, 320, 327, 339, 341, 343, 346, 348–350, 352–355, 361, 363, 364, 368, 370, 372, 374, 375
MODIS, 5, 69, 73–77, 82, 90, 103, 107, 209–212, 217, 220, 221, 256, 286, 289, 290, 297, 311–314
monitoring, 1, 3, 5, 6, 14, 16, 20, 32, 35, 38, 66, 69, 85, 90, 104, 105, 107, 109, 162, 164, 169, 170, 185, 204, 207, 221, 223, 248, 252, 256, 257, 270–272, 288–290, 297, 312, 319, 329, 337–339, 369–371
moraine, 37, 47
moraine-dammed, 51
morphological, 33, 167, 170
morphometric, 6, 19, 22, 24, 25, 30, 33, 70
morphometric parameters, 6, 19, 22, 24, 30, 70
mountain, 4, 14, 16–17, 32, 36–38, 40, 47, 48, 50, 71, 135, 138, 161, 163, 166, 191, 220, 277
multicriteria, 55, 68, 227

N

NCL, 355
NDBI, 268, 269
NDVI, 90, 106, 211, 260, 261, 268, 269, 275, 277, 280, 281, 289, 291
near-future, 143, 150, 152, 157

near-real-time, 66, 73
Nepal, 21, 67, 70–72, 75–77, 83–86, 92, 97, 111–115, 131, 134, 135, 137–141, 144, 146–148, 157–164, 172, 182, 186, 204, 248, 275, 277, 283, 286–291, 301, 302, 304, 306, 346, 347, 361, 371, 373, 374, 376
Nepal-Climate, 370
Nepal-Sikkim, 47
NetCDF, 115
network-centric, 181
nighttime, 286, 289, 369
nitrogen dioxide, 293, 294, 312, 314, 338, 369, 371
NOAA, 140, 141, 256, 359, 360, 371
nonparametric statistical, 35, 216
non-snow-covered, 57, 60
northwestern, 50, 192, 244, 361

O

object-based, 49
oceanography, 373
offset tracking, 20
outbreak, 294, 296, 315
outburst, 35, 36, 47–50

P

paleoclimate, 67
Panjshir, 215–217, 219
pan-sharpening, 39
PAN sharpening method, 39
peak ablation, 7, 14
polarimetric, 203, 248
policy(ies), 14, 48, 82, 83–85, 113, 131, 137, 139, 158, 182, 190, 197, 201, 203, 251, 257, 267–269, 283, 318, 327, 337, 338, 340
policymakers, 113, 131, 158, 201, 219, 297, 310
pollutants, 172, 252, 270, 283, 294–299, 301, 306, 307, 309, 310, 318, 319, 325, 335, 338, 339, 350, 354, 359, 361, 363–367, 369, 370, 372, 374–376
pollution–monitoring, 319
population(s), 41, 70, 72, 73, 80, 82, 85, 90, 103, 105, 135, 171–173, 176, 178, 179, 180, 190, 201, 206, 207, 221, 222, 229, 231, 246, 250–252, 263, 265, 273, 276, 278, 283, 286, 291, 293, 294, 297–300, 307, 308, 310–312, 315, 319–325, 327, 328, 331, 332, 334, 336, 337, 351, 356, 357
post-disaster, 70
postglacial, 250
post-impact, 174, 175
post-monsoon, 115, 126, 139, 146, 149, 150, 155–157
precipitation, 4–8, 10, 14–15, 21, 22, 38, 46, 54, 56–61, 63, 64, 66, 70, 71, 73, 74, 85, 86, 89, 92–94, 97, 112, 114, 116–118, 124, 130, 133, 134, 136, 141, 142, 158, 161–164, 190, 192, 196, 203, 206, 208, 209, 218, 228, 229, 241, 248, 250, 289, 310, 344, 350, 364, 365, 374

preindustrial, 369
pre-lockdown, 296
pro-glacial, 25, 40
pro-glacial lakes, 25
projected, 39, 111–113, 117, 119–122, 126–130,
132, 133, 138, 146–148, 150–154, 157, 158,
161, 163, 197, 199–201, 206, 207, 227, 246,
323, 351
projecting, 58, 117, 163
projection, 58, 111, 116, 117, 123, 130, 137, 139,
141, 142, 146, 148, 150, 158, 161–163, 225,
287, 355
p-value, 43, 44, 212

Q

quasi-global, 73

R

radar, 20, 31, 33, 39, 70, 86, 89–91, 96, 104–106, 209
rate of glacial lake changes, 39
receding rate, 4
reflectance values, 7
remote sensing, 4, 14, 17, 19, 20, 31–33, 45,
47–50, 55, 66–68, 84, 86, 90, 91, 104–106,
136, 162, 170, 183–186, 202–205, 207, 219–
222, 247, 248, 255, 256, 270–273, 287–291,
313–315, 370, 375
runoff, 3–17, 70, 76, 118, 124, 133, 134, 139, 157,
158, 161, 162, 190, 196, 199–201, 204, 206,
207, 217, 220, 221, 226, 231, 236, 237

S

SAR, 20–22, 26–33, 82, 85, 86, 89–93, 96, 97,
103–107, 290
sensitivity, 20, 50, 101, 118, 123, 124, 189, 199,
220, 346, 350, 373, 374
settlements, 103, 114, 251, 276, 286
Shillong, 225, 227–240, 244–246, 299, 303, 306,
307, 309
shrubland, 278, 281, 283
Siachen Glacier, 20, 21, 24–26, 28, 30, 348,
367, 368
Sikkim Himalaya, 37, 38, 40, 44, 46–48, 50, 203
slope, 8, 11, 22, 24, 25, 30, 38, 54, 55, 58–60,
63, 64, 70, 97, 113, 117, 163, 199, 205, 210,
212–214, 216, 217, 225, 227, 229, 231, 234,
237, 239, 241–243, 277, 361
snowcapped, 120
snow-clad, 113
snow-cover, 3, 5, 7, 10, 11, 54, 55, 61, 138, 158,
192, 209, 217, 218
snow cover area SCA, 214
snow-cover dynamics, 3, 54, 192
Snowfall–Snowmelt, 135

snowmelt, 4, 5, 7, 13, 14, 17, 31, 113, 118, 124,
126, 133, 134, 150, 157, 158, 164, 202, 206,
209, 212, 217, 220
spatio-temporal, 33, 49, 86, 106, 207, 219, 271,
273, 289
stochastic, 204
subsurface runoff, 3, 8, 12
surface temperature, 4, 7, 48, 112, 210, 211,
249, 251, 252, 254, 255, 257, 259, 261, 264,
265, 267, 270–273, 275, 276, 281, 284,
287–291, 313
susceptibility, 36, 37, 47, 49, 53–55, 58, 60, 61,
63–68, 84, 204
sustainability, 84–86, 105, 106, 138, 164, 184,
201, 204, 246, 248, 276, 291, 341

T

temperature, 4–10, 13–15, 17, 46, 48, 54, 56, 60,
69, 70, 106, 111, 112, 114–118, 120, 121, 123,
126, 130–134, 137–146, 150–154, 157–163,
167, 190, 193, 194, 204, 206, 210, 211, 218,
220, 221, 228, 249–255, 257, 259–261, 264,
265, 267, 270–273, 275–277, 280, 281, 284,
286–291, 313, 344, 345, 353, 355, 361–363
terrain parameters, 19, 22–25, 30, 54, 240
topography, 32, 39, 70, 72, 105, 113, 115, 117, 209,
237, 256, 277, 318, 353, 363

U

urban, 47, 54, 82, 86, 105, 203, 223, 225–231,
233, 234, 244–253, 256–258, 261–265,
267–273, 275–278, 281, 283, 286–291, 293,
294, 297, 311–313, 318, 322, 323, 325, 329,
331, 333, 337, 338, 340–342, 351, 356, 370,
374–376

V

validation, 11, 53, 55, 56, 61, 64, 66, 103, 111, 118,
123, 125, 134, 136, 189, 195, 199–201, 203,
227, 312, 313, 344, 361, 363, 364
velocity difference, 28
velocity zones, 31
vulnerability, 14, 60, 70, 85, 90, 181, 203

W

western Himalayas, 17, 24, 36, 60, 85, 106, 220,
222, 248
WHO, 29, 227, 254, 255, 294, 315

Z

Zanskar valley, 3, 5, 6, 10, 12–14
Zemu Glacier, 21, 22, 26–30

For Product Safety Concerns and Information please contact our EU
representative GPSR@taylorandfrancis.com
Taylor & Francis Verlag GmbH, Kaufingerstraße 24, 80331 München, Germany

www.ingramcontent.com/pod-product-compliance
Lightning Source LLC
Chambersburg PA
CBHW060752220326
41598CB00022B/2412